T0329884

UAV Communications for 5G and Beyond

UAV Communications for 5G and Beyond

Edited by

Yong Zeng
Southeast University, China
Jiangsu, China

and

Purple Mountain Laboratories
Jiangsu, China

Ismail Guvenc
North Carolina State University
NC, USA

Rui Zhang
National University of Singapore
Singapore

Giovanni Geraci
Universitat Pompeu Fabra
Barcelona, Spain

David W. Matolak
University of South Carolina
SC, USA

This edition first published 2021
© 2021 John Wiley & Sons Ltd

The right of Yong Zeng, Ismail Guvenc, Rui Zhang, Giovanni Geraci, and David W. Matolak to be identified as the authors of this editorial work has been asserted in accordance with law.

Registered Offices
John Wiley & Sons, Inc., 111 River Street, Hoboken, NJ 07030, USA
John Wiley & Sons Ltd, The Atrium, Southern Gate, Chichester, West Sussex, PO19 8SQ, UK

Editorial Office
The Atrium, Southern Gate, Chichester, West Sussex, PO19 8SQ, UK

For details of our global editorial offices, customer services, and more information about Wiley products visit us at www.wiley.com.

Wiley also publishes its books in a variety of electronic formats and by print-on-demand. Some content that appears in standard print versions of this book may not be available in other formats.

Library of Congress Cataloging-in-Publication Data

Names: Zeng, Yong (Professor at Southeast University), author. |
  Guvenc, Ismail (Professor at North Carolina State University) | Zhang, Rui (Professor at National
  University of Singapore), author. | Geraci, Giovanni (Assistant Professor at Universitat Pompeu Fabra),
  author. | Matolak, David W., author. | John Wiley & Sons, Inc., publisher.
Title: UAV communications for 5G and beyond / Yong Zeng, Ismail Guvenc, Rui
  Zhang, Giovanni Geraci, David W. Matolak.
Description: Hoboken, NJ : Wiley-IEEE Press, [2021] | Includes
  bibliographical references and index.
Identifiers: LCCN 2020030506 (print) | LCCN 2020030507 (ebook) | ISBN
  9781119575696 (hardback) | ISBN 9781119575672 (adobe pdf) | ISBN
  9781119575726 (epub)
Subjects: LCSH: Drone aircraft–Control systems. |
  Aeronautics–Communication systems. | Mobile communication systems. | 5G
  mobile communication systems.
Classification: LCC TL589.4 .Z465 2021 (print) | LCC TL589.4 (ebook) |
  DDC 629.135/5–dc23
LC record available at https://lccn.loc.gov/2020030506
LC ebook record available at https://lccn.loc.gov/2020030507

Cover Design: Wiley
Cover Image: © Waitforlight/Getty Images

Set in 9.5/12.5pt STIXTwoText by SPi Global, Chennai, India
Printed and bound by CPI Group (UK) Ltd, Croydon, CR0 4YY

10 9 8 7 6 5 4 3 2 1

Contents

List of Contributors

Rafhael Medeiros de Amorim
Nokia Bell Labs
Denmark

Chethan Kumar Anjinappa
Department of Electrical and Computer
Engineering
North Carolina State University
NC
USA

M. Mahdi Azari
Department of Electrical Engineering
KU Leuven
Belgium

Morteza Banagar
Wireless@VT
Department of ECE
Virginia Tech
Blacksburg
VA
USA

Arupjyoti Bhuyan
Idaho National Laboratory
Idaho Falls
ID
USA

Vishnu V. Chetlur
Wireless@VT
Department of ECE
Virginia Tech
Blacksburg
VA
USA

Huaiyu Dai
Department of Electrical and Computer
Engineering
North Carolina State University
NC
USA

Harpreet S. Dhillon
Wireless@VT
Department of ECE
Virginia Tech
Blacksburg
VA
USA

Fatih Erden
Department of Electrical and Computer
Engineering
North Carolina State University
NC
USA

Martins Ezuma
Department of Electrical and Computer
Engineering
North Carolina State University
NC
USA

Uwe-Carsten Fiebig
Institute of Communications and
Navigation
German Aerospace Center (DLR)
Wessling
Germany

Robert W. Heath
Electrical and Computer Engineering
Department
University of Texas at Austin
USA

Lorenzo Galati Giordano
Nokia Bell Labs
Dublin
Ireland

Adrian Garcia-Rodriguez
Nokia Bell Labs
Dublin
Ireland

Giovanni Geraci
Universitat Pompeu Fabra
Barcelona
Spain

Nuria González-Prelcic
Electrical and Computer Engineering
Department
University of Texas at Austin
USA

Ismail Guvenc
Department of Electrical and Computer
Engineering
North Carolina State University
NC
USA

Tianwei Hou
School of Electronic and Information
Engineering
Beijing Jiaotong University
PR China

Wahab Khawaja
Department of Electrical and Computer
Engineering
North Carolina State University
NC
USA

Aldebaro Klautau
Computer and Telecommunication
Engineering Department
Universidade Federal do Pará
Brazil

István Z. Kovács
Nokia Bell Labs
Denmark

Abhaykumar Kumbhar
Department of Electrical and Computer
Engineering
Florida International University
Miami
USA

Liang Liu
Department of Electronic and Information
Engineering
The Hong Kong Polytechnic University
Hong Kong

Yuanwei Liu
School of Electronic Engineering and
Computer Science
Queen Mary University of London
UK

David López-Pérez
Nokia Bell Labs
Dublin
Ireland

David W. Matolak
Department of Electrical Engineering
University of South Carolina
SC
USA

Helka-Liina Määttänen
Ericsson Research
Finland

Kamesh Namuduri
University of North Texas
USA

Ozgur Ozdemir
Department of Electrical and Computer
Engineering
North Carolina State University
NC
USA

Sofie Pollin
Department of Electrical Engineering
KU Leuven
Belgium

Fernando Rosas
Data Science Institute
Department of Brain Sciences
and Center for Complexity Science
Imperial College London
UK

Nadisanka Rupasinghe
Department of Electrical and Computer
Engineering
North Carolina State University
NC
USA

and

DOCOMO Innovations, Inc.
Palo Alto
CA
USA

Cristian Rusu
LCSL
Istituto Italiano di Tecnologia (IIT)
Liguria
Italy

Nicolas Schneckenberger
Institute of Communications and
Navigation
German Aerospace Center (DLR)
Wessling
Germany

Troels B. Sørensen
Aalborg University
Denmark

Xin Sun
School of Electronic and Information
Engineering
Beijing Jiaotong University
PR China

Jeroen Wigard
Nokia Bell Labs
Denmark

Qingqing Wu
State Key Laboratory of Internet of Things
for Smart City
University of Macau
China

Jie Xu
Future Network of Intelligence Institute
(FNii) and School of Science and
Engineering
The Chinese University of Hong Kong
Shenzhen
PR China

Yavuz Yapici
Department of Electrical and Computer
Engineering
North Carolina State University
NC
USA

Chiya Zhang
School of Electronic and Information
Engineering
Harbin Institute of Technology
Shenzhen
China

and

Peng Cheng Laboratory (PCL)
Shenzhen
China

Rui Zhang
Department of Electrical and Computer
Engineering
National University of Singapore
Singapore

Wei Zhang
School of Electrical Engineering and
Telecommunications
University of New South Wales
Sydney
Australia

Yong Zeng
National Mobile Communications
Research Laboratory
Southeast University
China

and

Purple Mountain Laboratories
Jiangsu
China

Acronyms

3GPP	3rd/third generation partnership project
5G	5th/fifth generation
5pSE	5th/fifth percentile spectral efficiency
AA	air-to-air
AG	air-to-ground
AG-HetNet	air–ground heterogeneous cellular network
ASE	area spectral efficiency
ASTA	arrivals see time averages
AWGN	additive white Gaussian noise
B5G	beyond 5th/fifth generation
b/s/Hz	bits per second per hertz
BER	bit error rate
BHCA	busy hour call attempts
BPP	binomial point process
BPSK	binary phase shift keying
BR	bandwidth reservation
BS	base station
BSs/km^2	base stations per square kilometer
b.u.	bandwidth unit(s)
BVLoS	beyond-visual-line-of-sight
BW	bandwidth
C2	command and control
CAC	call/connection admission control
CBP	call blocking probability(-ies)
CCDF	complementary cumulative distribution function
CCS	centum call seconds
CDF	cumulative distribution function
CDTM	connection dependent threshold model
CE2R	curved Earth two-ray
CFO	carrier frequency offset
CI	close-in
CIR	channel impulse response
CNPC	control and non-payload communications

CRE	cell range expansion
CS	complete sharing
CSF	coordinated radio subframe
CSI	channel state information
CTF	channel transfer function
CW	continuous wave
DBS	drone base station
DiffServ	differentiated services
DME	distance-measuring equipment
DPP	Doppler power profile
DS	dual slope
DSB-AM	double-sideband amplitude modulation
DS-SS	direct sequence spread spectrum
EMLM	Erlang multirate loss model
eICIC	enhanced inter-cell interference coordination
erl	the Erlang unit of traffic-load
FAA	federal aviation administration
FBMC	filter bank multicarrier
FCC	federal communications commission
FeICIC	further-enhanced inter-cell interference coordination
FI	floating intercept
FIFO	first in-first out
FMCW	frequency-modulated continuous wave
Freq.	frequency
FSPL	free-space path loss
GA	genetic algorithm
GBSCM	geometrically based stochastic channel model
GMSK	Gaussian minimum shift keying
GPS	global positioning system
GS	ground station
GSa/s	gigasamples per second
GSM	global system for mobile communication
GUE	ground user / ground user equipment
HAP	high-altitude platform
HD	high definition
HetNet	heterogeneous network
ICI	inter-carrier interference
ICIC	inter-cell interference coordination
IMPC	intermittent multipath component
Infs.	infrastructure
IS-GBSCM	irregular-shaped geometric-based stochastic channel model
ITU	International Telecommunication Union
kbps	kilobits per second
LAP	low-altitude platform
LDACS	L-band digital aeronautical communications systems

LDPLM	log-distance path-loss model
LoS / LOS	line-of-sight
LTE	long-term evolution
LUI	Lisbon University Institute
mAh	milli-amp hour
Mbps	megabits per second
MBS	macro base station
mgf	moment generating function
MIMO	multiple input–multiple output
MISO	multiple input–single output
mmWave	millimeter wave
Mod. sig.	modulated signal
MOI	MBS cell of interest / macro base station cell of interest
MPC	multipath component
mph	miles per hour
MSK	minimum shift keying
MUE	MBS GUE / macro base station ground user equipment
N/A	not applicable / not available
NGSCM	non-geometric stochastic channel model
NLoS / NLOS	non-line-of-sight
OFDM	orthogonal frequency-division multiplexing
OHPLM	Okumura–Hata path-loss model
OLOS	obstructed line-of-sight
PAPR	peak-to-average-power ratio
PBS	pico base station
PDF	probability density function
PDP	power delay profile
PG	path gain
pgfl	probability generating functional
PL	path loss
PLE	path-loss exponent
PPP	Poisson point process
PRN	pseudo-random number
PSC	public safety communications
PSD	power spectral density
QoS	quality of service
RED	random early detection
RF	radio frequency
RHS	Right hand side
RMa	rural macro
RMS-DS	root-mean-square delay spread
RS-GBSCM	regular-shaped geometric-based stochastic channel model
RSRP	reference symbol received power
RSRQ	reference signal receive quality
RSS	received signal strength

RSSI	received signal strength indicator
RTT	round-trip time
r.v.	random variable(s)
RW	random walk
RWP	random waypoint
RX	receiver
Satel.	satellite
SDMA	space-division multiple access
SE	spectral efficiency
SIMO	single input–multiple output
SINR	signal-to-interference-plus-noise ratio
SIR	signal-to-interference ratio
SIRO	service in random order
SISO	single input–single output
SNR	signal-to-noise ratio
TDL	tapped delay line
TDMA	time division multiple access
Terres.	terrestrial
TOA	time of arrival
TX	transmitter
UABS	unmanned aerial base station
UAS	unmanned aircraft system / unmanned aerial system
UAV	unmanned aerial vehicle
UDM	user-dependent model
UE	user / user equipment
UIM	user-independent model
UMa	urban macro
UMi	urban micro
UMTS	universal mobile telecommunications service
UOI	UABS cell of interest / unmanned aerial base station cell of interest
USF	uncoordinated radio subframe
UUE	UABS GUE / unmanned aerial base station ground user equipment
UWB	ultra-wideband
V2V	vehicle-to-vehicle
Vehic.	vehicular
VHF	very high frequency
WSS	wide-sense stationary

Part I

Fundamentals of UAV Communications

1

Overview

Qingqing Wu[1], Yong Zeng[3,4], and Rui Zhang[2]

[1]*State Key Laboratory of Internet of Things for Smart City, University of Macau, China*
[2]*Department of Electrical and Computer Engineering, National University of Singapore, Singapore*
[3]*National Mobile Communications Research Laboratory, Southeast University, China*
[4]*Purple Mountain Laboratories, Jiangsu, China*

1.1 UAV Definitions, Classes, and Global Trend

Unmanned aerial vehicles (UAVs), also commonly known as drones, are aircraft piloted by remote control or embedded computer programs without a human on-board. Historically, UAVs were mainly used in military applications deployed in hostile territory for remote surveillance and armed attack, to reduce pilot losses. In recent years, enthusiasm for using UAVs in civilian and commercial applications has skyrocketed, thanks to the advancement of UAVs' manufacturing technologies and their reducing cost, making them more easily accessible to the public. Nowadays, UAVs have found numerous applications in a proliferation of fields, such as aerial inspection, photography, precision agriculture, traffic control, search and rescue, package delivery, and telecommunications, among others. In June 2016, the US Federal Aviation Administration (FAA) released operational rules for routine civilian use of small unmanned aircraft systems (UASs) with aircraft weight less than 55 pounds (25 kg) [9]. In November 2017, the FAA further launched a national program, namely the "Drone Integration Pilot Program," to explore the expanded use of drones, including beyond-visual-line-of-sight (BVLoS) flights, night-time operations, and flights above people [6]. It is anticipated that these new guidelines and programs will spur the further growth of the global UAV industry in the coming years. The scale of the industry of UAVs is potentially enormous, with realistic predictions in the realm of 80 billion US dollars for the US economy alone, expected to create tens of thousands of new jobs within the next decade [1]. Therefore, UAVs have emerged as a promising technology to offer fertile business opportunities in the next decade.

There are many types of UAVs due to their numerous and diversified applications. UAVs can be practically sorted into different categories according to criteria such as functionality, weight/payload, size, endurance, wing configuration, control method, cruising range, flying altitude, maximum speed, and energy-supplying method. For example, in terms of wing configuration, fixed-wing and rotary-wing UAVs are the two main types of UAVs that

UAV Communications for 5G and Beyond, First Edition.
Edited by Yong Zeng, Ismail Guvenc, Rui Zhang, Giovanni Geraci, and David W. Matolak.
© 2021 John Wiley & Sons Ltd. Published 2021 by John Wiley & Sons Ltd.

Table 1.1 Characteristics of different types of UAVs. Source: From Fotouhi et al. [10].

	Micro	Small	Medium	Large
Example model	Kogan Nano Drone	DJI Spreading Wings S900	Scout B-330 helicopter	Predator B
Weight	16 g	3.3 kg	90 kg	2223 kg
Payload	N/A	4.9 kg	50 kg	1700 kg
Flying mechanism	Multi-rotor	Multi-rotor	Multi-rotor	Fixed wing
Range	50–80 m	N/A	N/A	1852 km
Altitude	N/A	N/A	3 km	5 km
Endurance	6–8 min	18 min	180 min	1800 min
Maximum speed	N/A	57.6 km h^{-1}	100 km h^{-1}	482 km h^{-1}
Power supply	160 mAh Li battery	12000 mAh LiPo battery	21 kW gasoline	712 kW 950 shaft horsepower turboprop engine
Application	Recreation	Professional aerial photography; suitable to carry cellular base stations or user equipment	Data acquisition, HD video live stream; can carry cellular base stations or user equipment	Reconnaissance, airborne surveillance,target acquisition

have been widely used in practice. Typically, fixed-wing UAVs have higher maximum flying speed, greater payloads, and longer endurance as compared to rotary-wing UAVs, while their disadvantages lie in their inability to hover and the fact that a runway or launcher is needed for take-off/landing. In contrast, rotary-wing UAVs are able to take off/land vertically and hover at prescribed locations. Such different characteristics of these two types of UAVs thus have a great impact on their suitable use cases. Another common UAV classification method is based on size. Table 1.1 summarizes several key characteristics of four typical UAVs based on their size. A more comprehensive classification has been provided in [13]. In general, selecting a suitable UAV type is crucial for accomplishing their mission efficiently, which needs to take into account their specifications as well as the requirements of practical applications.

1.2 UAV Communication and Spectrum Requirement

An essential enabling technology of UAS is wireless communication. On the one hand, UAVs need to exchange safety-critical information with various parties such as remote pilots, nearby aerial vehicles, and air traffic controllers, to ensure safe, reliable, and efficient flight operation. This is commonly known as control and non-payload communication (CNPC) [11]. On the other hand, depending on their missions, UAVs may need to transmit and/or receive in a timely manner mission-related data such as aerial images, high-speed video, and data packets for relaying to/from various ground entities such as UAV operators, end-users, or ground gateways. This is known as payload communication.

Table 1.2 UAV communication requirements specified by 3GPP. Source: Data from 3GPP TR 36.777 [2].

	Data type	Data rate	Reliability	Latency
DL (ground station to UAV)	Command and control	60–100 kbps	10^{-3} packet error rate	50 ms
UL (UAV to ground station)	Command and control	60–100 kbps	10^{-3} packet error rate	N/A
	Application data	Up to 50 Mbps	N/A	Similar to ground user

Enabling reliable and secure CNPC links is a necessity for the large-scale deployment and wide usage of UAVs. The International Telecommunication Union (ITU) has classified the required CNPC to ensure safe UAV operations into three categories [11].

- *Communication for UAV command and control*: This includes the telemetry report (e.g., flight status) from the UAV to the ground pilot, the real-time telecommand signaling from ground to UAVs for non-autonomous UAVs, and regular flight command update (such as waypoint update) for (semi-)autonomous UAVs.
- *Communication for air traffic control (ATC) relay*: It is critical to ensure that UAVs do not cause any safety threat to traditional manned aircraft, especially for operations approaching areas with a high density of aircraft. To this end, a link between the air traffic controller and the ground control station via the UAV, called ATC relay, is required.
- *Communication supporting "sense and avoid"*: The ability to support "sense and avoid" ensures that the UAV maintains sufficient safety distance from nearby aerial vehicles, terrain, and obstacles.

The specific communication and spectrum requirements in general differ for CNPC and payload communications. Recently, the 3rd generation partnership project (3GPP) has specified the communication requirements for these two types of links [2], which are summarized in Table 1.2. CNPC is usually of low data rate, say, in the order of kilobits per second (kbps), but has rather stringent requirement on high reliability and low latency. For example, as shown in Table 1.2, the data rate requirement for UAV command and control is only in the range of 60–100 kbps for both downlink (DL) and uplink (UL) directions, but a reliability of less than 10^{-3} packet error rate and a latency less than 50 ms (milliseconds) are required. While the communication requirements of CNPC links are similar for different types of UAVs due to their common safety considerations, those for payload data are highly application-dependent. In Table 1.3, we list several typical UAV applications and their corresponding data communication requirements based on [4].

Since the loss of CNPC link may cause catastrophic consequences, the International Civil Aviation Organization (ICAO) has determined that CNPC links for UAVs must operate over protected aviation spectrum [8, 12]. Furthermore, ITU studies have revealed that, to support CNPC for the forecast number of UAVs in coming years, 34 MHz (megahertz) terrestrial

Table 1.3 Communication requirements for typical UAV applications. Source: Data from China mobile technical report [4].

UAV application	Height coverage (m)	Payload traffic latency (ms)	Payload data rate (DL/UL)
Drone delivery	100	500	300 kbps/200 kbps
Drone filming	100	500	300 kbps/30 Mbps
Access point	500	500	50 Mbps/50 Mbps
Surveillance	100	3000	300 kbps/10 Mbps
Infrastructure inspection	100	3000	300 kbps/10 Mbps
Drone fleet show	200	100	200 kbps/200 kbps
Precision agriculture	300	500	300 kbps/200 kbps
Search and rescue	100	500	300 kbps/6 Mbps

spectrum and 56 MHz satellite spectrum are needed for supporting both line-of-sight (LoS) and beyond-LoS UAV operations [11]. To meet such requirements, the C-band spectrum at 5030–5091 MHz was made available for UAV CNPC at the 2012 World Radiocommunication Conference (WRC-12). More recently, the WRC-15 has decided that geostationary Fixed Satellite Service (FSS) networks may be used for UAS CNPC links.

Compared to CNPC, UAV payload communication usually has a much higher data rate requirement. For instance, to support the transmission of full high-definition (FHD) video from the UAV to the ground user, the transmission rate is about several megabits per second (Mbps), while for 4K video, it is higher than 30 Mbps. The rate requirement for UAV serving as an aerial communication platform can be even higher, e.g., up to dozens of gigabits per second (Gbps) for data forwarding/backhauling applications.

1.3 Potential Existing Technologies for UAV Communications

To support the CNPC and payload communication in multifarious UAV applications, proper wireless technologies need to be selected for achieving seamless connectivity and high reliability/throughput for both air-to-air and air-to-ground wireless communications in 3D space. Towards this end, four candidate communication technologies are listed and compared in Table 1.4, including (i) direct link, (ii) satellite, (iii) ad-hoc network, and (iv) cellular network.

1.3.1 Direct Link

Due to its simplicity and low cost, the direct point-to-point communication between a UAV and its associated ground node over the unlicensed band (e.g., the Industrial Scientific Medical (ISM) 2.4 GHz band) was most commonly used for commercial UAVs in the past, where the ground node can be a joystick, remote controller, or ground station. However, it is usually limited to LoS communication, which significantly constrains its operation range and hinders its applications in complex propagation environments. For example, in

Table 1.4 Comparison of wireless technologies for UAV communication.

Technology	Description	Advantages	Disadvantages
Direct link	Direct point-to-point communication with ground node	Simple, low cost	Limited range, low data rate, vulnerable to interference, non-scalable
Satellite	Communication and Internet access via satellite	Global coverage	Costly, heavy/bulky/energy-consuming communication equipment, high latency, large signal attenuation
Ad-hoc network	Dynamically self-organizing and infrastructure-free network	Robust and adaptable, support for high mobility	Costly, low spectrum efficiency, intermittent connectivity, complex routing protocol
Cellular network	Enabling UAV communications by using cellular infrastructure and technologies	Almost ubiquitous accessibility, cost-effective, superior performance and scalability	Unavailable in remote areas, potential interference with terrestrial communications

urban areas, the communication can be easily blocked by, e.g., trees and high-rise buildings, which results in poor reliability and low data rate. In addition, such a simple solution is usually insecure and vulnerable to interference and jamming. Due to the above limitations, the simple direct-link communication is not a scalable solution for supporting large-scale deployment of UAVs in the future.

1.3.2 Satellite

Enabling UAV communications by leveraging satellites is a viable option due to their global coverage. Specifically, satellites can help relay data communicated between widely separated UAVs and ground gateways, which is particularly useful for UAVs above oceans and in remote areas where terrestrial network (WiFi or cellular) coverage is unavailable. Furthermore, satellite signals can also be used for navigation and localization of UAVs. In WRC-15, the conditional use of satellite communication frequencies in the Ku/Ka band has been approved to connect drones to satellites, and some satellite companies such as Inmarsat have launched a satellite communication service for UAVs [5]. However, there are also several disadvantages of satellite-enabled UAV communications. First, the propagation loss and delay are quite significant due to the long distances between satellite and low-altitude UAVs/ground stations. This thus poses great challenges for meeting ultra-reliable and delay-sensitive CNPC for UAVs. Second, UAVs usually have stringent size, weight, and power (SWAP) constraints, and thus may not be able to carry the heavy, bulky, and energy-consuming satellite communication equipment (e.g., dish antenna) required. Third, the high operational cost of satellite communication also hinders its wide use for densely deployed UAVs in consumer-grade applications.

1.3.3 Ad-Hoc Network

Mobile ad-hoc network (MANET) is an infrastructure-free and dynamically self-organizing network for enabling peer-to-peer communications among mobile devices such as laptops, cellphones, and walkie-talkies. Such devices usually communicate over bandwidth-constrained wireless links using, e.g., IEEE 802.11 a/b/g/n. Each device in a MANET can move randomly over time; as a result, its link conditions with other devices may change frequently. Furthermore, for supporting communications between two far-apart nodes, some other nodes in between need to help forward the data via multi-hop relaying, thus incurring more energy consumption, low spectrum efficiency, and long end-to-end delay. Vehicular ad-hoc network (VANET) and flying ad-hoc network (FANET) are two applications of MANET, for supporting communications among high-mobility ground vehicles and UAVs in 2D and 3D networks, respectively [7].

The topology or configuration of a FANET for UAVs may take different forms, such as a mesh, ring, star, or even a straight line, depending on the application scenario. For example, a star network topology is suitable for UAV swarm applications, for which UAVs in a swarm all communicate through a central hub UAV that is responsible for communicating with the ground stations. Although FANET is a robust and flexible architecture for supporting UAV communications in a small network, it is generally unable to provide a scalable solution for serving massive UAVs deployed in a wide area, due to the complexities and difficulties for realizing a reliable routing protocol over the whole network with dynamic and intermittent link connectivities among the flying UAVs.

1.3.4 Cellular Network

It is evident that the aforementioned technologies generally cannot support large-scale UAV communications in a cost-effective manner. On the other hand, it is also economically nonviable to build new and dedicated ground networks for achieving this goal. As such, there has been significantly growing interest recently in leveraging the existing as well as future-generation cellular networks for enabling UAV–ground communications [17]. Thanks to the almost ubiquitous coverage of the cellular network worldwide as well as its high-speed optical backhaul and advanced communication technologies, both CNPC and payload communication requirements for UAVs can be potentially met, regardless of the density of UAVs as well as their distances from the corresponding ground nodes. For example, the forthcoming fifth-generation (5G) cellular network is expected to support a peak data rate of 10 Gbps with only 1 ms round-trip latency, which in principle is adequate for high-rate and delay-sensitive UAV communication applications such as real-time video streaming and data relaying.

Despite the promising advantages of cellular-enabled UAV communications, there are still scenarios where the cellular services are unavailable, e.g., in remote areas such as sea, desert, and forest. In such scenarios, other technologies such as direct link, satellite, and FANET can be used to support UAV communications beyond the terrestrial coverage of cellular networks. Therefore, it is envisioned that the future wireless network for supporting large-scale UAV communications will have an integrated 3D architecture consisting of UAV-to-UAV, UAV-to-satellite, and UAV-to-ground communications, as shown in

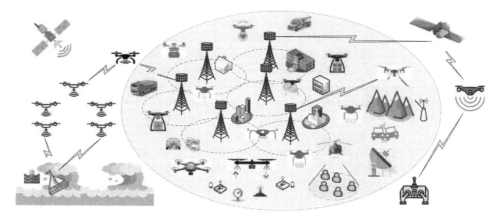

Figure 1.1 Supporting UAV communications with an integrated network architecture. Source: From Zeng et al. [19].

Figure 1.1, where each UAV may be enabled with one or more communication technologies to exploit the rich connectivity diversity in such a hybrid network.

1.4 Two Paradigms in Cellular UAV Communications

In this section, we further discuss the aforementioned new paradigm of integrating UAVs into the cellular network, to provide their full horizon of applications and benefits. In particular, we partition our discussion into two main categories. On the one hand, UAVs are considered as new aerial users that access the cellular network from the sky for communications, which we refer to as *cellular-connected UAVs*. On the other hand, UAVs are used as new aerial communication platforms such as base stations (BSs) and relays, to assist in terrestrial wireless communications by providing data access from the sky, thus called *UAV-assisted wireless communications*.

1.4.1 Cellular-Connected UAVs

By incorporating UAVs as new user equipment in the cellular network, the following benefits can be achieved [17]. First, thanks to the almost worldwide accessibility of cellular networks, a cellular-connected UAV makes it possible for the ground pilot to remotely command and control the UAV with virtually unlimited operation range. Besides, it also provides an effective solution to maintain wireless connectivity between UAVs and various other stakeholders, such as the end-users and air traffic controllers, regardless of their locations. This thus opens up many new UAV applications in the future.

Second, with the advanced cellular technologies and authentication mechanisms, a cellular-connected UAV is expected to achieve significant performance improvement over the other technologies introduced in Section 1.3, in terms of reliability, security, and data throughput. For instance, the current fourth-generation (4G) long-term evolution (LTE) cellular network employs a scheduling-based channel access mechanism, where multiple users can be served simultaneously by assigning them orthogonal resource blocks (RBs). In contrast, WiFi (e.g., 802.11g employed in FANET) adopts contention-based channel access with a random backoff mechanism, where users are allowed to access only channels that are sensed to be idle. Thus, multiuser transmission with centralized scheduling/control enables the cellular network to make a more efficient use of the spectrum than WiFi, especially when the user density is high. In addition, UAV-to-UAV communication can also be realized by leveraging the available device-to-device (D2D) communications in LTE and 5G systems.

Third, a cellular-assisted localization service can provide UAVs with a new and complementary means in addition to the conventional satellite-based global positioning system (GPS) for achieving more robust UAV navigation performance. Last, but not least, a cellular-connected UAV is a cost-effective solution since it reuses the millions of cellular BSs worldwide without the need to build new infrastructure dedicated for UAS only. Thus, the cellular-connected UAV is expected to be a win–win technology for both UAV and cellular industries, with rich business opportunities to explore in the future.

1.4.2 UAV-Assisted Wireless Communications

Thanks to the continuous cost reduction in UAV manufacturing and device miniaturization in communication equipment, it has become more feasible to mount compact and small-size BSs or relays on UAVs to enable flying aerial platforms to assist in terrestrial wireless communications. For instance, commercial LTE BSs with light weight (e.g., less than 4 kg) are already available in the market, which are suitable to be mounted on UAVs with moderate payload.

Compared to conventional terrestrial communications with typically static BSs/relays deployed at fixed locations, UAV-assisted communications bring the following main advantages [21]. First, UAV-mounted BSs/relays can be swiftly deployed on demand. This is especially appealing for application scenarios such as temporary or unexpected events, emergency response, and search and rescue, among others. Second, thanks to their high altitude above the ground, UAV BSs/relays are more likely to have LoS connection with their ground users as compared to their terrestrial counterparts, thus providing more reliable links for communication as well as multiuser scheduling and resource allocation. Third, thanks to the controllable high mobility of UAVs, UAV BSs/relays possess an additional degree of freedom (DoF) for communication performance enhancement, by dynamically adjusting their locations in 3D to cater for the terrestrial communication demands.

For 5G wireless networks, the three most representative commercial scenarios are enhanced mobile broadband (eMBB), massive machine-type communications (mMTC), and ultra-reliable and low-latency communications (URLLC) (also known as mission-critical communications), which are particularly appealing for UAV communications. Specifically, eMBB supports reliable connections with very high peak data rates, as well as moderate rates for cell-edge users; mMTC supports a massive number of Internet-of-Things (IoT) devices, which are only sporadically active and send small data payloads; and URLLC supports low-latency transmissions of small payloads with very high reliability from a limited set of terminals, which are active according to patterns typically specified by outside events, such as alarms. As such, the advantages of UAV-assisted communication make it a promising technology to support the main 5G applications with ever-increasing and highly dynamic wireless data traffic.

1.5 New Opportunities and Challenges

The integration of UAVs into cellular networks, as either aerial users or communication platforms, brings new design opportunities as well as challenges. Both cellular-connected UAV communication and UAV-assisted wireless communication are significantly different from their terrestrial counterparts, due to the high altitude and high mobility of UAVs, the high probability of UAV–ground LoS channels, the distinct communication quality of service (QoS) requirements for CNPC versus mission-related payload data, the stringent SWAP constraints of UAVs, as well as the new design DoF by jointly exploiting the UAV mobility control and communication scheduling/resource allocation. Table 1.5 summarizes the main design opportunities and challenges of cellular communications with UAVs, which are further elaborated as follows.

1.5.1 High Altitude

Compared with conventional terrestrial BSs/users, UAV BSs/users usually have much higher altitude. For instance, the typical height of a terrestrial BS is around 10 m for Urban Micro (UMi) deployment and 25 m for Urban Macro (UMa) deployment [2], whereas the current regulation already allows UAVs to fly up to 122 m [9]. For cellular-connected UAVs, the high UAV altitude requires cellular BSs to offer 3D aerial coverage for UAV users, in contrast to the conventional 2D coverage for terrestrial users. However, existing BS antennas are usually tilted downwards, either mechanically or electronically, to cater for the ground coverage as well as suppressing inter-cell interference. Nevertheless, preliminary

Table 1.5 New opportunities and challenges in UAV communications.

Characteristic	Opportunities	Challenges
High altitude	Wide ground coverage as aerial BS/relay	Requires 3D cellular coverage for aerial user
High LoS probability	Strong and reliable communication link; high macro-diversity; slow communication scheduling and resource allocation	Severe aerial–terrestrial interference; susceptible to terrestrial jamming/eavesdropping
High 3D mobility	Traffic-adaptive movement; QoS-aware trajectory design	Handover management; wireless backhaul
SWAP constraint	N/A	Limited payload and endurance; energy-efficient design; compact and lightweight BS/relay and antenna design

field measurement results have demonstrated satisfactory aerial coverage to meet the basic communication requirements by the antenna sidelobes of BSs for UAVs below 400 feet (122 m) [3]. However, as the altitude further increases, weak signal coverage is observed, which thus calls for new BS antenna designs and cellular communication techniques to achieve satisfactory UAV coverage up to the maximum altitude of 300 m as currently specified by 3GPP [2]. On the other hand, for UAV-assisted wireless communications, the high UAV altitude enables the UAV BS/relay to achieve wider ground coverage as compared to their terrestrial counterparts.

1.5.2 High LoS Probability

The high UAV altitude leads to unique air–ground channel characteristics as compared to terrestrial communication channels. Specifically, compared to the terrestrial channels, which generally suffer severe path loss due to shadowing and multipath fading effects, the UAV–ground channels, including both the UAV–BS and UAV–user channels, typically experience limited scattering and thus have a dominant LoS link with high probability. The LoS-dominant air–ground channel brings both opportunities and challenges to the design of UAV communications as compared to traditional terrestrial communications. On the one hand, it offers more reliable link performance between the UAV and its serving/served ground BSs/users, as well as a pronounced macro-diversity in terms of more flexible UAV–BS/user associations. Moreover, as LoS-dominant links have less channel variations in time and frequency, communication scheduling and resource allocation can be more efficiently implemented at a much slower pace as compared to that over terrestrial fading channels. On the other hand, however, it also causes strong air–ground interference, which is a critical issue that may severely limit the cellular network capacity with coexisting aerial and terrestrial BSs/users. For example, in the UL communication of a UAV user, it may pose severe interference to many adjacent cells in the same frequency band due to its high-probability LoS channels with their BSs; while in the DL communication, the UAV user also suffers strong interference from these co-channel BSs. Interference mitigation is crucial for both frameworks of cellular-connected UAVs and UAV-assisted terrestrial communications. Furthermore, the LoS-dominant air–ground links also make UAV communications more susceptible to jamming/eavesdropping attacks by malicious ground nodes as compared to the terrestrial communications over fading channels, thus imposing a new security threat at the physical layer [14].

1.5.3 High 3D Mobility

Different from the terrestrial networks where the BSs/relays are usually at fixed locations and the users move sporadically and randomly, UAVs can move at high speed in 3D space with partially or fully controllable mobility. On the one hand, the high mobility of UAVs generally results in more frequent handovers and time-varying wireless backhaul links with ground BSs/users. On the other hand, it also leads to an important new design approach of communication-aware UAV mobility control, such that the UAV's position, altitude, speed, heading direction, etc., can be dynamically changed to better meet its communication objectives. For example, in UAV-assisted wireless communication, UAV BSs/relays can design their trajectories (i.e., locations and speeds over time) either off-line or in real time to adapt to the locations and communication channels of their served ground users. Similarly, for cellular-connected UAVs, they can also adjust their trajectories based on the locations of the ground BSs to find the best route to fulfill their mission requirements, and in the meanwhile ensure that it is covered by at least one BS along its flight path to satisfy its communication needs. Furthermore, UAV 3D placement/trajectory design can be jointly considered with communication scheduling and resource allocation for further performance improvement.

1.5.4 SWAP Constraints

Different from terrestrial communication systems where the ground BSs/users usually have a stable power supply from the grid or a rechargeable battery, the SWAP constraints of UAVs pose critical limits on their endurance and communication capabilities. For example, in the case of UAV-assisted wireless communications, customized BSs/relays, generally of smaller size and lighter weight as well as with more compact antenna and power-efficient hardware as compared to their terrestrial counterparts, need to be designed to cater for the limited payload and size of UAVs. Furthermore, besides the conventional communication transceiver energy consumption, UAVs need to spend additional propulsion energy to remain aloft and move freely in the air [16, 18], which is usually much more significant than the communication-related energy (e.g., in the order of kilowatt versus watt). Thus, the energy-efficient design of UAV communication is more involved than that for the conventional terrestrial systems considering the communication energy only [15, 20].

1.6 Chapter Summary and Main Organization of the Book

In this chapter, we have given an overview of UAV communications, which have found numerous applications and are expected to bring fertile business opportunities in the next decade. The spectrum requirement was first discussed for both CNPC and payload communication in different applications. We then discussed the potential wireless technologies for supporting UAV communications, while a hybrid use of them was anticipated for achieving seamless connectivity and high reliability/throughput for UAV communications in the future. Then we discussed two promising research and application frameworks of UAV communications in cellular networks, namely *UAV-assisted wireless communications* and *cellular-connected UAVs*, where UAVs serve as aerial communication platforms and users, respectively. Compared to the conventional terrestrial communications, UAVs' communications face new opportunities and challenges due to their high altitude above the ground and great flexibility of movement in 3D space. Several critical issues arise, including the LoS-dominant UAV–ground channels and resultant strong aerial–terrestrial network interference, the distinct communication quality of service (QoS) requirements for UAV control messages versus payload data, the stringent constraints imposed by the SWAP limitations of UAVs, as well as the exploitation of the new design DoF brought by the highly controllable 3D UAV mobility.

This book brings together experts around the world working in related areas to discuss the various aspects of UAV communications. Including this one, it contains a total of

18 chapters, which are organized into four parts. Part I aims to introduce some basics about UAV communications. Besides the current chapter (Chapter 1), which gives an overview of integrating UAVs into 5G and beyond, Part I also include Chapter 2, which provides a comprehensive review on the air–ground propagation channel measurements and modelling, and Chapter 3, which discusses the detection and classification techniques of UAVs.

Part II mainly focuses on the first paradigm of integrating UAVs into 5G and beyond, namely cellular-connected UAVs, and it contains five chapters. Chapter 4 gives the performance analysis of UAV systems supported by cellular networks, in terms of coverage probability, spectral efficiency, as well as throughput. Chapter 5 studies LTE-connected UAVs via both experiments and simulations, and also discusses various performance enhancement techniques from the terminal side and network side, respectively. In Chapter 6, the recent standardization efforts of supporting UAVs via cellular networks by the 3rd generation partnership project (3GPP) are introduced. Chapter 7 takes a step further, by first studying the performance of UAVs supported by current cellular network infrastructure based on the latest 3GPP models, and then demonstrating the enhanced performance achieved by massive multiple input–multiple output (MIMO) technology. In Chapter 8, high-capacity UAV communications via millimeter wave (mmWave) is discussed, where the topics covered include the aerial mmWave channel modeling, channel estimation, beam training and tracking, as well as hybrid precoding.

Part III of this book mainly focuses on the paradigm of UAV-assisted wireless communications, and consists of six chapters. Chapter 9 gives a performance analysis of cellular networks with UAV BSs to study the impact of UAV mobility by using the powerful tool of stochastic geometry. Chapter 10 studies the heterogeneous cellular network consisting of the conventional ground macro-BSs and small cells formed by UAV BSs, where the UAV placement and interference coordination are optimized. In Chapter 11, the highly controllable UAV mobility is exploited for UAV-assisted communications, where the main techniques for joint UAV trajectory design and communication resource optimization are discussed. Chapter 12 tackles the limited on-board energy of UAVs, where a new framework for energy-efficient UAV communications is introduced by taking into account the UAV's propulsion energy consumption. In Chapter 13, the various trade-offs among throughput, communication delay, and energy consumption are discussed for single- and multi-UAV-assisted wireless communications. Chapter 14 focuses on the spectrum sharing problem between the small cells formed by UAV BSs and the traditional cellular networks to maximize network throughput.

Part IV of this book contains four chapters to discuss some other advanced technologies for UAV communication. In particular, Chapter 15 investigates the utilization of non-orthogonal multiple access (NOMA) technique for UAV communications to enhance the spectrum and energy efficiency. In Chapter 16, the physical layer security of UAV communications is discussed. Chapter 17 studies the radio-frequency wireless power transfer with flying wireless chargers mounted on UAVs, where the UAV trajectory is optimized to maximize the minimum received energy among all the energy receivers. Last, but not least, Chapter 18 discusses the ad-hoc networking in the sky as an efficient solution to achieve the beyond-radio-line-of-sight (BRLOS) connectivity among UAVs. Besides presenting the fundamental concepts, challenges and solutions for creating ad-hoc networks with UAVs, the chapter also discusses the current standards and products available for mesh networking of UAVs.

References

1 Inc.com (2016). With 1 announcement, the FAA just created an $82 billion market and 100,000 new jobs. https://www.inc.com/yoram-solomon/with-one-rule-the-faa-just-created-an-82-billion-market-and-100000-new-jobs.html (accessed 18 February 2019).

2 3GPP TR 36.777 (2017). *Technical specification group radio access network: study on enhanced LTE support for aerial vehicles*, v.15.0.0.

3 Qualcomm (2016). Paving the path to 5G: optimizing commercial LTE networks for drone communication. https://www.qualcomm.com/news/onq/2016/09/06/paving-path-5g-optimizing-commercial-lte-networks-drone-communication (accessed 18 February 2019).

4 China Mobile (2017). China Mobile technical report: Internet of drones (in Chinese). http://www.jintiankansha.me/t/AE9FsWW9tc (accessed 18 February 2019).

5 Inmarsat (2017). Launch of Inmarsat SwiftBroadband unmanned aerial vehicle service to provide operational capability boost. https://www.inmarsat.com/press-release/launch-inmarsat-swiftbroadband-unmanned-aerial-vehicle-service-provide-operational-capability-boost/ (accessed 18 February 2019).

6 FAA (2018). UAS Integration Pilot Program Resources. https://www.faa.gov/uas/programs_partnerships/integration_pilot_program/ (accessed 18 February 2019).

7 I. Bekmezci, O. K. Sahingoz, and Ş. Temel (2013). Flying ad-hoc networks (FANETs): a survey. *Ad Hoc Netw.*, 11 (3): 1254–1270.

8 C. Carlos (2017). Spectrum management issues for the operation of commercial services with UAVs. https://ssrn.com/abstract=2944132 or http://dx.doi.org/10.2139/ssrn.2944132 (accessed 18 February 2019).

9 FAA (2016). Summary of small unmanned aircraft rule. https://www.faa.gov/uas/media/Part_107_Summary.pdf (accessed 18 February 2019).

10 A. Fotouhi, H. Qiang, M. Ding et al. (2019). Survey on UAV cellular communications: practical aspects, standardization advancements, regulation, and security challenges. *IEEE Commun. Surveys Tuts.*, early access.

11 ITU (2009). Characteristics of unmanned aircraft systems and spectrum requirements to support their safe operation in non-segregated airspace. ITU Tech. Rep. M.2171.

12 R. J. Kerczewski, J. D. Wilson, and W. D. Bishop (2013). Frequency spectrum for integration of unmanned aircraft. In *Proceedings of the IEEE/AIAA Digital Avionics Systems Conference (DASC)*.

13 K. P. Valavanis and G. J. Vachtsevanos (2015). *Handbook of Unmanned Aerial Vehicles*. Springer Netherlands.

14 Q. Wu, W. Mei, and R. Zhang (2019). Safeguarding wireless network with UAVs: a physical layer security perspective. *IEEE Wireless Commun.*, June, to appear.

15 Y. Chen, S. Zhang, S. Xu, and G. Y. Li (2011). Fundamental trade-offs on green wireless networks. *IEEE Commun. Mag.* 49 (6): 30–37.

16 Y. Zeng and R. Zhang (2017). Energy-efficient UAV communication with trajectory optimization. *IEEE Trans. Wireless Commun.* 16 (6): 3747–3760.

17 Y. Zeng, J. Lyu, and R. Zhang (2019). Cellular-connected UAV: potentials, challenges and promising technologies. *IEEE Wireless Commun.* 26 (1): 120–127.

18 Y. Zeng, J. Xu, and R. Zhang (2019). Energy minimization for wireless communication with rotary-wing UAV. *IEEE Trans. Wireless Commun.*, 18 (4): 2329–2345.

19 Y. Zeng, Q. Wu, and R. Zhang (2019). Accessing from the sky: a tutorial on UAV communications for 5G and beyond. *Proc. IEEE*, submitted, arXiv/1903.05289.

20 Z. Hasan, H. Boostanimehr, and V. K. Bhargava (2011). Green cellular networks: a survey, some research issues and challenges. *IEEE Commun. Surveys Tuts.* 13 (4): 524–540.

21 Y. Zeng, R. Zhang, and T. J. Lim (2016). Wireless communications with unmanned aerial vehicles: opportunities and challenges. *IEEE Commun. Mag.* 54 (5): 36–42.

2

A Survey of Air-to-Ground Propagation Channel Modeling for Unmanned Aerial Vehicles

Wahab Khawaja[1], Ismail Guvenc[1], David W. Matolak[2], Uwe-Carsten Fiebig[3], and Nicolas Schneckenberger[3]*

[1]Department of Electrical and Computer Engineering, North Carolina State University, Oval Drive, Raleigh, NC 27606, USA
[2]Department of Electrical Engineering, University of South Carolina, Main Street, Columbia, SC 29208, USA
[3]Institute of Communications and Navigation, German Aerospace Center (DLR), Wessling 82234, Germany

2.1 Introduction

Unmanned aerial vehicles (UAVs), also commonly referred to as "drones," have long been used for military and specialized applications [25, 67, 115–117]. Owing to recent technological advancements, they have attracted major attention from industry for uses ("use cases") from package delivery to communications, surveillance, inspection, transportation, search and rescue, and others [1–6, 23, 42, 66, 96]. These UAVs can vary in size from small toys that fit in the palm of a human hand to large military aircraft such as the General Atomics MQ-9 Reaper (commonly termed Predator) [136], with a wingspan over 15 m. The small battery-powered toys can typically fly for up to 15 minutes, whereas the larger UAVs can be designed for long-endurance (30 hours) and high-altitude operations (higher than 15 km). According to Goldman Sachs, there exists a $100 billion market opportunity for UAVs in the commercial, civil government, and military sectors combined till 2020 [52].

Wireless connectivity with UAVs is a requirement for their integration into any national airspace and for facilitating new usages. However, UAV operating environments and scenarios introduce unique technical challenges, and these are presently being investigated by telecommunications companies such as AT&T [19], Vodafone [132], Ericsson [84], Nokia [18, 77], and Qualcomm [104], among others. A major challenge is obtaining realistic air-to-ground (AG) propagation models for various UAV operating environments and scenarios. Having accurate characterization of the AG channel is of paramount importance for designing robust and effective waveforms, modulation, resource allocation, link adaptation, and multiple antenna techniques.

The AG channel for UAVs has not been studied as extensively as terrestrial propagation channels. The available AG propagation channel models for high-altitude aeronautical communications generally cannot be employed directly for low-altitude UAV

---

* Corresponding Author: Wahab Khawaja

UAV Communications for 5G and Beyond, First Edition.
Edited by Yong Zeng, Ismail Guvenc, Rui Zhang, Giovanni Geraci, and David W. Matolak.

communications due to their differences in the channel scattering environment. Small UAVs may also possess distinct structural and flight characteristics such as different airframe shadowing features due to unique body shapes and materials, and potentially sharper rates of change of pitch, roll, and yaw during flight. In this chapter, we provide a comprehensive, unified review of the existing work on UAV AG propagation channels. We discuss recent channel measurement campaigns and modeling efforts to characterize the AG channel for UAVs. We also describe future research challenges and possible enhancements relating to UAV propagation channels.

This chapter is organized as follows. Section 2.2 provides a review of various use cases for UAVs to provide some context, and gives a brief literature review on existing efforts for modeling propagation characteristics of aerial links. Section 2.3 explains some of the unique propagation channel characteristics for UAV AG channels such as operating frequencies, scattering geometries, antenna effects, and Doppler, all in comparison with terrestrial channels. A review of some key considerations for AG channel measurements is provided in Section 2.4, including actual measurement frequencies, platform configurations, measurement environments, unique challenges for AG measurements, sounding waveform types for AG measurements, and the effects of elevation angle on measurement results.

Section 2.5 completes the review of existing UAV AG propagation measurements and some representative simulation studies in the literature. In particular, this section reviews the path loss and shadowing, delay dispersion, narrowband fading, Doppler spread, throughput and bit error rate characteristics, and the effects of different measurement environment types. Section 2.6 discusses AG propagation channel models, including models based on deterministic and stochastic models, their combination, and ray tracing simulations. After classification and review of different channel model types, we provide a survey of path-loss and large-scale fading models, airframe shadowing, small-scale fading models, modeling of intermittent multipath components (MPCs), effects of the choice of frequency band, and multiple input–multiple output (MIMO) propagation models. We also review the recent 3GPP UAV AG channel models, and provide a comparison of existing AG propagation models with each other and with traditional cellular and satellite channel models. Finally, Section 2.7 provides some concluding remarks.

The acronyms used in this chapter are given (along with others used throughout the book) in the list of acronyms in the front matter; they are given in full herein on first mention. The variables used throughout this chapter are given in Table 2.1. Multiple tables and figures are provided to list, classify, and review existing aerial channel modeling studies in the literature. Table 2.2 provides a review of AG channel measurements in the literature along with their measurement configurations, Table 2.3 classifies related measurement studies with respect to five different measurement environments, Table 2.4 summarizes the existing literature on large-scale AG propagation and key path-loss parameters, and finally, Table 2.5 summarizes the small-scale AG model parameters documented in the literature. On the other hand, Figure 2.2 provides a taxonomy of measurement scenarios and related literature in terms of aerial vehicle type and measurement environment, while Figure 2.4 classifies the literature on UAV AG channel models in terms of deterministic and stochastic models.

Table 2.1 Variables used in this chapter.

Variable	Definition	Variable	Definition
A_m	Median attenuation as compared to FSPL	a	Amplitude of the MPC
b	Correction factor	c	Speed of light
d	Link distance between TX and RX	d_h	Horizontal distance between UAV and base station
d_0	Reference distance between TX and RX	f	Frequency instance
f_c	Carrier frequency	f_d	Doppler frequency shift of the MPC
f_{MHz}	Frequency in MHz	Δh	Height difference between UAV and GS
h_G	Ground station height	h_{RX}	Height of receiver
h_S	Height of scatterer	h_{TX}	Height of transmitter
h_U	UAV altitude above ground	G_A	Gain from the environment
G_{Gr}	Antenna gain for the ground-reflected component	G_{LOS}	Antenna gain for LOS component
K-factor	Ricean K-factor	L_{rts}	Loss due to rooftop to street diffraction and scattering
L_{msd}	Loss due to multi-screen diffraction	M	Total number of MPCs
n	Frequency-dependent factor	p	MPC persistence coefficient
P_{d_0}	Received power at d_0	P_R	Received power
P_T	Transmit power	PL	Modified FSPL
PL_0	Reference path loss	PL_b	Base propagation loss
PL_{CI}	Close-in PL	PL_{DS}	Dual-slope PL
PL_F	FSPL	PL_{FI}	Floating-intercept PL
PL_{LOS}	PL for LOS	PL_m	Median propagation loss
PL_{NLOS}	PL for NLOS	r	Horizontal distance between UAV and GS
r_1	LOS component path length	r_2	Ground-reflected component path length
t	Time instant	υ	Velocity of UAV
υ_{max}	Maximum speed	X	Shadowing random variable
X_{FS}	Random variations in the CI PL model	X_{FI}	Random variations in the FI PL model
X_{DS}	Random variations in the DS PL model	γ_s	Propagation loss slope
Γ	Ground reflection coefficient	α_0	Adjustment factor
α	Slope of linear least-squares regression fit	β	Y-intercept point for the linear least-squares regression fit
Θ	Aggregated phase angles	θ	Elevation angle
θ_{Gz}	Grazing angle	ϕ	Phase of the MPC
$\Delta\psi$	Phase difference between the LOS and ground-reflected MPC	γ	Path-loss exponent
τ	Delay of the MPC	λ	Wavelength of the radio wave
σ	Standard deviation of shadow fading	ς	Ratio of built-up area to total area
ξ	Mean number of buildings per unit area	Ω	Height distribution of buildings

2.2 Literature Review

Various organizations have developed classifications for UAVs according to size, with designations large, medium, and small being typical. In the USA, the Federal Aviation Administration (FAA) has issued rules for small UAVs weighing less than 55 pounds (25 kg) [46]. Highlights of these rules include the requirement for a visual line-of-sight (LOS) from pilot to aircraft, flight under daylight or during twilight (within 30 minutes of official sunrise/sunset) with appropriate lighting for collision avoidance, a maximum flight ceiling of 400 feet (122 m) above the ground (higher if the UAV is within 122 m of a construction site), and a maximum speed of 100 mph (87 knots, or 161 km h^{-1}). Restrictions also apply regarding proximity to airports, and, generally, a licensed pilot must operate or supervise UAV operation. In this chapter our focus is on the smaller UAVs, and specifically on the AG channel between these UAVs and the ground stations (GSs) with which they communicate. These GSs are also usually the UAV control stations.

2.2.1 Literature Review on Aerial Propagation

The available research on the UAV-based AG wireless propagation channel can be largely classified into two categories. The first one is payload communications, which can be narrowband or wideband and is mostly application-dependent. The second one is control and non-payload communications (CNPC) for telemetric control of UAVs; CNPC is largely synonymous with command and control (C2). Most of the CNPC employs the unlicensed bands, e.g., 2.4 GHz and 5.8 GHz, which is not preferred by the aviation community, as these bands can be congested and easily jammed. In the USA, CNPC is potentially planned for a portion of L-band (0.9–1.2 GHz) and C-band (5.03–5.091 GHz), though use of these bands is still being negotiated [63, 71]. On the other hand, channel measurements and modeling for UAVs are (other than bandwidth and carrier frequency) largely independent of whether signaling is for payload or CNPC.

AG communications can be traced back to approximately 1920 [135], when manually operated radio telegraphs were being used. Lower frequency bands were used in the early 1930s and these links did not support simultaneous voice communications in both directions (AG and ground-to-air). From the early 1940s, double-sideband amplitude modulation (DSB-AM) in the very high frequency (VHF) band (118–137 MHz) was adopted for voice communications between pilots and ground controllers. This system supported a maximum of 140 channels until 1979. Multiplexing and multiple access were frequency division with *manual* channel assignment by air traffic control. In more dense air traffic spaces, to enable a larger number of simultaneous transmissions, 25 kHz DSB-AM channels were subdivided into three channels each of bandwidth 8.33 kHz.

The civilian aeronautical AG communications continues to use the reliable analog DSB-AM system today, although since 1990 some small segments of the VHF band in some geographic locations are being upgraded to a digital VHF data link that can in principle support 2280 channels [86, 114]. This system employs both time division and frequency division, with single-carrier phase-shift keying modulation. Military AG communications uses different frequency bands (ultra-high frequency) and modulation schemes for short and long ranges [105]. Due to very low data rates, the civil aviation systems cannot support modern AG communication requirements. In 2007, use of a portion of the L-band was suggested for new civil aviation systems, and two such systems known as L-band digital aeronautical communications systems (LDACS) were developed [114]. Due to its compatibility with numerous existing systems that operate in the L-band, the LDACS system is still being refined. LDACS-1 is currently undergoing standardization by the International Civil Aviation Organization.

There are numerous studies available in the literature on the characterizations of aeronautical channels [49, 55, 82, 86, 89]. Aeronautical communications can be broadly classified into communications between the pilot or crew with the ground controller and wireless data communication for passengers. Both of these types of communication are dependent on the flight route characteristics. In [55] the propagation channel is divided into three main phases of flight, termed parking and taxiing, en-route, and take-off and landing. Each phase of flight was described by different channel characteristics (type of fading, Doppler spread, and delay), but this relatively early paper was neither comprehensive nor fully supported by measurements.

There are also studies on long-distance AG propagation channel for satellites and high-altitude platforms (HAPs), which can also be considered as UAV communication channels. However, due to the long distances from the GS on the Earth's surface, normally greater than 17 km, modeling of these links may also need to take into account upper-atmospheric and very-low-elevation-angle effects. Depending on frequency and UAV altitude, long-distance AG channels may also be much more vulnerable to lower-tropospheric effects such as fading from hydrometeors [83]. For most of these HAPs, a LOS component is required because of power limitations; hence the AG channel amplitude fading is typically modeled as Ricean [43].

2.2.2 Existing Surveys on UAV AG Propagation

There are several surveys available on UAV AG propagation channel measurements and modeling [76, 89, 93]. In [89], a survey on wideband AG propagation channels was provided, with focus on the L-band and C-band that are envisioned as possible candidate bands for future CNPC AG communications. A tapped delay line model for a time-varying channel was provided. A general finding was that no accurate and comprehensive wideband AG propagation channel model based on empirical data existed for these bands. However, the literature review content in that survey is now outdated and new research has since appeared in the literature. A similar short description of overall propagation channel characteristics for UAVs was provided in [93] with analysis of a two-ray geometrical model and its applications in different scenarios. Limitations of existing AG channel models for UAVs were also discussed.

A recent survey on AG propagation channel modeling was provided in [76]. The survey discussed measurement and analytical channel models available in the literature. The AG propagation channel measurements were divided into three parts. The first part covered narrowband and wideband channel measurements, the second part discussed channel measurements using 802.11 radios, and the third part covered the cellular infrastructure-based AG propagation channel measurements. A table of AG propagation channel measurements available in the literature was also provided. Additionally, large-scale and small-scale fading statistics and their respective models from the literature were provided. The survey also covered analytical AG channel models from the literature. These analytical channel models were divided into three categories, namely, deterministic, stochastic (based on time delay line (TDL)), and geometric. Additionally, the survey discussed air-to-air (AA) channel characterization studies from the literature; a common result here is the small path-loss exponent (PLE), essentially that of free space, in contrast to larger PLEs found in terrestrial or AG propagation. Important issues relating to AG propagation were also briefly discussed, including the airframe shadowing, stationarity intervals, and diversity gain.

2.3 UAV AG Propagation Characteristics

In this section, salient characteristics of UAV AG propagation channels are described. A common AG propagation scenario is shown in Figure 2.1 in the presence of terrestrial obstacles, which are also commonly referred to as *scatterers*, even if the propagation mechanism they produce is different (e.g., reflection, diffraction). In the figure, h_G, h_S, and h_U represent the heights of the GS, scatterers, and UAV above the ground, respectively, d is the slant range between the UAV antennas and the GS, and θ is the elevation angle between the GS and UAV antennas. We note that airborne scatterers may be present as well; however, in this chapter, for the AG link, we neglect this condition.

2.3.1 Comparison of UAV AG and Terrestrial Propagation

The AG channel exhibits distinctly different characteristics from those well-studied terrestrial communication channels, e.g., the urban channel. There is one inherent advantage over terrestrial communications with aerial platforms, i.e., a higher likelihood of LOS propagation. This reduces required transmit power and can translate to higher link reliability as well. In cases where only non-line-of-sight (NLOS) paths exist, when the elevation angle to the UAV is large enough, the AG channel may incur smaller diffraction and shadowing losses than near-ground terrestrial links.

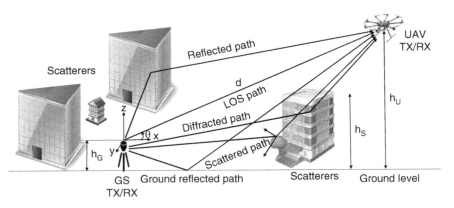

Figure 2.1 A typical air-to-ground propagation scenario with a UAV.

On the other hand, the AG channel can exhibit significantly higher rates of change than typical terrestrial communication channels because of UAV mobilities. When the channel is modeled statistically, this implies that the channel's statistics are approximately constant (the channel is wide-sense stationary (WSS)) for only a small spatial extent and a short time. This is often loosely termed "non-stationarity." If the UAV is not in the direct vicinity of scattering objects or the GS, the characteristics of the channel could instead actually change very slowly, especially for hovering UAVs. In such a case, adverse propagation conditions, e.g., deep fades of the received signal, may last several seconds or even minutes; hence common communication techniques of interleaving or averaging may not be affective. In many cases, when UAV altitudes are well above scattering objects, the AG channel's "non-stationarity" will be attributable to the direct surroundings of the GS, e.g., nearby buildings or obstacles around the GS.

Additionally, AG communications with UAVs face many other challenges, due to three-dimensional arbitrary mobility patterns and diverse communication application requirements [11, 20, 140, 146]. As an aerial node, some of the UAV-specific characteristics that need to be taken into account include airframe shadowing, mechanical and electronic noise from UAV electronics and motors, and antenna characteristics, including size, orientation, polarization, and array operation (e.g., beam steering) for MIMO systems. For UAVs in motion, the effect of Doppler shifts and spread must also be considered for specific communication applications [44, 57]. For a given setting, an optimum UAV height may need to be considered, e.g., for maintaining LOS in that environment [14].

2.3.2 Frequency Bands for UAV AG Propagation

As with all communication channels, a fundamental consideration is the frequency band, since propagation characteristics can vary significantly with frequency. There are typically two popular bands, namely 2.4 GHz and 5.8 GHz, used by commercial UAVs for CNPC operations during flight. However, other frequency bands may be used for additional features, e.g., for transmitting videos from the UAV to the GS at 3.4 GHz. The 5.8 GHz band is a better choice than the 2.4 GHz for the majority of scenarios, because of lower interference (at least at present). For the L- and C-bands envisioned for CNPC, and for the currently popular unlicensed bands for payload communications, tropospheric attenuations from atmospheric gases and hydrometeors are mostly negligible. This is not true for operation at higher frequency bands, e.g., at Ku, Ka, and other millimeter wave (mmWave) bands, which may be as high as 100 GHz. These higher frequency bands can hence suffer both larger free-space path loss (FSPL) as well as tropospheric attenuations. Because of this, these frequency bands will generally be used for short-range AG links.

Although the lower frequency bands have smaller attenuation, mmWave bands offer a wider bandwidth, which is their primary appeal for 5G cellular systems. Wider bandwidths can be more robust to the larger values of Doppler shift and Doppler spread encountered with UAVs moving at high speed. The mmWave bands may be supported by UAVs in the future for payload data communications that are typically of high rate requirement. However for CNPC, mmWave bands may not be a good option because of their larger attenuation and weaker diffraction, making these bands unreliable even in the presence of small blockages.

2.3.3 Scattering Characteristics for AG Propagation

In an AG propagation channel using UAVs, the MPCs appear due to reflections from the Earth's surface, from terrestrial objects (ground scatterers), and sometimes from the airframe of the UAV itself. The characteristics of the channel will be dependent on the material, shape, and electrical size of the scattering objects. The strongest MPC apart from the LOS component in an AG propagation scenario is often the single reflection from the Earth's surface. This gives rise to the well-known two-ray model.

For high enough frequencies, the scatterers on the ground and around the UAV can be modeled as point scatterers on the surface of two respective cylinders, spheres [34, 78], or ellipsoids, and these can be bounded (truncated) by the intersection of the elliptical planes on the ground [53, 54]. These topologies can help in deriving geometrical characteristics of the AG propagation channel. The distribution of scattering objects, on land or water, can be modeled stochastically, and this concept can be used to create the so-called geometry based stochastic channel models (GBSCMs). For aircraft moving through an area above such a distribution, this gives rise to intermittent MPCs [92], as also seen in vehicle-to-vehicle (V2V) channels.

For propagation over water, in the presence of an LOS component, the path loss (PL) is similar to that of free space [39], with a strong surface reflection. Any other MPCs from the water surface due to rough surface scattering are weaker, and often of approximately equal power and time of arrival (TOA), whereas MPCs from obstacles on the water surface, e.g., large ships or construction platforms, can be stronger.

2.3.4 Antenna Configurations for AG Propagation

The aircraft antenna is a critical component for AG communications due to limited space, and limitations of the aerodynamic structure [69, 111]. Factors that affect AG link performance include the number, type, and orientation of the antennas used, as well as the UAV shape and material properties.

The majority of AG channel measurements employ a single antenna at both the aircraft and the GS, whereas in [99] an antenna array is used. There are some single input–multiple output (SIMO) and MIMO antenna configurations described in the literature for AG propagation measurements [31, 37]. Omnidirectional antennas are most popular for vehicular communications due to their easier use in the presence of motion, whereas directional antennas (having better range via directional gain) can perform poorly during motion due to misalignment losses. With high maneuverability of UAVs during flight, omnidirectional antennas are generally better suited than directional antennas. A potential major drawback of any antenna on-board UAVs is the shadowing from the body of the UAV. Similarly, orientation of antennas on-board UAVs can affect the communication performance [33, 142]. A better throughput performance was reported with horizontal–horizontal orientation than with vertical–vertical orientation in [33], whereas in [142], it was observed that horizontal antenna orientation can help to overcome yaw misalignment; vertical orientation was found to perform better during tilting (change of pitch) of the UAV.

The use of multiple antennas to enable diversity can yield spatial diversity gains even in sparse multipath environments [40, 137]. It was demonstrated that using MIMO and

SIMO in AG propagation can yield spatial diversity dependent on the antenna geometry and surrounding environment. Similarly, multiple antennas can be used for spatial selectivity such as beamforming/steering. However, due to limited space of UAVs, space diversity using multiple antennas is often difficult to achieve, especially for lower carrier frequencies or very small UAVs. Beamforming using antenna arrays operating at mmWave frequencies, for example, can be used to overcome fading and improve coverage, but array processing will require high computational resources on-board. The employment of MIMO systems for enhancing the channel capacity of the AG propagation channel has been suggested in [48, 147]. By changing the diameter of a circular antenna array and the UAV flying altitude, different values of MIMO channel capacity were obtained [48]. With another approach, in [147], optimizing the distance between the antenna elements using linear adaptive antenna arrays was proposed to increase MIMO channel capacity.

2.3.5 Doppler Effects

Doppler shifts can introduce carrier frequency offset (CFO) and inter-carrier interference (ICI) in multicarrier signaling, e.g., orthogonal frequency-division multiplexing (OFDM). There are several studies that consider modeling of Doppler spread [26, 44, 55, 57, 118, 131, 139, 141] for AG scenarios. Some channel access algorithms, e.g., multicarrier code-division multiple access, have been shown to be robust against Doppler spread in AG channels [129]. For OFDM, Doppler spread can be mitigated by adjusting the carrier spacing.

Doppler frequency shifts depend on the velocity of the UAV and the geometry. If the different signal paths are associated with largely different Doppler frequencies, this yields large Doppler spread. This can happen if the aircraft is relatively close to the GS. If the aircraft is far away from the GS, and at sufficiently high altitude, the paths should all have a very similar Doppler frequency, as the objects in the close surroundings of the GS causing MPCs are seen all under similar angles from the aircraft, i.e., the angular spread is small. The effect of a large Doppler frequency that is constant for all MPCs can be mitigated by accurate frequency synchronization.

In order to describe the statistical characteristics of a fading channel, typically first- and second-order fading statistics are used. The majority of the AG propagation literature discusses first-order fading statistics. The second-order statistics of envelope level crossing rate and average fade duration are discussed in [34, 118], but many authors address other second-order properties, primarily correlation functions in the time or frequency domains.

2.4 AG Channel Measurements: Configurations, Challenges, Scenarios, and Waveforms

Several AG channel measurement campaigns using piloted aircraft and UAVs have been recently reported in the literature. These measurements were conducted in different environments and with different measurement parameters. In this section, we provide a brief classification of these measurements based on environmental scenario, sounding signal, carrier frequency, bandwidth, and antenna specifications and placement. Whenever available, we also provide UAV type and speed, heights of UAV and GS from the terrain surface, link distance between transmitter (TX) and receiver (RX), elevation angle, and the channel statistics provided by the authors. These channel measurement parameters are summarized in Table 2.2.

In the reported AG propagation measurements, either the TX or the RX on the UAV/GS is stationary. Measurements for AG propagation with both the TX and the RX moving are rare. A notable contribution of wideband AG propagation measurements is available in the form of multiple campaigns conducted in the L- and C-bands using a SIMO antenna configuration for different terrain types and over water/sea [37–39, 90–92, 94, 95, 121, 122]. The rest of the cited channel measurements are conducted in different frequency bands ranging from narrowband to ultra-wideband (UWB) with various types of sounding signals.

2.4.1 Channel Measurement Configurations

These channel measurements used different types and configurations of antennas. The most commonly used antenna type is omnidirectional and the most commonly used configuration is single input–single output (SISO). The positioning of an antenna on the UAV is important to avoid both shadowing from the airframe and disruption of the aircraft's aerodynamics. In the majority of measurements the antennas were mounted on the bottom of the aircraft's fuselage or wings. The orientation of antennas on UAV and ground can also affect the signal characteristics [12, 33, 142, 143]. One characteristic, namely airframe shadowing, is most important during banking turns, and when the aircraft's pitch angle deviates from horizontal. The elevation angle between the TX and the RX antennas is dependent on the height of the UAV and GS, which often continuously varies during flight.

In many of the communication applications envisioned for UAVs, the aerial node is expected to be stationary (or mostly so) in space for a given time. For UAVs operating at higher velocities, the coherence time of the channel decreases, and this translates into a larger Doppler spread. For connections to multiple UAVs, where handovers are required, this means that the number of handovers will also generally increase with velocity, and this will require additional processing. Additionally, higher velocities will result in increased air friction and mechanical turbulence that generally result in increased noise levels. Many of the AG channel measurements in the literature have been conducted with fixed-wing aircraft with maximum speeds varying from 17 to 293 m s^{-1}. The speed of rotorcraft and air balloons is much less than that of fixed-wing aircraft, ranging from 8 to 20 m s^{-1}.

The height of the UAV above ground is an important channel parameter that will also affect the channel characteristics. For example, increasing the height of the UAV can result in reduced effects of MPCs [72] from surrounding scatterers if the power of the LOS component becomes dominant. Another benefit of higher UAV altitude is larger coverage area on the ground. Similarly, the height of GS will also affect the channel characteristics. For a given environmental scenario, there may be an optimal height of the GS [73], e.g., this might be a balancing of attenuation and multipath diversity.

Example propagation measurements using rotorcraft and air balloons during flight and hovering are available in [12, 73, 118, 142]. These AG propagation measurements were obtained at different UAV heights ranging from 16 m to 11 km, and link distances from 16.5 m to 142 km. The UAV latitude, longitude, yaw, pitch, and roll readings are typically obtained from Global Positioning System (GPS) RXs and often stored on-board.

Table 2.2 Review of important empirical AG channel measurement studies in the literature.

Ref.	Scenario	Sound. sig.	Freq. (GHz)	BW (MHz)	Antenna and mounting	P_T (dBm)	UAV, v_{max} (m s^{-1})	h_U, h_G, d (m)	θ (deg.)	Channel statistics
[118]	Urban	CW	2	0.0125	1 monopole on UAV for TX, 4 on GS for RX	27	Air balloon, 8	170, 1.5, 6000	1–6	P_R, autocorrelation of direct and diffuse components
[73]	Open field, suburban	PRN	3.1–5.3	2200	1 dipole on UAV for TX, 1 on GS for RX	−14.5	Quadcopter, 20	16, 1.5, 16.5	—	PL, PDP, RMS-DS, TOA of MPCs, PSD of sub-bands
[37, 38, 39, 90, 91, 92, 94, 95, 121, 122]	Urban, suburban, hilly, desert, fresh water, harbor, sea	DS-SS	0.968, 5.06	5, 50	1 directional antenna on GS for TX, 4 monopoles on UAV for RX	40	Fixed wing, 101	520–1952, 20, 1000–54 390	1.5–48	PL, PDP, RMS-DS, K-factor, tap probability and statistics (power, delay, duration) in TDL model
[113]	Rural, suburban	OFDM	0.97	10	1 monopole antenna on GS for TX, 1 monopole on aircraft for RX	37	Fixed wing, 235	11 000, 23, 350 000	0–45	PL, PDP, DPP
[124]	Rural, suburban, urban, forest	FMCW	5.06	20	1 monopole on UAV for TX, 1 patch antenna on GS for RX	30	Fixed wing, 50	—, 0, 25 000	—	CIR, PG, RSS
[100]	Urban	MSK	2.3	6	1 whip antenna on UAV as transceiver, 1 patch antenna as transceiver on GS	33	Fixed wing, 50	800, 0.15, 11 000	4.15–86	RSS
[51]	Urban, suburban, rural	GSM, UMTS	0.9, 1, 9–2, 2	—	Transceiver on balloon and GS	41.76	Captive balloon	450, —, —	—	RSSI, handover analysis
[31]	Urban, hilly, ocean	OFDM	2.4	4.375	4 whip antennas on AV for TX, 4 patch antennas on GS for RX	—	Fixed wing, 120	3500, —, 50 000	—	Eigenvalues, beamforming gain
[137]	Rural	PRN, BPSK	0.915	10	2 helical antennas on AV for TX, 8 at GS for RX	44.15	Fixed wing, 36	200, —, 870	13–80	CIR, P_R, RMS-DS, spatial diversity
[33]	—	OFDM	5.28	—	4 omnidirectional on UAV for TX, 2 on GS for RX	18	Fixed wing, 17.88	45.72, 4.26, —	—	P_R, RSSI

Table 2.2 (Continued)

Ref.	Scenario	Sound. sig.	Freq. (GHz)	BW (MHz)	Antenna and mounting	P_T (dBm)	UAV, v_{max} (m s^{-1})	h_U, h_G, d (m)	θ (deg.)	Channel statistics
[142]	Urban, open field	OFDM	5.24	—	2 omnidirectional on UAV for TX, 2 on GS for RX	20	Quadcopter, 16	120, 2, 502.5	—	RSSI
[143]	Open field	OFDM	5.24	—	3 omnidirectional on UAV for TX, 3 on GS for RX	20	Quadcopter, 16	110, 3, 366.87	10–85	RSS
[12]	—	IEEE 802.15.4	2.4	—	On-board inverted F transceiver antenna on UAV and GS	0	Hexacopter, 16	20, 1.4, 120	—	RSSI
[11]	Suburban	WiFi, 3G/4G	—	—	Transceiver on UAV and GS	—	Hexacopter, 8	100, —, —	—	P_R, RTT of packets
[36]	Forest (anechoic chamber)	—	8–18	—	Spiral antennas on TX and RX	—	—	2.3, 0.6, 2.85	26–45	P_R
[110]	Open area	Mod. sig.	5.8	—	2 Monopole, 1 horn on UAV for TX, 2 on GS for RX	—	Fixed wing, —	150, 0, 500	—	P_R
[79]	Open area/foliage	802.11b/g	5.8	—	1 omnidirectional on GS for TX, 4 on UAV for RX	—	Fixed wing, 20	75..2, —	—	Diversity performance
[145]	Urban/suburban, open field, foliage	CW	2.00106, 2.00086	—	2 monopoles on UAV for TX, 2 on GS for RX	27	Gondola airship, 8.3	50 and above, 1.5, 2700	1	P_R
[126]	Urban, rural, open field	—	0.915	—	1 omnidirectional antenna on UAV for TX, 1 on GS for RX	—	Quadcopter, —	—, 13.9, 500	—	RSSI, PL
[97]	Sea	PRN	5.7	—	Omnidirectional on AV for TX, 2 directional antennas at GS for RX	40	Fixed wing AV, —	1830, 2.1, 7.65,95 000	—	PL
[99]	Urban	CW	2.05	—	1 monopole on AV for TX, 4 on GS for RX	—	Aerial platform, —	975, —, —	7.5–30	PDP, RMS-DS, MPCs count, K-factor, PL
[129]	Near airport	CW	5.75	—	Directional antenna on GS for TX, omnidirectional on AV for RX	33	Fixed wing AV, —	914, 20,85 000	80	P_R, fading depth, K-factor, PL
[80]	Urban, hilly	Chirp	5.12	20	1 monopole antenna on GS for TX, 1 omnidirectional on AV for RX	40	Fixed wing AV, 293	11 000, 18,142 000	(−16) − 5	PDP

Apart from conventional AG channel sounding, there are some indirect UAV AG channel measurements available from use of radios employing different versions of protocols of the IEEE 802.11 standards [12, 33, 142, 143]. The IEEE 802.11-supported devices offer a very flexible platform and may provide insight for UAV deployments in different topologies and applications, e.g., UAV swarms. Yet, because of the specific features of 802.11, the resulting measurements are applicable to particular protocol setup and radio configuration, and rarely provide detailed propagation channel characteristics.

AA communications with UAVs has not been studied extensively in the literature [133]. The AA communications is particularly important for scenarios where multiple drones communicate among a swarm. This swarm then usually communicates with one or more GS via a backhaul link from one or several of the UAVs. The AA communications is similar to free space with a strong LOS and often a weak ground reflection, but this is dependent on the flight altitude and environment. The communication channel is mostly non-dispersive for higher altitudes but can be rapidly time-varying, dependent on the relative velocities of the UAVs and the scattering environment [50].

2.4.2 Challenges in AG Channel Measurements

There are many challenges in AG channel measurement campaigns as compared to terrestrial measurements. Two big challenges are the payload limitation of the UAVs, and the operating range and height of UAVs, which in the USA is set by the FAA [45]. Larger UAVs also incur larger test costs. Due to restrictions on the height of UAVs above ground, UAVs at lower altitudes have lower LOS probability and are hence more susceptible to shadowing, especially in suburban and urban areas. Due to limitations on payload, higher transmit power measurements on-board the UAVs are difficult to achieve, and, similarly, complex RX processing on-board UAVs can consume a prohibitive amount of power.

Other challenges include precise frequency synchronization for channel measurements, varying conditions of the terrain during flight, meteorological conditions (wind and rain), antenna positioning on the UAV, precise location measurement of UAVs in space over time, diverse telemetry control for different types of UAVs having specific latencies, available bandwidth, potential reliability issues, and limited flight time for most small UAVs due to limited battery life [11, 20, 146]. Due to the motion of UAVs in three-dimensional space, it is challenging to precisely measure the distance between the UAV and the GS. Momentary wind gusts that cause sudden shifts in UAV position can make it difficult to accurately track the UAV path. The most common technique of measuring the instantaneous distance is by using GPS traces on both the UAV and GS, but of course GPS devices have accuracy limitations and navigation signals may also be susceptible to interference in different flying zones.

2.4.3 AG Propagation Scenarios

A typical terrestrial channel sounding instrument, a vector network analyzer, cannot be used for UAV-based AG channel sounding due to payload constraints, physical synchronization link requirements, and UAV mobility [130]. Therefore, channel sounding for both narrowband and wideband channels using impulse, correlative, or chirp sounding techniques are employed, where the RX is often on the ground due to payload and processing constraints.

Proper selection of channel measurement parameters in a given environment is critical for obtaining accurate channel statistics for a given application. The AG propagation environment is generally classified on the basis of the terrain type, e.g., flat, hilly, mountainous, and over water. A particular terrain can have a given cover type as well, e.g., grass, forest, or buildings. The most widely accepted terrain cover classification is provided by the International Telecommunication Union (ITU) [59]. In this survey we classify the cited measurement scenarios as open (flat), hilly/mountainous, and over water. Each scenario can be subdivided on the basis of the terrain cover as shown in Figure 2.2.

For any environment, different types of radio-controlled UAVs can be used. Balloons or dirigibles are simple to operate but do not have robust movement characteristics. The non-balloon UAVs can be broadly classified as fixed wing and rotorcraft. The fixed-wing UAVs can glide and attain higher air speeds and generally travel farther than the rotorcraft, but rotorcraft are more agile, e.g., most can move straight vertically. Rotorcraft also have the ability to hover, which is not possible for nearly all fixed-wing UAVs. The UAV AG propagation scenarios in different environments with particular characteristics are described in Table 2.3. In the rest of this subsection, we review the different AG measurement scenarios depicted in Figure 2.2.

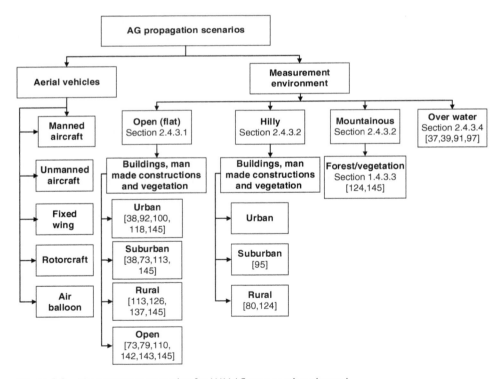

Figure 2.2 Measurement scenarios for UAV AG propagation channel.

Table 2.3 Literature on UAV AG propagation characteristics for five different flight environments.

Ref.	Scenario	Characteristics of scenario	Important factors
[11, 38, 73, 92, 99, 100, 118]	Urban/suburban	Ratio of land area versus ratio of open to built-up area, distribution of building sizes and heights, distribution of ground terminals (vehicles, pedestrians), distribution and characteristics of vegetation, water bodies, etc.	Material of buildings and rooftops
[79, 110, 113, 126, 129, 137, 142, 143]	Rural/open field	Type and density of vegetation, distribution and sizes of the sparse buildings	Surface roughness, soil type, and moisture content
[80, 95, 124]	Hilly/mountainous	Terrain heights and slopes, distribution and type of vegetation, distribution and sizes of buildings	Ground slope, ground roughness
[36, 124, 145]	Forest	Density and types of foliage, and height distributions	Leaves and branches distributions
[37, 39, 97, 121]	Over water	Water type (sea or fresh), distributions and sizes of water surface objects (boats, platforms, etc.), distributions of littoral objects (buildings, water tanks, etc.), and water surface variation (e.g., sea state)	Modified reflection coefficient as compared to ground, ducting effect in case of over sea

2.4.3.1 Open Space

A major part of the literature on AG propagation covers open (flat) terrain. This open terrain can have different terrain covers that affect the channel characteristics. One of the major terrain cover types is buildings. The distribution of building sizes, heights, and their area-wise densities allows sub-classification into urban, suburban, and rural areas as depicted in Figure 2.2. In the case of urban and suburban areas, there is a higher concentration of manmade structures in a given space, e.g., buildings, roads, bridges, large signs, etc. The distribution (and composition) of these complex scatterer structures can strongly influence the channel characteristics. In rural areas, typically buildings are sparse, and of lower height than in urban settings, although large warehouses and other structures could yield strong MPCs.

2.4.3.2 Hilly/Mountainous

The hilly/mountainous terrain is characterized by uneven ground height or, equivalently, a large standard deviation of terrain height. The propagation PL in hilly and mountainous areas will mostly follow the two-ray model with adjustments due to surface roughness, and potentially reflections from smooth sections of mountain slopes or an occasional large building. The PL for links that encounter terrain obstructions can employ established models for diffraction, e.g., [58], but with first Fresnel zone clearance between the TX and the RX, PL is close to free space [95, 122]. Channel dispersion, typically quantified by the root-mean-square delay spread (RMS-DS), is generally smaller than in urban/suburban environments [122] but can be large if a strong reflection occurs from a large and distant mountain slope. Generally, hilly and mountainous settings present fewer reflections than more populated regions because of the absence of large numbers of nearby scatterers.

2.4.3.3 Forest

There are few comprehensive studies covering AG propagation in forests, especially with UAVs, although there are numerous publications for *roadside shadowing* for satellite chan-nels, e.g., [10, 70, 81]. In these studies, propagation effects – typically attenuation – from particular volumes of trees, along with temporal fade statistics, are analyzed for long-range AG communications. Generally for AG propagation with a GS within a forest, the channel characteristics are dominated by the type and density of trees. Small UAVs within a forest experience different scattering characteristics depending upon height, e.g., the scattering near the tree trunk will be different from that near the tree crown [36]. The scattering is also dependent on the type and density of leaves and branches of the trees, and hence, for deciduous trees, can vary seasonally.

2.4.3.4 Water/Sea

The AG propagation channel for over-water settings is similar to that for open settings, with different surface reflectivity and roughness than ground. The PL can be represented using a two-ray model, with variations attributable to surface roughness (see small-scale fading in Sections 2.6.4 and 2.6.8.2). The RMS-DS in this case is generally smaller than in environments with a large number of obstacles (urban, suburban), although if large objects are on- or just offshore, these may produce significant reflections and large delay spreads depending upon geometry.

In the case of propagation over sea, the height of waves in a rough sea can introduce additional scattering and even diffraction or obstruction for very low-height stations on the sea. An interesting propagation phenomenon that can also occur over sea is ducting, where anomalous index-of-refraction variation with height results in propagation loss less than that of free space [97]. This phenomenon is dependent on frequency and meteorological conditions, and is thus typically addressed statistically [60].

2.4.4 Elevation Angle Effects

It is important to consider the effect of elevation angle for UAV AG communications, as it is for satellite communications. This is in contrast to the case of terrestrial communications, where the effect of the elevation angle is less significant due to (1) smaller heights of TX and RX, and (2) NLOS communication for a substantial fraction of users. The effect of elevation angle can also vary with the type of antennas used. If the communication is directional with LOS, and the UAV and GS beams are aligned, then the effect of the elevation angle is negligible. However, if the communication is omnidirectional, then the effect of the antenna pattern as a function of elevation angle can be significant.

In [32, 75], the effect of elevation angle at different UAV altitudes for different antenna orientations was discussed. The gain of the omnidirectional dipole antenna in the elevation plane was modeled as a trigonometric function of physical elevation angle between the UAV and the GS. It was shown that the effect of the elevation angle on the received power was essentially deterministic when the UAV is hovering, but was not as easily modeled when the UAV was in circular motion around the ground RX.

The effect of elevation angle on UAV AG omnidirectional communications at 28 GHz was discussed in [74]. It was observed that received power was mainly dependent on the elevation angles of the LOS and ground-reflected components (as they were the strongest). The received power was a function of link distance and elevation angle: for some distances and angles, antenna gain effects dominated, but at larger distances the antenna gain was less significant than the effect of attenuation (FSPL) with distance.

Similarly, in [72], the received power was shown to follow a two-ray model. It was also shown that the received power and RMS-DS dependence on UAV height is a function of the specific propagation environment. For example, in a rural area, the effect of ground scatterers was negligible at higher UAV altitudes, whereas for an urban area, the effect of scatterers was notable even at higher UAV altitudes. Similar results were reported for TOA, angle of arrival, and angle of departure of MPCs in [74], where larger temporal and angular spreads were observed at UAV heights comparable to the scatterer heights.

2.5 UAV AG Propagation Measurement and Simulation Results in the Literature

Several types of channel statistics are useful for characterizing the channel for different applications. For AG propagation, the channel statistics are similar to those gathered for terrestrial channels. In general, propagation channels are linear and time-varying, but can sometimes be approximated or modeled as time-invariant. For linearly time-varying channels, the channel impulse response (CIR) or its Fourier transform, the time-varying channel transfer function (CTF), completely characterizes the channel [37–39, 80, 90–92, 94, 95, 99, 118, 121, 122, 137]. As noted, due to relative motion of the UAV, the AG channel may be statistically stationary only for small distances [37]. Thus, *stationarity distance* may need to be accounted for when estimating the channel statistics [39, 101, 108].

Another higher-level parameter that has been used by some researchers to characterize the quality of the AG propagation channel is throughput, but, of course, this is highly dependent upon the TX and the RX implementation, and parameters of the air interface, such as bandwidth, transmit power, and the number of antennas. Hence this measure is of limited use for assessing the AG channel itself. Similarly, for MIMO channels, beamforming gain, diversity, and capacity of the channel are often estimated. Some commonly reported channel characteristics for AG propagation channels are given in the following subsections.

2.5.1 Path Loss/Shadowing

Most of the AG propagation campaigns address PL and, if present, shadowing, in different scenarios. For AG channels with a LOS component, PL modeling begins with FSPL;

when the Earth's surface reflection is present (not blocked or suppressed via directional antennas), PL can be described by the well-known two-ray model. Parallel to the developments in terrestrial settings, most of the measurements employ the log-distance PL model, where the loss increases with distance and is indicated by the PLE. In [73], PL is calculated for open-field and suburban areas for different UAV and GS heights for a small hovering UAV. Comprehensive PL measurements in L- and C-bands were carried out in different propagation scenarios [37–39, 90–92, 94, 95, 121, 122], as summarized in Table 2.2. The values of PLE were found to be slightly different for urban, suburban, hilly, and over-water scenarios, but are generally close to the free-space value of 2, with standard deviation around the linear fit typically less than 3 dB.

In [142], it was observed that the PLEs for IEEE 802.11 communications were different during UAV hovering and moving due to different orientations of the on-board UAV antennas. Therefore, antenna patterns can distort the true channel PL characteristics, and removing their effect is not always easy or possible. On the other hand, for the specific UAV configuration used, the resulting PL model is still useful. Typically, PL for LOS and NLOS conditions are provided separately, e.g., [134], where for the NLOS case, there is an additional small-scale (often modeled as Rayleigh) fading term, and potentially other terms in addition to the LOS PL. Analogously, the LOS models for L- and C-bands can incorporate Ricean small-scale effects [37].

In [13], the reported PL is described as a function of the elevation angle, θ, between the low-altitude platform and GS, given as

$$PL = 20\log\left(\frac{\Delta h}{\sin\theta}\right) + 20\log(f_{\text{MHz}}) - 27.55, \tag{2.1}$$

where $\Delta h = h_{\text{U}} - h_{\text{G}}$ is the difference between the height of the low-altitude platform (UAV) and the GS, and f_{MHz} is the operating frequency expressed in MHz. The argument $\Delta h / \sin\theta$ is simply the link distance expressed as a function of the elevation angle.

Path loss results including the effects of shadowing were reported in [47, 56, 73, 99, 118, 129]. Note that in LOS cases without actual obstruction of the first Fresnel zone, the physical mechanism causing PL variation is not actually shadowing but either small-scale or antenna effects. In [118], PL and its associated shadowing were attributed to buildings only when the UAV was flying near the ground, whereas, when flying higher, actual shadowing was not present but variation still occurred. One can also estimate losses due to "partial" shadowing by conventional methods. For example, the shadowing in [56] was found to be a function of the elevation angle, where the shadowing magnitude was estimated by using the uniform theory of diffraction.

In Figure 2.3(a) we show an example for the variation of the LOS signal power due to ground-reflected MPCs versus the link distance d. Specifically, this is the *combined* effect of the LOS component and the unresolved ground reflection. The measurements were taken in a rural environment using a 10 MHz signal bandwidth. The GS height h_{G} was 23 m. The UAV trajectory is shown in Figure 2.3(b). During the measurements, the specular reflection point first passed over the roof of a building and then over open grassy fields [112]. From Figure 2.3(a) we observe a periodic variation of the received power: an attenuation of the signal by more than 10 dB is common. For an increasing link distance, the frequency of the variation decreases – a direct manifestation of the two-ray model. Thus in such a

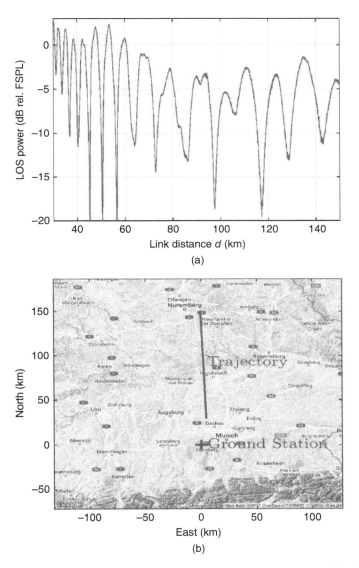

Figure 2.3 (a) The LOS signal power variation due to ground multipath propagation. The power is normalized to FSPL. (b) Measurement scenario environment. Source: Based on Schneckenburger et al. [112].

channel, even for a UAV flying at a high speed, a fade can easily last several seconds. It is essential to note that a ground MPC may not always be present, e.g., for the case when the ground is a poorly reflecting ground surface, or the surface is very rough relative to the signal wavelength.

The PL provides complete information on link attenuation, but another indirect parameter often used for channel attenuation estimation is the received signal strength (RSS). In [33, 142, 143], RSS indicator data for an AG propagation channel based on IEEE 802.11a transmissions with different antenna orientations was provided. Data on fluctuations in

RSS due to multipath fading from tall building reflections was provided in [124], where the RSS was found to decrease due to polarization mismatch between the TX and the RX antennas when the aerial vehicle made a banking turn. The accuracy of RSS values in commercial products can vary considerably, so, when these are used, care should be taken in calibration.

2.5.2 Delay Dispersion

The power delay profile (PDP) is the "power version" of the CIR. This can be computed "instantaneously" or, more traditionally, as an average over a given spatial volume (where the channel can be considered WSS). Various AG propagation studies in different environments have measured PDPs, and via the PDP the most common estimate of the delay-domain dispersion is estimated: the RMS-DS. Other dispersion measures such as the delay window or delay interval are also sometimes reported. Statistics for the RMS-DS itself are often computed, e.g., mean RMS-DS values for different elevation angles were reported in [99]. As generally expected from geometry, the RMS-DS was found to decrease as elevation angle increases. In [73], the PDPs were measured for open areas, suburban areas, and areas covered with foliage.

The Saleh–Valenzuela model, originally developed for indoor channels, is sometimes used to model the PDP when MPCs appear grouped or "clustered" in delay. This model specifies the MPCs by such clusters, and the number of clusters is different for different environmental scenarios. PDPs were measured for different environments in [38, 39, 90, 94, 95, 121, 122], and resulting RMS-DS statistics were provided. As expected, the delay spread was found to be dependent on the terrain cover, with maximum delay spread values of 4 μs for urban and suburban settings. The largest RMS-DS values generally occur when there are large buildings that can provide strong MPC reflections. For hilly and mountainous terrain, maximum RMS-DS values of 1 μs for hilly regions and 180 ns for mountainous terrain have been reported. In over-water settings, the maximum RMS-DS value reported was 350 ns. Again, in all the settings cited here, a LOS component was present between GS and UAV; hence, for the majority of the time, RMS-DS was small, on the order of a few tens of nanoseconds. The maximum RMS-DS values occurred intermittently. In [127], a finite-difference time-domain model for the electric field propagating at *very* low heights over sea was developed. An RMS-DS model for VHF to 3 GHz was presented, with RMS-DS a function of wave height.

2.5.3 Narrowband Fading and Ricean *K*-factor

Small-scale amplitude fading in AG propagation channels usually follows a Ricean distribution due to the presence of a LOS component. The Ricean K-factor is defined as the ratio of dominant channel component power to the power in the sum of all other received components. The K-factor is often used to characterize the AG channel amplitude fading. In [99], as generally expected, the authors found that the K-factor increased with increasing elevation angle. The Ricean K-factor as a function of link distance was given in [129], during multiple phases of flight (parking and taxiing, take-off and landing, and en-route). The en-route phase showed the largest K-factor, followed by take-off and landing, and parking and taxiing. In [36], it was observed that the K-factor will differ with different types of scattering trees: values of K ranging from 2 dB to 10 dB were reported.

The K-factor was measured for both L-band and C-band AG propagation in [37, 90, 92, 122] for urban, suburban, hilly, and mountainous settings, and also for over-freshwater and over-seawater scenarios. The mean values of K-factor for urban areas were reported to be 12 dB and 27.4 dB for L-band and C-band, respectively. The mean K-factor values for hilly and mountainous terrain were reported to be 12.8 dB and 29.4 dB for L-band and C-band, respectively, whereas for over-sea settings, mean values of K-factor for L-band and C-band were found to be 12.5 dB and 31.3 dB, respectively. It is worth pointing out that in these "strong LOS" channels, the K-factor does not strongly depend on the GS environment. Also observed was that the K-factor in C-band was larger than that in L-band in all environments. This is attributable to two causes: first, the C-band measurement signal bandwidth was larger than that of L-band, ameliorating fading; and second, for any given incident angle and surface roughness (e.g., ground or ocean), as carrier frequency increases, the surface roughness with respect to the wavelength also increases, and hence incident signals are scattered in multiple directions rather than being reflected as a primarily specular component in a single direction (toward the RX). With fewer and/or weaker MPCs at the higher frequency, the K-factor is larger.

2.5.4 Doppler Spread

The Doppler effect is a well-known phenomenon for wireless mobile communications. Considering AG propagation with UAVs in a multipath environment, if we let ϕ_i represent the angle between the aircraft velocity vector and the direction from which the ith MPC is received, the Doppler frequency shift of this ith MPC is $f_d^i = (v \cos \phi_i)/\lambda$, where v is the UAV velocity, and λ is the wavelength of the radio wave. We assume here that the GS is motionless; otherwise a more general formulation for the Doppler shift must be used. If the MPCs are received with different Doppler frequencies, this phenomenon produces spectral broadening, called Doppler spread.

In [55, 129], simulations were used to find the Doppler shift and its effect on the channel at different phases of flight (parking and taxiing, en-route, and take-off and landing). Doppler spread in a multipath environment implementing OFDM systems was considered in [26], where arriving MPCs were observed to have different frequency offsets. In such a case, if the RX CFO synchronizer cannot mitigate the effect of these different frequency offsets, it results in ICI. In [44], a mitigation technique for Doppler shift was proposed for the case where the UAV is relaying between two communication nodes. The UAV acts as a repeater that provides the required frequency shift to mitigate the Doppler effect. A three-dimensional AG Doppler delay spread model was provided in [57] for high scattering scenarios. Doppler spread for AG propagation is also discussed in [73, 80, 91, 137, 141].

2.5.5 Effects of UAV AG Measurement Environment

The different AG propagation channel measurement campaigns can be broadly classified based on terrain, terrain cover, and sounding signal characteristics. In this subsection, we provide a brief overview and comparison of different approaches.

2.5.5.1 Urban/Suburban

UWB AG propagation channel measurements were provided in [73], using pseudo-random number (PRN) sounding pulses. These measurements are unique, as large-bandwidth AG channel measurements are not common in the literature. However, the link distance and UAV heights considered are small due to the very small transmit power allowed by the Federal Communications Commission (FCC) in the United States. Buildings over a flat terrain in a suburban area with an average height of 12 m resulted in more reflections than in open areas. In [118], a continuous wave (CW) narrowband sounding signal with a center frequency of 2 GHz was used in an urban scenario. There were uniformly built buildings in this urban area with an average height of 22 m. Second-order channel fading statistics of level crossing rate and average fade duration were analyzed for different heights and horizontal distances of the aerial platform from the GS. This work is unique, as the second-order channel fading statistics for AG propagation using UAVs are rarely available in the literature. However, the velocity of the aerial platform considered was very small. Additionally, it would have been more interesting if second-order statistics were compared at different UAV velocities in this environment.

In [38, 92], wideband AG propagation channel measurements were reported for suburban and urban areas in the L-band and C-band. It was observed that reflections from high-rise buildings increased the RMS-DS. The large-scale and small-scale fading at two frequency bands in a similar environment were found to be different. In [99], a wideband sounding signal at a center frequency of 2.05 GHz was used in an urban-like environment of a university campus. There were four- to six-story buildings and the terrain was rolling. It was observed that the RMS-DS increased with decreasing elevation angle, whereas the number of MPCs remained the same, suggesting that at lower elevation angles the MPCs have larger power. In both [99] and [118], measurements were performed in approximately similar environments and at the same center frequency, only with different bandwidths. However, the multipath fading distribution in [99] was found to be Rayleigh/Ricean, whereas for [118], it was found to better follow the Loo distribution.

2.5.5.2 Rural/Open Field

In [113], a rural environment similar to an airport with large and small buildings and open grassy fields was considered for channel measurements. Wideband OFDM channel sounding was performed in the L-band at a center frequency of 970 MHz. These channel measurements were performed at an aerial height up to 11 km. At these higher altitudes, tropospheric effects were also considered. In addition, interference effects such as from distance measuring equipment (DME) were also taken into account. A link distance up to 350 km was used, and that is much larger than in most reported measurements available in the literature. In [129], channel measurements were carried out near airports using a wideband signal centered at 5.75 GHz. Channel measurements were obtained for different flight scenarios of parking and taxiing, en-route, and take-off and landing. Received power and small-scale fading statistics for different flight scenarios were analyzed, where during taxiing and take-off, larger values of RMS-DS, *K*-factor, and Doppler shift were observed. In [126], measurements were conducted in order to explore the feasibility of using the fixed cellular network for telemetry and control of UAVs, particularly considering radio propagation at shorter distances in the sky than for terrestrial. The center frequency of operation was 0.915 GHz. Comparisons of AG measurement results with the COST 231-Walfisch–Ikegami (COST-231-WI) model were provided showing that received power from measurements was overestimated by the COST-231-WI model.

In [110], channel measurements were carried out in an open field using a modulated signal centered at 5.8 GHz, with two TXs operating at slightly different frequencies. A MIMO configuration was used for channel measurements with directional and omnidirectional antennas. It was confirmed that multipath interference can be reduced using directional antennas. Additionally, using multiple antennas on-board UAVs can provide robustness against received power fluctuations due to varying antenna orientations on the aerial platform. However, no propagation model was provided. In [137], a wideband PRN, binary phase shift keying (BPSK) sounding signal was used at a center frequency of 0.915 GHz. Analysis of spatial diversity for MIMO signaling was carried out. Additionally, a near-field scattering region analysis around the UAV and the GS was carried out. It was observed that additional spatial diversity could be obtained from objects near the GS. The MPCs were reported to be sparse, similar to the observation in [74], though at a lower frequency.

2.5.5.3 Mountains/Hilly, Over Sea, Forest

In [80], AG propagation channel measurements were performed in a near-mountainous area using a wideband chirp signal at a center frequency of 5.12 GHz. The overall terrain was flat with some nearby mountains resulting in moderate MPCs. The channel characteristics were largely dictated by the LOS component. The effects of airframe shadowing were also observed. In [95, 122] channel measurements were performed in the L-band and C-band in an urban hilly area. Curved Earth two-ray (CE2R) modeling was used, with weaker reflections and scattering from the hilly terrain and terrain cover observed. Additionally, multiple clusters of MPCs due to reflections from hills and buildings on hills were observed. In [124], a wideband frequency-modulated continuous wave (FMCW) signal centered at 5.06 GHz was used for channel measurements in a hilly area. RSS fluctuations due to MPCs from nearby buildings were reported. In the measurement campaigns of [95, 122, 124], the frequency bands were near this, with measurements in approximately similar environments. However, different results were observed, e.g., in [122], a better fitting with the CE2R model was observed in the C-band, whereas a free-space attenuation fitting was observed in [124] for one of the flight tracks.

In [97], wideband channel measurements were performed using a PRN sequence at a center frequency of 5.7 GHz over the sea. Multipath channel statistics were analyzed at different heights of the aerial platform. It was observed that the CIR can be represented using a three-ray model. Elevation and evaporation ducting effects were observed that resulted in reduction of the attenuation. Over-sea measurements were also carried out in [37, 39, 121], for over-sea and near-harbor areas. A dominant two-ray model was observed for all the cases. No ducting effect was observed over the sea for these measurements as opposed to [97].

In [36], an AG propagation scenario through a forest was imitated in an anechoic chamber using different heights of the TX and the RX and with different species of trees. Channel measurements were carried out in the X-band and Ku-bands. Different diffuse scattering regions from different parts and respective species of the trees were observed, resulting in corresponding small-scale fading statistics. Similar results were observed in [73], where foliage in the form of a medium-sized tree obstructed the direct LOS path between the UAV and the GS, resulting in small-scale fading due to diffraction and scattering from different parts of the tree. In [145], channel measurements were carried out using a CW sounding signal centered at 2 GHz. Received power was measured in different propagation environments, including woods, where the shadowing from the woods was found to be significantly different than obtained from the buildings.

2.5.6 Simulations for Channel Characterization

Apart from measurement campaigns for AG propagation channel modeling, some simulation-based channel characterizations are also available in the literature, where the real scenarios are imitated using computer simulations. Simulations in urban/suburban areas were performed in [13, 24, 47, 131]. The antenna considered in these environments was omnidirectional. Different carrier frequencies, namely 200 MHz, 700 MHz, 1 GHz, 2 GHz, 2.5 GHz, 5 GHz, and 5.8 GHz, were covered for AG channel characterization in the urban/suburban environments, and different heights of UAVs, ranging from 200 m to 2000 m were considered. The PL (from simulated RSS) was estimated. Over-sea-based channel simulations were carried out in [127], where a channel simulator imitating the sea environment was developed. Carrier frequencies from 3 kHz to 3 GHz were used, with the TX and the RX placed 3.75 m above the sea surface. The main goal of the study was to quantify sea surface shadowing for the marine communication channel using UAVs. The channel characteristics of PL and RMS-DS were modeled based on the sea surface height.

In [64], simulations were conducted in environmental scenarios consisting of over-sea, hilly, and mountainous terrain. Performance of AG communications using filter bank multicarrier (FBMC) modulation systems and LDACS were compared. The results showed that FBMC has better performance than LDACS, especially in the presence of interference from DME signals. In the presence of the AG channel, the FBMC and LDACS performances are comparable. Other simulations of communication systems employed over AG propagation channels, for particular simulation scenarios, are also available in the literature [26, 44, 65].

In [22], the effect of the UAV height for optimal coverage radius was considered. It is observed that, by adjusting UAV altitude, outage probability can be minimized: a larger "footprint" is produced with a higher UAV altitude, but, of course, increased altitude can increase PL. An optimum UAV height is evaluated that maximizes the coverage area for a given signal-to-noise ratio (SNR) threshold. The Ricean K-factor was found to increase exponentially with elevation angle between UAV and GS, given as $K = c_1 \exp(c_2 \theta)$, where c_1 and c_2 are constants that depend on the environment and system parameters. The relation between minimizing outage probability or maximizing coverage area for a given SNR threshold is solved only based on PL without considering the effect of scatterers in the environment. The consideration of geometry of scatterers in the analysis would, of course, make it more robust and realistic.

2.6 UAV AG Propagation Models

The UAV AG propagation measurements discussed in the previous section are useful for developing models for different environments. In the literature, UAV AG propagation channel models have been developed using deterministic or statistical approaches, or their combination. These channel models can be for narrowband, wideband, or even UWB communications. Complete channel models include both large-scale and small-scale effects. In this section, we categorize AG propagation channel models in the literature as shown in Figure 2.4, and review some of the important channel models.

2.6.1 AG Propagation Channel Model Types

Time-variant channel models can be obtained via deterministic or stochastic methods, or by their combination. The deterministic methods often use ray tracing (or geometry) to estimate the CIR in a given environment. These deterministic channel models can have very high accuracy but require extensive data to characterize any real environment. This includes the sizes, shapes, and locations of all obstacles in the environment, along with the electrical properties (permittivity, conductivity) of all materials. Hence such models are inherently site-specific. They also tend to require adjustment of parameters when comparing with measurement data. Since ray-tracing-based techniques employ high-frequency approximations, they are not always accurate. They are not as accurate as full wave electromagnetic solutions, e.g., the method of moments and finite-difference time-domain methods, for solving the Maxwell equations [144], but ray tracing methods are, of course, far less complex

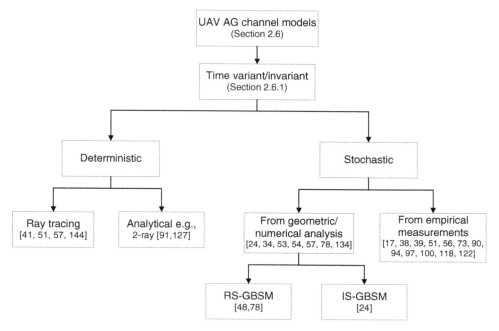

Figure 2.4 AG channel model characterization.

than these full-wave solutions. Such deterministic simulators are also very complex when they are used to model time-varying channels. Ray tracing was used in [13, 41, 47, 51, 57, 138] for different fully deterministic AG propagation scenarios.

The models in [38, 39, 122] are a mix of deterministic and stochastic models (sometimes termed quasi-deterministic). Specifically, the LOS and Earth's surface reflection are modeled deterministically via geometry, and the remaining MPCs are modeled stochastically, with parameter distributions (for MPC amplitude, delay, and duration) for each environment based on a large set of measurement data.

Purely stochastic channel models either can be obtained from geometric and numerical analysis without using measurements or they can be wholly empirical. Early cellular radio channel models, e.g., the COST 207 models, are examples of the latter. These types of models are becoming less and less common over time though, as incorporation of known physical information is shown to improve accuracy, and the greater model complexity is no longer prohibitive because of continuing advances in computer memory capacity and computational power. Geometry-based channel models for AG propagation generally require three spatial dimensions to be accurate. The associated velocity vector for UAV motion in space also requires three dimensions, although two-dimensional approximations can often be very accurate. In order to model the scatterers around the GS, two elliptical planes intersecting a main ellipsoid were considered in [24, 53, 54, 134], where the MPCs are defined by the ellipsoid and the two elliptical planes. Scatterers are considered to be randomly distributed on two spheres surrounding the TX and the RX in [78]. In [34, 48], the distribution of scatterers around the GS is modeled using a three-dimensional cylinder.

The GBSCMs can be further classified into regular-shaped GBSCMs (RS-GBSCMs) or irregular-shaped GBSCMs (IS-GBSCMs). For RS-GBSCMs, the scatterers are assumed to be distributed on regular shapes, e.g., ellipsoids, cylinders, or spheres. These models often result in closed-form solutions, but are, of course, generally unrealistic. In contrast, the IS-GBSCM distributes the scatterers at random locations through some statistical distribution. The properties of the scatterers in both cases are generally defined beforehand. In some cases, authors assume a large number of scatterers a priori, and, via the central limit theorem, obtain a Ricean amplitude distribution to obtain estimates of the CIR based upon some geometry. Alternatively, signal interaction from randomly distributed scatterers can be estimated directly, or with the help of ray tracing software [13, 41, 47]. A non-geometric stochastic channel model (NGSCM) based on a Markov process is provided in [141]. The ground-to-air fading channel was described by a Markov process that switches between the Ricean and Loo models, depending on the flight altitude.

2.6.2 Path-Loss and Large-Scale Fading Models

As noted, in mostly LOS AG channels, large-scale fading only occurs when the LOS path between UAV and GS gets obstructed by an object that is large relative to the wavelength. Some models for this attenuation mechanism exist (e.g., terrain diffraction, tree shadowing), but not much measurement data for UAV channels obstructed by buildings has been reported. When the LOS path does not get obstructed, the only other truly large-scale effect is the two-ray variation from the Earth's surface MPC. There are numerous measurement campaigns in the literature for PL estimation in different environments, as summarized in Table 2.4. Large-scale fading models in the literature cover both the PL and shadowing.

2.6.2.1 Free-Space Path-Loss Model

In the majority of the literature, the well-known terrestrial-based log-distance PL model with FSPL reference ("close-in," CI) is used:

$$PL_{CI}(d) = PL_0 + 10\gamma \log_{10}(d/d_0) + X_{FS}, \tag{2.2}$$

where $PL_{CI}(d)$ is the model PL as a function of distance, PL_0 is the PL at reference distance d_0 in free space given by $10\log[(4\pi d_0/\lambda)^2]$, γ is the PLE obtained using a least-squares error best fit, and X_{FS} is a random variable to account for shadowing, or, in the case of LOS channels, the variation about the linear fit. In free space the value of PLE is 2, but, as seen from Table 2.4, measured values of PLE vary from approximately 1.5 to 4. One might conceptually divide the path between the UAV and the GS into two components: the free-space component above the ground and the remaining *terrestrial-influenced* components. When the GS antenna height is well above surrounding obstacles, we expect the terrestrial components to have smaller effect and the PLE is near that of free space.

2.6.2.2 Floating-Intercept Path-Loss Model

Another PL model used in the literature for large-scale fading is floating intercept (FI) [62]. This model is similar to Eq. (2.2), but the FSPL at a reference distance is removed and the model's intercept is allowed to vary based on the least-squares fit without any intercept restriction. This model thus has two parameters denoted α and β [17], where α is the slope (PLE) and β represents the (floating) intercept. The FI model is given as

$$PL_{FI}(d) = 10\alpha \log_{10}(d) + \beta + X_{FI}, \tag{2.3}$$

where X_{FI} is again a random variable representing the PL variation.

2.6.2.3 Dual-Slope Path-Loss Model

The two PL models discussed above employ a single slope, or PLE. These models hold in areas where the rate of change of PL with distance does not change significantly over the complete range of distances. The two PL models in Eqs. (2.2) and (2.3) address this. At short distances, the model in Eq. (2.3) provides an underestimated PL, whereas it provides an overestimate at larger distances [123]; therefore, use of the CI model in Eq. (2.2) is recommended for all the cited scenarios. However, in some settings with NLOS paths and complex geometries resulting in higher-order reflections and diffractions, these single-slope models can have large regression errors. In such cases, a dual-slope (DS) PL model is sometimes used [29, 106]. This model is similar to the FI model, but has two different slopes for different link distance ranges, and can be represented as

$$PL_{DS}(d) = \begin{cases} 10\alpha_{d_1} \log_{10}(d) + \beta_{d_1} + X_{DS}, & d \leq d_1, \\ 10\alpha_{d_1} \log_{10}(d_1) + \beta_{d_1} + 10\alpha_{d_2} \log_{10}(d/d_1) + X_{DS}, & d > d_1, \end{cases} \tag{2.4}$$

where α_{d_1} and α_{d_2} are the slopes of the fits for the two link distance ranges separated by threshold d_1, β_{d_1} is the intercept, and X_{DS} is a random variable representing the variation in the fit.

Table 2.4 Review of existing literature on large-scale AG propagation and the path-loss parameters in those papers.

Ref.	Scenario	Path (LOS/NLOS)	Model type	PLE (γ) or (α, β) parameters	Intercept PL_0 (dB)	σ (dB)
[73]	Suburban, open field	LOS, OLOS	Log-distance PL, eqs. (1), (2)	γ: 2.54–3.037	21.9–34.9	2.79–5.3
[17]	Lightly hilly rural (for $h_U = 120$ m); for other values of the height, see table II in the paper	LOS	Log-distance PL (alpha–beta model), eq. (2)	$\alpha = 2.0, \beta = -35.3$	—	3.4
[124]	Urban, suburban, rural	—	FSPL	—	—	—
[100]	Urban	—	FSPL, eq. (1)	—	—	—
[38]	Urban, suburban	LOS	Log-distance PL, two-ray model, eqs. (1), (2)	γ: 1.7 L-band, 1.5–2 C-band	98.2–99.4 L-band, 110.4–116.7 C-band	2.6–3.1 L-band, 2.9–3.2 C-band
[56]	Urban, suburban	LOS, NLOS	Modified FSPL, eqs. (5), (6)	—	—	—
[142]	Urban, open field	LOS	Log-distance PL, eq. (1)	γ: 2.2–2.6	—	—
[119]	Urban	LOS, NLOS	Modified FSPL, eqs. (1), (9), (10), (13)	—	—	—
[125]	Urban	—	Modified LUI model, eq. (1)	—	—	—
[99]	Urban, rural	LOS	Log-distance PL, eq. (7)	γ: 4.1	—	5.24
[129]	Near airports	LOS	Log-distance PL, eq. (1)	γ: 2–2.25	—	—
[143]	Open field	—	Log-distance PL, eq. (1)	γ: 2.01	—	—
[12]	—	LOS	Log-distance PL	γ: 2.32	—	—
[122]	Hilly, mountainous	LOS	Log-distance PL, eq. (3)	γ: 1.3–1.8 L-band, 1–1.8 C-band	96.1–106.5 L-band, 115.4–123.9 C-band	3.2–3.9 L-band, 2.2–2.8 C-band
[36]	Forest/foliage	—	—	—	—	—
[37]	Over sea	LOS	Two-ray PL, eq. (1)	—	—	—
[39]	Over water, sea	LOS	Log-distance PL, two-ray PL, eqs. (15), (16), (17)	γ: 1.9, 1.9 over water and sea for L-band, 1.9, 1.5 over water and sea for C-band	104.4, 100.7 over water and sea for L-band, 116.3, 116.7 over water and sea for C-band	3.8–4.2 over water and sea for L-band, 3.1–2.6 for over water and sea for C-band
[97]	Over sea	LOS	Two-ray PL, FSPL, eqs. (2), (3)	γ: 0.14–2.46	19–129	—
[29]	Ensemble of containers, see table II in the paper	LOS	Dual slope, eqs. (1), (2), (3)	—	—	—

2.6.2.4 Log-Distance Path-Loss Model

PL estimates using log-distance models (Eq. (2.2)) are given in [12, 33, 35, 38, 51, 73, 92, 94, 97, 122, 125, 127, 129, 134, 142, 143]. There are other PL models that consider shadowing for NLOS paths, and additional losses incurred from other obstacles [56, 100, 119]. In [56], shadowing loss was considered in the modeling and evaluated as a function of the elevation angle for NLOS paths. The shadowing loss was calculated based on the uniform theory of diffraction. The distribution of the shadowing was found to be normal. Strong shadowing was observed in [119], in an urban area, mostly due to diffraction from the surrounding buildings. In [100] additional losses at the GS and the UAV for the overall PL modeling were considered.

2.6.2.5 Modified FSPL Model

Due to the potential three-dimensional motion of UAVs, modified FSPL models accounting for UAV altitude can also be developed; several that are a function of elevation angle are considered in [13, 47, 74, 102, 125]. In [74], the effect of the three-dimensional antenna radiation pattern is included in the received power calculation. At higher elevation angles with omnidirectional antennas at both the TX and the RX, the low antenna gain resulted in reduction of the received power. This loss, due to small antenna radiation gain in the elevation plane at higher UAV heights, was reduced as the UAV moved away from the GS (decreasing elevation angle). Similar observations on antenna radiation patterns were made in [125]. In [13], a modified FSPL model was provided, taking into account the height of the aerial platform. Specifically, the distance d obtained from the geometry of the setup in the FSPL expression was modified as $d = \Delta h / \sin \theta$. In [47], PL considering the elevation angle was modeled for $\theta > 10$. The PL was modeled for both LOS and NLOS paths using intrinsic coefficients, and this was claimed to be independent of the antenna heights at the TX and the RX. A similar three-dimensional PL model was provided in [102].

2.6.2.6 Two-Ray PL Model

The two-ray PL model described earlier in Section 2.3.3 is provided in [38, 39, 89–91, 94, 95, 97, 121]. In the case of two-ray PL modeling, the variation of the PL with distance has distinctive peaks due to destructive superposition of the dominant LOS and surface-reflected components. In the majority of PL models, PL variation is approximated as a log-normal random variable. This variation can be due to shadowing from the UAV body (see Section 2.6.2.7), or from obstacles, or from MPCs attributable to terrestrial scatterers such as buildings [12, 38, 39, 56, 73, 89, 92, 95, 97, 99, 122, 129, 134].

2.6.2.7 Log-Distance FI Model

In [17], log-distance FI models for the PLE and shadowing for the AG radio channel between airborne UAVs and cellular networks were presented for 800 MHz and UAV heights from 1.5 to 120 m above ground. In [29], the low-altitude UAV AG wireless channel has been investigated for a scenario where a UAV was flying above an ensemble of containers at 5.76 GHz. Narrow- and wideband measurements were carried out. The paper presents a modified PL model and example PDPs. Most interesting is that, in this particular environment, delay dispersion actually increases with altitude as the UAV rises above metallic structures.

2.6.2.8 LOS/NLOS Mixture Path-Loss Model

Another model used in the literature [14–16, 21, 28, 68, 98] represents the PL in both LOS and NLOS conditions probabilistically, as follows [14, 61]:

$$PL_{avg} = P(LOS)\, PL_{LOS} + [1 - P(LOS)]\, PL_{NLOS}, \tag{2.5}$$

where PL_{LOS} and PL_{NLOS} are the PLs in LOS and NLOS conditions, respectively, and $P(LOS)$ denotes the probability of having a LOS link between the UAV and the ground node, given by [14, 61]

$$P(LOS) = \prod_{n=0}^{m}\left[1 - \exp\left(-\frac{[h_U - (n + \frac{1}{2})(h_U - h_G)/(m + 1)]^2}{2\Omega^2}\right)\right]. \tag{2.6}$$

Here we have $m = \text{floor}(r\sqrt{\varsigma\xi} - 1)$, r is the horizontal distance between the UAV and the ground node, h_U and h_G are as shown in Figure 2.1, ς is the ratio of built-up land area to the total land area, ξ is the mean number of buildings per unit area (in km^2), and Ω characterizes the height (denoted by H) distribution of buildings, which is based on a Rayleigh distribution, i.e., $P(H) = (H/\Omega)^2 \exp(-H/2\Omega^2)$. In [14], for a specific value of θ in figure 2 therein, a sigmoid function is also fitted to Eq. (2.6) for different environments (urban, suburban, dense urban, and high-rise urban) to enable analytical tractability of UAV height optimization. Since Eq. (2.5) averages the PL over a large number of potential LOS/NLOS link possibilities, it should be used carefully if employed with system-level analysis for calculating end metrics such as throughput and outage. Similarly, PL variability should be added to the model of Eq. (2.5).

Overall, a comparison of different PL models in Table 2.4 shows that the two-ray PL model is a better choice for open field, rural areas, and over sea, whereas higher-order ray models, e.g., three- or four-ray models, can be employed for environments with a larger number of scatterers whose heights are comparable to the height of the UAV. This has been validated by channel measurements carried out in [38, 39, 122], where the CE2R model was found to provide better fitting for over-water, harbor, and mountainous settings; however, the log-distance PL model was found to have a better fit for the environments with taller scatterers, e.g., urban and suburban. Additionally, the two-ray PL fitting for L-band was better than for C-band. However, in [97], the measured data obtained over sea was compared with FSPL and two-ray model, and it was observed that both models overestimated the PL. This was hypothesized to be due to ducting over the sea surface causing reduction in the PL. In [124], the RSS obtained from wideband measurements at 5.06 GHz was found to closely fit the FSPL for one location, yet the FSPL yielded an upper bound for another location. However, in [122], with similar propagation environment at C-band, it was observed that the CE2R model provided a better fit than the flat-Earth two-ray model.

Selection of a suitable PL model for a given AG propagation scenario is pivotal. In most of the literature, the PL model of Eq. (2.2) is used due to its simplicity and provision of a standard platform based on reference distance FSPL for comparison of measurements in different environments. A reference distance of 1 m is often taken as a standard for short-range systems, but larger values are also used. However, in some scenarios, where the reference FSPL is not available, the FI model (Eq. (2.3)) may be used. Yet, due to lack of any standard physical reference, the FI slope will be dependent on the environment.

Additionally, the variability of the PL is generally a zero-mean Gaussian random variable that has approximately similar values for both the CI and FI model types.

A general recommendation for selection of AG PL model for a given measurement scenario from Table 2.4 is as follows. For an open flat or hilly area with light suburban, rural, or no terrain cover, and for over water, the two-ray PL model or free-space reference log-distance model (Eq. (2.2)) may be preferred. This is due to the small number of MPCs reaching the aerial platform and the high probability of the presence of a dominant LOS and ground-reflected component only in these environments. For complex geometrical environments with large numbers of scatterers and with a LOS path between the GS and the UAV, a DS PL (Eq. (2.4)) model may be best if two distinct regions can be identified: one is where persistent components arise from the LOS and the ground reflection paths, and the other where MPCs arise from the surrounding scatterers. The FI model in Eq. (2.3) may be preferred in certain specialized environments, e.g., [29], due to their ease in applicability and PL model predictions only for a given area. In Table 2.4, the model types denoted log-distance refer to the general log-distance equation for PL with different reference distances and additional parameters.

2.6.3 Airframe Shadowing

Airframe shadowing occurs when the body of the aircraft itself obstructs the LOS to the GS. This impairment is somewhat unique to AG communications, and there are not many studies in the literature on this effect. One reason for this is that such shadowing can be largely (but not always completely) alleviated by using multiple spatially separated antennas: airframe shadowing on one antenna can be made unlikely to occur at the same time as shadowing on the other(s). In addition to frequency and antenna placement, shadowing results also depend on the exact shape, size, and material of the aircraft. For small rotorcraft, depending on frequency and antenna placement, airframe shadowing could be minimal. Example measurement results, as well as models for airframe shadowing, for a fixed-wing medium-sized aircraft, were provided in [120].

For these results, at frequencies of 970 and 5060 MHz, wing shadowing attenuations were generally proportional to aircraft roll angle, with maximum shadowing depths exceeding 35 dB at both frequencies. Shadowing durations depend upon flight maneuvers, but for long, slow banking turns, can exceed tens of seconds.

An illustration of airframe shadowing is shown in Figure 2.5 where received power is plotted against time for a wideband (50 MHz) signal in C-band before, during, and after the medium-sized aircraft made a banking turn. The received power on two aircraft antennas (denoted C1 and C2), bottom-mounted and separated by approximately 1.2 m, is shown. Attenuations due to airframe shadowing, along with the polarization mismatch that occurs during the aircraft maneuver, exceed approximately 30 dB in this case.

2.6.4 Small-Scale Fading Models

Small-scale fading models apply to narrowband channels or to individual MPCs, or *taps* in tapped delay line wideband models, with bandwidth up to some maximum value (i.e., small-scale fading may not pertain to MPCs in a UWB channel). The depth of small-scale

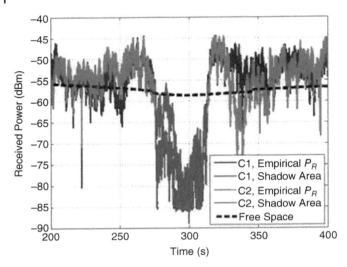

Figure 2.5 Received power versus time for illustration of shadowing before, during, and after the banking turn of a medium-sized aircraft at C-band.

amplitude fades on a given signal also generally varies inversely with signal bandwidth [87]. Stochastic fading models are obtained through analysis, empirical data, or through geometric analysis and simulations [24, 34, 48, 54, 78, 134]. As noted in Section 2.6.1, the GBSCMs can be subdivided into RS-GBSCM and IS-GBSCM. In [24], a time-variant IS-GBSCM was used with a Ricean distribution for small-scale fading. Time-variant RS-GBSCMs were provided in [48, 78], and these also illustrated Ricean small-scale fading.

A NGSCM was provided in [141], where ground-to-air fading was described using Ricean and Loo models. The Loo model was derived based on the assumption that the amplitude attenuation of the LOS component due to foliage in a land mobile satellite link follows a log-normal distribution, and that the fading due to MPCs follows a Rayleigh distribution. The switching between Ricean and Loo models was controlled by a Markov process dependent on flight height. In [54], a GBSCM for MPCs was provided in the form of shape factors describing angular spread, angular compression, and direction of maximum fading using the probability density function (PDF) of angle of arrival.

Table 2.5 provides measured small-scale AG fading characteristics reported in the literature for various environments. As previously noted, the most common small-scale fading distribution for AG propagation is the Ricean. As in terrestrial channels, for the NLOS case, the Rayleigh fading distribution typically provides a better fit [65, 99, 118, 129, 131, 147, 148], and, of course, other distributions such as the Nakagami-m and Weibull distributions might also be employed. Small-scale fading rates depend upon velocity, and these rates are proportional to the Doppler spreads of the MPCs [121, 124, 129, 139].

2.6.5 Intermittent MPCs

Another AG characteristic that may be of interest in high-fidelity and long-term channel models is the intermittent nature of MPCs. From geometry, it is easy to deduce that, for a

Table 2.5 Review of the existing literature on small-scale AG propagation channel fading characteristics.

Ref.	Scenario	Time-variant/ invariant	Modeling type	Frequency spectrum	Center frequency (GHz)	Doppler spread (Hz)	Fading distribution	K-factor (dB)
[118]	Urban/suburban	Time-invariant	Statistical	Narrowband	2	—	Ricean	—
[73]	Suburban/open field	Time-invariant	Statistical	Ultra-wideband	3.1–5.3	—	Nakagami	—
[124]	Suburban/open field	Time-variant	—	—	5.06	833	—	—
[100]	Urban/suburban	—	—	Narrowband	2.3	—	—	—
[99]	Urban/suburban	Time-invariant	Statistical	Wideband	2	—	Rayleigh, Ricean	—
[129]	Urban/suburban	—	Statistical	Wideband	5.75	1400	Ricean	–5 to 10
[38]	Urban/suburban	Time-variant	Statistical	Wideband	0.968, 5.06	—	Ricean	12–27.4 in L- and C-bands
[80]	Hilly	Time-variant	—	Wideband	5.12	5000	—	—
[36]	Forest/foliage	—	Statistical	Ultra-wideband	8–18	—	Ricean, Nakagami	2–5
[94]	Sea/freshwater	Time-variant	Statistical	Wideband	0.968, 5.06	—	Ricean	12, 28 for L- and C-bands
[26]	—	Time-variant	Statistical	Wideband	5.135	5820	—	—

given vehicle trajectory in some environment, individual MPCs will persist only for some finite span of time [39]. This has been noted in V2V channels as well, but with UAVs and their potentially larger velocities, the intermittent MPC (IMPC) dynamics can be greater. These IMPCs arise (are "born") and disappear ("die") naturally in GBSCMs. They may also be modeled using discrete-time Markov chains. The IMPCs can significantly change the CIR for some short time span, hence yielding wide variation in RMS-DS (another manifestation of so-called "non-stationarity"). Example models for the IMPCs – their probability of occurrence, duration, delay, and amplitude – appear in [37, 39, 74, 90]. In these studies it was found that IMPCs follow a random process which is highly dependent on the geometry and distribution of the scatterers for a given UAV trajectory.

In Figure 2.6, from [89], the fading of MPCs as a function of time and delay are shown. The amplitude of MPCs generally decay with excess delay for any given time instant. Additionally, there may be a continuous birth and death process of MPCs at different instants of time. This can be represented using CIR as [89]

$$h(t, \tau) = \sum_{i=0}^{M(t)-1} p_i(t) a_i(t) \exp(j\phi_i(t)) \, \delta(\tau - \tau_i(t)), \tag{2.7}$$

where $h(t, \tau)$ is the time-variant CIR, $M(t)$ is the total number of MPCs at time instant t, and $p_i(t)$ represents the multipath persistence process coefficient and can take binary values [0, 1]. The amplitude, phase, and delay of the ith MPC at time instant t are represented as $a_i(t)$, $\phi_i(t)$, and $\tau_i(t)$, respectively. The phase term is given as $\phi_i(t) = 2\pi f_d^i(t)(t - \tau_i(t)) - f_c(t)\tau_i(t)$, where $f_d^i(t) = v(t)f_c(t)\cos(\Theta_i(t))/c$ is the Doppler frequency of the ith MPC, $\Theta_i(t)$ is the aggregate phase angle in the ith delay bin, c is the speed of light, and f_c represents the carrier frequency. The channel transfer function $H(f, t)$ from (2.7) is then given

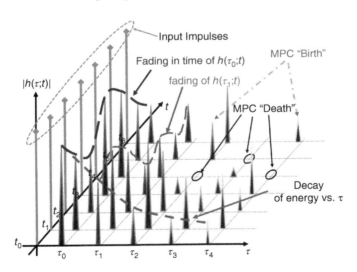

Figure 2.6 Fading and birth and death process of intermittent MPCs, from [89].

as follows:

$$H(f,t) = \sum_{i=0}^{M(t)-1} p_i(t)a_i(t)\exp[j2\pi f_d^i(t)(t - \tau_i(t))]$$

$$\times \exp(-j2\pi f_c \tau_i(t))\exp(-j2\pi f \tau_i(t)). \tag{2.8}$$

The effect of the Doppler frequency term in the phase is typically negligible compared to the carrier frequency term, especially at lower velocities. Therefore, the carrier frequency term will dominate the variation of the transfer function.

Figure 2.7(a) shows a sequence of PDPs versus link distance for a near-urban AG link near Cleveland, OH. Flight parameters can be found in [38]. In this figure, the IMPCs are clearly visible, here caused by reflections from obstacles near the Lake Erie shoreline. In Figure 2.7(b) RMS-DS versus link distance for a hilly environment in Palmdale, CA, is shown. The intermittent nature of the MPCs produces "spikes" and "bumps" in the RMS-DS values, illustrating the potential rapid time variation of AG channels.

2.6.6 Effect of Frequency Bands on Channel Models

The selection of frequency bands for CNPC and payload communications using AG links for UAVs is discussed in Section 2.3.2. In the literature, most of the AG propagation channel measurements were carried out using wideband signals, whereas some narrowband channel measurements are also available. Several current and future communications are expected to deploy OFDM technology. Therefore, narrowband characteristics are of value for their effect on individual OFDM subcarriers; frequency correlation is also important.

UWB AG propagation channel measurements in the frequency band 3.1–5.3 GHz were performed in open and suburban areas in [73]. A narrow sub-band frequency analysis was provided, where higher mean attenuation and larger variance of received power were observed in the higher frequency sub-bands. A narrowband frequency measurement campaign for three different frequencies of 2 GHz, 3.5 GHz, and 5.5 GHz was carried out in an urban area in [56]. Higher attenuation was observed in the higher frequency bands, whereas the standard deviation did not show much difference for the three different frequency bands. Wideband channel measurements at L-band and C-band were provided in [38, 39, 122] for different propagation environments. Different attenuation and small-scale effects were observed at the two different frequency bands in a given propagation environment; e.g., in [122], channel measurements were performed in a hilly/mountainous area. Similarly, higher K-factor and PLE were observed for C-band than for L-band, whereas a larger shadowing variation was reported for L-band than C-band.

Some mmWave AG propagation channel characteristics using UAVs at 28 GHz and 60 GHz in different environments were provided in [72] using ray tracing simulations. The RSS generally followed a two-ray model at both frequencies in different propagation environments; however, a higher rate of maxima/minima at 60 GHz was observed than at 28 GHz. In [100], the two-ray model was used for PL representation at 2 GHz in an

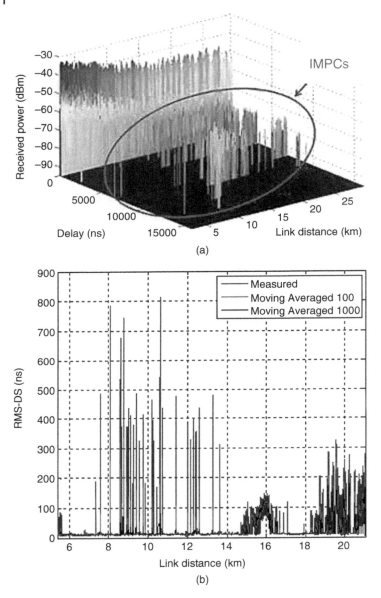

Figure 2.7 (a) Sequence of PDPs versus link distance for a near-urban AG link near Cleveland, OH. (b) RMS-DS versus link distance for a hilly environment in Palmdale, CA.

urban environment, whereas channel measurements at 2 GHz obtained in [99] in a similar environment were found to be better fitted with a log-distance PL model.

2.6.7 MIMO AG Propagation Channel Models

The use of MIMO systems for UAV AG communications has been gaining popularity. The rationale, increased throughput and reliability, is the same one that increased the use of

MIMO in other terrestrial systems. In [27], it was shown that it is possible to attain higher spatial multiplexing gains in LOS channels by properly selecting the antenna separation and orientation as a function of carrier wavelength and link distance. This careful alignment is not always practicable or possible with UAVs, especially when mobile.

The advantages of spatial diversity and multiplexing gains in MIMO are often only moderate due to limited scattering available near UAVs or GSs. In [109], it was demonstrated that, due to limited spatial diversity in the AG channel, only moderate capacity gains are possible. In order to obtain better spatial multiplexing gains, larger antenna separations are required, and this requires large antenna arrays that are not feasible on-board small UAVs. Use of higher carrier frequencies makes it possible to use electrically large antenna arrays, but higher frequencies yield higher PL (this can be mitigated somewhat by beamforming, at the expense of the complexity required for beam steering). Moreover, accurate channel state information (CSI) is important for MIMO systems for higher performance, but, in a rapidly varying AG propagation channel, it can be difficult to provide accurate CSI and hence MIMO gains can be limited. The use of MIMO on airborne platforms also incurs additional cost, computational complexity, and power consumption.

There are a limited number of studies available in the literature for MIMO AG propagation channel measurements. In [137], a detailed measurement analysis of the AG MIMO propagation channel was provided. It was observed that a considerable spatial decorrelation of the received signal at the GS was achieved due to the interaction of non-planar wavefronts. These wavefronts are generated due to near-field effects from the measurement vehicle, on which the GS antennas were mounted. Spatial diversity from antennas located on the UAV was also observed, interestingly at *higher* elevation angles. The authors suggest that having scatterers near the GS can yield larger spatial diversity. The received signal in [110] was analyzed for multiple input–single output (MISO) and MIMO systems, and it was observed that the use of MIMO systems enables a more robust channel for changes in antenna orientations arising from UAV maneuvering. In [145], MIMO system performance was tested in different scenarios of the outdoor environment, including urban, rural, open field, and forest. The effect of terrain cover on the received power was analyzed for these different scenarios, with the result that the propagation channel in the open field is mostly influenced by the ground reflections, whereas in the case of forests, the reflection and shadowing from the trees is a major contributor to the propagation channel characteristics. In rural and urban cases, the reflections from the walls and surfaces of building structures play an important role.

Time-variant GBSCMs for MIMO systems provided in [24, 30, 48, 78] were explored through simulations, where different propagation geometries and scatterer distributions were used in order to analyze the capacity for MIMO AG channels. A common – and expected – observation was that higher capacity is achievable with MIMO AG systems. A simulation-based AG MIMO channel propagation model was provided in [134] for a hilly area. The results indicate increased throughput from spatial multiplexing and higher SNR from the MIMO system in comparison to SISO, as expected. A stochastic model for a mobile-to-mobile AG MIMO propagation channel was presented in [78]. These results show that there was considerable capacity increase and reduction in outage probabilities using MIMO systems if perfect instantaneous CSI is available. In [30], geometry-based simulations were conducted for a massive MIMO implementation for a UAV AG propagation channel. The simulation results illustrate the expected result of a significant capacity increase when a large number of antennas are used at the GS.

2.6.8 Comparison of Different AG Channel Models

In this subsection, we briefly analyze and compare different UAV AG propagation channel models in the literature.

2.6.8.1 Large-Scale Fading Models

The large-scale fading models for AG propagation available in the literature generally can be fitted with a modified FSPL model. In [73], PL was measured for open and suburban scenarios in the presence and absence of foliage. The PL was found to be highest for the foliage due to obstruction. In addition, the PL was found to be dependent on the height of the GS, apart from the height of the UAV. The PLE reported was above 2.5 for all the propagation scenarios involving open and suburban areas; the PLE for suburban scenarios was slightly larger than that for the open area scenario. In [36], PL due to diffraction and scattering from different species and parts of the tree was provided. It was observed that the loss from the trunk of the tree is due to diffuse scattering, whereas it was mostly due to diffraction on the edges of the crown of the tree. In [37–39, 92, 95, 122], PLs for channel measurements in the urban, suburban, hilly, mountainous, and over-sea scenarios were provided. A comparison of PL obtained from free space, analytical CE2R model, and measurements was provided. The PLE for C-band and L-band for different measurement scenarios was found to be approximately equal to or less than the FSPL. For over-sea scenario, ducting was not observed, whereas in [97], ducting was hypothesized as the cause of the reduction of the PL as compared to FSPL. A study for PL during different flight scenarios of take-off, en-route, landing, taxiing, and parking were provided in [129]. Higher PLE was observed for take-off and en-route as compared to other flight scenarios. In [119], excess PL was provided for an urban scenario. It was found that excess loss was dependent on different diffractions from the edges of the surrounding buildings.

Antenna orientation effects on the RSS were provided in [142], where the PLE was found to be close to FSPL in different antenna orientations in urban area and open field. The PLE was found to be larger in the urban area than in the open field. A similar study taking into account the antenna radiation effects is available in [74, 125], where it was found that minimum received power was observed when the UAV was on the top of the base station due to minimum antenna gain in the elevation plane at that point.

2.6.8.2 Small-Scale Fading Models

In the literature, there are limited small-scale models for AG propagation using UAVs. The L-band and C-band measurement and modeling campaign in [38, 39, 122] provides the bulk of the small-scale measurement-based modeling information for AG propagation channels. A TDL model was used to represent the channel response in all the scenarios. A two-ray model in addition to varying numbers of intermittent MPCs in different environments constituted the TDLs. In all the scenarios, the K-factor for C-band was found to be higher than the L-band. The highest K-factor was observed for the C-band for the over-sea scenario, followed by hilly/mountainous, and suburban/urban scenarios. The K-factor in the L-band was found to vary less in different propagation environments than in the C-band.

A small-scale propagation channel model for UWB was provided in [73], for suburban and open field scenarios. The small-scale fading amplitudes were found to be Nakagami distributed. A modified Saleh–Valenzuela model was used to model the CIR. Different numbers of MPC clusters were observed in the open field as compared to the suburban environment. It was observed that the RMS-DS changed with the height of the UAV in the suburban scenario, whereas it was approximately flat for the open field scenario. Similarly, higher TOA of MPCs was observed for suburban scenario as compared to open field for different UAV heights. In [118], second-order channel statistics of average fade duration and level crossing rate were provided for narrowband AG signal propagation. The amplitude of the MPCs was found to be log-normally distributed. A time-series generator was used for emulation of RSS based on the analytical model to simulate the measurement results.

2.6.9 Comparison of Traditional Channel Models with UAV AG Propagation Channel Models

The UAV AG propagation channel naturally has similarities with outdoor terrestrial propagation channels, at least for elevated base stations. We do not provide a comprehensive comparison between UAV AG channels and terrestrial channels here, as there are numerous terrestrial channel models for a variety of terrestrial environments [8] that differ based on specific environment, center frequency of operation, bandwidth, antenna type, and configuration, among other factors. Here we provide some remarks for context.

Among other applications, UAVs are likely to be used in future cellular communication networks either as a base station [85, 107] or as a user equipment node [51]. When a UAV is used as a base station, the corresponding propagation channel *may* exhibit cellular base station characteristics if they are hovering (with no mobility) at similar heights/environments as terrestrial base stations, and assuming they communicate with similarly placed user equipment as in terrestrial cellular networks. In such cases, terrestrial cellular channel models might be applicable for AG channels [8].

On the other hand, many times the UAVs may be mobile, they might operate at significantly higher altitudes than terrestrial base stations, and the operational environment may also be very different, all of which should be taken into account for the channel model. A major difference in the AG channel would be that, for typical situations, the vast majority of scatterers are located around the GS and not the UAV. This, of course, has a direct influence on the MPC characteristics, mainly the Doppler frequencies of the MPCs. In particular, if the UAV is moving while the GS is stationary, all MPCs usually have a very similar Doppler frequency, in strong contrast to the terrestrial channel.

When the UAV is used as a user equipment node (communicating with a terrestrial base station), the terrestrial cellular channel models cannot be directly applied [75]. This is true because, when aloft, hovering, or moving, the UAV experiences large- and small-scale fading characteristics different from ground-based users due to the different spatial distribution of nearby obstacles. Channel models for a user equipment node as a

UAV in a cellular communication network are provided in the recent version of the 3GPP documentation covering UAV communications [7]. Details are discussed in Section 2.6.11.

One can also compare UAV channel models with satellite channel models [88, 103]. Overall, both satellite and UAV communication links have higher probability of LOS than terrestrial cellular links [128]. One of the main differences is that, for most satellite applications, we are not interested in small elevation angles (e.g., smaller than 5°). This is due to the negative effects of the troposphere and/or ionosphere on the propagation. However, in the AG channel we often end up with very low elevation angles and long link distances: this implies that the effects of the troposphere can be significantly larger than for the typically high elevation angles used in satellite applications. Many satellite communication links are directional and point-to-point, whereas UAV communications need not be directional.

2.6.10 Ray Tracing Simulations

In the literature, in addition to measurements, channel characterization for AG propagation is also carried out using simulations. These simulators are either based on customized channel environments on a given software platform or can be realized using ray tracing simulations. There are PL models available for these simulated environments [13, 14, 47, 102, 127, 148]. Urban environmental scenarios for LOS and NLOS paths were considered in [13, 14, 47] where log-distance and modified FSPL models were suggested. In [127] a log-distance path model was provided for LOS and NLOS paths for over-sea settings in a simulated environment. However, to the best knowledge of the authors, there are no specific experimental studies available in the literature that experimentally validate the channel models proposed using geometrical analysis and simulations in [13, 14, 47, 102, 127, 148].

Ray tracing was used for mmWave channel characterization for 28 GHz and 60 GHz frequency bands for UAV AG propagation in [72]. Different environments were realized, namely urban, suburban, rural, and over sea. It was observed that the RSS followed the two-ray model and is, of course, affected by the presence of scatterers in the surroundings. The RMS-DS was also affected by the presence of scatterers in the surrounding environment and the UAV height in the given environment. If the height of the scatterers was comparable to the UAV height, larger RMS-DS was observed due to multiple reflections from the randomly distributed scatterers. In contrast, if the height of the scatterers is small, we have smaller RMS-DS at higher UAV heights due to fewer significant MPCs reaching the UAV. This phenomenon was verified at 28 GHz and 60 GHz, where at 60 GHz we have smaller RMS-DS than at 28 GHz due to higher attenuation of MPCs.

Ray tracing simulations using the Wireless InSite software were carried out to estimate PL for an over-sea scenario as shown in Figure 2.8. The channel measurement parameters were set according to [39], and the simulated PL results were compared with the measured values. Figure 2.9 shows the simulated PL results. In this simulated environment, we have buildings as scatterers near the TX. Due to reflections and diffractions from these scatterers, we observe additional fluctuations on top of the two-ray propagation model. The deviations are due to MPCs reflected and diffracted from the different-shaped scatterers at

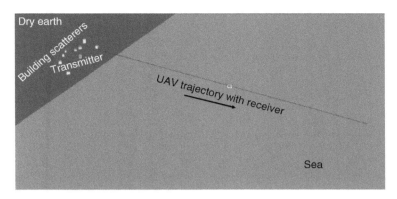

Figure 2.8 Ray tracing simulation scenario for over-sea scenario, where the UAV flies over a straight line.

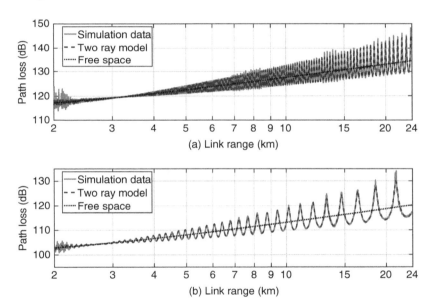

Figure 2.9 Ray tracing PL results for over-seawater settings: (a) C-band (5.03–5.091 GHz) and (b) L-band (0.9–1.2 GHz).

different angles. These weak MPCs reach the UAV RX at different link distances, resulting in variations from the two-ray model as shown in Figure 2.10 at a link distance between 13 and 14 km.

Similarly, in Figure 2.11(a), the effect of MPCs from the scatterers around the TX for link distances of 100 m to 2 km are shown. It can be observed that, without the scatterers and seawater (with ground only), we have a perfect two-ray PL model. Yet in the presence of the scatterers around the TX, superimposed upon this effect is a variation from additional MPCs from the scatterers. This effect can be modeled as a random PL component on top

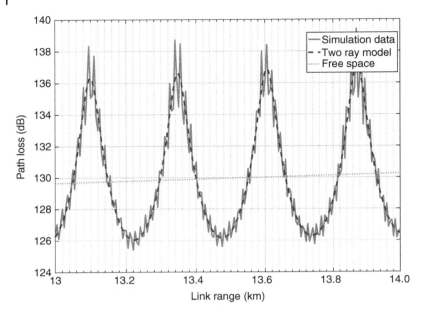

Figure 2.10 Zoomed-in results of PL for the over-seawater simulations of Figure 2.9 for C-band at link distances of 13–14 km.

of the two-ray model, or in effect a small-scale fading. This effect is, of course, dependent on the geometry of the scenario and will cause the PL to vary along the trajectory of the UAV. A similar effect at the larger link distance range of 13–13.5 km in Figure 2.11(a) can be observed in Figure 2.11(b).

There is a good match between the ray tracing simulation results and analytical results for this over-sea scenario in Figure 2.9. However, when comparing measurement data, e.g., in [39], with simulation data, we observe more fluctuations in measurements due to several factors: ambient noise, measurement equipment variation, and in particular, scattering from the rough sea surface, which is not as easily modeled with the basic ray tracing.

2.6.11 3GPP Channel Models for UAVs

The latest version of the 3GPP model [7] provides channel modeling details for UAV AG communications, considering the user equipment on the UAV communicating with the fixed base station. These details include LOS probability, PL models, and small-scale fading models. The height of the user equipment on-board the UAV in the air can be smaller or greater than the base station height.

LOS probability for rural macro (RMa), urban macro (UMa), and urban micro (UMi) are provided for different aerial user heights. The LOS probability is smaller for all the scenarios when the UAV height is small due to obstruction from scatterers on the ground. As the height of the aerial user increases, the LOS probability also increases; e.g., for RMa scenario,

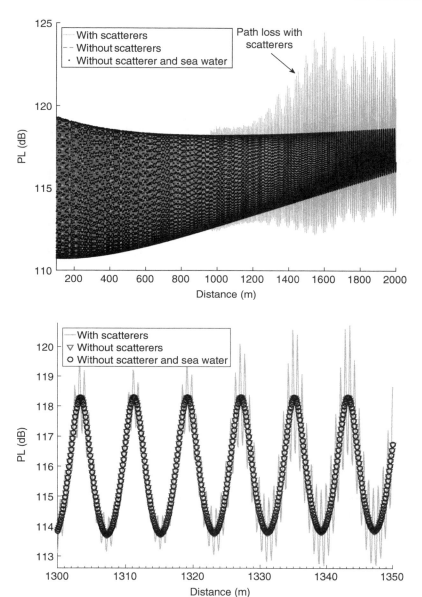

Figure 2.11 Path loss versus distance with/without scatterers, and without the sea surface: (a) 100 m to 2 km range, and (b) 1300 m to 1350 m range.

we have 100% LOS probability after 50 m of aerial user height, whereas this threshold distance is 100 m for UMa scenario. The LOS probability expression is dependent on the height of the UAV. For example, for a RMa scenario, with a UAV height in the range of 10–40 m, and for a UMa and UMi scenario with a UAV height range of 22.5–100 m, the LOS probability expression is given as follows [7]:

$$P_{LOS} = \begin{cases} 1, & \text{for } d_h \leq d_1, \\ \dfrac{d_1}{d_h} + \exp\left(-\dfrac{d_h}{p_1}\right)\left(1 - \dfrac{d_1}{d_h}\right), & \text{for } d_h > d_1, \end{cases} \tag{2.9}$$

where d_h is the horizontal distance of the UAV from the base station. The values of variables d_1 and p_1 are dependent on the UAV height and the scenario considered.

The PL models are modified FSPL models taking into account the height of the user with respective constants. For the three scenarios of RMa, UMa, and UMi, the lower bound height of the user is 1.5 m and goes up to 300 m. The PL models using UAVs are divided into two categories based on the height of the aerial user. For a RMa scenario, with height of aerial user in the range 1.5 m to 10 m, the PL model from section 7.4 of [9] were recommended for both LOS and NLOS paths, whereas for aerial user heights above 10 m, additional PL models are provided for LOS and NLOS paths [7]. Similarly, for the UMa and UMi scenarios, instead of 10 m, an upper bound height of 22.5 m is used for separation of the two PL model categories. The distribution of the shadow fading in all the scenarios is represented as log-normal. The standard deviation of the shadow fading for LOS scenarios for aerial user height of more than 10 m for RMa and 22.5 m for UMa and UMi is a function of height of the aerial user. On the other hand, for aerial user height of less than 10 m for RMa, and 22.5 m for UMa and UMi, and NLOS scenarios, it is assigned a constant value.

Fast (small-scale) fading models are also provided in [7] for aerial users. The models are provided for aerial user height of 10 m to 300 m for RMa and 22.5 m to 300 m for UMa and UMi. Three different alternatives are provided for evaluation of fast fading models. In each alternative, specific parameters are provided or parameters from [9] are used for fast fading modeling for RMa, UMa, and UMi scenarios.

2.7 Conclusions

In this chapter, we have provided a comprehensive survey for AG propagation channels for UAVs. The measurement campaigns in the literature for AG propagation were summarized, with information provided on the type of channel sounding signal, its center frequency, bandwidth, transmit power, UAV speed, height of UAV and GS, link distance, elevation angle, and local GS environment characteristics. AG channel statistics from the literature were also provided. Various UAV propagation scenarios and important implementation factors for these measurements were also discussed. Large-scale fading, small-scale fading, MIMO channel characteristics and models, and channel simulations were all described. Finally, future research directions and challenges were highlighted.

References

1 Amazon Prime Air Drone Delivery. https://www.amazon.com/Amazon-Prime-Air/b? ie=UTF8&node=8037720011.

2 Flytrex. http://www.flytrex.com/.

3 Google Project Loon. https://loon.co/.

4 Google Project Wing. https://x.company/projects/wing/.

5 Uber Elevate. https://www.uber.com/us/en/elevate/.

6 Zipline. http://www.flyzipline.com/.

7 3GPP (2017). Specification Number 36.777. 3GPP Specifications, RAN 78. https://portal .3gpp.org/desktopmodules/Specifications/SpecificationDetails.aspx?specificationId=3231 (accessed 17 May 2018).

8 3GPP (2018). *3GPP, The mobile broadband standard.* http://www.3gpp.org/ specifications.

9 3GPP (2017). ETSI TR 138 901. *Study on channel model for frequencies from 0.5 to 100 GHz.* http://www.etsi.org/deliver/etsi_tr/138900_138999/138901/14.00.00_60/tr_ 138901v140000p.pdf (accessed 17 May 2018).

10 A. Aboudebra, K. Tanaka, T. Wakabayashi et al. (1999). Signal fading in land–mobile satellite communication systems: statistical characteristics of data measured in Japan using ETS-VI. *Proc. IEEE Microw. Antennas Propag.* 146 (5): 349–354. doi: 10.1049/ip-map:19990655.

11 L. Afonso, N. Souto, P. Sebastiao et al. (2016). Cellular for the skies: exploiting mobile network infrastructure for low altitude air-to-ground communications. *IEEE Aerosp. Electron. Syst. Mag.* 31 (8): 4–11. doi: 10.1109/MAES.2016.150170.

12 N. Ahmed, S. S. Kanhere, and S. Jha (2016). On the importance of link characterization for aerial wireless sensor networks. *IEEE Commun. Mag.* 54 (5): 52–57.

13 A. Al-Hourani, S. Kandeepan, and A. Jamalipour (2014). Modeling air-to-ground path loss for low altitude platforms in urban environments. *Proceedings of the IEEE Global Communications (GLOBECOM) Conference*, pp. 2898–2904.

14 A. Al-Hourani, S. Kandeepan, and S. Lardner (2014). Optimal LAP altitude for maximum coverage. *IEEE Wireless Commun. Lett.* 3 (6): 569–572.

15 Mohamed Alzenad, Amr El-Keyi, Faraj Lagum, and Halim Yanikomeroglu. 3D placement of an unmanned aerial vehicle base station (UAV-BS) for energy-efficient maximal coverage. *IEEE Wireless Commun. Lett.*, 2017a.

16 Mohamed Alzenad, Amr El-Keyi, and Halim Yanikomeroglu. 3D placement of an unmanned aerial vehicle base station for maximum coverage of users with different QoS requirements. *IEEE Wireless Commun. Lett.*, 2017b.

17 R. Amorim, H. Nguyen, P. Mogensen et al. Radio channel modeling for UAV communication over cellular networks. *IEEE Wireless Commun. Lett.*, 6 (4): 514–517, Aug. 2017a. doi: 10.1109/LWC.2017.2710045.

18 R. Amorim, H. Nguyen, P. Mogensen et al. Radio channel modeling for UAV communication over cellular networks. *IEEE Wireless Commun. Lett.*, 6 (4): 514–517, 2017b.

19 Art Pregler, AT&T. *When COWs Fly: AT&T Sending LTE Signals from Drones.* https:// about.att.com/innovationblog/cows_fly, Feb. 2017.

20 M. Asadpour, B. Van den Bergh, D. Giustiniano et al. Micro aerial vehicle networks: An experimental analysis of challenges and opportunities. *IEEE Commun. Mag.*, 52 (7): 141–149, Jul. 2014. doi: 10.1109/MCOM.2014.6852096.

21 Dasun Athukoralage, Ismail Guvenc, Walid Saad, and Mehdi Bennis. Regret based learning for UAV assisted LTE-U/WiFi public safety networks. *Proceedings of the IEEE Global Communications (GLOBECOM) Conference*, pp. 1–7, 2016.

22 M. M. Azari, F. Rosas, K. C. Chen, and S. Pollin. Optimal UAV positioning for terrestrial–aerial communication in presence of fading. *Proceedings of the IEEE Global Communications (GLOBECOM) Conference*, pages 1–7, Dec. 2016. doi: 10.1109/GLO-COM.2016.7842099.

23 Lana Bandoim. Uber plans to launch food-delivery drones. *Forbes*, Oct. 2018.

24 Steve Blandino, Florian Kaltenberger, and Michael Feilen. Wireless channel simulator testbed for airborne receivers. *Proceedings of the IEEE Global Communications (GLOBECOM) Workshops*, pages 1–6, 2015.

25 John David Blom. *Unmanned Aerial Systems: A Historical Perspective*, volume 45. Combat Studies Institute Press, 2010.

26 Christian Bluemm, Christoph Heller, Bertille Fourestie, and Robert Weigel. Air-to-ground channel characterization for OFDM communication in C-band. *Proceedings of the International Conference on Signal Processing and Communication Systems (ICSPCS)*, pp. 1–8, 2013.

27 Frode Bohagen, Pal Orten, and Geir E Oien. Design of optimal high-rank line-of-sight MIMO channels. *IEEE Trans. Wireless Commun.*, 6 (4), 2007.

28 R Irem Bor-Yaliniz, Amr El-Keyi, and Halim Yanikomeroglu. Efficient 3-D placement of an aerial base station in next generation cellular networks. *Proceedings of the IEEE International Conference on Communications (ICC)*, pages 1–5, 2016.

29 X. Cai, A. Gonzalez-Plaza, D. Alonso et al. Low altitude UAV propagation channel modelling. *Proceedings of the European Conference on Antennas and Propagation (EUCAP)*, pages 1443–1447. IEEE, 2017.

30 P. Chandhar, D. Danev, and E. G. Larsson. Massive MIMO as enabler for communications with drone swarms. *Proceedings of the International Conference on Unmanned Aircraft Systems (ICUAS)*, pages 347–354, Jun. 2016. doi: 10.1109/ICUAS.2016.7502655.

31 J. Chen, B. Daneshrad, and Weijun Zhu. MIMO performance evaluation for airborne wireless communication systems. *Proceedings of the Military Communications Conference (MILCOM)*, pages 1827–1832, Nov. 2011. doi: 10.1109/MILCOM.2011.6127578.

32 J. Chen, D. Raye, W. Khawaja et al. Impact of 3D UWB antenna radiation pattern on air-to-ground drone connectivity. *Proceedings of the Vehicular Technology Conference (VTC)*, Fall, pages 1–5, 2018, 2018.

33 Chen-Mou Cheng, Pai Hsiang Hsiao, HT Kung, and Dario Vlah. Performance measurement of 802.11 a wireless links from UAV to ground nodes with various antenna orientations. *Proceedings of the International Conference on Computer Communications and Networks (ICCCN)*, pages 303–308, 2006.

34 Xiang Cheng, C-X Wang, David I Laurenson, and Athanasios V Vasilakos. Second order statistics of non-isotropic mobile-to-mobile Ricean fading channels. *Proceedings of the IEEE International Conference on Communications (ICC)*, pages 1–5, 2009.

35 JR Child. Air-to-ground propagation at 900 MHz. *Proceedings of the IEEE Vehicular Technology Conference (VTC)*, volume 35, pages 73–80, 1985.

36 E. Lemos Cid, A. V. Alejos, and M. Garcia Sanchez. Signaling through scattered vegetation: empirical loss modeling for low elevation angle satellite paths obstructed by isolated thin trees. *IEEE Vehic. Technol. Mag.*, 11 (3): 22–28, Sep. 2016. doi: 10.1109/MVT.2016.2550008.

37 D. W. Matolak and R. Sun. Antenna and frequency diversity in the unmanned aircraft systems bands for the over-sea setting. *Proceedings of the IEEE Digital Avionics Systems Conference (DASC)*, pages 6A4-1–6A4-10, Oct. 2014. doi: 10.1109/DASC.2014.6979495.

38 D. W. Matolak and R. Sun. Air–ground channel characterization for unmanned aircraft systems – Part III: The suburban and near-urban environments. *IEEE Trans. Vehic. Technol.*, 2017a.

39 D. W. Matolak and R. Sun. Air–ground channel characterization for unmanned aircraft systems – Part I: Methods, measurements, and models for over-water settings. *IEEE Trans. Vehic. Technol.*, 66 (1): 26–44, Jan. 2017b. doi: 10.1109/TVT.2016.2530306.

40 D. W. Matolak, H. Jamal and R. Sun. Spatial and frequency correlations in two-ray SIMO channels. *Proceedings of the IEEE International Conference on Communications (ICC)*, May, 2017.

41 Kai Daniel, Markus Putzke, Bjoern Dusza, and Christian Wietfeld. Three dimensional channel characterization for low altitude aerial vehicles. *Proceedings of the International Symposium on Wireless Communication Systems (ISWCS)*, pages 756–760, 2010.

42 Jeff Desjardins. Amazon and UPS are betting big on drone delivery. *Business Insider*, March 2018.

43 G. M. Djuknic, J. Freidenfelds, and Y. Okunev. Establishing wireless communications services via high-altitude aeronautical platforms: A concept whose time has come? *IEEE Commun. Mag.*, 35 (9): 128–135, Sep. 1997. doi: 10.1109/35.620534.

44 R. Essaadali and A. Kouki. A new simple unmanned aerial vehicle doppler effect RF reducing technique. *Proceedings of the Military Communications Conference (MILCOM)*, pages 1179–1183, Nov. 2016. doi: 10.1109/MILCOM.2016.7795490.

45 Federal Aviation Administration. FAA rules for UAVs. https://www.faa.gov/uas/beyond_the_basics (accessed 25 February 2017).

46 Federal Aviation Administration. FAA small unmanned aircraft regulations. https://www.faa.gov/news/fact_sheets/news_story.cfm?newsId=20516 (accessed 3 July 2017).

47 Qixing Feng, Joe McGeehan, Eustace K Tameh, and Andrew R Nix. Path loss models for air-to-ground radio channels in urban environments. *Proceedings of the IEEE Vehicular Technology Conference (VTC), volume 6, pages* 2901–2905, 2006.

48 Xijun Gao, Zili Chen, and Yongjiang Hu. Analysis of unmanned aerial vehicle MIMO channel capacity based on aircraft attitude. *WSEAS Trans. Inform. Sci. Appl.*, 10: 58–67, 2013.

49 B. G. Gates. Aeronautical communications. *Electr. Eng. – Part IIIA: Radiocommun. J.*, 94 (11): 74–81, Mar. 1947. doi: 10.1049/ji-3a-2.1947.0009.

50 N. Goddemeier and C. Wietfeld. Investigation of air-to-air channel characteristics and a UAV specific extension to the Rice model. *Proceedings of the IEEE Global Communications (GLOBECOM) Workshops*, pages 1–5, Dec. 2015. doi: 10.1109/GLOCOMW.2015.7414180.

51 Niklas Goddemeier, Kai Daniel, and Christian Wietfeld. Coverage evaluation of wireless networks for unmanned aerial systems. *Proceedings of the IEEE Global Communications (GLOBECOM) Workshops*, pages 1760–1765, 2010.

52 Goldman Sachs. Drones: Reporting for Work. https://www.goldmansachs.com/insights/technology-driving-innovation/drones/.

53 S. M. Gulfam, S. J. Nawaz, M. N. Patwary, and M. Abdel-Maguid. On the spatial characterization of 3-D air-to-ground radio communication channels. *Proceedings of the IEEE International Conference on Communications (ICC)*, pages 2924–2930, 2015.

54 S. M. Gulfam, S. J. Nawaz, A. Ahmed, and M. N. Patwary. Analysis on multipath shape factors of air-to-ground radio communication channels. *Proceedings of the IEEE Wireless Telecommunications Symposium (WTS)*, pages 1–5, 2016.

55 E. Haas. Aeronautical channel modeling. *IEEE Trans. Vehic. Technol.*, 51 (2): 254–264, Mar. 2002. doi: 10.1109/25.994803.

56 Jaroslav Holis and Pavel Pechac. Elevation dependent shadowing model for mobile communications via high altitude platforms in built-up areas. *IEEE Trans. Antennas Propag.*, 56 (4): 1078–1084, 2008.

57 M. Ibrahim and H. Arslan. Air–ground Doppler-delay spread spectrum for dense scattering environments. *Proceedings of the Military Communications Conference (MILCOM)*, pages 1661–1666, Oct. 2015. doi: 10.1109/MILCOM.2015.7357683.

58 International Telecommunication Union. Propagation by diffraction. http://www.itu.int/dms_pubrec/itu-r/rec/p/R-REC-P.526-13-201311-I!!PDF-E.pdf (accessed 5 July 2017).

59 International Telecommunication Union. Terrain cover types. https://www.itu.int/oth/R0A04000031/en (accessed 5 July 2017).

60 International Telecommunication Union. Ducting over sea calculation. http://www.itu.int/md/dologin_md.asp?id=R03-WRC03-C-0025!A27-L188!MSW-E. (accessed 5 July 2017).

61 International Telecommunication Union (2003). Propagation data and prediction methods required for the design of terrestrial broadband millimetric radio access systems. http://www.catr.cn/catr/catr/itu/itur/iturlist.jsp?docplace=P&vchar1=P.1410-2 (accessed 27 November 2017).

62 IST-4-027756 WINNER II. D1.1.2 V1.0 WINNER II channel models. http://www2.tu-ilmenau.de/nt/generic/paper_pdfs/Part, 2003.

63 B. R. P. Jackson (2015). Telemetry, command and control of UAS in the National Airspace. *Proceedings of the International Telemetering Conference* International Foundation for Telemetering.

64 Hosseinali Jamal and David W Matolak. FBMC and LDACS performance for future air to ground communication systems. *IEEE Trans. Vehic. Technol.*, 2016.

65 F. Jiang and A. L. Swindlehurst (2012). Optimization of UAV heading for the ground-to-air uplink. *IEEE J. Sel. Areas Commun.* 30 (5): 993–1005.

66 J. Johnsson and A. Levin. *Boeing is getting ready to sell flying taxis*. Bloomberg, Mar. 2018.

67 George Pierce Jones IV, Leonard G Pearlstine, and H Franklin Percival. An assessment of small unmanned aerial vehicles for wildlife research. *Wildlife Soc. Bull.*, 34 (3): 750–758, 2006.

68 E. Kalantari, I. Bor-Yaliniz, A. Yongacoglu, and H. Yanikomeroglu (2017). User association and bandwidth allocation for terrestrial and aerial base stations with backhaul considerations. *Proceedings of the IEEE Annual International Symposium on Personal, Indoor, and Mobile Radio Communications (PIMRC)*, pp. 1–6.

69 S. Kaul, K. Ramachandran, P. Shankar et al. (2007). Effect of antenna placement and diversity on vehicular network communications. *Proceedings of the 4th Annual IEEE Communications Society Conference on Sensor, Mesh and Ad Hoc Communications and Networks (SECON)*, pages 112–121, Jun. 2007. doi: 10.1109/SAHCN.2007.4292823.

70 F. Kawamata. Optimum frame size for land mobile satellite communication channels. *Proceedings of the IEEE Global Communications (GLOBECOM) Conference*, pages 583–587 vol. 1, Nov. 1993. doi: 10.1109/GLOCOM.1993.318148.

71 Bob Kerczewski. Spectrum for UAS control and non-payload communications. *Proceedings of the IEEE Integrated Communications, Navigation and Surveillance Conference (ICNS)*, pages 1–21, 2013.

72 Wahab Khawaja, Ozgur Ozdemir, and Ismail Guvenc. UAV air-to-ground channel characterization for mmWave systems. *Proceedings of the IEEE Vehicular Technology Conference (VTC) Sep.* 2017.

73 Wahab Khawaja, Ismail Guvenc, and David W. Matolak. UWB channel sounding and modeling for UAV air-to-ground propagation channels. *Proceedings of the IEEE Global Communications (GLOBECOM) Conference*, pages 1–7, Dec. 2016. doi: 10.1109/GLO-COM.2016.7842372.

74 Wahab Khawaja, Ozgur Ozdemir, and Ismail Guvenc. Temporal and spatial characteristics of mmwave propagation channels for UAVs. *Proceedings of the IEEE Global Symposium on Millimeter Waves (GSMM)*, pages 1–6, 2018.

75 W. Khawaja, O. Ozdemir, F. Erden et al. (2019). UWB Air-to-ground propagation channel measurements and modeling using UAVs. *Proceedings of the IEEE Aerospace Conference*, 2019.

76 A. A. Khuwaja, Y. Chen, N. Zhao et al. (2018). A survey of channel modeling for UAV communications. *IEEE Commun. Surveys Tuts* 20 (4): 2804–2821.

77 I. Kovacs, R. Amorim, H. C. Nguyen et al. (2017). Interference analysis for UAV connectivity over LTE using aerial radio measurements. *Proceedings of the IEEE Vehicular Technology Conference (VTC)*, pages 1–6.

78 Alexander Ksendzov. A geometrical 3D multi-cluster mobile-to-mobile MIMO channel model with Rician correlated fading. *Proceedings of the IEEE International Congress on Ultra Modern Telecommunications (ICUMT) Conference*, pages 191–195, 2016.

79 H. T. Kung, C.-K. Lin, T.-H. Lin et al. (2010). Measuring diversity on a low-altitude UAV in a ground-to-air wireless 802.11 mesh network. *Proceedings of the IEEE Global Communications (GLOBECOM) Workshops*, pages 1799–1804, 2010.

80 J. Kunisch, I. De La Torre, A. Winkelmann et al. (2011). Wideband time-variant air-to-ground radio channel measurements at 5 GHz. *Proceedings of the European Conference on Antennas and Propagation (EUCAP)*, pp. 1386–1390.

81 Milan Kvicera, Fernando Pérez Fontán, Jonathan Israel, and Pavel Pechac. A new model for scattering from tree canopies based on physical optics and multiple scattering theory. *IEEE Trans. Antennas Propag.*, 65 (4): 1925–1933, 2017.

82 D. F. Lamiano, K. H. Leung, L. C. Monticone et al. (2009). Digital broadband VHF aeronautical communications for air traffic control. *Proceedings of the Integrated*

Communications, Navigation and Surveillance Conference (ICNS), pages 1–12, May 2009. doi: 10.1109/ICNSURV.2009.5172856.

83 Curt Levis, Joel T Johnson, and Fernando L Teixeira. *Radiowave Propagation: Physics and Applications*. Wiley, 2010.

84 X. Lin, V. Yajnanarayana, S. D. Muruganathan et al. (2018). The sky is not the limit: LTE for unmanned aerial vehicles. *IEEE Commun. Mag.* 56 (4): 204–210.

85 J. Lyu, Y. Zeng, R. Zhang, and T. J. Lim (2017).. Placement optimization of UAV-mounted mobile base stations. *IEEE Commun. Lett.*, 21 (3): 604–607.

86 M. S. Ben Mahmoud, C. Guerber, A. Pirovano et al. (2014). *Aeronautical Air–Ground Data Link Communications*. Wiley.

87 W. Q. Malik, B. Allen, and D. J. Edwards. Impact of bandwidth on small-scale fade depth. *Proceedings of the IEEE Global Communications (GLOBECOM) Conference*, pages 3837–3841, Nov. 2007. doi: 10.1109/GLOCOM.2007.729.

88 A. Matese, P. Toscano, S. F. Di Gennaro et al. (2015). Intercomparison of UAV, aircraft and satellite remote sensing platforms for precision viticulture. *Remote Sensing*, 7 (3): 2971–2990.

89 D. W. Matolak. Air–ground channels models: comprehensive review and considerations for unmanned aircraft systems. *Proceedings of the IEEE Aerospace Conference*, pages 1–17, Mar. 2012. doi: 10.1109/AERO.2012.6187152.

90 David W Matolak. Channel characterization for unmanned aircraft systems. *Proceedings of the European Conference on Antennas and Propagation (EUCAP)*, pages 1–5, 2015.

91 David W. Matolak and Ruoyu Sun. Air-ground channel measurements & modeling for UAS. *Proceedings of the IEEE Integrated Communications, Navigation and Surveillance Conference (ICNS)*, pages 1–9, 2013.

92 D. W. Matolak and R. Sun (2015). Air–ground channel characterization for unmanned aircraft systems: the near-urban environment. *Proceedings of the Military Communications Conference (MILCOM)*, pp. 1656–1660.

93 D. W. Matolak and R. Sun (2015). Unmanned aircraft systems: air–ground channel characterization for future applications. *IEEE Vehic. Technol. Mag.* 10 (2): 79–85.

94 David W. Matolak and Ruoyu Sun. Air–ground channels for UAS: summary of measurements and models for L-and C-bands. *Proceedings of the IEEE Integrated Communications, Navigation and Surveillance Conference (ICNS)*, pages 8B2–1, 2016.

95 David W. Matolak and Ruoyun Sun. Air–ground channel characterization for unmanned aircraft systems: the hilly suburban environment. *Proceedings of the IEEE Vehicular Technology Conference (VTC)*, pages 1–5, 2014.

96 Matt McFarland. UPS drivers may tag team deliveries with drones. CNN News Article, Feb. 2017.

97 Yu Song Meng and Yee Hui Lee. Measurements and characterizations of air-to-ground channel over sea surface at C-band with low airborne altitudes. *IEEE Trans. Vehic. Technol.* 60 (4): 1943–1948, 2011.

98 M. Mozaffari, W. Saad, M. Bennis, and M. Debbah (2017). Mobile unmanned aerial vehicles (UAVs) for energy-efficient internet of things communications. *IEEE Trans. Wireless Commun.* 16 (11): 7574–7589.

99 W. G. Newhall, R. Mostafa, C. Dietrich et al. (2003). Wideband air-to-ground radio channel measurements using an antenna array at 2 GHz for low-altitude operations. *Proceedings of the Military Communications Conference (MILCOM)*, volume 2, pages 1422–1427.

100 F. Ono, K. Takizawa, H. Tsuji, and R. Miura (2015). S-band radio propagation characteristics in urban environment for unmanned aircraft systems. *Proceedings of the International Symposium on Antennas and Propagation (ISAP)*, pages 1–4, 2015.

101 A. Paier, T. Zemen, L. Bernado et al. (2008). Non-WSSUS vehicular channel characterization in highway and urban scenarios at 5.2 GHz using the local scattering function. *Proceedings of the International Workshop on Smart Antennas*, pages 9–15, Feb. 2008. doi: 10.1109/WSA.2008.4475530.

102 Pyung Joo Park, Sung-Min Choi, Dong Hee Lee, and Byung-Seub Lee. Performance of UAV (unmanned aerial vehicle) communication system adapting WiBro with array antenna. *Proceedings of the International Conference on Advanced Communication Technology (ICACT)*, volume 2, pages 1233–1237, 2009.

103 Land Point. Satellite versus UAV mapping: How are they different. http://www .landpoint.net/satellite-versus-uav-mapping-how-are-they-different/.

104 Qualcomm (2017). LTE Unmanned Aircraft Systems. https://www.qualcomm.com/ media/documents/files/lte-unmanned-aircraft-systems-trial-report.pdf, May 2017.

105 RadioReference. VHF/UHF military monitoring. http://wiki.radioreference.com/index .phpVHF/UHF_Military_Monitoring (accessed 31 May 2017).

106 T. S. Rappaport (1996). *Wireless Communications: Principles and Practice*, vol. 2. Prentice-Hall.

107 V. V. C. Ravi and H. S. Dhillon (2016). Downlink coverage probability in a finite network of unmanned aerial vehicle (UAV) base stations. *Proceedings of the IEEE Signal Processing Advances in Wireless Communications (SPAWC) Conference*, pp. 1–5.

108 O. Renaudin, V. M. Kolmonen, P. Vainikainen, and C. Oestges (2010). Non-stationary narrowband MIMO inter-vehicle channel characterization in the 5-GHz band. *IEEE Trans. Vehic. Technol.* 59 (4): 2007–2015. doi: 10.1109/TVT.2010.2040851.

109 D. Rieth, C. Heller, D. Blaschke, and G. Ascheid (2015). On the practicability of airborne MIMO communication. *Proceedings of the IEEE Digital Avionics Systems Conference (DASC)*, p. 2C1–1.

110 J. Romeu, A. Aguasca, J. Alonso et al. (2010). Small UAV radiocommunication channel characterization. *Proceedings of the European Conference on Antennas and Propagation (EUCAP)*, pp. 1–5.

111 A. R. Ruddle. Simulation of far-field characteristics and measurement techniques for vehicle-mounted antennas. *IEE Colloquium on Antennas for Automotives*, (Ref. No. 2000/002), pages 7/1–7/8, 2000. doi: 10.1049/ic:20000007.

112 N. Schneckenburger, T. Jost, D. Shutin, and U. C. Fiebig. Line of sight power variation in the air to ground channel. *Proceedings of the European Conference on Antennas and Propagation (EUCAP)*, Davos, Switzerland, 2016a.

113 N. Schneckenburger, T. Jost, D. Shutin et al. (2016). Measurement of the L-band air-to-ground channel for positioning applications. *IEEE Trans. Aerosp. Electron. Syst.* 52 (5): 2281–2297. doi: 10.1109/TAES.2016.150451.

114 M. Schnell, U. Epple, D. Shutin, and N. Schneckenburger. LDACS: future aeronautical communications for air-traffic management. *IEEE Commun. Mag.*, 52 (5): 104–110, May 2014. doi: 10.1109/MCOM.2014.6815900.

115 C. S. Sharp, O. Shakernia, and S. S. Sastry (2001). A vision system for landing an unmanned aerial vehicle. *Proceedings of the IEEE International Conference on Robotics and Automation (ICRA)*, volume 2, pages 1720–1727.

116 D. H. Shim, H. J. Kim, and S. Sastry (2000). Control system design for rotorcraft-based unmanned aerial vehicles using time-domain system identification. *Proceedings of the IEEE International Conference on Control Applications*, pages 808–813.

117 Hyunchul Shim. Hierarchical flight control system synthesis for rotorcraft-based unmanned aerial vehicles. PhD Dissertation, Harvard University, 2000.

118 M. Simunek, F. P. Fontán, and P. Pechac (2013). The UAV low elevation propagation channel in urban areas: statistical analysis and time-series generator. *IEEE Trans. Antennas Propag.* 61 (7): 3850–3858. doi: 10.1109/TAP.2013.2256098.

119 Michal Simunek, Pavel Pechac, and Fernando P Fontán. Excess loss model for low elevation links in urban areas for UAVs. *Radioengineering*, 2011.

120 R. Sun, D. W. Matolak, and W. Rayess (2017). Air–ground channel characterization for unmanned aircraft systems 8212 – Part IV: Airframe shadowing. *IEEE Trans. Vehic. Technol.* 66 (9): 7643–7652. doi: 10.1109/TVT.2017.2677884.

121 Ruoyu Sun and David W. Matolak. Over-harbor channel modeling with directional ground station antennas for the air–ground channel. *Proceedings of the Military Communications Conference (MILCOM)*, pages 382–387, 2014.

122 Ruoyu Sun and David W. Matolak. Air–ground channel characterization for unmanned aircraft systems – Part II: Hilly and mountainous settings. *IEEE Trans. Vehic. Technol.*, 2016.

123 S. Sun and T. Rappaport. Investigation of prediction accuracy, sensitivity, and parameter stability of large-scale propagation path loss models from 500 MHz to 100 GHz. http://wireless.engineering.nyu.edu/presentations/NTIA-propagation-presentation-JUNE-15-2016_v1.

124 K. Takizawa, T. Kagawa, S. Lin et al. (2014). C-band aircraft-to-ground (A2G) radio channel measurement for unmanned aircraft systems. *Proceedings of the Wireless Personal Multimedia Communications (WPMC) Conference*, pages 754–758, 2014.

125 T. Tavares, P. Sebastiao, N. Souto et al. (2015). Generalized LUI propagation model for UAVs communications using terrestrial cellular networks. *Proceedings of the IEEE Vehicular Technology Conference (VTC)*, pages 1–6.

126 E. Teng, J. D. Falcao, C. R. Dominguez et al. (2016). Aerial sensing and characterization of three-dimensional RF fields. *Proceedings of the Second International Workshop on Robotic Sensor Networks* (accessed September 2016).

127 Ian J Timmins and Siu O'Young. Marine communications channel modeling using the finite-difference time domain method. *IEEE Trans. Vehic. Technol.* 58 (6): 2626–2637, 2009.

128 T. C. Tozer and D. Grace (2001). High-altitude platforms for wireless communications. *Electron. Commun. Eng. J.* 13 (3): 127–137.

129 H. D. Tu and S. Shimamoto. A proposal of wide-band air-to-ground communication at airports employing 5-GHz band. *Proceedings of the IEEE Wireless Communications and Networking Conference (WCNC)*, pages 1–6, 2009.

130 Universitat Politècnica de Catalunya. Vector network analyzer specifications. http://www.upc.edu/sct/en/documents_equipament/d_160_id-655-2.pdf (accessed 18 May 2017).

131 Vahid Vahidi and Ebrahim Saberinia. Orthogonal frequency division multiplexing and channel models for payload communications of unmanned aerial systems. *Proceedings of the International Conference on Unmanned Aircraft Systems (ICUAS)*, pages 1156–1161, 2016.

132 Vodafone. Beyond visual line of sight drone trial report. https://www.vodafone.com/content/dam/vodafone-images/media/Downloads/Vodafone_BVLOS_drone_trial_report.pdf, Nov. 2018.

133 Michael Walter, Snjezana Gligorević, Thorben Detert, and Michael Schnell. UHF/VHF air-to-air propagation measurements. *Proceedings of the European Conference on Antennas and Propagation (EUCAP)*, pages 1–5, 2010.

134 Michael Wentz and Milica Stojanovic. A MIMO radio channel model for low-altitude air-to-ground communication systems. *Proceedings of the IEEE Vehicular Technology Conference (VTC)*, pages 1–6, 2015.

135 F. White (1973). Air–ground communications: history and expectations. *IEEE Trans. Commun.*, 21 (5): 398–407. doi: 10.1109/TCOM.1973.1091709.

136 Wikipedia. General atomics MQ-9 Reaper. https://en.wikipedia.org/wiki/General_Atomics_MQ-9_Reaper (accessed 3 July 2017).

137 T. J. Willink, C. C. Squires, G. W. K. Colman, and M. T. Muccio (2016). Measurement and characterization of low-altitude air-to-ground MIMO channels. *IEEE Trans. Vehic. Technol.* 65 (4): 2637–2648. doi: 10.1109/TVT.2015.2419738.

138 Y. Wu, Z. Gao, C. Chen et al. (2015). Ray tracing based wireless channel modeling over the sea surface near Diaoyu Islands. *Proceedings of the International Conference on Computational Intelligence Theory, Systems and Applications (CCITSA)*, pages 124–128.

139 Zhiqiang Wu, Hemanth Kumar, and Asad Davari. Performance evaluation of OFDM transmission in UAV wireless communication. *Proceedings of the Southeastern Symposium on Systems Theory (SSST)*, pages 6–10, 2005.

140 Z. Xiao, P. Xia, and X. G. Xia (2016). Enabling UAV cellular with millimeter-wave communication: potentials and approaches. *IEEE Commun. Mag.* 54 (5): 66–73. doi: 10.1109/MCOM.2016.7470937.

141 J. Yang, P. Liu, and H. Mao (2011). Model and simulation of narrowband ground-to-air fading channel based on Markov process. *Proceedings of the Network Computing and Information Security Conference (NCIS)*, volume 1, pages 142–146, 2011.

142 Evşen Yanmaz, Robert Kuschnig, and Christian Bettstetter. Channel measurements over 802.11a-based UAV-to-ground links. *Proceedings of the IEEE Global Communications (GLOBECOM) Workshops*, pages 1280–1284, 2011.

143 Evşen Yanmaz, Robert Kuschnig, and Christian Bettstetter. Achieving air-ground communications in 802.11 networks with three-dimensional aerial mobility. *Proceedings of the IEEE International Conference on Computer Communications (INFOCOM)*, pages 120–124, 2013.

144 Z. Yun and M. F. Iskander (2015). Ray tracing for radio propagation modeling: principles and applications. *IEEE Access*, 3: 1089–1100.

145 Jan Zelený, Fernando Pérez-Fontán, and Pavel Pechač. Initial results from a measurement campaign for low elevation angle links in different environments. *Proceedings of the European Conference on Antennas and Propagation (EUCAP)*, pages 1–4, 2015.

146 Y. Zeng, R. Zhang, and T. J. Lim (2016). Wireless communications with unmanned aerial vehicles: opportunities and challenges. *IEEE Commun. Mag.* 54 (5): 36–42. doi: 10.1109/MCOM.2016.7470933.

147 Chao Zhang and Yannian Hui. Broadband air-to-ground communications with adaptive MIMO datalinks. *Proceedings of the IEEE Digital Avionics Systems Conference (DASC)*, pages 4D4–1, 2011.

148 Yi Zheng, Yuwen Wang, and Fanji Meng. Modeling and simulation of pathloss and fading for air-ground link of HAPs within a network simulator. *Proceedings of the International Conference on Cyber-Enabled Distributed Computing and Knowledge Discovery*, pages 421–426, 2013.

3

UAV Detection and Identification

Martins Ezuma[1]*, Fatih Erden[1], Chethan Kumar Anjinappa[1], Ozgur Ozdemir[1], Ismail Guvenc[1], and David Matolak[2]

[1]*Department of Electrical and Computer Engineering, North Carolina State University, Raleigh NC 27606, USA*
[2]*Department of Electrical Engineering, University of South Carolina, Columbia SC 29208, USA*

3.1 Introduction

Unmanned aerial vehicles (UAVs) are expected to play a major role in the fifth-generation (5G) systems. Emerging civil applications include package delivery, search and rescue operations, law enforcement, precision agriculture, mobile or temporary cellular base stations, and infrastructure inspection [32]. The UAV communication network is an integral part of the 5G Internet of Things (IoT) in the sky, which will be an essential component of future smart cities [29]. Moreover, there are proposals to integrate unmanned aerial systems (UAS) into national air traffic systems, following air traffic management (ATM) procedures. Such integration will make it possible for UAVs to access the national airspaces occupied by human-piloted aircraft.

Although there are many potential benefits of UAVs in the future 5G ecosystem, these devices can also be exploited for malicious purposes. In recent times, some UAV pilots have intentionally violated no-fly-zone restrictions around sensitive national facilities like airports and nuclear reactors. In 2015, the White House surveillance system, designed to detect aircraft, failed to detect a small hobby drone that crash-landed on the presidential lawn of the White House. In the same year, a similar drone, carrying traces of radioactive materials (cesium from Fukushima nuclear site) landed on the residence of the Japanese prime minister in Tokyo [34]. Events like this show the potential threat that UAVs pose to modern society. In addition, UAVs have been identified, along with cruise missiles, as new threats to homeland security [19]. Therefore, there is an urgent need to develop an effective system for detecting and classifying UAVs.

Figure 3.1 shows a typical UAV detection scenario in a 5G ecosystem [17]. In this scenario, the airspace around a facility is classified as either restricted (protected areas) or non-restricted. If the trajectory of a small UAV remains outside the protected area, it is considered to be safe and so no action needs to be taken against the UAV. On the other

---

* *Corresponding Author: Martins Ezuma, mcezuma@ncsu.edu

UAV Communications for 5G and Beyond, First Edition.
Edited by Yong Zeng, Ismail Guvenc, Rui Zhang, Giovanni Geraci, and David W. Matolak.
© 2021 John Wiley & Sons Ltd. Published 2021 by John Wiley & Sons Ltd.

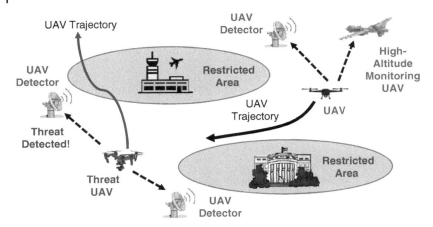

Figure 3.1 Detection scenarios of unauthorized small UAVs around restricted areas [17].

hand, a UAV is considered a threat if its trajectory crosses through a protected area. In such a situation, it is important to detect and classify the UAV.

Several techniques have been investigated for UAV detection and classification in the literature. These techniques can be categorized under the following headings: radar-based techniques, computer-vision-based techniques, acoustic-based techniques, and radio-frequency-based (RF-based) techniques. Most of the recent developments in radar-based techniques have focused on the extraction of the radar micro-Doppler signatures of UAVs. In [12], the radar micro-Doppler signature of a quadcopter UAV was extracted using a 25 GHz millimeter wave (mmWave) radar. Afterward, the radar micro-Doppler signature of the quadcopter UAV was compared with those extracted from a walking man and a three-blade helicopter. The results show that the quadcopter UAV can be distinguished from the other objects through its unique radar micro-Doppler signature.

In [25], micro-UAVs and small birds are classified based on the eigenpairs extracted from the decomposition of their micro-Doppler signatures. Larger birds can be readily recognized and distinguished from small UAVs because of the frequency modulation induced by their flapping wings. Therefore, using the radar micro-Doppler signature, it is possible to distinguish UAVs from other airborne objects such as birds, fixed-wing airplanes, and helicopters. However, research has shown that radar-based detection techniques are not very effective in detecting many UAVs. This is because most UAVs have very small radar cross-section (RCS) [16]. The RCS of a target is a measure of the amount of signal energy reflected by the target towards the radar receivers. Therefore, targets with small RCS, such as many commercial UAVs, are difficult to detect by radar.

Lately, computer-vision- and acoustic-based approaches to UAV detection and classification have gained attention. The recent surge is probably due to the availability of relatively inexpensive sensors used in these detection techniques [16]. The computer vision techniques use video cameras as sensors, whereas the acoustic techniques use microphones as sensors. In [14], a computer-vision-based technique was described for UAV detection. The approach used two different electro-optic cameras to capture images of a UAV flying in different orientations and background environments. Afterward, several features such as Haar-like, the histogram of gradients (HOG), and local binary patterns (LBP) were extracted from the UAV images. These features were fed into cascaded boosted classifiers for UAV detection. The results show that the UAV can be detected in near-real time with a detection rate of about 0.96 *F*-score. Furthermore, the distance of the UAV can be estimated with a camera sensor using the boosted classifiers. The experimental results show that distance estimation of UAVs is possible in about 60 ms indoors (1032×778 resolution) and 150 ms outdoors (1280×720 resolution) per frame.

Similar machine learning (ML) techniques are employed in acoustic-based UAV detection. In this case, microphones are used to record the unique audio signatures of UAVs. These audio signatures are mainly sounds generated by the direct current (DC) motors used in the UAVs. Thereafter, features extracted from these audio data are used for the detection and classification of a UAV. In [20], the authors investigate the effectiveness of the Gaussian mixture model (GMM) and deep learning algorithms for drone sound detection. The detection problem, which is modeled as a binary classification problem, is based on the detection of sound events. The study concludes that a long short-term memory (LSTM) recurrent neural network (RNN) shows the best performance for acoustic-based UAV detection.

Computer-vision- and acoustic-based detection techniques are very sensitive to environmental conditions. For instance, computer-vision-based techniques suffer severely from the changes in the ambient lighting conditions [11]. Similarly, acoustic-based detection techniques are susceptible to noise in the environment [11]. These limitations affect the efficiency and reliability of these UAV detection techniques, especially in outdoor urban environments.

As should be apparent from earlier discussions, the different technologies available for detection and tracking of UAVs have various trade-offs related to cost, accuracy, precision, range, energy efficiency (critical if sensors operate on batteries), portability (e.g., sensors deployed at other UAVs), and complexity (as summarized in Table 3.1). For example, while some sensor technologies can only operate very well in line-of-sight (LOS) conditions (e.g., radar and computer vision), some others can also operate in non-line-of-sight (NLOS) environments (e.g., RF-based approaches). For accurate and quick detection/tracking of UAVs, data fusion techniques that can simultaneously use information from multiple different types of sensors carry critical importance (see, e.g., [8] and [23] for joint use of 94 GHz mmWave radar and acoustic sensors, respectively, with high-resolution optical cameras), and this constitutes an open research area.

Once detected, several counter-actions like interdiction and jamming could be instituted against a UAV. Depending on the threat capability and the vulnerability of the UAV, different techniques can be employed to counter the threat. Figure 3.2 shows some of the counter-actions that can be initiated against a UAV. One way to interdict a drone is to jam its remote control signal (commonly in the 2.4 GHz ISM (industrial, scientific, and medical) band) and/or GPS signal using a jammer gun, as illustrated in Figure 3.2(a). In another approach, as shown in Figure 3.2(b), an operator fires a canister with large nets, and as soon as the drone's rotors become tangled, a parachute brings the drone safely down to the ground. As discussed previously, certain types of drones may be vulnerable to deauthentication attacks as in Figure 3.2(c), which can be used for breaking the communication link between the drone and its operator.

As shown in Figure 3.2(d), metropolitan police in Scotland are training eagles to take down suspicious drones. This low-cost idea does not require use of any other devices and has no danger to civilians. Eagles were trained to consider drones as prey so that they catch them

Table 3.1 Comparison of advantages/disadvantages for different drone detection and tracking techniques.

Detection technique	Advantages	Disadvantages
Ambient RF signals (e.g., [18, 27])	Low-cost RF sensors (e.g., software-defined radios (SDRs)), works in NLOS, long detection range. May allow deauthentication attack for taking control of drone by mimicking a remote controller or spoofing a Global Positioning System (GPS) signal	Need prior training to identify/classify different drones. Fail for fully/partially autonomous drone flights due to no/limited signal radiation from a drone/controller
Radar (e.g., [8, 9, 13, 22])	Low-cost frequency-modulated continuous wave (FMCW) radars, does not get affected by fog/clouds/dust, as opposed to vision-based techniques, can work in NLOS (more sophisticated). Higher (mmWave) frequencies allow capturing micro-Doppler/range accurately at the cost of higher path loss. Does not require active transmission from the drone	Small RCS of drone makes identification/classification difficult. Further research needed for accurate drone detection/ classification and machine learning techniques, considering different radar/drone geometries and different drone types, which all affect micro-Doppler signatures. Higher path loss at mmWave bands limits drone detection range
Acoustic signals (e.g., [2, 3, 7])	Low cost for simple microphones (cost depends on the quality of microphones). Can work in NLOS as long as the drone is audible	Need to develop database of acoustic signature for different drones. Knowledge of current wind conditions and background noise is needed. May operate poorly under high ambient noise such as in urban environments
Computer vision (e.g., [8, 23, 26])	Low cost of basic optical sensors. The cameras of many commercial drones can be used as sensors	Higher cost for thermal, laser-based, and wide field-of-view (FOV) cameras. Requires LOS. Level of visibility impacted by fog, clouds, and dust
Sensor fusion (e.g., [8, 23])	Can combine advantages of multiple different techniques for wider application scenario, high detection accuracy, and long-distance operation	Higher cost and processing complexity. Need effective sensor fusion algorithms

and place them in a safe area. One other way in which a drone could be brought down is by using another drone. For example, as shown in Figure 3.2(e), Tokyo police created a drone squad for privacy breaches. The police department was provided with net-carrying drones that will trap suspicious drones flying in the vicinity. The GPS spoofing in Figure 3.2(f) is another possible cyber attack that has been shown to work effectively on drones [21], which can be implemented even with low-cost SDRs. The communication links in drones include incoming signals from GPS satellites, signals notifying the drone's presence, and a two-way link between the ground station and the drone. The basic idea in GPS spoofing is to transmit

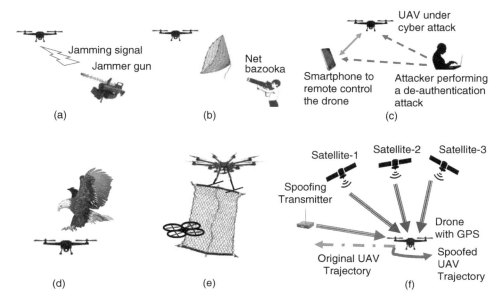

Figure 3.2 Different approaches for interdicting unauthorized UAVs: (a) use of a drone jammer gun; (b) bazooka that fires giant nets to a drone; (c) deauthentication cyber attack to a drone; (d) eagle trained to catch a drone and place in a safe zone; (e) net traps used by large drones to catch smaller drones; and (f) taking control of a drone using GPS spoofing.

fake GPS coordinates to the control system of the drone. This will hijack the drone and subsequently it will be in complete control of the attacker. A successful attack is conducted when the attacker is very close to the drone, or equivalently by using a directional antenna with narrow beamwidth aiming at the drone. Due to weak authentication mechanism, civilian drones can be attacked easily by delaying signals. However, attacks on military drones are complicated due to the use of more advanced authentication mechanisms.

In the remainder of this chapter, we will focus on RF-based techniques for detection and classification of drones. This chapter is organized as follows. First, Section 3.2 provides a brief survey of RF-based UAV detection techniques. Section 3.3 describes a multistage UAV RF signal detection technique in the presence of noise and radio interference. Feature extraction and the RF fingerprinting-based UAV classification system are explained in Section 3.4. The experimental setup together with the results of the detection and classification of UAV RF signals are described in Section 3.5. The conclusion is presented in Section 3.6.

3.2 RF-Based UAV Detection Techniques

UAV detection and classification through RF signals can be grouped into two major headings: RF fingerprinting techniques and WiFi fingerprinting techniques. These techniques use an RF sensing device to capture the RF communication signal between a UAV and its controller. In RF fingerprinting techniques, first, the physical layer features or the

RF signatures of the captured signals are extracted and then fed to ML algorithms for detection and classification of the UAV or its controller. The WiFi fingerprinting techniques extract the medium access control (MAC) and network layer features of the captured UAV RF transmission for the detection and classification of the UAVs.

3.2.1   RF Fingerprinting Technique

RF fingerprinting techniques rely on the unique characteristics of the RF signal waveforms captured from different UAVs or the UAV controllers. Experimental investigations show that most of the commercial UAVs have unique RF signatures, which is due to the circuitry design and modulation techniques employed. Therefore, RF fingerprints extracted from the signal waveform captured from the UAV or its remote controller can be used as the basis for detection and classification of the UAVs.

In [36], RF fingerprints captured from the UAVs' wireless control signals were extracted by computing the amplitude envelope of the signal. The dimensionality of the processed signal is further reduced by performing principal component analysis (PCA). The result is a set of lower-dimensional data which are fed into an auxiliary classifier Wasserstein generative adversarial networks (AC-WGANs). The AC-WGANs achieved an overall classification rate of 95% when four different types of UAVs are considered.

In [1], drones were detected by analyzing the RF background activities along with the RF signals emitted when the drones are operated in different modes, such as flying, hovering, and video recording. Afterward, the RF spectrum of the drone signal is computed using the discrete Fourier transform. The drone classification system is designed by training a deep neural network (DNN) with the RF spectrum data of different drones. The system shows an accuracy of 99.7% when two drones are classified, 84.5% with four drones, and 46.8% with 10 drones. The classification system performs best when only a few drone controllers are considered.

In [5], an industry integrated counter-drone solution was described. The solution is based on a network of distributed RF sensors. In this system, RF signals from different UAV controllers are detected using an energy detector. Afterward, the signals of interest are classified using RF spectral shape correlation features. The distributed RF sensors make it possible to localize the UAV controller using time-difference-of-arrival (TDoA) or multilateration techniques. However, this industrial solution is quite expensive. Therefore, cost-effective RF-based drone detection systems need to be investigated.

3.2.2   WiFi Fingerprinting Technique

These techniques are motivated by the fact that some UAVs use WiFi links for control and video streaming. The RF sensing system consists primarily of a WiFi packet-sniffing device which can intercept the WiFi data traffic between a UAV and its remote controller. A UAV can be detected and identified by analyzing the WiFi data and the associated WiFi fingerprints. In [4], unauthorized WiFi-controlled UAVs were detected by a patrolling drone using a set of WiFi statistical features. The extracted features include MAC addresses, root mean square (RMS) of the WiFi packet length, packet duration, average packet inter-arrival time, and the like. These features are used to train different ML algorithms that perform the UAV classification task. In [4], random tree and random forest (RandF) classifiers achieve the best performance as measured by the true positive and false positive rates.

In [28], drone presence was detected by eavesdropping on WiFi channels between the drone and its controller. The system detects drones by analyzing the impact of their unique vibration and body shifting motions on the WiFi signals transmitted by the drone. The system achieves accuracy above 90% at 50 m.

In general, a major concern with the WiFi fingerprinting techniques is privacy. That is, the same WiFi detection system can spoof the WiFi traffic data from a smartphone user or a private WiFi network. In addition, only a limited number of commercial drones employ WiFi links for video streaming and control. Most commercial drones use proprietary communication links.

In the rest of this chapter, we will describe in detail a new method for the detection and identification of UAVs based on the RF signals and the corresponding RF fingerprints. We will also present representative experimental results. The proposed system first considers a multistage detector to capture raw data and decide if the source is a UAV controller, a wireless interference device, or thermal noise. If the captured signal is detected to be a UAV controller signal, then the classification process is invoked to identify the UAV. The design of the multistage detector is described next.

3.3 Multistage UAV RF Signal Detection

We consider the scenario shown in Figure 3.3. Here, a passive RF surveillance system listens for the control signals transmitted between a UAV and its remote controller. The main hardware components of the surveillance system are 2.4 GHz RF antenna and a high-frequency oscilloscope, which is capable of sampling the captured data at 20 GSa/s (gigasamples per second). The high sampling rate enables the surveillance system to capture high-resolution waveform transient features of any detected RF signal. The waveform transient features of

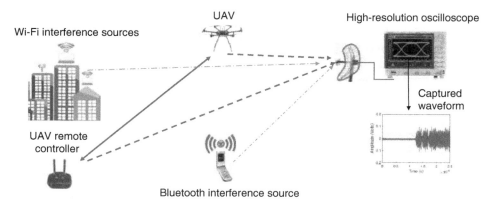

Figure 3.3 The scenario considered in the RF-based UAV detection system. The passive RF surveillance system listens for the signal transmitted between the controller and the UAV. The environment contains signals from WiFi and Bluetooth interference devices which operate in the same frequency band as the UAV and the remote controller.

different UAV controllers are unique. This is a useful property for detecting and classifying RF signals from different UAV controllers. Figure 3.4 shows examples of RF signals captured from eight different UAV controllers and four different UAVs. As it is clear from the figure, each signal has a unique waveform transient or shape, which can be exploited for identifying the source UAV controller.

Furthermore, since most commercial UAVs operate in the 2.4 GHz band, the passive RF surveillance system is designed to operate in this frequency band. However, this also corresponds to the operational band of WiFi and mobile Bluetooth devices. Therefore, in the wireless environment, signals from these wireless sources will act as interference to the detection of the UAV control signals. In this study, we focus on detecting only the signals from the UAV controller.

Given the scenario in Figure 3.3, the passive RF surveillance system has to decide if the captured data comes from a UAV controller, an interference source, or background noise. In the case where the captured data comes from a UAV controller, the detection system should be able to correctly identify the UAV controller. However, if the detected signal is from an interference source, the detection system should be able to correctly identify the source. Therefore, the detection problem is a multi-hypothesis problem. For such problems, it is well known that computational complexity increases as the number of hypotheses increases. Consequently, the multi-hypothesis detection problem can be simplified by using a multistage sequential detector. In this system, the first few steps involve simple binary hypothesis tests, which are much easier to solve.

Figure 3.5 is a flowchart that provides a high-level graphical description of the entire system. From the flowchart, we see that the first step in detecting and identifying a UAV controller is data capture. Usually, the captured raw dataset is large and samples are often very noisy. Therefore, before detection and classification, the captured raw data are first preprocessed using wavelet-based multiresolution analysis. Next, the data are transferred to a multistage detection system, which consists of two stages. In the first stage, the detector employs a Bayesian hypothesis test in deciding if the captured signal is an RF signal or noise. If an RF signal is detected, the second-stage detector is activated. This detector decides if the captured RF signal comes from an interference source (WiFi or Bluetooth device) or a UAV controller. This detector uses bandwidth analysis and modulation-based features for interference detection. Therefore, if the detected RF signal is not from a WiFi or Bluetooth interference source, it is presumed to be a signal transmitted by a UAV controller. Consequently, the detected signal is transferred to a ML-based classification system for the identification of the UAV controller.

3.3.1 Preprocessing Step: Multiresolution Analysis

The captured RF data are preprocessed by means of wavelet-based multiresolution analysis. It has been established that multiresolution decomposition using discrete wavelet transform (DWT) like the Haar wavelet transform is effective for analyzing the information content of signals and images [24].

In this work, multiresolution decomposition of the captured RF data is carried out using the two-level Haar transform as shown in Figure 3.6. Using this transform, the raw input signal is decomposed and important time–frequency information can be extracted at different

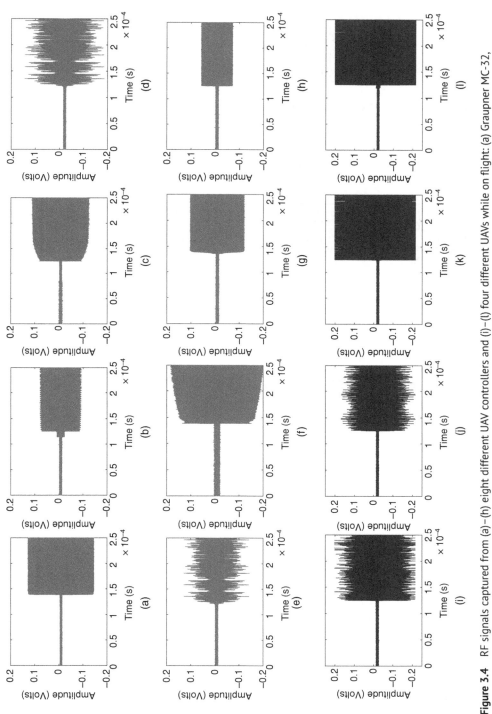

Figure 3.4 RF signals captured from (a)–(h) eight different UAV controllers and (i)–(l) four different UAVs while on flight: (a) Graupner MC-32, (b) Spektrum DX6e, (c) Futaba T8FG, (d) DJI Phantom 4 Pro, (e) DJI Inspire 1 Pro, (f) JR X9303, (g) Jeti Duplex DC-16, (h) FlySky FS-T6, (i) DJI Matrice 600 UAV, (j) DJI Phantom 4 Pro UAV, (k) DJI Inspire 1 Pro UAV, and (l) DJI Mavic Pro.

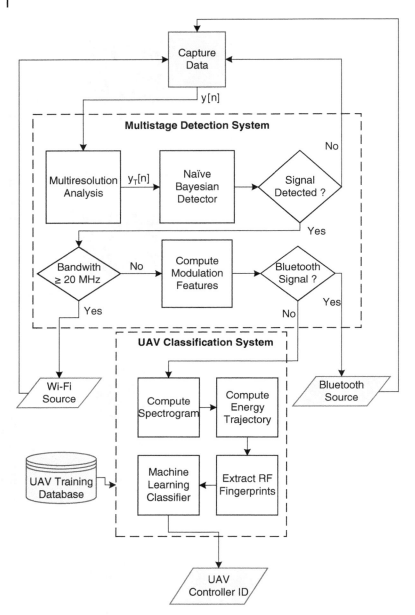

Figure 3.5 The system flowchart providing a graphical description of information processing and flow of data through the system. The system consists of two main subsystems: the multistage UAV detection system and the UAV classification system.

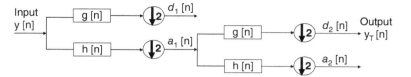

Figure 3.6 The two-level discrete Haar wavelet transform for preprocessing of the captured raw data. The transform consists of a set of low-pass ($h[n]$) and high-pass ($g[n]$) filters.

resolution levels. In the first level, the input RF data are split into low- and high-frequency components (sub-bands) by means of the half-band low-pass ($h[n]$) and high-pass ($g[n]$) filters, respectively. This is followed by a dyadic decimation (downsampling) of the filter outputs to produce the approximate coefficients, $a_1[n]$, and detail coefficients, $d_1[n]$. In the second level, the $a_1[n]$ coefficients are further decomposed in a similar manner and the generated $d_2[n]$ coefficients are taken as the output RF data ($y_T[n]$). Thereafter, $y_T[n]$ is input to the multistage detection system. Therefore, moving from left to right in Figure 3.6, we get coarser representation of the captured RF data. This coarse–fine strategy is useful for ML-based signal classification because it reveals the signal's features at different resolution levels [6]. Moreover, the output RF data will have fewer samples due to the successive downsampling of the input RF data. This reduces the computational complexity of the algorithms.

In addition, multiresolution analysis is useful for the detection of weak signals in the presence of background noise. This approach to noisy signal detection is based on wavelet denoising (wavelet decomposition and thresholding). This leads to a higher detection accuracy, which is required in applications like UAV threat detection.

Figure 3.7 shows the effect of the Haar wavelet decomposition on a signal captured from the controller of the DJI Phantom 4 Pro UAV. It is clear from the figure that the wavelet transform removes the bias in the signal alignment and reduces the number of samples in

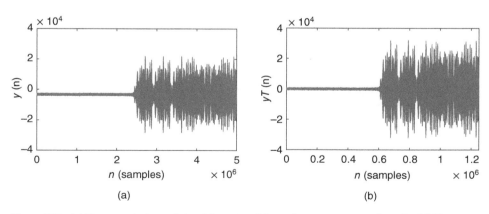

(a) (b)

Figure 3.7 (a) The sampled raw data $y[n]$ captured from the remote controller of a DJI Phantom 4 Pro UAV using an oscilloscope with a sampling rate of 20 Gsa/s, and (b) the transformed data $y_T[n]$ obtained at the output of the two-level Haar wavelet filter. Due to successive downsampling, $y_T[n]$ has about 3.8×10^6 fewer data samples than $y[n]$.

the original signal. Moreover, the transformation preserves the characteristics of the original waveform. This is important in ML-based signal classification.

After the preprocessing step, transformed RF data are transferred to the first stage of the detection system. In this stage, we decide if the captured data belong to an RF signal or noise samples using a Markov models-based Bayes classifier.

3.3.2 The Naive Bayesian Decision Mechanism for RF Signal Detection

In this stage, a decision is made as to whether the captured RF data represent an RF signal of interest or the background noise. First, we model the wavelet-transformed RF signal, $y_T[n]$, using a two-state Markov model. This allows us to compute the likelihood of the captured data coming from either the signal or noise class. According to the Bayesian decision theory, the optimum detector, which minimizes the probability of an error, is one that maximizes the posterior probability. Mathematically, let $C \in \{0, 1\}$ be an index denoting the class of the preprocessed RF signal $y_T[n]$, where $C = 1$ when the captured raw signal $y[n]$ is an RF signal, and $C = 0$ otherwise. Let $S_{y_T} = [S_{y_T}(1), S_{y_T}(2), \ldots, S_{y_T}(N)]^T$ be the state vector representation of the given test data $y_T[n]$ containing N samples, with $S_{y_T}(i) \in \{S_1, S_2\}$, $i = 1, 2, \ldots, N$, and S_1 and S_2 being the two states in the Markov models. Then, the posterior probability of the RF signal class given S_{y_T} is

$$P(C = 1 \mid S_{y_T}) = \frac{P(S_{y_T} \mid C = 1) \, P(C = 1)}{P(S_{y_T})}, \tag{3.1}$$

where $P(S_{y_T} \mid C = 1)$ is the likelihood function conditioned on $C = 1$, $P(C = 1)$ is the prior probability of the RF signal class, and $P(S_{y_T})$ is the evidence. A similar expression holds for the posterior probability $P(C = 0 | S_{y_T})$.

In practice, since the evidence does not depend on the C, it is a constant and can be ignored. Therefore, we are interested in maximizing the numerator in Eq. (3.1). That is,

$$\hat{C} = \arg\max_{C} P(S_{y_T} \mid C) \, P(C). \tag{3.2}$$

In other words, we decide that the captured data belong to an RF signal (i.e., $C = 1$), if the following condition is satisfied:

$$P(S_{y_T} \mid C = 1) \, P(C = 1) \geq P(S_{y_T} \mid C = 0) \, P(C = 0). \tag{3.3}$$

If we assume that the prior probabilities of the signal and noise class are equal, then the naive Bayesian decision rule in Eq. (3.3) becomes

$$P(S_{y_T} \mid C = 1) \geq P(S_{y_T} \mid C = 0). \tag{3.4}$$

Therefore, for given test data, we need to compute and compare the likelihood probabilities $P(S_{y_T} \mid C = \{0, 1\})$. In order to compute the likelihood probability for the RF signal and noise classes, we use a large amount of training data captured from multiple UAV controllers, WiFi routers, mobile Bluetooth emitters, and background noise. This training dataset is stored in a database as shown in Figure 3.5. Since the captured RF signal (after sampling) has a discrete time-varying waveform, we can model it as a stochastic sequence of states/events. The likelihood of such a state sequence belonging to any class can be computed using the state transitions in the Markov models.

First, we define a two-state Markov model for $y_T[n]$. This is achieved by representing $y_T[n]$ in terms of two states (S_1 and S_2) using thresholding. Therefore, the samples in $y_T[n]$ whose absolute amplitudes are less than or equal to a predetermined threshold value (δ) are considered as in state S_1, while the samples with absolute amplitude greater than δ are considered as in state S_2. Mathematically, the state transformation is given as

$$
S_{y_T}(n) = \begin{cases} S_1, & |y_T[n]| \le \delta, \\ S_2, & |y_T[n]| > \delta. \end{cases} \tag{3.5}
$$

Based on the above rule, it is straightforward to transform $y_T[n]$ into the state vector, S_{y_T}. Once S_{y_T} is obtained, the probability of a transition between any two states is calculated.

Note that the state vector is generated based on the amplitude of the signal samples in the wavelet domain because of the benefits of reduced noise power. The choice of δ in Eq. (3.5) depends on the operating signal-to-noise ratio (SNR) of the system and will be discussed in Section 3.5.2. The transition probability matrix is generated based on the transitions from adjacent indexed states. The transition count matrix, T_N, and the transition probability matrix, T_P, are defined as follows:

$$
T_N = \begin{bmatrix} N_{11} & N_{12} \\ N_{21} & N_{22} \end{bmatrix}, \qquad T_P = \begin{bmatrix} p_{11} & p_{12} \\ p_{21} & p_{22} \end{bmatrix} = \frac{T_N}{\sum_{i,j} N_{ij}}. \tag{3.6}
$$

Here N_{ij} is the number of samples transiting from state S_i to S_j, T_N is a matrix whose elements are the number of state transitions in y_T, and the matrix T_P contains the state transition probabilities, which are obtained by normalizing the T_N matrix with the total number of samples in $y_T[n]$. That is, the element $p_{ij} = P(S_i \to S_j)$ is the transition probability from state S_i to S_j. It is expected that the transition probabilities generated for the signal class (UAV, WiFi, and Bluetooth) and the noise class will be significantly different at modest SNR levels. Also, the choice of δ dictates the transition probabilities for both the signal and noise classes. In Section 3.5.2, the threshold δ is expressed in terms of standard deviation (σ) of the preprocessed noise data captured from the environment.

During the experiments, the data is captured within a short time window (0.25 ms), and thus we assume that the environmental noise is stationary during this interval. Figure 3.8 shows the two-state Markov model for the signal and noise class obtained from the training data using $\delta = 3.5\sigma$. From Figure 3.8, we see that, for the signal class, p_{22} is significantly higher than p_{11}, p_{12}, and p_{21}. On the other hand, from Figure 3.8, we see that, for the noise class, p_{11} is significantly higher than p_{22}, p_{12}, and p_{21}. Therefore, for any given test data, using the two-state Markov model, an appropriate δ, and a naive Bayesian classifier, it is possible to accurately determine if the data is an RF signal or noise.

For a given test data $y_T[n]$, the likelihood of the data being a signal is calculated as follows:

$$
\begin{aligned}
P(S_{y_T} \mid C = 1) &= \prod_{n=1}^{N-1} p(S_{y_T}(n) \to S_{y_T}(n+1) \mid C = 1) \\
&= \prod_{i,j=\{1,2\}} T_{P_{C=1}}^{T_N(i,j)}(i,j) \\
&= \prod_{i,j=\{1,2\}} p_{ij;\,C=1}^{N_{ij}}.
\end{aligned} \tag{3.7}
$$

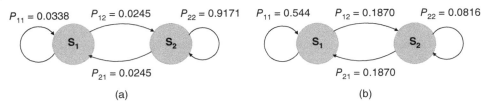

Figure 3.8 Two-state Markov model and associated state transition probabilities using $\delta = 3.5\sigma$ for (a) the signal class and (b) the noise class.

The product of the conditional transition probabilities in the above equation gives the likelihood of obtaining the state vector S_{y_T} given that the hypothesis $C = 1$ is true. The log-likelihood of the expression in Eq. (3.7) is given by

$$\log(P(S_{y_T} \mid C = 1)) = \sum_{i,j=\{1,2\}} N_{ij} \log(p_{ij; \, C=1}). \tag{3.8}$$

Similarly, the log-likelihood of the signal coming from a noise class is calculated by

$$\log(P(S_{y_T} \mid C = 0)) = \sum_{i,j=\{1,2\}} N_{ij} \log(p_{ij; \, C=0}). \tag{3.9}$$

The decision will be favored to $C = 1$ if $\log(P(S_{y_T} \mid C = 1)) \geq \log(P(S_{y_T} \mid C = 0))$; otherwise, $C = 0$. If the captured RF data come from the signal class (i.e., if $C = 1$), then the second-stage detector is invoked. Otherwise, the system continues sensing the environment for the presence of signals as shown in Figure 3.5.

3.3.3 Detection of WiFi and Bluetooth Interference

The second-stage detector decides if the signal detected in the first stage is an emission from a wireless interference source or a UAV controller. The wireless interference sources of interest are WiFi and mobile Bluetooth devices.

In recent times, there has been interest in detecting WiFi and Bluetooth signals [30]. However, in the context of RF-based UAV detection in an urban environment, where these wireless signals are considered as interference, there has been very little research. This is a practical problem. Fortunately, these wireless interference signals are well standardized. Therefore, they can be detected and classified by using the knowledge of their specifications. Table 3.2 provides a brief summary of the specifications for WiFi and Bluetooth transmissions. From the table, it is obvious that the signal bandwidth and the modulation type are two important features for identifying the WiFi and Bluetooth signals. Therefore, the second-stage detector exploits these features for detecting these interference signals/-sources.

The first step in deciding if the detected signal is a wireless interference signal or not is to perform bandwidth analysis. This is particularly important because WiFi signals are easily identified by their bandwidth. According to Table 3.2, Bluetooth 2.0 signals have a bandwidth of 1 or 2 MHz, WiFi signals have a bandwidth of 20 MHz (or more), while all the UAV controller signals in our database have bandwidth less than 10 MHz. Therefore, if the detected RF signal has a bandwidth equal to or greater than 20 MHz, it is classified as a WiFi signal. Bandwidth analysis is performed by taking the Fourier transform of the

Table 3.2 Specifications of the WiFi and Bluetooth standards.

Standard	Bluetooth (IEEE 802.15.1 WPAN)	WiFi (IEEE 802.11 WLAN)
Center frequency (GHz)	2.4	2.4 / 5
Bandwidth (MHz)	1	20 / 40 / 80 / 160
PHY modulation	GFSK / FSK / DPSK	DSSS / OFDM
Range (m)	variable	> 50
Data rate (Mbps)	2	variable

resampled signal. Figure 3.9 shows the result of the bandwidth analysis of a WiFi signal, a Bluetooth signal from a Motorola e5 cruise, and a UAV remote controller signal from a Spektrum DX5e controller.

On the other hand, if the detected signal has a bandwidth less than 20 MHz, it is assumed to be transmitted either from a Bluetooth interference source or a valid UAV controller. Since most mobile Bluetooth devices employ Gaussian frequency shift keying (GFSK/FSK) modulation, it is reasonable to detect and discriminate these devices by means of modulation features. In this study, two GFSK/FSK modulation features, namely, frequency deviation and symbol duration, will be used to discriminate Bluetooth signals. The frequency deviation is a measure of the maximum difference between the peak frequency in the GFSK/FSK signal and the center frequency. On the other hand, the symbol duration is the minimum time interval in the observed Bluetooth waveform or pulse. Therefore, using a GFSK/FSK demodulator, these features can be extracted and used as the basis for Bluetooth signal detection.

In this study, a zero-crossing GFSK/FSK demodulator is considered. It is known that the Bluetooth GFSK/FSK signal is transmitted in bursts consisting of M data bits $d_m \in \{-1, +1\}$, each bit having a period T_b and average energy per bit E_b [31]. A general model for such a signal is given as

$$s(t) = \sqrt{\frac{2E_b}{T_b}} \cos(2\pi f_o t + \varphi(t, \alpha) + \varphi_o) + n(t), \tag{3.10}$$

where $\varphi(t, \alpha)$ is a phase modulating function, φ_o is an arbitrary phase constant, f_o is the operational frequency, and $n(t)$ is the channel noise. The zero-crossing demodulator considered herein for Bluetooth interference detection is able to detect the time instants at which the signal $s(t)$ is equal to zero and has a positive slope, i.e., the zero-crossings. When a Bluetooth device transmits at the basic rate using the standard GFSK/FSK modulation, one symbol represents one bit. Therefore, the time interval between consecutive zero-crossings is a measure of the symbol duration of the Bluetooth signal. This is the basis for the detection of Bluetooth interference signal using the zero-crossing demodulator.

Figure 3.10 shows the results of the zero-crossing demodulation of a Bluetooth signal from a Motorola e5 cruise. The captured Bluetooth signal and its fast Fourier transform (FFT) are shown in Figure 3.10(a) and Figure 3.10(b), respectively. From Figure 3.10(b),

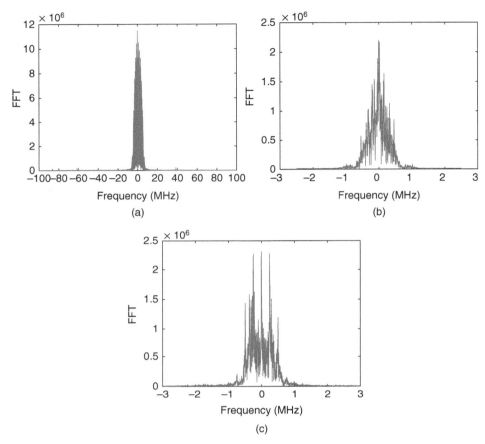

Figure 3.9 Bandwidth analysis of different signals: (a) WiFi signal, (b) Bluetooth signal from a Motorola e5 cruise, and (c) Spektrum DX5e UAV controller signal.

we see that the transmit frequency of the Bluetooth device is 2.4 GHz. Afterward, the signal is shifted and resampled by 1/2000. The FFT of the resampled signal is shown in Figure 3.10(c). This figure shows that the bandwidth of the Bluetooth signal is around 2 MHz, much smaller than 20 MHz. Next, the resampled signal is demodulated by taking the derivative of its phase angle, and the start point of the demodulated signal is estimated using the Higuchi algorithm [10]. The Higuchi algorithm detects the start point of the signal by measuring the fractal dimensions of the signal. Once the start point is detected, the frequency deviation is estimated as one half of the peak-to-peak frequency of the demodulated signal. Figure 3.10(d) shows a plot of the demodulated signal and the estimated start point. The peak-to-peak frequency of the demodulated signal is estimated as 551.12 kHz, and therefore the frequency deviation is 275.56 kHz.

In order to estimate the symbol duration, the demodulated signal is converted to a binary signal using the mean as a threshold. Figure 3.10(e) shows the binary signal, where a binary 1 represents a positive frequency deviation and a binary 0 represents a negative frequency deviation. Then, we compute the derivative of the binary signals to locate the

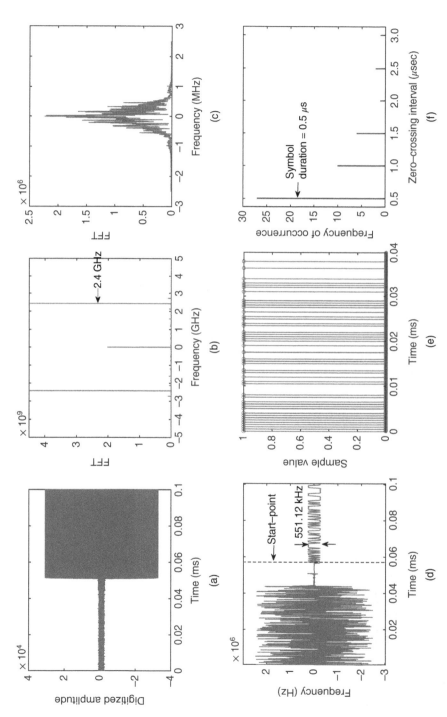

Figure 3.10 Extraction of the modulation features of a Bluetooth interference signal from a Motorola e5 cruise mobile device using zero-crossing demodulation technique: (a) raw signal, (b) FFT of the raw signal, (c) FFT of the shifted and resampled signal (by 1/2000), (d) the demodulated signal showing a peak-to-peak frequency of 551.12 kHz (the frequency deviation is one half of the peak-to-peak value), (e) binary signal, and (f) histogram of the time interval between consecutive zero-crossings in the modulated signal.

zero-crossings. To accurately compute the symbol duration, we take the histogram of the time intervals between consecutive zero-crossings. This is necessary because channel distortions will cause some deviation in these intervals. Figure 3.10(f) shows the histogram of the time intervals between consecutive zero-crossings for the Bluetooth signal extracted from a Motorola e5 cruise mobile device. From the histogram, the zero-crossing interval with the highest frequency of occurrence is a measure of the symbol duration of the Bluetooth signal. The estimated symbol duration is 0.5 μs.

By estimating the modulation-based features, as illustrated in Figure 3.10, it is possible to detect and classify a Bluetooth interference signal. In order to validate the joint discriminating ability of these modulation features, Bluetooth signals from six mobile phones and signals from nine UAV controllers were collected. The mobile phones are iPhone 7, iPhone XR, LG X Charger, Motorola GXX Play, Motorola e5 cruise, and Samsung Galaxy. The UAV controllers considered are Jeti Duplex DC16, Spektrum DX5e, Spektrum DX6e, Spektrum DX6i, Fly-SKy FS-T6, Graupner MC-32, HK-T6, Turnigy 9X, and JR X9303. The UAV signals are frequency modulated as well. Therefore, all the collected signals are demodulated using the zero-crossing technique. Figure 3.11 shows the feature space of the demodulated Bluetooth and UAV controller signals. The figure shows a clear clustering pattern in the feature space. The Bluetooth signals from different mobile devices are clustered together. In addition, Figure 3.11 shows that the Bluetooth signals all have the symbol duration of 0.5 μs and frequency deviation less than 350 kHz. Therefore, the frequency deviation and symbol duration can be used as features in a simple maximum likelihood classifier for detecting and classifying Bluetooth interference signals. If the detected signal is not from a Bluetooth interference source, it is presumed to be an emission from a UAV controller. In this case, the signal is transferred to the UAV classification system for the identification process.

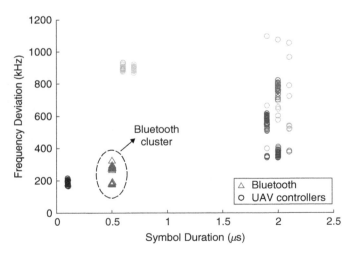

Figure 3.11 Feature space showing the symbol duration and frequency deviation of the signals from several mobile Bluetooth devices and UAV controllers. Each UAV controller is represented by a circular marker of a unique color. The feature clustering validates the hypothesis that the symbol duration and frequency deviation are good features for detecting Bluetooth interference signals.

3.4 UAV Classification Using RF Fingerprints

Once we detect the presence of a UAV signal, we identify the UAV controller using ML techniques. The inputs into the ML classifiers are the RF fingerprints extracted from the energy–time–frequency domain representation of the signals. For the representation of RF signals in the energy–time–frequency domain, we use the spectrogram method. The spectrogram of any signal can be computed as the squared magnitude of the short-time Fourier transform (STFT)

$$
\text{spectrogram}(m, \omega) = \left| \sum_{k=-\infty}^{\infty} y_{\text{T}}[k] w[k - m] e^{-j\omega k} \right|^2, \tag{3.11}
$$

where $y_{\text{T}}[n]$ is the preprocessed signal captured by the surveillance system, m is discrete time, ω is the frequency, and $w[n]$ is a sliding window function that acts as a filter. The spectrogram analysis of the captured RF signals can reveal the transmit frequency of the signal as well as the frequency hopping patterns. The spectrogram of the signal captured from the remote controller of the DJI Phantom 4 Pro UAV (the signal in Figure 3.7) is shown in Figure 3.12. In computing the spectrogram, the signal is divided into segments of length 128 with an overlap of 120 samples between adjoining segments. Then, a Hamming window is used, followed by a 256-point discrete Fourier transform (DFT). The spectrogram in Figure 3.12(a) shows that the transmit frequency of the signal is 2.4 GHz.

The spectrogram, by definition, displays the energy/intensity distribution of the signal on time–frequency axes. Therefore, the energy trajectory can be computed from the spectrogram by taking the maximum energy values along the time axis. From this distribution, we estimate the energy transient by searching for the most abrupt change in the mean or variance of the normalized energy trajectory. The energy transient defines the transient characteristics of the signal in the energy domain. For the RF signal in Figure 3.7, the normalized energy trajectory computed from the spectrogram and the corresponding energy transient are shown in Figure 3.12.

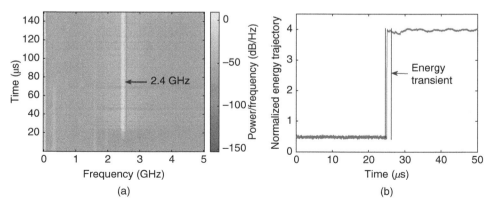

Figure 3.12 (a) The spectrogram and (b) the energy trajectory of the UAV controller signal shown in Figure 3.7. The energy transient is detected from the energy trajectory by searching for the most abrupt change in the mean.

Once the energy transient is detected, RF fingerprints (a set of 15 statistical features) are extracted. These RF fingerprints are the statistical moments that characterize the energy transient. That is, they are the physical descriptors of the energy transients and provide vital information for pattern recognition/ML-based classification of the signals captured from different UAV controllers. Table 3.3 gives the list of the extracted features used in this study.

For this study, the RF fingerprints extracted from 17 UAV controllers were used to train five different ML algorithms: k nearest neighbors (kNN), RandF, discriminant analysis (DA), support vector machine (SVM), and neural networks (NN) [33]. Since some of the RF fingerprints may be correlated, and therefore redundant, we also perform feature selection to reduce the computational cost of the classification algorithm.

Table 3.3 Statistical features.

Features	Formula	Measures
Mean, μ	$\frac{1}{N}\sum_{i=1}^{N} x_i$	Central tendency
Absolute mean, $\bar{x}$	$\frac{1}{N}\sum_{i=1}^{N} |x_i|$	Central tendency
Standard deviation, σ_T	$\left[\frac{1}{N-1}\sum_{i=1}^{N}(x_i-\bar{x})^2\right]^{1/2}$	Dispersion
Skewness, γ	$\frac{\sum_{i=1}^{N}(x_i-\bar{x})^3}{(N-1)\sigma_T^3}$	Asymmetry/shape descriptor
Entropy, H	$-\sum_{i=1}^{N} x_i\log_2 x_i$	Uncertainty
Root mean square, x_{rms}	$\left[\frac{1}{N}\sum_{i=1}^{N} x_i^2\right]^{1/2}$	Magnitude/average power
Root, x_{r}	$\left[\frac{1}{N}\sum_{i=1}^{N} |x_i|^{1/2}\right]^2$	Magnitude
Kurtosis, k	$\frac{\sum_{i=1}^{N}(x_i-\bar{x})^4}{(N-1)\sigma_T^4}$	Tail/shape descriptor
Variance	$\frac{1}{N}\sum_{i=1}^{N}(x_i-\mu)^2$	Dispersion
Peak value, x_{pv}	$\max(x_i)$	Amplitude
Peak-to-peak, x_{ppv}	$\max(x_i)-\min(x_i)$	Waveform amplitude
Shape factor, x_{sf}	$\frac{x_{\text{rms}}}{\bar{x}}$	Shape descriptor
Crest factor	$\frac{x_{\text{max}}}{x_{\text{rms}}}$	Peak extremity
Impulse factor	$\frac{x_{\text{max}}}{x_{\text{rms}}}$	Impulse
Clearance factor	$\frac{x_{\text{max}}}{\bar{x}}$	Spikiness

3.4.1 Feature Selection Using Neighborhood Components Analysis (NCA)

The NCA algorithm is a nearest-neighbor-based feature weighting algorithm, which learns a feature weighting vector by maximizing a leave-one-out classification accuracy using a gradient-based optimizer. It is a non-parametric, embedded, and supervised learning method for feature selection. NCA learns the weighting vector by which the primary data are transformed into a lower-dimensional space [15]. In this lower-dimensional space, the features are ranked according to a weight metric, with the more important features receiving higher weights.

Let U be a set of training samples representing the different UAV controllers,

$$U = \{(\boldsymbol{x}_i, Y_i),\ i = 1, 2, \ldots, n\}, \tag{3.12}$$

where $\boldsymbol{x}_i$ is a p-dimensional feature vector extracted from the energy transient, $Y_i \in \{1, 2, \ldots, C\}$ are the corresponding class labels, and C is the number of classes. Then the NCA learns the feature weighting vector $\boldsymbol{w}$ by maximizing a regularized objective function $f(\boldsymbol{w})$ with respect to the feature weights w_r. The regularized objective function is defined as

$$f(\boldsymbol{w}) = \frac{1}{n} \sum_{i=1}^{n} \left[\sum_{j=1, i \neq j}^{n} p_{ij} Y_{ij} - \lambda \sum_{r=1}^{p} w_r^2 \right], \tag{3.13}$$

where

$$p_{ij} = \begin{cases} \dfrac{k(d_{\mathrm{w}}(\boldsymbol{x}_i, \boldsymbol{x}_j))}{\sum_{j=1, i \neq j}^{n} k(d_{\mathrm{w}}(\boldsymbol{x}_i, \boldsymbol{x}_j))}, & \text{if } i \neq j, \\ 0, & \text{if } i = j, \end{cases}$$

$$Y_{ij} = \begin{cases} 1, & \text{if } Y_i = Y_j, \\ 0, & \text{otherwise.} \end{cases}$$

Here n is the number of samples in the feature set, λ is the regularization term, w_r is a weight associated with the rth feature, and p_{ij} is the probability with which each point $\boldsymbol{x}_i$ selects another point $\boldsymbol{x}_j$ as its reference neighbor and inherits the class label of the latter [35]. The parameter Y_{ij} is an indicator function, $d_{\mathrm{w}}(\boldsymbol{x}_i, \boldsymbol{x}_j) = \sum_{r=1}^{p} w_r^2 |x_{ir} - x_{jr}|$ is a weighted distance function between $\boldsymbol{x}_i$ and $\boldsymbol{x}_j$, and $k(a) = \exp(a/\sigma)$ is some kernel function. Thus, NCA is a kernel-based feature selection algorithm that selects the most descriptive and informative features by optimizing Eq. (3.13) using gradient update techniques.

Figure 3.13 shows the results of the NCA ranking of 15 features extracted from the 17 UAV controllers considered in this study. The experimental setup and data structure of the captured data are described in Section 3.5.1. In Figure 3.13, we see that NCA ranks the RF fingerprints according to their weights. As can be seen, the shape factor has the highest weight and thus is the most discriminative feature in the feature set. The next significant feature is the kurtosis, which describes the tailedness of the energy trajectory curve. Next are variance and standard deviation, which measure the dispersion of the energy trajectory curve. Most pattern recognition and ML algorithms rely on shape descriptors in the datasets. On the other hand, entropy, which measures the uncertainty in the dataset, is the least significant feature. Consequently, for training and testing, the ML algorithms used in this study can safely discard the less important features and still achieve good classification

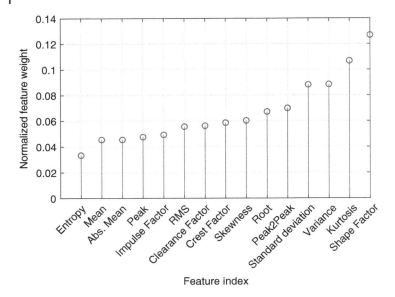

Figure 3.13 NCA ranking of all 15 features extracted from 17 UAV controllers. The shape factor is the most significant RF fingerprint for the dataset, while entropy is the least significant.

performance. This is because discarding the less significant features reduces the chance of overfitting. In addition, for large-scale classification problems, there can be huge computational savings in training and testing the classifiers with fewer features. The next section describes the experimental setup and presents the results.

3.5 Experimental Results

3.5.1 Experimental Setup

In this study, RF signals were captured from 17 UAV controllers, six mobile Bluetooth devices (smartphones), and a WiFi router. Table 3.4 gives the catalogue of the UAV controllers used in this study. The catalogue includes UAV controllers from eight different manufacturers. All the UAV controllers transmit control signals in the 2.4 GHz frequency band. In particular, a pair of UAV controllers from DJI Matrice 600 and DJI Phantom 4 Pro models are used, but only one controller from the other models was used. This is important for forensic and security analysis to investigate the confusion that would arise when the ML classifiers or other automatic target recognition (ATR) system attempts to distinguish between two UAV controllers of the same make and model. For the remaining part of the study, we will refer to the pair of DJI Matrice 600 as DJI M600 Mpact and DJI M600 Ngat. Similarly, the pair of Phantom 4 Pro controllers will be referred to as DJI Phantom 4 Pro Mpact and DJI Phantom 4 Pro Ngat.

Figure 3.14 shows the indoor and outdoor experimental scenarios. In either case, the RF passive surveillance system can detect signals transmitted by the UAV controllers and the wireless interference sources. In both indoor and outdoor environments, the dominant

Table 3.4 UAV catalogue.

Make	Model
DJI	Inspire 1 Pro
	Matrice 100
	Matrice 600[a]
	Phantom 4 Pro[a]
	Phantom 3
Spektrum	DX5e
	DX6e
	DX6i
	JRX9303
Futaba	T8-FG
Graupner	MC-32
HobbyKing	HK-T6A
FlySky	FS-T6
Turnigy	FS-T6
Jeti Duplex	DC-16

a) A pair of these controllers is used in this study. For all other controllers, only one of each type is considered.

interference signals come from WiFi and Bluetooth devices. In this chapter, due to space limitations, only the results from the indoor experiment will be reported.

The experimental RF passive surveillance system used in this study consists of a 6 GHz bandwidth Keysight MSOS604A oscilloscope with a maximum sampling frequency of 20 Gsa/s, a 2 dBi omnidirectional antenna (for short-distance detection), and a 24 dBi WiFi grid antenna (for longer-distance detection). The antennas operate in the 2.4 GHz frequency band, which is the operational band of most commercial and hobby-grade UAVs. Detection range for the far-field scenario can be further improved by using a combination of high-gain receive antennas and low-noise power amplifiers (LNAs).

The receiver antenna senses the environment for the presence of signals from UAV controllers. However, since the environment is noisy, captured RF signals could be emissions from interference sources or background noise. Once the signal is captured, it is fed into the oscilloscope (central receiver system). The collected signals are automatically saved in a cloud database for postprocessing. For each controller, 100 RF signals were collected. Each RF signal is a vector of size $5\,000\,000 \times 1$ and has a time span of 0.25 ms. The database is partitioned with the ratio $p = 0.2$. That is, 80% of the saved data is randomly selected for training, and the remaining 20% is used for testing (4 : 1 partitioning).

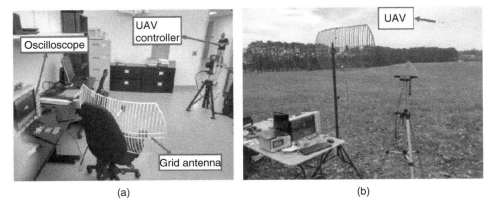

Figure 3.14 Experimental scenario for UAV signal detection in (a) indoor and (b) outdoor environments. Source: Guvenc, Matolak.

3.5.2 Detection Results

Once the RF data has been detected by the receiver, the first-stage detector is invoked. This detector decides if the captured data is a signal or noise. Figure 3.15 shows the detection accuracy versus SNR for different thresholds. The selected thresholds are functions of the standard deviation (σ) of the preprocessed noise data and the false alarm rate (FAR) specification. The value of σ is estimated after performing multiresolution analysis (wavelet preprocessing) of the concatenation of several noise data captured from the environment. On the other hand, FAR, also known as the probability of false detection, is the percentage of false alarms per the number of non-events.

From Figure 3.15, we see that, at very low SNR, such as -10 dB, the detection accuracy is generally very low irrespective of the threshold. As a result, in the case of low-level signals (where signals are completely buried in the noise), the probability of missed detection is high. Besides, for a given SNR, it is observed that the set threshold determines the performance of the detection system. For instance, when the system operates at an SNR of 2 dB, a threshold of $\delta = 0.1\sigma$ will achieve a detection accuracy of above 99%. This high detection accuracy implies that almost all the test signals are correctly detected when the threshold is set at $\delta = 0.1\sigma$. However, the threshold $\delta = 0.1\sigma$ yields a FAR of 100%. Therefore, a very low threshold value will result in a high percentage of misclassification of the noise data as signals. Furthermore, for the given SNR of 2 dB, an increase in the threshold value to $\delta = 1.1\sigma$ will reduce the detection accuracy and FAR to 96.6% and 14.8%, respectively. Further increase in the threshold to $\delta = 2.5\sigma$ will greatly reduce the detection accuracy and FAR to 40.4% and 3.2%, respectively. Therefore, the optimum threshold depends on the operating condition and the requirements on the FAR. Besides, the input impedance of the oscilloscope places a fundamental limit on the sensitivity of the passive detection system used in this study. The relationship between the input impedance and the detection accuracy will be investigated in future work.

In addition, Figure 3.15 shows that better detection performance (with low FAR) can be achieved if the detector operates at high SNR (above 8 dB) and with $\delta \in [2.5\sigma, 4.1\sigma]$. For

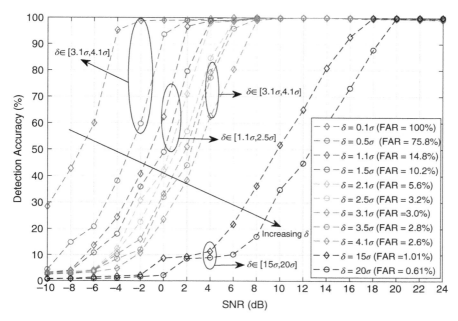

Figure 3.15 The signal detection accuracy of the Markov model-based naive Bayesian detector versus SNR for different values of δ. For each value of δ, there is an associated FAR. Moreover, for a given SNR, the detection accuracy and FAR reduce as the value of δ increases. The choice of δ will depend on the operational requirement of the system.

instance, when the receiver operates at an SNR of 10 dB with a threshold $\delta = 2.5\sigma$, the detection accuracy is 100% and FAR is 3.2%. Although a continuous increase in the threshold will further reduce the FAR, it will not always guarantee a better detection accuracy especially when the receiver operates at an SNR less than 18 dB. This is obvious in the case where δ is set to 15σ or 25σ. This is because the level of δ determines the state transition probability matrix (T_P) of the signal and noise classes. The dissimilarity between the transition matrices of the two classes reduces as δ increases beyond some optimum value. Therefore, there is a high chance of detection error as δ increases indefinitely. In the context of detection theory, a much higher level of δ will lead to an increase in missed detection, which reduces the overall detection accuracy.

Once a signal has been detected, the bandwidth and the modulation features are estimated as described in Section 3.3.3. This information is used to decide if the signal comes from a UAV controller or any of the known interference sources (WiFi and Bluetooth sources). Given that the detected signal comes from a UAV controller, it is sent to the ML classification system for proper identification. The results of the ML classifiers is discussed next.

3.5.3 UAV Classification Results

For the classification problem, the 15 statistical features given in Table 3.3 were extracted. Feature selection was performed using the NCA algorithm as described in Section 3.4.1.

To validate the result of the NCA algorithm and the ML classifiers, 10 Monte Carlo simulations were run on the test dataset. On the one hand, all the 15 features are used for the UAV controller classification problem. On the other hand, only the three most significant features are used according to the NCA weight ranking shown in Figure 3.13. These are the shape factor, kurtosis, and variance. The classification experiments were run separately for the case of 15 and 17 UAV controllers. Here, the number of controllers represents the number of classes considered. In the case of 15 controllers, all the controllers are of a different model. However, in the case of 17 controllers, a pair of DJI Matrice 600 (labeled as DJI Matrice 600 Mpact and DJI Matrice 600 Ngat) and a pair of DJI Phantom 4 Pro controllers (labeled as DJI Phantom 4 Pro Mpact and DJI Phantom 4 Pro Ngat) were considered in addition to 13 different models.

Table 3.5 provides the classification accuracy of all five ML algorithms. With the exception of the kNN classifier, the classification accuracy is only slightly higher when all the features are used as compared to when only the three selected features are used. Therefore, there is little performance loss in using only the selected features. In addition, there are computational savings in time and storage when only the selected features are used for the classification. These savings in time and computational resources are important in aerial surveillance system, where the response time to effectively neutralize a threat is very small. In addition, as the number of UAV classes increases, the savings in time will become more substantial. Therefore, the results in Table 3.5 validate the decision to perform feature selection using the NCA algorithm.

In addition, Table 3.5 shows that the RandF classifier has the highest classification accuracy when all the features are used. For the case of 15 and 17 controllers, RandF achieves an accuracy of 98.53% and 96.32%, respectively. Therefore, when all the features are used, RandF is the best-performing classifier. This is followed by the kNN classifier,

Table 3.5 Performance of the ML classification algorithms at 25 dB SNR. A total of 100 sample signals from each UAV controller are captured, with 80% used for training and 20% for testing (partition ratio = 0.2). The selected RF fingerprints are: shape factor, kurtosis, and variance.

No. of controllers	Classifier	Accuracy (%) [a]		Computational time (s) [a]	
		All feat.	Selected feat.	All feat.	Selected feat.
15	kNN	97.30	98.13	24.85	24.57
	DA	96.30	94.43	19.42	18.58
	SVM	96.47	91.67	119.22	111.02
	NN	96.73	96.13	38.73	38.14
	RandF	98.53	97.73	21.37	20.89
17	kNN	95.62	95.53	26.16	25.13
	DA	92.77	88.12	19.36	18.90
	SVM	93.82	87.88	139.94	141.68
	NN	92.88	93.03	46.04	43.33
	RandF	96.32	95.18	24.71	24.84

a) Both the accuracy and total computation time are the average of the 10 Monte Carlo simulations.

which achieves an accuracy of 97.30% and 95.62% with 15 and 17 controllers, respectively. The DA classifier is the least optimal when all the features are utilized. On the other hand, when only the three selected RF features are used, the kNN classifier achieves the best accuracy with 98.13% and 95.53% with 15 and 17 controllers, respectively. This is followed by the RandF classifier, which achieves an accuracy of 97.73% and 95.18% with 15 and 17 controllers, respectively. When only the three most significant features are used, the least performing classifier is SVM. Furthermore, Table 3.5 shows that the DA classifier has the shortest computational time, and the SVM classifier has the longest computational time.

It is important to note that Table 3.5 only provides the average classification accuracy. A more detailed summary can be obtained from a box plot analysis. Figure 3.16 provides a box plot comparison of the classifiers. Each box plot corresponds to a different classifier. The box plot gives a summary of the performance of a classifier in terms of the minimum, first quartile, median (red horizontal line), third quartile, and the maximum accuracy values. In terms of these metrics, we see that kNN is the optimal classifier for the classification problem when only the selected features are used. Comparing the box plots in Figure 3.16(a) and Figure 3.16(b), we see that the box plot metrics for each classifier are lower in the case of 17 controllers as compared to the case of 15 controllers. This will be further investigated with the help of the confusion matrices. In addition, the box plots reveal the presence of outliers in the performance of the SVM and NN classifiers. These outliers suggest that, for a given test signal, SVM and NN classifiers could produce accuracy values well below the average values reported in Table 3.5. This raises concerns about the reliability of these classifiers for the UAV controller classification problem.

Furthermore, the SNR of the detected signal is an important factor that influences the accuracy of the classifiers. Figure 3.17 shows the accuracy versus SNR for the kNN, RandF, and DA classifiers. For signals with SNR in the interval between 15 and 25 dB, the kNN performs slightly better than the RandF for the case of 15 controllers. In the same SNR region, the RandF performs best for the case of 17 controllers. In this SNR region, the DA classifier has the worst performance. On the other hand, for SNR between 4 and 15 dB, the performance of the DA classifier improves significantly, outperforming the kNN and RandF classifiers when 15 controllers are considered. This is an interesting observation, since DA is known to have the shortest computational time. However, for SNR between

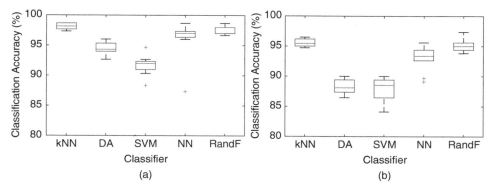

Figure 3.16 Box plot analysis of the performance of the ML classifiers using the three selected features (shape factor, kurtosis, and variance): classification accuracy with (a) 15 controllers and (b) 17 controllers.

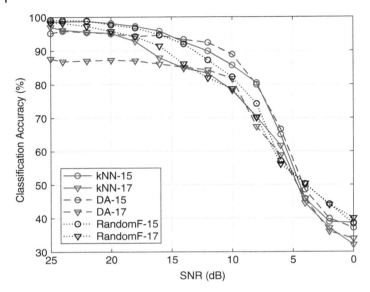

Figure 3.17 Classification accuracy versus SNR for kNN, RandF, and DA classifiers using the three selected RF fingerprints (shape factor, kurtosis, and variance) as features for training and testing the ML classifiers.

0 and 4 dB, the RandF classifier has the best performance. In general, the accuracy of all the classifiers increases with SNR. Therefore, to ensure accurate identification of the UAV controller, it is best to operate the receiver at SNR above 15 dB, in which case, kNN and RandF are the best classifiers for the datasets. In addition, Figure 3.17 shows that, for all SNR, the accuracy plot is slightly lower when 17 controllers are considered as compared to the case of 15 controllers.

The confusion matrix gives an idea of what a classifier is getting right and the type of errors it makes. Figure 3.18 shows the confusion matrices of the classifiers: kNN, RandF, and DA for the case of 15 and 17 remote controllers. The vertical axis of each confusion matrix is the output class or the prediction of the classifier, while the horizontal axis is the target class or the true label. From the confusion matrices in Figure 3.18, we observe that, in the case of 17 controllers, the degree of confusion around the DJI controllers is relatively higher as compared to the case of 15 controllers. This is because, in the former, we intentionally included two pairs of identical DJI controllers (DJI Matrice 600 MPact, DJI Matrice 600 Ngat, DJI Phantom 4 Pro Mpact, and DJI Phantom 4 Pro Ngat). Consequently, there is some confusion among these four controllers, leading to a slight reduction in the classification accuracy in the case of 17 controllers. However, the kNN and RandF classifiers still achieve an average accuracy of 95.53% and 95.18%, respectively. Therefore, these classifiers are robust in identifying UAV controllers of the same make and model. On the other hand, the DA classifier is characterized by more confusion among different controllers, which reduces its average accuracy to 88.12% in the case of 17 remote controllers. Therefore, while the kNN and RandF seem to be the best classifiers, the DA classifier still performs well for the given dataset.

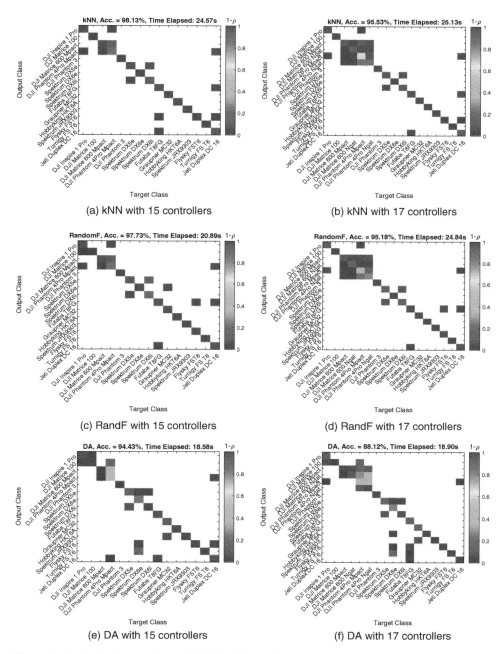

Figure 3.18 Confusion matrices of kNN, RandF, and DA classifiers using the three selected RF fingerprints (shape factor, kurtosis, and variance). In the confusion matrices, the colorbar is used to specify the degree of confusion in terms of the confusion probability ρ. Moving down the colorbar, the degree of confusion increases with increasing value of ρ.

3.6 Conclusion

In this chapter, the problem of detecting and classifying UAVs with a focus on RF fingerprinting techniques was investigated. The proposed detection system operates in the presence of wireless interference from WiFi and Bluetooth sources. These interference signals are detected using a multistage detector that estimates the bandwidth and modulation features of the detected signals. Once the signal from a UAV controller is detected, it is identified using different ML algorithms. Using only three features selected by the NCA for the ML classifiers, the study shows that kNN achieves the best classification performance, with an accuracy of 98.13% in the case of 15 different controllers. However, the RandF classifier is the best when all 15 features are used, achieving an accuracy of 98.53% in the case of 15 different controllers. On the other hand, the DA classifier has the shortest computation time and performed well for some range of SNR. In addition, the confusion matrices show that, in the presence of multiple controllers of the same make and model, the average classification accuracy is slightly reduced. However, even in this situation, the kNN and RandF classifiers achieve an average accuracy of 95.53% and 95.18%, respectively, using only the three selected RF fingerprints. Future studies will investigate the outdoor experimental scenario and we will consider the potential of sensor fusion for improved UAV detection.

Acknowledgments

This work was supported by the National Aeronautics and Space Administration under Federal Award ID number NNX17AJ94A. The authors would like to thank their project co-investigators Paul Davis and Benjamin Boisevert from Architecture Technology Corporation for their feedback.

References

1 M. F. Al-Sa'd, A. Al-Ali, A. Mohamed et al. (2019). RF-based drone detection and identification using deep learning approaches: an initiative towards a large open source drone database. *Future Gener. Comput. Syst.* 100: 86–97.

2 M. Benyamin and G. H. Goldman (2014). Acoustic Detection and Tracking of a Class I UAS with a Small Tetrahedral Microphone Array. Technical Report, DTIC Document, September.

3 G. C. Birch, J. C. Griffin, and M. K. Erdman (2015). UAS detection, classification, and neutralization: market survey. Prepared by Sandia National Laboratories.

4 I. Bisio, C. Garibotto, F. Lavagetto et al. (2018). Unauthorized amateur UAV detection based on WiFi statistical fingerprint analysis. *IEEE Commun. Mag.* 56 (4): 106–111.

5 T. Boon-Poh (2017). RF techniques for detection, classification and location of commercial drone controllers. Technical Report, KeySight Technologies, July.

6 A. Bultan and R. A. Haddad (2000). System identification with denoising. *Proceedings of the IEEE International Conference on Acoustics, Speech, and Signal Processing (ICASSP)*, Istanbul, Turkey, June, vol. 1, pp. 576–579.

7 J. Busset, F. Perrodin, P. Wellig et al. (2015). Detection and tracking of drones using advanced acoustic cameras. *Proc. SPIE Security + Defense* 9647: 1–8.

8 M. Caris, W. Johannes, S. Stanko, and N. Pohl (2015). Millimeter wave radar for perimeter surveillance and detection of MAVs (micro aerial vehicles). *Proceedings of the IEEE International Radar Symposium (IRS)*, Dresden, Germany, June, pp. 284–287.

9 J. Drozdowicz, M. Wielgo, P. Samczynski et al. (2016). 35 GHz FMCW drone detection system. *Proceedings of the IEEE International Radar Symposium (IRS)*, Krakow, Poland, May, pp. 1–4.

10 R. Esteller, G. Vachtsevanos, J. Echauz, and B. Litt (2001). A comparison of waveform fractal dimension algorithms. *IEEE Trans. Circuits Syst. I, Fundam. Theory Appl.* 48 (2): 177–183.

11 M. Ezuma, F. Erden, C. K. Anjinappa et al. (2019). Micro-UAV detection and classification from RF fingerprints using machine learning techniques. *Proceedings of the IEEE Aerospace Conference*, Big Sky, MT, March, pp. 1–13.

12 M. Ezuma, O. Ozdemir, C. Kumar et al. (2019). Micro-UAV detection with a low-grazing angle millimeter wave radar. *Proceedings of the IEEE Radio Wireless Week (RWW) Conference*, Orlando, FL, January, pp. 1–4.

13 F. Fioranelli, M. Ritchie, H. Griffiths, and H. Borrion (2015). Classification of loaded/unloaded micro-drones using multistatic radar. *IET Electron. Lett.* 51 (22): 1813–1815.

14 F. Gökçe, G. Üçoluk, E. Şahin, and S. Kalkan (2015). Vision-based detection and distance estimation of micro unmanned aerial vehicles. *Sensors* 15 (9): 23 805–23 846.

15 Jacob Goldberger, Geoffrey E Hinton, Sam T Roweis, and Ruslan R Salakhutdinov. Neighbourhood components analysis. *Proceedings of Conference on Neural Information Processing Systems (NeurIPS)*, pages 513–520, Vancouver, Canada, December 2004.

16 İsmail Güvenç, Ozgur Ozdemir, Yavuz Yapici, Hani Mehrpouyan, and David Matolak. Detection, localization, and tracking of unauthorized uas and jammers. *Proceedings of the 2017 IEEE/AIAA 36th Digital Avionics Systems Conference (DASC)*, pages 1–10, St. Petersburg, FL, September 2017.

17 Ismail Guvenc, Farshad Koohifar, Simran Singh, Mihail L Sichitiu, and David Matolak. Detection, tracking, and interdiction for amateur drones. *IEEE Commun. Mag.*, 56(4):75–81, Apr. 2018.

18 Michal Haluza and Jaroslav Čechák. Analysis and decoding of radio signals for remote control of drones. *Proceedings of the IEEE Meeting on New Trends in Signal Processing (NTSP)*, pages 1–5, Demanovska Dolina, Slovakia, Oct. 2016.

19 Brian A Jackson, David R Frelinger, Michael Lostumbo, and Robert W Button. *Evaluating Novel Threats to the Homeland: Unmanned Aerial Vehicles and Cruise Missiles*. Rand Corporation, Santa Monica, CA, 2008.

20 S. Jeon, J. Shin, Y. Lee, W. Kim, Y. Kwon, and H. Yang. Empirical study of drone sound detection in real-life environment with deep neural networks. *Proceedings of the European Signal Processing Conference (EUSIPCO)*, pages 1858–1862, Aug. 2017.

21 Andrew J Kerns, Daniel P Shepard, Jahshan A Bhatti, and Todd E Humphreys. Unmanned aircraft capture and control via GPS spoofing. *J. Field Robotics*, 31(4):617–636, 2014.

22 Jens Klare, Oliver Biallawons, and Delphine Cerutti-Maori. Detection of UAVs using the MIMO radar MIRA-CLE Ka. *Proceedings of the European Conference on Synthetic Aperture Radar*, pages 1–4, Hamburg, Germany, June 2016.

23 H. Liu, Z. Wei, Y. Chen, J. Pan, L. Lin, and Y. Ren. Drone detection based on an audio-assisted camera array. *Proceedings of the IEEE International Conference on Multimedia Big Data (BigMM)*, pages 402–406, Apr. 2017.

24 Stephane G Mallat. A theory for multiresolution signal decomposition: the wavelet representation. *IEEE Trans. Pattern Anal. Mach. Intell.*, 11 (7):674–693, July 1989.

25 Pavlo Molchanov, Ronny IA Harmanny, Jaco JM de Wit, Karen Egiazarian, and Jaakko Astola. Classification of small UAVs and birds by micro-Doppler signatures. *Int. J. Microw. Wireless Technol.*, 6(3-4): 435–444, 2014.

26 Thomas Müller. Robust drone detection for day/night counter-UAV with static VIS and SWIR cameras. *Proc. SPIE Defense + Security*, 10190, id 1019018, May 2017.

27 Phuc Nguyen, Mahesh Ravindranatha, Anh Nguyen, Richard Han, and Tam Vu. Investigating cost-effective RF-based detection of drones. *Proceedings of the ACM Workshop on Micro Aerial Vehicle Networks, Systems, and Applications for Civilian Use*, pages 17–22, Singapore, June 2016.

28 Phuc Nguyen, Hoang Truong, Mahesh Ravindranathan, Anh Nguyen, Richard Han, and Tam Vu. Matthan: drone presence detection by identifying physical signatures in the drone's RF communication. *Proceedings of the ACM International Conference on Mobile Systems, Applications, and Services (ACM MobiSys)*, pages 211–224, Niagara Falls, NY, June 2017.

29 Fei Qi, Xuetian Zhu, Ge Mang, Michel Kadoch, and Wei Li. UAV network and IoT in the sky for future smart cities. *IEEE Netw.*, 33(2):96–101, Mar. 2019.

30 Shravan Rayanchu, Ashish Patro, and Suman Banerjee. Airshark: detecting non-WiFi RF devices using commodity WiFi hardware. *Proceedings of the ACM Internet Measurement Conference (ACM IMC)*, pages 137–154, Berlin, Germany, Nov. 2011.

31 T Scholand and P Jung. Bluetooth receiver with zero-crossing zero-forcing demodulation. *IET Electron. Lett.*, 39(17):1275–1277, Aug. 2003.

32 H. Shakhatreh, A. H. Sawalmeh, A. Al-Fuqaha et al. (2019). Unmanned aerial vehicles UAVs: a survey on civil applications and key research challenges. *IEEE Access* 7: 48 572–48 634.

33 Sergios Theodoridis and Konstantinos Koutroumbas. *Pattern Recognition*. Academic Press, Burlington, MA, Oct. 2008.

34 Ryan J Wallace and Jon M Loffi. Examining unmanned aerial system threats and defenses: a conceptual analysis. *Int. J. Aviation, Aeronaut., Aerosp.*, 2(4):1, Sept. 2015.

35 Wei Yang, Kuanquan Wang, and Wangmeng Zuo. Neighborhood component feature selection for high-dimensional data. *J. Comput.*, 7(1): 161–168, Jan. 2012.

36 Caidan Zhao, Caiyun Chen, Zhibiao Cai, Mingxian Shi, Xiaojiang Du, and Mohsen Guizani. Classification of small UAVs based on auxiliary classifier wasserstein GANs. *Proceedings of the IEEE Global Telecommunications (GLOBECOM) Conference*, pages 206–212, Abu Dhabi, UAE, Dec. 2018.

Part II

Cellular-Connected UAV Communications

4

Performance Analysis for Cellular-Connected UAVs

M. Mahdi Azari[1], Fernando Rosas[2,3,4], and Sofie Pollin[1]

[1]*Department of Electrical Engineering, KU Leuven, Belgium*
[2]*Data Science Institute, Imperial College London, UK*
[3]*Department of Brain Sciences, Imperial College London, UK*
[4]*Center for Complexity Science, Imperial College London, UK*

4.1 Introduction

Reliable wireless communication is a key enabler for UAV-enabled applications, allowing both mission control, real-time flight control, or even real-time access to the UAV data. The use of the cellular network is a promising candidate, as it is widely deployed and would allow also beyond-visual-line-of-sight UAV applications. In this chapter, a mathematical framework based on stochastic geometry is proposed to analyze the performance of UAV systems relying on the cellular network for their wireless communication. Closed-form expressions are derived for various important performance indicators, such as coverage probability, spectral efficiency, and throughput. Using this framework, the impact of various design parameters, such as ground base station density, UAV altitude, and its antenna tilt, is analyzed. Also, the coexistence of ground and aerial users, making use of the same cellular infrastructure and spectrum, is investigated. The main conclusions are that interference is significant for aerial users, and that it is important to properly dimension the network and antenna parameters to avoid interference to aerial users. Nevertheless, when sufficiently optimizing network and antenna parameters, networks with both ground and aerial users achieve a higher overall area spectral efficiency. As a result, the analysis and conclusions in this chapter confirm that cellular communication is a promising technology for connecting UAVs.

4.1.1 Motivation

A novel and promising feature of 5G networks is the potential for extending their reach towards the sky, this by integrating unmanned aerial vehicles (UAVs) – popularly known as drones – to the networked environment. UAVs have rapidly caught the attention of both academia and industry due to their flexibility, low flying altitude, and potential cost efficiency in comparison with conventional aircraft. These features make drones specially suitable for a range of applications, including surveillance and monitoring, search and rescue

UAV Communications for 5G and Beyond, First Edition.
Edited by Yong Zeng, Ismail Guvenc, Rui Zhang, Giovanni Geraci, and David W. Matolak.
© 2021 John Wiley & Sons Ltd. Published 2021 by John Wiley & Sons Ltd.

operations, remote sensing, product delivery, and many others [1]. As a matter of fact, it has been predicted that the sales of UAVs will surpass 12 billion US dollars per year by 2021 [2], and that the global UAV payload (including equipment carried by UAVs such as cameras, etc.) market value is expected to reach 3 billion US dollars by 2027 [3]. All this indicates that UAVs are here to stay, and therefore integrating them into 5G networks is an important step towards enabling them into new commercial opportunities and civil services.

Most of the novel roles and applications of UAVs depend on a reliable control of these devices, for which significant real-time data exchange is mandatory. Therefore, there is the need for adequate wireless technology which can guarantee an appropriate connectivity between drones and ground-based networks [4, 5]. This technology needs to serve two main purposes: command/control services, and data communication [6]. Associated with command and control functions are the following key features:

(1) *high coverage and continuous connectivity* to guarantee reliable control and tracking of autonomous or human-driven UAVs; and
(2) *low latency* to enable robust remote control or real-time applications such as event monitoring.

Additionally, in order to satisfy the requirements of applications that heavily rely on data exchange, this technology might need to have capabilities for

- *high throughput* to allow data exchange and video monitoring,
- *secure communication* for data protection and privacy,
- *location verification* for traffic management,
- *licensed spectrum* for mission-critical applications,
- *system scalability* for supporting rapid growth of UAVs, and
- *regulatory compliance* of UAV communication.

The wireless cellular network is a natural candidate for providing these services to UAVs, as it is a mature technology that is able to satisfy similar design objectives when serving more traditional ground users. In particular, using long-term evolution (LTE) technology and its existing infrastructure for this purpose could be particularly convenient, as it would provide important features such as flexible scheduling, resource management, and multiple access solutions [7]. This, in turn, could induce important reductions in the operational costs needed to enable the desired UAV connectivity.

Integrating UAVs into the existing LTE systems is not a straightforward "plug-and-play" procedure. In effect, the current instantiation of the cellular infrastructure and corresponding LTE technology have been designed for serving ground users, and some design choices might not be favorable for UAV devices. For example, base stations (BSs) usually tilt their antennas down, pointing towards the ground, which would induce a considerable antenna gain loss when serving aerial devices. Also, spatial reuse factors optimized for ground propagation conditions and inter-cell interference assumptions are unlikely to fit well the needs of UAVs. These concerns are motivated in a number of recent studies, which have highlighted the important difference between ground-to-UAV communications and conventional ground-to-ground systems in terms of propagation conditions [8–10]. In summary, there might be a significant degradation in the service received by UAVs with respect to ground users when served by the ground cellular infrastructure.

In order to allow a successful use of cellular technology for serving UAVs, it is crucial to consider potential coexistence conflicts between ground and aerial user equipment (UE), and study the effect of various design factors on the quality of service (QoS) of both communities. Moreover, such study requires a careful characterization of the corresponding ground-to-UAV link properties, and their relationship with the main network parameters.

4.1.2 Related Works

The first theoretical study on the cellular-connected UAVs can be found in [6, 11, 12]. In these works, the coverage probability of a UAV served by the ground cellular network is obtained where the spectrum is shared with the ground users. Several interesting insights are reported, and user-based solutions are proposed to enhance the efficiency of the network in the presence of UAVs. For instance, it is shown that inter-cell interference is a considerably more important limiting factor for aerial users; however, UAVs are capable of dealing with this issue by the proper use of their design degrees of freedom, such as their adjustable altitude or antenna tilting. Also, the impact of many ground-based parameters such as BS antenna tilt angle and height are investigated and general guidelines are presented.

According to [6] the association pattern of ground cellular networks is different in the sky such that an aerial user might be served by a much further BS due to having line-of-sight (LoS) condition and receiving the signal from higher BS antenna gain compared to the closer BSs. A detailed effect of BS antenna radiation pattern is also considered in [13], where UAV-to-UAV communication link performance is the main focus. In this study, Azari et al. provide a generic view on using the ground uplink cellular spectrum for UAV interconnections following a distance-proportional power control policy. It is concluded that when using the power control mechanism the effect of aerial interference on the existing ground networks is limited. Nevertheless, one should still say that the UAV-to-UAV link performance is notably affected by LoS interference received from the ground users, which can be controlled by their altitude and interconnection distances. Uplink cellular-connected UAVs are also studied in [14].

In parallel to academic efforts, several industrial-based research studies about cellular-connected UAVs are available [15–17]. The Qualcomm report [15] shows that, in general, UAV users are capable of detecting more cells as their height increases. This implies that the more favorable LoS propagation condition dominates the lower BS antenna gain towards the sky. Accordingly, a UAV user receives a higher reference signal strength and also higher level of interference. The same report shows that the performance of an omnidirectional UAV at high altitude is lower than at ground level but yet able to establish a communication link up to a certain examined altitude. The Bell Labs team in Dublin investigated the use of massive multiple input–multiple output (MIMO) for serving UAVs through simulation tools [16]. Their results demonstrate the advantages of massive MIMO BSs and shows that a reliable communication link can be set up for UAV operations.

A significant amount of literature considers UAVs acting as aerial BSs or relays [8, 9, 18–28]. In [25] a general overview and tutorial is given about UAV wireless communication and networking. The performance of UAVs in the role of aerial BSs serving ground users is investigated in [8, 9, 18]. In [8] a novel generic altitude- and elevation-angle-dependent path-loss and small-scale fading model is used for the analysis of the UAV-to-ground link with and without relaying. Static multiple aerial BSs are considered in [18], and [9] accounts for random location of UAVs as well. The authors in [9] optimized the UAV altitude, density, and antenna beamwidth for maximum coverage received by a typical ground user. The performance of aerial UAV BSs for ground node localization is considered in [26, 27], where the number and trajectory of UAVs are optimized for higher localization accuracy. In [27] the detailed energy consumption of UAVs is presented and the performance of the network is optimized under a given energy budget.

4.1.3 Contributions and Chapter Structure

In this chapter we introduce a generic framework that can be used to assess the performance of UAVs which are served by a ground cellular network. The framework includes a dual-slope line-of-sight/non-line-of-sight (LoS/NLoS) propagation model for both path loss and small-scale fading, which are combined according to their probability of occurrence. The model also uses a generic distance- and altitude-dependent LoS probability, which is able to reflect the characteristics of different types of urban environments. Therefore, following [8], this channel modeling choice is consistent with the fact that the path-loss exponent and fading effect is likely to be affected by the link's distance and altitude.

By exploiting the proposed framework, this chapter presents a unified view over the results reported in [6, 11, 12]. At the core of these findings is the derivation of a number of exact analytical expressions for important key performance metrics, including the link coverage probability, capacity, and network area spectral efficiency. Several approximations are also provided which facilitate the complex computation of the expressions. Using these expressions, the main findings reported in this chapter are listed in the following:

- We evaluate the capability of the existing ground networks to serve aerial users. For this, the impact of several important system parameters are investigated, which include the BS density, UAV altitude and antenna parameters, and different types of urban areas. Particularly, we consider the coexistence of ground and aerial users when sharing the same spectrum.
- We show that the altitude range of UAV operation can be extended significantly by tilting the UAV antenna towards the further BSs in order to block the LoS interfering BSs. However, this method is not beneficial for very dense networks.
- The UAV connectivity in heterogeneous and ultra-dense networks is addressed as well. Appropriate tier selection in serving UAVs is of utmost importance to mitigate the increased amount of aggregate interference in the sky. Furthermore, the optimal density of BSs from both user and network perspectives is analyzed when the network accommodates aerial users.

In the sequel, Section 4.2 introduces the system model, and then Section 4.3 presents the mathematical derivations for key performance metrics. Numerical and simulation results are presented in Section 4.4, which also includes a discussions on the impact of various system parameters on UAV connectivity. Finally, our main conclusions are summarized in Section 4.5.

4.2 Modelling Preliminaries

This section introduces the main modeling assumptions. First, a general overview on stochastic geometry is presented in Section 4.2.1. Then in Section 4.2.2 considerations regarding the network architecture are presented. Next, the channel model is elaborated in Section 4.2.3, and Section 4.2.4 describes the modeling of urban blockages and the resulting LoS probability. Finally, the user association method and the definition of instantaneous signal-to-interference-plus-noise ratio (SINR) is explained in Section 4.2.5.

4.2.1 Stochastic Geometry

A wireless network is a collection of nodes distributed over space, which successively act as transmitters or receivers. Due to the broadcast nature of the wireless media, the signal sent to an intended receiver might simultaneously generate interference to neighbouring nodes whose locations might be deterministic or unknown. Additionally, the distance between the transmitter and receiver might also be unknown when establishing a communication link.

For these reasons, it is rational to model the geometry of wireless networks using stochastic elements, which in turn provide statistical estimates of diverse quantities of interest including the SINR and other key performance metrics or indicators (KPIs). In particular, *stochastic geometry* (SG) [29] is an effective formalism to capture crucial collective properties of wireless networks that are related to the spatial distribution of their nodes. Moreover, adequate modeling choices allow stochastic geometry to provide tractable, compact, and even closed-form expressions of various KPIs. These expressions can be used to deepen our understanding of the behavior of wireless networks, in a more comprehensive manner than mere brute-force numerical simulations.

The use of SG in modeling wireless networks has a long history, which can be traced back to the late 1970s [30, 31]. An SG approach is useful not only for modeling the random positions of the wireless nodes, but also for considering scenarios where the geometry is deterministic but irregularly shaped – e.g., a cellular network with macro/micro BS locations. Using SG, the locations of the wireless nodes are modeled following the properties of a given stochastic process. By exploiting these properties, one can usually obtain statistical properties for various KPIs, which capture crucial dependencies between the network's performance and various system parameters. Although basic SG approaches focus on the nodes' locations, more advanced applications can include other sources of uncertainties in wireless networks, including large-scale shadow fading, small-scale fading, and also the effects of power control [32].

In this chapter we use SG to model cellular networks and analyze the effect of various system parameters on the user- and network-level performance. For modeling the locations of the BSs, our basic choice is to use a Poisson point process (PPP), in which the nodes are randomly, independently, and uniformly scattered in the spatial domain. This implies that the number of nodes in any bounded region is a Poisson random variable, which motivates the name of these processes.

It is important to note that, in contrast to ground-to-ground network modeling that relies on 2D PPP models, the positions of aerial users should be modeled via 3D PPP modeling. This difference can induce important differences in the results, being related to some of the most interesting features that distinguish aerial networks from their traditional ground-based counterparts.

4.2.2 Network Architecture

Let us consider a ground cellular network that includes BSs of height h_b serving UE in downlink. The locations of BSs follow a homogeneous Poisson point process (HPPP) with a fixed density of λ_b base stations per square kilometer (BSs/km^2). The users are located at altitude h_u, which can be either a ground UE (G-UE) at height 1.5 m or a UAV UE (U-UE) at a higher height. We denote by r the ground distance between a BS and a typical user whose 2D location on the ground is denoted by O (see Figure 4.1). Additionally, the 3D distance between them is denoted by $d = \sqrt{r^2 + \Delta_h^2}$, where $\Delta_h \triangleq h_u - h_b$.

We focus on the case that the density of users is significantly larger than the density of BSs λ_b, so that every BS can be assumed to be active over a given time/frequency resource

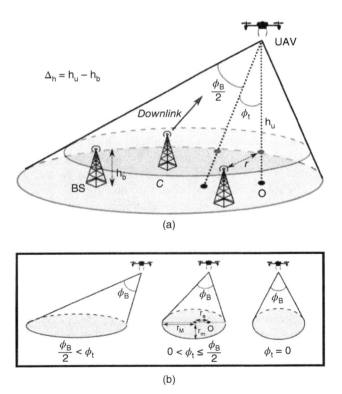

(a)

(b)

Figure 4.1 Downlink of a cellular-connected UAV. The UAV is equipped with a tilted directional antenna, with a tilt angle ϕ_t and antenna beamwidth ϕ_B. The area detectable by the UAV forms an ellipse denoted by C.

block. A maximum of one UE is assumed to be scheduled on each time/frequency resource block within each cell, so that one can neglect intra-cell interference.

In agreement with the 3GPP regulation, the antennas of the BSs are assumed to be directional on the vertical axis, and tilted down towards the ground [7]. As a consequence, G-UE is usually served via the main lobe, while U-UE tends to receive signals from the side lobes. The gain of the BS antenna is denoted by G_b where g_M and g_m represent the main- and side-lobe gains, respectively.

The UAVs are assumed to employ tilted directional antennas, while the ground users are equipped with omnidirectional antennas. The UAVs antenna beamwidths and tilt angles are denoted by ϕ_B and ϕ_t, respectively, and the main-lobe and side-lobe gains are approximated by $G_u = 29\,000/\phi_B^2$ and 0, respectively [33]. Due to the tilted antennas, a UAV can only detect BSs that are within its main lobe, which forms an elliptical contour on the ground (see Figure 4.1). The elliptical region is denoted by C and the location of BSs within this region forms a HPPP represented by Φ_C of the same density as λ_b [29].

4.2.3 Channel Model

The wireless link between two given nodes is assumed to undergo large-scale path loss and small-scale fading. For each, we employ different parameters for LoS and NLoS components. The path loss endured in a link between two nodes separated by a ground distance r and height difference Δ_h can be stated as

$$\zeta_v(r) = A_v d^{-\alpha_v} = A_v(r^2 + \Delta_h^2)^{-\alpha_v/2}, \quad v \in \{L, N\}, \tag{4.1}$$

where $v \in \{L, N\}$ represents the type of link, which is either LoS ($v = L$) or NLoS ($v = N$), α_v is the path-loss exponent, and A_v is the reference path loss at the distance $d = 1$.

For modeling the effects of small-scale fading, we use the well-known Nakagami-m model, which can be used to represent various types of fading environments, including Rayleigh or Ricean channels while providing a tractable mathematical structure [34]. Under Nakagami-m fading the fading power Ω_v follows a Gamma distribution,[1] whose cumulative distribution function (CDF) is

$$F_{\Omega_v}(\omega) \triangleq \mathbb{P}[\Omega_v < \omega] = 1 - \sum_{k=0}^{m_v-1} \frac{(m_v\omega)^k}{k!} e^{-m_v\omega}, \quad v \in \{L, N\}. \tag{4.2}$$

Above, m_v is the fading parameter, which is considered a positive integer for the sake of tractability. Note that a larger m_v corresponds to a lighter fading, with the limit being equivalent to an additive white Gaussian noise (AWGN) channel. Accordingly, $m_L > m_N$ holds.

Finally, the received power at a UE from a BS with a transmit power of P_{tx} is calculated as

$$P_{rx}(r) = P_{tx}\, G_{tot}\, \zeta_v(r)\, \Omega_v, \tag{4.3}$$

where $G_{tot} = G_b\, G_u$ indicates the total gain of transmitter and receiver antennas and v indicates the type of link being LoS or NLoS.

---

1 Without loss of generality, we follow the convention $\mathbb{E}\{\Omega_v\} = 1$.

4.2.4 Blockage Modeling and LoS Probability

Following [35], the LoS probability is modeled as

$$
\mathcal{P}_{\mathrm{L}}(r) = \prod_{n=0}^{M} \left[1 - \exp\left(-\frac{[\mathrm{h_b} - (n + \frac{1}{2})(\mathrm{h_b} - \mathrm{h_u})/(M+1)]^2}{2c^2} \right) \right],
\tag{4.4}
$$

where (a, b, c) are the parameters that determine the type of urban areas (e.g., suburban, urban, dense urban, or high-rise urban), and $M = \lfloor (r\sqrt{ab}/1000) - 1 \rfloor$. Conversely, the probability of NLoS is $\mathcal{P}_{\mathrm{N}}(r) = 1 - \mathcal{P}_{\mathrm{L}}(r)$.

In our modeling we assume that the LoS probabilities of different links are statistically independent of each other, and hence Φ_C can be decomposed into two independent inhomogeneous PPP processes:

(1) Φ_C^{L} of LoS BSs with respect to the typical UE, and
(2) Φ_C^{N} of NLoS BSs.

The densities of Φ_C^{L} and Φ_C^{N} might not be constant, and are denoted by $\lambda_{\mathrm{L}}(r) = \lambda \mathcal{P}_{\mathrm{L}}(r)$ and $\lambda_{\mathrm{N}}(r) = \lambda \mathcal{P}_{\mathrm{N}}(r)$, respectively. In this manner, $\lambda = \lambda_{\mathrm{L}} + \lambda_{\mathrm{N}}$ and $\Phi_C = \Phi_C^{\mathrm{L}} \cup \Phi_C^{\mathrm{N}}$.

4.2.5 User Association Strategy and Link SINR

For association strategy, we consider the case where a UE is served by the best BS that offers the highest signal strength, and hence SINR, at the UE receiver. Accordingly, due to the effect of blockages, a UE might associate with a LoS BS that is not the closest but still provides a higher SINR than another NLoS BS that could be closer. By denoting the serving BS distance as R_{S}, the aggregate received interference from the other co-channel cells can be written as

$$
I = \sum_{r \in \Phi_C^{\mathrm{L}} \backslash R_{\mathrm{S}}} P_{\mathrm{tx}} G_{\mathrm{tot}} \zeta_{\mathrm{L}}(r) \Omega_{\mathrm{L}} + \sum_{r \in \Phi_C^{\mathrm{N}} \backslash R_{\mathrm{S}}} P_{\mathrm{tx}} G_{\mathrm{tot}} \zeta_{\mathrm{N}}(r) \Omega_{\mathrm{N}}.
\tag{4.5}
$$

Above, the first summation term corresponds to LoS interfering BSs, and the second one is related to NLoS interfering BSs. By denoting the received useful signal power as $P_{\mathrm{rx}}(R_{\mathrm{S}}) = P_{\mathrm{tx}} G_{\mathrm{tot}} \zeta_v(R_{\mathrm{S}}) \Omega_v$, the instantaneous SINR at the UE can therefore be expressed as

$$
\mathrm{SINR} = \frac{P_{\mathrm{rx}}(R_{\mathrm{S}})}{I + N_0},
\tag{4.6}
$$

where N_0 represents the noise power at the UE receiver.

A list of variables and corresponding definitions is presented in Table 4.1.

4.3 Performance Analysis

In this section, the key performance metrics that correspond to the user and network perspectives are evaluated. The analysis provides a number of exact formal expressions, which – to facilitate numerical evaluations – are later simplified into several tight approximations. In the sequel, Section 4.3.1 presents the exact expressions for the link coverage

Table 4.1 Variables and definitions.

Variable	Definition
λ_b	BS density
h_u, h_b	User and BS height, respectively
g_M, g_m	Main-lobe and side-lobe gain of BS antenna
G_b, G_u	BS and UAV antenna gain, respectively
$G_{tot} = G_b G_u$	Total effect of transmitter and receiver antenna gains
ϕ_B, ϕ_t	UAV antenna opening and tilt angle, respectively
ζ_v, A_v	Total and reference path loss, $v \in \{L, N\}$
r, d	2D and 3D distances, respectively
α_v	Path-loss exponent, $v \in \{L, N\}$
Ω_v	Small-scale fading power
m_v	Nakagami-m fading parameter
P_{tx}	BS transmit power
$\mathcal{P}_L, \mathcal{P}_N$	LoS and NLoS probability
(a, b, c)	Urban type parameters
λ_L, λ_N	Densities of LoS and NLoS active BSs
I	Aggregate interference
N_0	Noise power
R_S	Ground distance between a given UE and its serving BS
$\mathcal{P}_{cov}$	Coverage probability
$\mathcal{R}$	Achievable throughput
$\mathcal{A}$	Area spectral efficiency
T	SINR threshold
ρ	Ratio of aerial UEs to total aerial and ground UEs

probability (i.e., the complementary cumulative distribution function (CCDF) of the SINR distribution), and Section 4.3.2 includes the approximate expressions. We further study the user- and network-level spectral efficiency in Section 4.3.3.

4.3.1 Exact Coverage Probability

The coverage probability, which is denoted by $\mathcal{P}_{cov}$, is the probability that an instantaneous SINR is greater than a given threshold T that depends on the system requirement, i.e.,

$$\mathcal{P}_{cov} \triangleq \mathbb{P}[\text{SINR} > T]. \tag{4.7}$$

Note that coverage probability is a function of UE altitude as well and hence one can write $\mathcal{P}_{cov} = \mathcal{P}_{cov}(h_u, T)$. This metric represents the reliability of the communication link and particularly can be used to evaluate the suitability of such a network for UAV command

and control (C&C), which includes important information such as real-time control for piloting, telemetry, identity, and so on. Safe UAV deployments and traffic management, and beyond-visual LoS UAV operations, will be enabled by establishing a reliable C&C link with the ground stations. It is worth noting that the SINR threshold T can relate to the target rate R_t and user bandwidth BW by $T = 2^{R_t/\mathrm{BW}} - 1$.

In the following theorem, we analytically obtain the serving BS distance distribution R_S, characterize the aggregate interference through its Laplacian, and derive the link coverage probability.

Theorem 4.1 *The coverage probability of a typical cellular-connected UAV in downlink can be expressed as*

$$P_{\mathrm{cov}} = 2 \sum_{v \in \{L,N\}} \int_{r_0}^{r_e + r_M} r_S \mathcal{P}_{\mathrm{cov}|R_S}^v (r_S) f_{R_S}^v (r_S) [\pi + \varphi_1(r_S) - \varphi_2(r_S)] \, dr_S, \tag{4.8}$$

where

$$r_M = \frac{\Delta_h \sin(\phi_B)}{2[\cos^2(\phi_t) - \sin^2(\phi_B/2)]}, \qquad r_m = \frac{\Delta_h \sin(\phi_B/2)}{\sqrt{\cos^2(\phi_t) - \sin^2(\phi_B/2)}}, \tag{4.9}$$

$$r_e = \Delta_h \tan(\phi_t - \phi_B/2) + r_M, \qquad r_0 = \max(0, r_e - r_M). \tag{4.10}$$

Here,

- *if* $r < (r_m/r_M)\sqrt{r_M^2 - r_e^2}$ *and* $\phi_t < \phi_B/2$ *then*

$$\varphi_1(r) = \varphi_2(r) = \frac{\pi}{2};$$

- *if* $(r_m/r_M)\sqrt{r_M^2 - r_e^2} \leq r \leq r_M - r_e$ *and* $\phi_t < \phi_B/2$ *then*

$$\varphi_1(r) = \cos^{-1}\left[\frac{r_e r_m^2 - \sqrt{r_e^2 r_m^4 - (r_m^2 - r_M^2)(r_e^2 r_m^2 + r^2 r_M^2 - r_m^2 r_M^2)}}{r(r_m^2 - r_M^2)}\right],$$

$$\varphi_2(r) = \cos^{-1}\left[\frac{r_e r_m^2 + \sqrt{r_e^2 r_m^4 - (r_m^2 - r_M^2)(r_e^2 r_m^2 + r^2 r_M^2 - r_m^2 r_M^2)}}{r(r_m^2 - r_M^2)}\right];$$

- *and if* $r > |r_M - r_e|$ *then*

$$\varphi_1(r) = \cos^{-1}\left[\frac{r_e r_m^2 - \sqrt{r_e^2 r_m^4 - (r_m^2 - r_M^2)(r_e^2 r_m^2 + r^2 r_M^2 - r_m^2 r_M^2)}}{r(r_m^2 - r_M^2)}\right],$$

$$\varphi_2(r) = \pi.$$

Above, $f_{R_S}^v(r_S)$ *indicates the probability density function (PDF) of* R_S, *i.e., the serving BS distance, which is obtained as*

$$f_{R_S}^v(r_S) = \lambda_b \mathcal{P}_v(r_S) \, e^{-2\lambda_b[\mathcal{I}_{\mathrm{IL}}^v + \mathcal{I}_{\mathrm{IN}}^v]}, \qquad v \in \{L,N\},$$

with

$$\mathcal{I}_{1\xi}^{\upsilon} \triangleq \int_{r_0}^{r_{\xi}^{\upsilon}} r P_{\xi}(r)[\pi + \varphi_1(r) - \varphi_2(r)] \, dr, \quad \xi \in \{L, N\},$$

and r_{ξ}^{υ} is as follows:

$$r_L^L = r_S, \qquad r_N^L = \sqrt{\max[r_0^2, (A_N/A_L)^{2/\alpha_N}(r_S^2 + \Delta_h^2)^{\alpha_L/\alpha_N} - \Delta_h^2]},$$

$$r_L^N = \min[r_e + r_M, \sqrt{(A_L/A_N)^{2/\alpha_L}(r_S^2 + \Delta_h^2)^{\alpha_N/\alpha_L} - \Delta_h^2}], \qquad r_N^N = r_S. \tag{4.11}$$

Moreover, $\mathcal{P}_{\text{cov}|R_S}^{\upsilon}$, which is the conditional coverage probability given the serving BS ground distance R_S and its condition υ, is expressed as

$$\mathcal{P}_{\text{cov}|R_S}^{\upsilon} = \sum_{k=0}^{m_{\upsilon}-1} (-1)^k q_k \frac{d^k}{dy_{\upsilon}^k} \mathcal{L}_{I|R_S}^{\upsilon}(y_{\upsilon}), \quad \upsilon \in \{L, N\} \tag{4.12}$$

where

$$q_k \triangleq \frac{e^{-N_0 y_{\upsilon}}}{k!} \sum_{j=k}^{m_{\upsilon}-1} \frac{N_0^{j-k} y_{\upsilon}^j}{(j-k)!}, \tag{4.13}$$

$$y_{\upsilon} \triangleq \frac{m_{\upsilon} T}{P_{\text{tx}} G_{\text{tot}} \zeta_{\upsilon}(r_S)}. \tag{4.14}$$

Furthermore, $\mathcal{L}_{I|R_S}^{\upsilon}(\cdot)$ represents the Laplace transform of the conditional aggregate interference $I|R_S$ for the serving BS of condition υ. The exact expression of $\mathcal{L}_{I|R_S}^{\upsilon}(\cdot)$ is

$$\mathcal{L}_{I|R_S}^{\upsilon}(y_{\upsilon}) = e^{-2\lambda[\mathcal{I}_{2L}^{\upsilon} + \mathcal{I}_{2N}^{\upsilon}]}, \tag{4.15}$$

where

$$\mathcal{I}_{2\xi}^{\upsilon} \triangleq \int_{r_{\xi}^{\upsilon}}^{r_e + r_M} r P_{\xi}(r)[1 - \Upsilon_{\xi}(r, y_{\upsilon})][\pi + \varphi_1(r) - \varphi_2(r)] \, dr$$

and

$$\Upsilon_{\xi}(r, y_{\upsilon}) \triangleq \left(\frac{m_{\xi}}{m_{\xi} + y_{\upsilon} P_{\text{tx}} G_{\text{tot}} \zeta_{\xi}(r)}\right)^{m_{\xi}}. \tag{4.16}$$

Proof: See Appendix 4.7. ∎

Theorem 4.1 expresses the coverage probability of a UAV equipped with a tilted directional antenna. In order to obtain the coverage probability of a UE that uses an omnidirectional antenna, one needs to replace $r_e = r_0 = 0$, $r_M = \infty$, and $\varphi_1(r) = \varphi_2(r)$.

4.3.2 Approximations for UAV Coverage Probability

In this section, we propose two approximations that significantly reduce the complexity in numerical computation of the UAVs coverage performance. By studying these approximations, one can identify some major factors that determine the UAV communications performance.

4.3.2.1 Discarding NLoS and Noise Effects

The number of detected and LoS BSs increases with the altitude of the UAVs. In fact, UAVs flying higher than the height of the BSs logically find several BSs in LoS condition. The locations of such BSs are likely to be closer to the UAV as the LoS probability increases when the ground distance decreases. Accordingly, LoS BSs can introduce much higher signal power at a UAV receiver compared to NLoS BSs due to more favorable propagation conditions. In addition, LoS interfering BSs impose a high amount of aggregate interference typically much higher than noise power. These facts motivate us to propose the following approximation for UEs in the sky, by eliminating NLoS BSs and noise impact from the mathematical derivations.

Proposition 4.1 The UAV UE coverage performance $\mathcal{P}_{\text{cov}}$ can be approximated and simplified by discarding NLoS BSs and noise effects as

$$\mathcal{P}_{\text{cov}} \approx 2 \int_0^{r_{\text{e}}+r_{\text{M}}} \mathcal{P}_{\text{cov}|R_{\text{S}}}^{\text{L}}(r_{\text{S}}) f_{R_{\text{S}}}^{\text{L}}(r_{\text{S}})[\pi + \varphi_1(r_{\text{S}}) - \varphi_2(r_{\text{S}})] r_{\text{S}} \, dr_{\text{S}},$$

where

$$f_{R_{\text{S}}}^{\text{L}}(r_{\text{S}}) \approx \lambda_{\text{b}} \mathcal{P}_{\text{L}}(r_{\text{S}}) \, e^{-2\lambda_{\text{b}} I_{1\text{L}}^{\text{L}}}, \tag{4.17}$$

$$\mathcal{P}_{\text{cov}|R_{\text{S}}}^{\text{L}} \approx \sum_{k=0}^{m_{\text{L}}-1} \frac{(-y_{\text{L}})^k}{k!} \frac{d^k}{dy_{\text{L}}^k} \, \mathcal{L}_{I|R_{\text{S}}}^{\text{L}}(y_{\text{L}}), \tag{4.18}$$

and

$$\mathcal{L}_{I|R_{\text{S}}}^{\text{L}}(y_{\text{L}}) \approx e^{-2\lambda I_{2\text{L}}^{\text{L}}}. \tag{4.19}$$

Proof: The desired result is obtained by replacing $\mathcal{P}_{\text{N}}$ with zero and $N_0 \approx 0$ in Theorem 4. ∎

Please note that several terms are eliminated in Proposition 4.1 in order to obtain the coverage probability. These terms include $I_{1\xi}^{\text{N}}$, $I_{1\text{N}}^{\text{L}}$, and $I_{2\text{N}}^{\text{L}}$. The accuracy of such an approximation is examined in Section 4.4.1.

Proposition 4.1 shows that, on the one hand, increasing the UAV height h_{u} from zero results in more LoS interfering BSs, and hence larger value of $I_{1\text{L}}^{\text{L}}$. This, in turn, degrades the coverage performance since $\mathcal{L}_{I|R_{\text{S}}}^{\text{L}}(y_{\text{L}})$ and $f_{R_{\text{S}}}^{\text{L}}$ in Eqs. (4.23) and (4.17), respectively, decrease. On the other hand, at higher UAV altitude, the UAV has a higher LoS probability with the serving BS, which leads to a better overall performance due to larger value of $f_{R_{\text{S}}}^{\text{L}}(r_{\text{S}})$ in Eq. (4.17). These two opposite effects might finally be balanced at an optimum altitude of UAV at which the performance is maximized. A similar trade-off can be found when looking into the other system parameters such as BS density.

4.3.2.2 Moment Matching

In order to further simplify the expressions of the UE performance, now we approximate the aggregate interference using a Gamma distribution [36], which enables us to express the Laplacian of the interference in closed form. For this, we employ a method of moments matching in order to estimate the interference distribution. In particular, the first and second moments of the interference are calculated and then used to determine the Gamma distribution. The following lemma yields the mean and variance of the aggregated interference.

Lemma 4.1 *The mean and variance of the aggregated interference can be calculated as*

$$\mu_{I|R_S} = 2\lambda P_{tx} G_{tot} \int_{r_S}^{r_e+r_M} r \mathcal{P}_L(r) \zeta_L(r) [\pi + \varphi_1(r) - \varphi_2(r)] \, dr \tag{4.20}$$

and

$$\sigma_{I|R_S}^2 = 2\lambda (P_{tx} G_{tot})^2 \left(\frac{m_L + 1}{m_L} \right) \int_{r_S}^{r_e+r_M} r \mathcal{P}_L(r) \zeta_L^2(r) [\pi + \varphi_1(r) - \varphi_2(r)] \, dr. \tag{4.21}$$

Proof: See Appendix 4.8. ∎

For the case that the UAV antenna is pointing directly down, we can further provide simpler equations by deriving the integrals as follows.

Corollary 4.1 *If $\phi_t = 0$, the first and second moments of the aggregate interference are calculated as*

$$\mu_{I|R_S} = \begin{cases} \dfrac{\pi \lambda P_{tx} G_{tot} A_L}{0.5\alpha_L - 1} \displaystyle\sum_{k=i}^{j} p_k [(r_k^2 + \Delta_h^2)^{1-0.5\alpha_L} - (r_{k+1}^2 + \Delta_h^2)^{1-0.5\alpha_L}], & \text{if } \alpha_L > 2, \\[2ex] \pi \lambda P_{tx} G_{tot} A_L \sum_{k=i}^{j} p_k \ln \left(\dfrac{r_{k+1}^2 + \Delta_h^2}{r_k^2 + \Delta_h^2} \right), & \text{if } \alpha_L = 2, \end{cases}$$

and

$$\sigma_{I|R_S}^2 = \frac{\lambda (P_{tx} G_{tot} A_L)^2}{\alpha_L - 1} \left(\frac{m_L + 1}{m_L} \right) \sum_{k=i}^{j} p_k [(r_k^2 + \Delta_h^2)^{1-\alpha_L} - (r_{k+1}^2 + \Delta_h^2)^{1-\alpha_L}],$$

in which $r_k = 1000(k+1)/\sqrt{ab}$ except that $r_i = r_S$ and $r_{j+1} = r_M$, $i = \lfloor (r_S \sqrt{ab}/1000) - 1 \rfloor$, $j = \lfloor (r_M \sqrt{ab}/1000) - 1 \rfloor$ and p_k is a fixed value that results from Eq. (4.4) by substituting $M = k$.

Proof: See Appendix 4.9. ∎

By looking at the mean interference, Corollary 4.1 shows that increasing the BS height results in

(1) an increase in p_k (since the LoS probability is an increasing function of h_b), and
(2) an increase in the second term of the summation (since this term is a decreasing function of $\Delta_h = h_u - h_b$, and hence an increasing function of h_b).

Accordingly, in total, the mean interference increases with the higher h_b. However, the LoS probability and the received useful signal power are higher when h_b is larger. All these suggest that an optimum value of BS height might exist, at which the increased interference and signal levels are optimally balanced, resulting in a maximum UAV performance.

Using the scale and shape parameters of a Gamma distribution, denoted as β_2 and β_1, respectively, we approximately characterize the distribution of the aggregate interference. We indeed apply the following relationship between these parameters and the interference's mean and variance: [36]

$$\beta_1 = \frac{\sigma^2_{I|R_s}}{\mu_{I|R_s}}, \qquad \beta_2 = \frac{\mu^2_{I|R_s}}{\sigma^2_{I|R_s}}. \tag{4.22}$$

Furthermore, the Gamma Laplacian transform is [36]

$$\mathcal{L}^{\text{L}}_{I|R_s}(y_{\text{L}}) = (1 + \beta_1 y_{\text{L}})^{-\beta_2}. \tag{4.23}$$

Using such Gamma approximation of the aggregate interference, we present the next proposition.

Proposition 4.2 Approximating the statistics of I with a Gamma distribution of scale and shape parameters β_2 and β_1, respectively, results in the conditional coverage performance of a UAV UE expressed as

$$\mathcal{P}^{\text{L}}_{\text{cov}|R_s} \approx \frac{1}{\Gamma(\beta_2)} \sum_{k=0}^{m_{\text{L}}-1} \frac{(\beta_1 y_{\text{L}})^k}{k!} \, \Gamma(\beta_2 + k) \, (1 + \beta_1 y_{\text{L}})^{-\beta_2-k}, \tag{4.24}$$

where $\Gamma(\cdot)$ is the complete Gamma function.[2]

Proof: The result is obtained by substituting Eq. (4.23) in Eq. (4.18). ∎

4.3.3 Achievable Throughput and Area Spectral Efficiency Analysis

The achievable throughput, denoted by $\mathcal{R}$, is the highest bit rate that a UE can receive from the network. This metric is calculated (in bits per second per hertz) as

$$\mathcal{R} \triangleq \mathbb{E}[\log_2(1 + \text{SINR})] \quad (\text{b/s/Hz}). \tag{4.25}$$

To compute $\mathcal{R}$, one can write

$$\mathcal{R} = \frac{1}{\ln 2} \int_0^\infty \frac{\mathcal{P}_{\text{cov}}(h_u, t)}{1+t} \, dt$$

$$\approx \frac{1}{\ln 2} \sum_{n=1}^{K} \frac{\mathcal{P}_{\text{cov}}(h_u, t_n)}{1 + t_n} \frac{\pi^2 \sin\left(\dfrac{2n-1}{2K}\pi\right)}{4K\cos^2\left[\dfrac{\pi}{4}\cos\left(\dfrac{2n-1}{2K}\pi\right) + \dfrac{\pi}{4}\right]}.$$

We used the last approximation in order to facilitate the numerical calculation of the main equation by eliminating one integral. This approximation follows the Gauss–Chebyshev quadrature (GCQ) rule, with a free parameter (K) that is chosen large enough to enable high accuracy [37]. Moreover, t_n is given by

$$t_n = \tan\left[\frac{\pi}{4}\cos\left(\frac{2n-1}{2K}\pi\right) + \frac{\pi}{4}\right]. \tag{4.26}$$

---

2 Note that using the property of the Gamma function, we have $\Gamma(\beta_2 + k)/\Gamma(\beta_2) = \prod_{i=0}^{k-1}(\beta_2 + i)$.

To evaluate the network-level performance in the presence of the UAVs, the *area spectral efficiency* (ASE) denoted by $\mathcal{A}$, is an adequate metric that defines the achievable network rate per square kilometer (in bits per second per hertz per square kilometer). Let us denote as ρ the ratio of aerial UEs over the total aerial and ground UEs. Therefore, one can write

$$\mathcal{A} \triangleq \lambda[(1 - \rho)\, \mathcal{R}(1.5, \lambda) + \rho\, \mathcal{R}(h_u, \lambda)] \quad (\text{b/s/Hz/km}^2), \tag{4.27}$$

where $\mathcal{R}(1.5, \lambda)$ and $\mathcal{R}(h_u, \lambda)$ are, respectively, related to ground users at altitude of 1.5 m and aerial users at altitude of h_u.

ASE allows us to study the impact of adding aerial users to the network when the spectrum is shared with ground users. We also aim to investigate how the network scales with the density of BSs in the presence of aerial users. This metric can be derived directly by replacing $\mathcal{R}(h_u, \lambda)$ from Eq. (4.26) into Eq. (4.27).

4.4 System Design: Study Cases and Discussion

In this section, by using the theoretical tools presented in Section 4.3, qualitative and quantitative understandings of important design parameters of the cellular-connected UAVs are provided. The default values of system parameters for numerical evaluations and simulations are listed in Table 4.2.

4.4.1 Analysis of Accuracy

Figure 4.2 reveals the accuracy of analysis and proposed approximations. In this figure, the CCDF of SINR is shown for both ground and aerial users equipped with omnidirectional antennas. The simulations are the result of 10^5 network realizations. Note that the approximations for ground users are not applicable, as their communication links are not

Table 4.2 Default values for simulation and numerical evaluation.

Parameter	Value
$(\alpha_L, \alpha_N, m_L, m_N)$	(2.09, 3.75, 3, 1)
(A_L, A_N)	(−41.1, −32.9) dB
P_{tx}	46 dBm
(a, b, c)	(0.3, 500, 15) for urban
λ_b	10 BSs/km²
(g_M, g_m)	(10, 0.5)
(h_u, h_b)	(100, 25) m
BW	200 kHz
R_t	100 kbps
T	$2^{R_t/BW} - 1 = -3.8$ dB

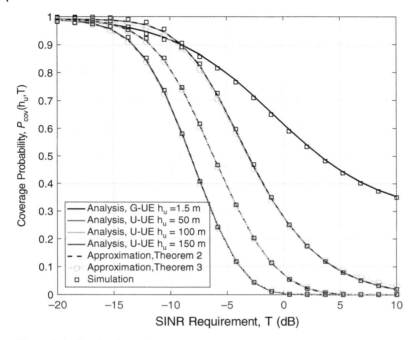

Figure 4.2 Results from simulation and analytical expressions are precisely matched. Moreover, the approximations are in a good conformity with the exact numerical results. This highlights the minority of NLoS influence for UAV UE.

LoS dominant. In other words, ground user links to the BSs are less likely to be in LoS conditions and hence the NLoS links cannot be ignored in performance evaluations.

In general, omnidirectional UAV communication performance is low. The performance penalty at increasing altitudes is bigger when the SINR threshold is larger. As can be seen, the aerial users' SINR distribution is more concentrated. This is due to the fact that multi-path scatters are less involved in LoS UAV communication links. However, the multipath effect for ground users is significant and hence cannot be avoided.

4.4.2 Design Parameters

This subsection provides an in-depth analysis of the impact of various design parameters on the quality of the UAV communication link with BSs. We study how each parameter affects the user-level performances, i.e., coverage probability and achievable throughput. In the sequel, we analyze the role of UAV altitude, antenna beamwidth, and tilt angle along with the network load effect. Finally, we discuss how the type of environment impacts the ground and UAV users' performance.

4.4.2.1 Impact of UAV Altitude

In general, there exists an optimum altitude at which the UAV performance is the highest, as is shown in Figure 4.3. Indeed, when the UAV altitude increases from zero, the links to the BSs transit from NLoS to LoS condition. This is beneficial up to a certain point, as

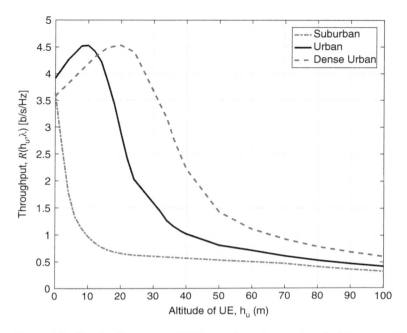

Figure 4.3 The feasible range of UAV operation when equipped with an omnidirectional antenna is limited due to the detrimental effect of LoS interfering BSs.

the serving BS finds itself in a LoS situation with the UE. However, as the altitude increases further, the interfering links also become more in LoS and hence the performance degrades. Such results show that the BSs antenna gain reduction is completely compensated by the more favorable LoS propagation condition for the sky.

Figure 4.3 also illustrates two important results: First, the UAV UE at the optimum altitude receives a higher quality of service compared to a ground UE, although the optimum altitude is in general very low. Second, the optimum altitude is higher in denser environments, which is due to having more interfering BSs in NLoS condition. Furthermore, as can be seen, in suburban areas the UAV has to fly as low as possible to reduce the detrimental impact of LoS interference.

4.4.2.2 Impact of UAV Antenna Beamwidth

The variation of UAV performance with its antenna beamwidth ϕ_B in Figure 4.4 shows that an appropriate choice of ϕ_B enhances a UAV's performance significantly, even higher than a ground user's performance. When increasing ϕ_B, the candidate serving BSs (to be selected) within the UAV's main lobe increase, which is beneficial for the UAV's performance. However, further increase in ϕ_B results in the inclusion of more interfering BSs, which is detrimental. These two opposite factors are in balance at an optimum antenna beamwidth.

We note that increasing ϕ_B reduces the antenna gain, as stated in our system modeling. However, the variation of UAV antenna gain has the same effect on useful signal and interference power. This basically means that, if a network is interference-limited, which

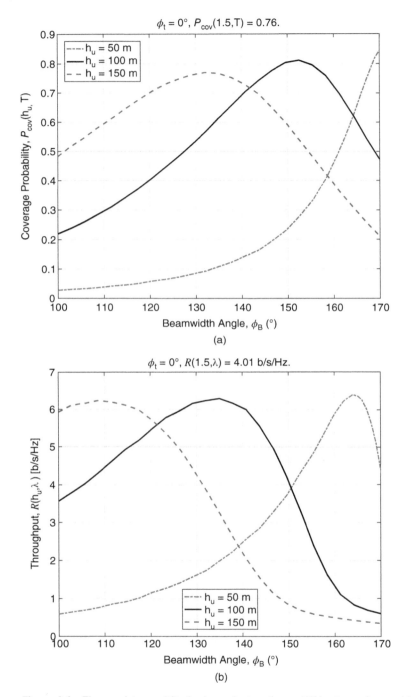

Figure 4.4 There exists an altitude-dependent optimum UAV antenna beamwidth for maximum performance.

is the case here as stated in Proposition 4.1, the variation of ϕ_B does not change the SINR distribution and hence the coverage probability.

Furthermore, we note that the optimal ϕ_B is different for coverage probability and achievable throughput. This is because throughput is a weighted average over all realizations of SINR distribution; however, coverage probability only considers a specific SINR threshold. Therefore, an adequate design should take into account such different optimal system factors and establish a proper coverage and throughput trade-off.

4.4.2.3 Impact of UAV Antenna Tilt

Figure 4.5 reveals the effect of UAV antenna tilting. In fact, increasing the tilt angle ϕ_t first increases the number of BSs within the main lobe, and second the BSs transit from the LoS to NLoS conditions. Each of these two can introduce beneficial or detrimental effects. The former can be beneficial as the number of candidate BSs goes up and can be detrimental as the number of interfering BSs also increases. The latter might be constructive since the interfering BSs become more NLoS; however, it might be destructive as the serving BS also goes to NLoS situation. All together, the numerical results show that, at a certain optimum angle, these effects are balanced and the UAV performance is maximized. The optimum tilt angle is nonzero for non-dense networks; however, the tilting approach is not helpful for very dense networks.

It is worth noting that tilting the antenna might introduce a BS misdetection problem, i.e., receiving very low reference signal. This is due to the fact that BS distances become longer and they are more likely to be in NLoS condition. However, we note that the coverage probability takes into account such an effect and accounts for both misdetection or outage issues caused by interference. In addition, when the tilting and beamwidth angles are small, there might be no BS within the UAV main lobe. This situation also corresponds to zero coverage in our framework.

4.4.2.4 Impact of Different Types of Environment

Table 4.3 reveals that the effect of UE altitude on the coverage performance is worse in less-obstructed environments. For example, in a suburban area, the link coverage goes down from 90% at ground to 4% at 150m, which is more severe than the corresponding fall from 76% to 10% in an urban environment. Moreover, a ground UE has the highest coverage performance in suburban areas, while a UAV UE is best served in urban environments. Accordingly, the network deployment strategies for ground and aerial users are to be different.

It is straightforward to see that, as the UAV link is interference-limited, the achievable throughput $\mathcal{R}$ grows linearly with the assigned bandwidth (BW). In contrast, the link coverage probability is affected by the BW in a non-trivial way. Table 4.3 illustrates this dependence through four sample altitudes in two different environments, focusing on a fixed target rate of $R_t = 100$kbps for UAV C&C [38]. The results show that UAVs with omnidirectional antenna endure considerably lower link coverage than a ground user particularly in less-obstructed areas. Doubling the bandwidth for a given target rate reduces the SINR threshold constraint and hence significantly improves the coverage probability; e.g., the coverage probability is improved almost four times for a UAV at 150 m in an urban setting. On the other hand, the same BW doubling enables a growth of only 8% for UE at ground

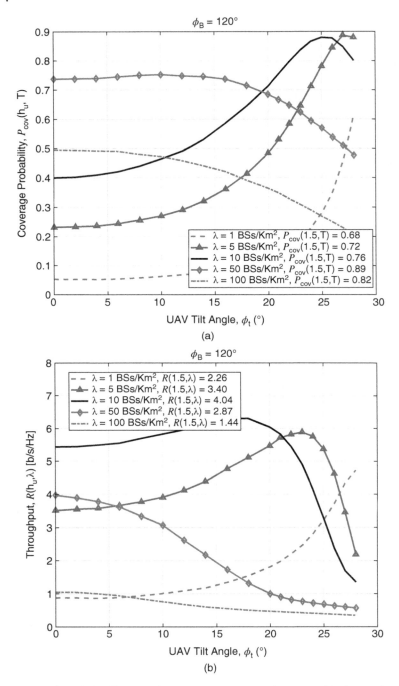

Figure 4.5 A significant performance enhancement is observed by leveraging UAV antenna tilt for sparse to medium-dense networks. However, tilting is not beneficial for a dense network due to the inclusion of significantly more interfering BSs.

Table 4.3 Link coverage of a cellular-connected UAV with an omnidirectional antenna in two types of urban areas. A target rate of $R_t = 100$kbps is assumed for coverage assessment.

UAV altitude,	Assigned BW	Coverage (%)	
h_u (m)	(kHz)	suburban	urban
1.5	200	90	76
50	200	34	54
100	200	20	30
150	200	4	10
1.5	400	97	85
50	400	60	82
100	400	48	60
150	400	28	39

level. Therefore, in general, our results suggest that an increase in bandwidth is more efficient for aerial than for ground users.

4.4.3 Heterogeneous Networks – Tier Selection

The use of different tiers in cellular architecture is a common approach to address several important problems. A homogeneous macro cellular network might not be able to provide a proper quality of service in hotspots or when the users are suffering from significant shadowing due to obstacles. Moreover, the considerable growth in data demand is another reason to consider a heterogeneous multi-tier cellular network. To do so, less-expensive micro BSs might be deployed to address such issues. As compared to macro BSs, micro BSs typically have lower height, smaller cell size, lower transmit power, and larger deployment densities. Furthermore, the operating frequency of different tiers can be orthogonal, as recommended by 3GPP [39]. Considering the different characteristics of the tiers and the fact that both tiers can be seen by a UAV, it is interesting to investigate the suitability of each tier in serving aerial users.

Indeed, it is not straightforward to conclude which tier is best to serve aerial users. The higher density of micro cells might impose higher interference; however, their lower BS height increases shadowing and limits interference propagation. Therefore, an investigation of such problems is required, which is provided through Figure 4.6. In the two panels of this figure, we have used the parameter values as specified in 3GPP recommendations [38]. These parameters include BS heights, densities, and transmit powers.

Figure 4.6 shows that, in general, aerial users are best served by micro cells at low altitudes. However, above a certain altitude, macro cells are the best candidate for UAV connectivity. Note that the switching point depends critically on the UAVs' antenna beamwidth and tilt angle. In effect, the larger ϕ_B results in lower switching altitude.

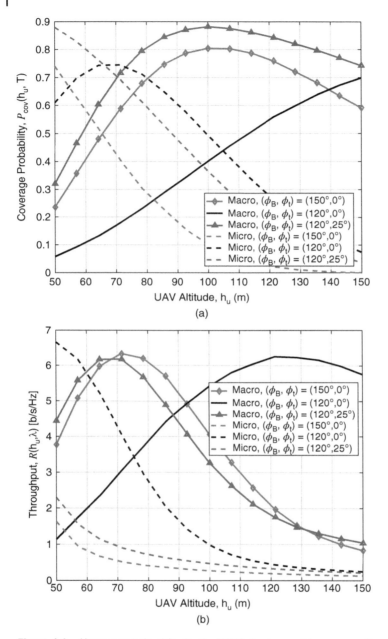

Figure 4.6 Above a certain altitude, the UAV is best served by macro cells.

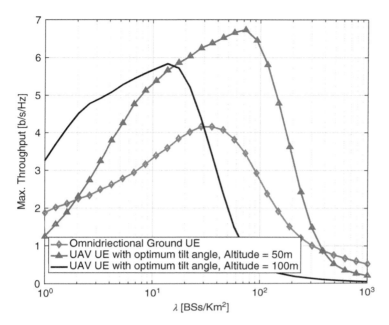

Figure 4.7 From the UAV perspective, network performance converges to zero as it becomes dense even though the optimal tilting angle is adopted. However, lowering the UAV altitude partially saves the UAV. Note that for these simulations ground UE is considered to be equipped with omnidirectional antenna.

4.4.4 Network Densification

In order to enhance the network capacity, network densification is another promising solution. In this manner, the number of cells grows to bring the BSs closer to users. However, due to the harmful effect of interfering BSs, this method is useful only up to a certain density, which should be determined. The optimal network density and the benefits of densification might be significantly influenced by the inclusion of aerial users. For this, in the following, we examine densification in the presence of aerial UEs. First we focus on user-level performance via Figure 4.7 and then we investigate network-level performance in Figure 4.8a.

Figure 4.7 shows the user achievable throughput versus BS density. The UAV performance at any given λ_b is obtained by employing the optimum tilt angle, i.e., $\phi_t^{opt}(\lambda)$. Indeed, as discussed before, there is an optimum tilt angle at which the UAV performance is maximized. As can be seen from the figure, aerial users are capable of receiving significantly higher rate, thanks to their flexibility in choosing their antenna parameters. Moreover, at a certain density, the users' performance is the highest, indicating the existence of an optimal network density. Also, when the network is dense, lowering the UAV altitude considerably improves its performance. This is due to the fact that lowering the UAV altitude blocks the interfering BSs.

The scaling property with respect to λ_b is studied, when the network is composed only of ground users, or only UAV users, or when aerial and ground UEs are present together. As shown in Figure 4.8a, when the network includes only aerial UEs, three scaling regimes can be realized. The first is where the increased signal power dominates the higher interference, named *signal-dominating regime*. In this region, the overall ASE increases by including

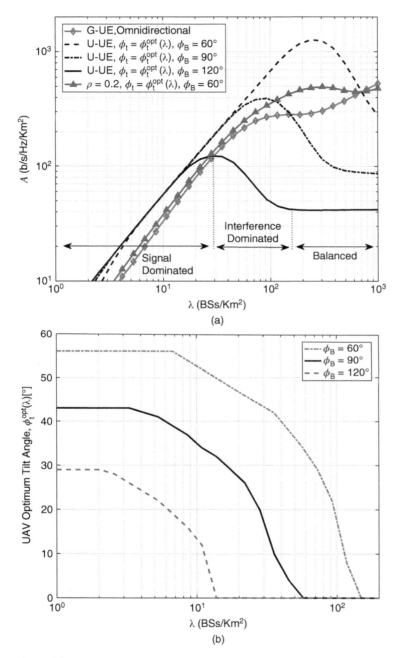

Figure 4.8 The inclusion of aerial users might enhance the area spectral efficiency up to a certain BSs density. Moreover, the optimum tilt angle is lower at higher network density.

the aerial users with appropriate ϕ_t. In other words, sharing resources with UAVs is beneficial as the network capacity increases. The second regime is where the higher interference dominates and therefore the overall performance degrades. This region is named *interference-dominating regime*. Such impact degrades the ASE of networks with UAV UEs. Finally, the *balanced regime* is where ASE is constant. In this region, the UAV UE achievable throughput is inversely proportional to λ_b. As can be seen, when the network is very dense, including more UAV users is detrimental from the network perspective. Finally, Figure 4.8b illustrates that the optimal tilt angle is lower for higher density, which is also significantly depends on the UAV antenna beamwidth.

4.5 Conclusion

In this chapter we studied the feasibility of integrating UAVs into existing and future 5G cellular networks. For this, a generic 3D framework based on stochastic geometry was developed that allowed us to build analytic expressions for various user and network KPIs. A number of design insights were found through the analysis, which might serve to guide a smooth integration of UAVs into 5G cellular networks.

The results suggest that we can be optimistic in that *current cellular networks might be capable of providing effective support to drone UEs*. In particular, our main findings can be summarized as follows:

- The BS antenna gain reduction is made up by the more favorable free-space propagation condition in the sky. Accordingly, the received signal strength is higher at altitudes compared to at ground level.
- The performance of a UAV user equipped with an omnidirectional antenna is limited due to the increased number of LoS interfering nodes detected from the sky. This in turn restricts the feasible range of the UAV's operation.
- There exists an optimum altitude at which a UAV's performance is higher compared to a ground user's, although this optimum altitude is typically very low and may not be feasible for the UAV's flight.
- The UAV's antenna configuration plays a significant role. The proper design of its beamwidth and tilt angle overcomes the adverse effect of LoS interference in the sky and hence allows UAVs to operate safely and reliably. Such a result, however, is not valid for very dense ground networks, and hence other solutions relying on ground networks are needed.
- Network operators can benefit by the inclusion of aerial users, as the area spectral efficiency can be enhanced.
- When micro BSs are available in addition to macro BSs, a proper tier selection for the UAVs significantly improves the network and UAV efficiency. We note that the UAVs are capable of detecting both tiers thanks to the high LoS probability.

In summary, due to the unique properties of air-to-ground wireless links, the most critical aspect for serving drones is to manage their extreme vulnerability to interference. The results suggest that the interference level might be successfully controlled by designing the network with appropriate BS height and down-tilt angle, drone antenna parameters, and altitude.

However, there are still many unsolved issues related to the integration of UAVs and 5G networks. In particular, their integration into future ultra-dense networks might still be challenging due to the high level of interference that characterize these systems. Although this could be mitigated by choosing low flying altitudes and optimized drone antenna beamwidth, good integration eventually might require novel interference compensation techniques. Other interesting lines of future research may include the following.

(1) *Handover of cellular-connected UAVs.* UAVs connected to ground cellular networks are normally capable of detecting several LoS BSs which are radiating their signals through their antennas' main lobe or side lobes. Accordingly, the pattern of the cells is different in the air. This fact will result in a different handover characteristic for aerial UEs. Such characteristics are highly dependent on the blockage distribution, height, and mobility pattern of the flying UAVs. In fact, the handover pattern can be changed where some of them might fail due to a low received signal power from side lobes. In order to establish reliable and safe cellular-connected UAVs, a qualitative and quantitative understanding of such network behavior is of utmost importance.

(2) *UAV mmWave communications.* mmWave as an important element in 5G and beyond can be used in UAV communications as well. Indeed, this technology even sounds more promising for UAVs due to their LoS communication. The framework proposed in this chapter allows one to analyze the performance of such a scenario and provide general guidelines by comparing mmWave and sub-6 GHz technologies. However, due to the movement of UAVs, in order to be able to use mmWave for UAV communication efficiently, signal blockages, fast beamforming training and tracking, and Doppler effects should be addressed properly.

(3) *Recognition of UAV users and scheduling schemes.* UAV users in cellular networks define a new service with different requirements than ground users. For instance, UAVs in downlink need highly reliable, low-data-rate, and low-latency communication links for safe command and control. To satisfy such demand and to accommodate UAVs reliably, aerial users should first be distinguished. This can be done by developing new signaling overhead. Alternative solutions can be developed for low-latency communication, which is an open problem. Moreover, new user scheduling should be adopted due to the different requirements for UAV links.

Appendix A
Proof of Theorem 1

Using Eq. (4.7) one sees that

$$P_{\text{cov}} = \sum_{v \in \{\text{L,N}\}} \int_C P^v_{\text{cov}|R_S}(r_S) f^v_{R_S}(r_S) r_S \, d\varphi \, dr_S, \tag{4.28}$$

where

$$P^L_{\text{cov}|R_S} = \mathbb{P}[\text{SINR} > T \mid R_S = r_S, \text{ LoS}], \tag{4.29}$$

$$P^N_{\text{cov}|R_S} = \mathbb{P}[\text{SINR} > T \mid R_S = r_S, \text{ NLoS}], \tag{4.30}$$

are the conditional coverage probabilities given the type of the serving BS (being LoS and NLoS, respectively) and also given the location of the serving BS in $\mathbb{R}^2$. Moreover, we note that $f_{R_s}^L(r_s)$ and $f_{R_s}^N(r_s)$ are the probability distributions of the serving BS distance and type at a specific location r_s.

The received signal (either LoS or NLoS) experiences interference from both the aggregate of LoS and NLoS interfering BSs, which is expressed in Eq. (4.5). Using the properties of PPP, the function $f_{R_s}^L(r_s)$ can be obtained as

$$f_{R_s}^L(r_s) = \lambda_b \mathcal{P}_L(r_s) P_{noL}^L(r_s) P_{noN}^L(r_s), \qquad (4.31)$$

where the unconditional PDF of having an LoS BS is $\lambda_b \mathcal{P}_L(r_s)$, $P_{noL}^L(r_s)$ denotes the probability that there is no LoS BS with stronger signal for the typical UE, and $P_{noN}^L(r_s)$ is the probability of no NLoS BS with stronger signal strength. Let us assume that the set of $\mathcal{A}_{noL}^L(r_s)$ includes all the ground distance at which a LoS BS could provide a higher signal power. Accordingly, the probability $P_{noL}^L(r_s)$ can be expressed as

$$P_{noL}^L(r_s) = \exp\left(-2\int_{\mathcal{A}_{noL}^L} \lambda_b \mathcal{P}_L(r) r \, d\varphi \, dr\right). \qquad (4.32)$$

Similarly, if we define the set $\mathcal{A}_{noN}^L(r_s)$ as the locations from which a stronger NLoS signal can be received, one can write

$$P_{noN}^L(r_s) = \exp\left(-2\int_{\mathcal{A}_{noN}^L} \lambda_b \mathcal{P}_N(r) r \, d\varphi \, dr\right). \qquad (4.33)$$

The sets $\mathcal{A}_{noL}^L$ and $\mathcal{A}_{noN}^L$ are geometry-dependent.

Let us denote by C the area within the main lobe of a UAV with the antenna parameters specified in the system model, i.e., beamwidth angle ϕ_B and tilt angle ϕ_t. This area is an elliptical section, which can be characterized by its semi-major and semi-minor axes denoted by r_M and r_m, respectively. Also the origin of this region is $(r_e, 0)$, as shown in Figure 4.9. Using

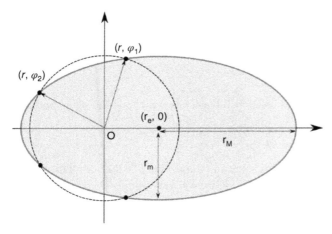

Figure 4.9 Illustration of the parameters φ_1 and φ_2 and the region C.

Table 4.4 Parameter values.

$$r_M = \frac{\Delta_h \sin(\phi_B)}{2[\cos^2(\phi_t) - \sin^2(\phi_B/2)]}$$

$$r_m = \frac{\Delta_h \sin(\phi_B/2)}{\sqrt{\cos^2(\phi_t) - \sin^2(\phi_B/2)}}$$

$$r_e = \Delta_h \tan(\phi_t - \phi_B/2) + r_M$$

$$r_0 = [r_e - r_M]_0^+$$

$$r_L^L = r_S, \quad r_N^L = \sqrt{[(A_N/A_L)^{2/\alpha_N}(r_S^2 + \Delta_h^2)^{\alpha_L/\alpha_N} - \Delta_h^2]_{r_0^2}^+}$$

$$r_L^N = [\sqrt{(A_L/A_N)^{2/\alpha_L}(r_S^2 + \Delta_h^2)^{\alpha_N/\alpha_L} - \Delta_h^2}]_{r_e+r_M}^-, \quad r_N^N = r_S$$

$$\varphi_1(r) = \varphi_2(r) = \frac{\pi}{2}, \text{ if } r < \frac{r_m}{r_M}\sqrt{r_M^2 - r_e^2} \text{ and } \phi_t < \frac{\phi_B}{2}$$

$$\varphi_1(r) = \cos^{-1}\left[\frac{r_e r_m^2 - \sqrt{r_e^2 r_m^4 - (r_m^2 - r_M^2)(r_e^2 r_m^2 + r^2 r_M^2 - r_m^2 r_M^2)}}{r(r_m^2 - r_M^2)}\right],$$

$$\varphi_2(r) = \cos^{-1}\left[\frac{r_e r_m^2 + \sqrt{r_e^2 r_m^4 - (r_m^2 - r_M^2)(r_e^2 r_m^2 + r^2 r_M^2 - r_m^2 r_M^2)}}{r(r_m^2 - r_M^2)}\right],$$

$$\text{if } \frac{r_m}{r_M}\sqrt{r_M^2 - r_e^2} \le r \le r_M - r_e \text{ and } \phi_t < \frac{\phi_B}{2}$$

$$\varphi_1(r) = \cos^{-1}\left[\frac{r_e r_m^2 - \sqrt{r_e^2 r_m^4 - (r_m^2 - r_M^2)(r_e^2 r_m^2 + r^2 r_M^2 - r_m^2 r_M^2)}}{r(r_m^2 - r_M^2)}\right],$$

$$\varphi_2(r) = \pi, \text{ if } r > |r_M - r_e|$$

[40], these parameters are obtained and listed in Table 4.4, in which the following notation is used:

$$[y]_x^+ \triangleq \max(x, y), \qquad [y]_x^- \triangleq \min(x, y). \tag{4.34}$$

In order to obtain A_{noL}^L, we note that this set is an intersection of a disk of radius r_S centered at the origin and C, which is visualized in Figure 4.9. Accordingly

$$A_{noL}^L = \{(r, \varphi) \mid r < r_L^L, \ \varphi \in [0, \pi] \backslash [\varphi_1(r), \varphi_2(r)]\}, \tag{4.35}$$

where $r_L^L = r_S$. Furthermore $\varphi_1(r)$ and $\varphi_2(r)$ are expressed in Table 4.4.

The integral in Eq. (4.32), therefore, is

$$\int_{\mathcal{A}_{\text{noL}}^{\text{L}}} \lambda_{\text{b}} \mathcal{P}_{\text{L}}(r) r \, d\varphi \, dr = \int_0^{r_{\text{S}}} \int_0^{\varphi_1(r)} \lambda_{\text{b}} \mathcal{P}_{\text{L}}(r) r \, d\varphi \, dr + \int_0^{r_{\text{S}}} \int_{\varphi_2(r)}^{\pi} \lambda_{\text{b}} \mathcal{P}_{\text{L}}(r) r \, d\varphi \, dr$$

$$= \lambda_{\text{b}} \int_0^{r_{\text{S}}} r \varphi_1(r) \mathcal{P}_{\text{L}}(r) \, dr + \lambda \int_0^{r_{\text{S}}} r[\pi - \varphi_2(r)] \mathcal{P}_{\text{L}}(r) \, dr$$

$$= \lambda_{\text{b}} \int_0^{r_{\text{S}}} r \mathcal{P}_{\text{L}}(r)[\varphi_1(r) + \pi - \varphi_2(r)] \, dr$$

$$\triangleq \lambda_{\text{b}} \, \mathcal{I}_{1\text{L}}^{\text{L}}.$$

We further proceed with the derivation of the set $\mathcal{A}_{\text{noN}}^{\text{L}}$, which is

$$\mathcal{A}_{\text{noN}}^{\text{L}} = \{(r, \varphi) \mid r < r_{\text{N}}^{\text{L}}, \ \varphi \in [0, \pi] \backslash [\varphi_1(r), \varphi_2(r)]\}, \tag{4.36}$$

where r_{N}^{L} is expressed in Table 4.4. Thus,

$$\int_{\mathcal{A}_{\text{noN}}^{\text{L}}} \lambda_{\text{b}} \mathcal{P}_{\text{N}}(r) r \, d\varphi \, dr = \lambda_{\text{b}} \int_0^{r_{\text{N}}^{\text{L}}} r \varphi_1(r) \mathcal{P}_{\text{N}}(r) \, dr + \lambda_{\text{b}} \int_0^{r_{\text{N}}^{\text{L}}} r[\pi - \varphi_2(r)] \mathcal{P}_{\text{N}}(r) \, dr$$

$$= \lambda_{\text{b}} \int_0^{r_{\text{N}}^{\text{L}}} r \mathcal{P}_{\text{N}}(r)[\varphi_1(r) + \pi - \varphi_2(r)] \, dr \tag{4.37}$$

$$\triangleq \lambda_{\text{b}} \, \mathcal{I}_{1\text{N}}^{\text{L}}. \tag{4.38}$$

Finally, by using Eqs. (4.31)–(4.37) we have

$$f_{R_{\text{S}}}^{\text{L}}(r_{\text{S}}) = \lambda_{\text{b}} \mathcal{P}_{\text{L}}(r_{\text{S}}) \, e^{-2\lambda_{\text{b}}[\mathcal{I}_{1\text{L}}^{\text{L}} + \mathcal{I}_{1\text{N}}^{\text{L}}]}. \tag{4.39}$$

Similarly one sees that

$$f_{R_{\text{S}}}^{\text{N}}(r_{\text{S}}) = \lambda_{\text{b}} \mathcal{P}_{\text{N}}(r_{\text{S}}) \, e^{-2\lambda_{\text{b}}[\mathcal{I}_{1\text{L}}^{\text{N}} + \mathcal{I}_{1\text{N}}^{\text{N}}]}, \tag{4.40}$$

where the integrals and related parameters are specified in Theorem 4.1.

The conditional coverage probability, i.e., $\mathcal{P}_{\text{cov}|R_{\text{S}}}^{\upsilon}$, can be obtained as

$$\mathcal{P}_{\text{cov}|R_{\text{S}}}^{\upsilon} = \mathbb{P}\left[\frac{P_{\text{tx}} G_{\text{tot}} \zeta_{\upsilon} \Omega_{\upsilon}}{N_0 + I} > T | R_{\text{S}} = r_{\text{S}}\right]$$

$$= \mathbb{E}_I\left\{\mathbb{P}\left[\Omega_{\upsilon} > \frac{T}{P_{\text{tx}} G_{\text{tot}} \zeta_{\upsilon}} (N_0 + I) | R_{\text{S}} = r_{\text{S}}\right]\right\}$$

$$\overset{(a)}{=} \mathbb{E}_I\left\{\sum_{k=0}^{m_{\upsilon}-1} \frac{y_{\upsilon}^k}{k!} (N_0 + I)^k \exp[-y_{\upsilon}(N_0 + I)] | R_{\text{S}} = r_{\text{S}}\right\}$$

$$= \mathbb{E}_I\left\{\sum_{k=0}^{m_{\upsilon}-1} \frac{y_{\upsilon}^k}{k!} e^{-N_0 y_{\upsilon}} \sum_{j=0}^{k} \binom{k}{j} N_0^{k-j} I^j \exp[-y_{\upsilon} I] | R_{\text{S}} = r_{\text{S}}\right\}$$

$$= \sum_{k=0}^{m_{\upsilon}-1} q_k \, \mathbb{E}_I\{I^k \exp(-y_{\upsilon} I) \mid R_{\text{S}} = r_{\text{S}}\}$$

$$= \sum_{k=0}^{m_{\upsilon}-1} (-1)^k q_k \, \frac{d^k}{dy_{\upsilon}^k} \, \mathcal{L}_{I|R_{\text{S}}}^{\upsilon}(y_{\upsilon}),$$

where the Gamma distribution of Ω_v is used in expression (a), and q_k and y_v are stated in Eqs. (4.13) and (4.14), respectively.

The Laplacian of the aggregate interference, i.e., $\mathcal{L}_{I|R_S}(y_v)$, is

$$\mathcal{L}_{I|X_S}(y_v) = \mathbb{E}_I\{e^{-y_v I} \mid R_S = r_S\}$$

$$= \mathbb{E}_{\Phi,\Omega}\left\{ \prod_{r \in \Phi \setminus r_S} e^{-y_v P_{tx} G_{tot} \zeta_\xi(r) \Omega_\xi} \right\}$$

$$= \mathbb{E}_{\Phi}\left\{ \prod_{r \in \Phi \setminus r_S} \mathbb{E}_\Omega\{e^{-y_v P_{tx} G_{tot} \zeta_\xi(r) \Omega_\xi}\} \right\}, \tag{4.41}$$

which can be written as

$$\mathcal{L}_{I|R_S}(y_v) = \mathbb{E}_{\Phi_L}\left\{ \prod_{r \in \Phi_L \setminus r_S} \mathbb{E}_\Omega\{e^{-y_v P_{tx} G_{tot} \zeta_L(r) \Omega_L}\} \right\}$$

$$\times \mathbb{E}_{\Phi_N}\left\{ \prod_{r \in \Phi_N \setminus r_S} \mathbb{E}_\Omega\{e^{-y_v P_{tx} G_{tot} \zeta_N(r) \Omega_N}\} \right\}$$

$$\overset{(a)}{=} \exp\left\{ -2 \int_{\overline{\mathcal{A}}_{noL}^v} \lambda_L(r)[1 - \Upsilon_L(r, y_v)] r \, d\varphi \, dr \right\}$$

$$\times \exp\left\{ -2 \int_{\overline{\mathcal{A}}_{noN}^v} \lambda_N(r)[1 - \Upsilon_N(r, y_v)] r \, d\varphi \, dr \right\}.$$

Above, the probability generating functional (PGFL) of PPP is used in expression (a). Furthermore, the complementary of the sets $\mathcal{A}_{no\xi}^v$ over C is denoted as $\overline{\mathcal{A}}_{no\xi}^v$. Mathematically speaking $\overline{\mathcal{A}}_{no\xi}^v = C \setminus \mathcal{A}_{no\xi}^v$. Therefore,

$$\int_{\overline{\mathcal{A}}_{noL}^L} \lambda_L(r)[1 - \Upsilon_L(r, y_L)] \, dx$$

$$= \int_{r_S}^{r_e + r_M} \int_0^{\varphi_1(r)} \lambda_L(r)[1 - \Upsilon_L(r, y_v)] r \, d\varphi \, dr$$

$$+ \int_{r_S}^{r_e + r_M} \int_{\varphi_2(r)}^{\pi} \lambda_L(r)[1 - \Upsilon_L(r, y_L)] r \, d\varphi \, dr$$

$$= \lambda \int_{r_S}^{r_e + r_M} \varphi_1(r) P_L(r)[1 - \Upsilon_L(r, y_L)] r \, dr$$

$$+ \lambda \int_{r_S}^{r_e + r_M} [\pi - \varphi_2(r)] P_L(r)[1 - \Upsilon_L(r, y_L)] r \, dr$$

$$= \lambda \int_{r_S}^{r_e + r_M} r P_L(r)[1 - \Upsilon_L(r, y_L)][\varphi_1(r) + \pi - \varphi_2(r)] \, dr$$

$$\triangleq \lambda \, \mathcal{I}_{2L}^L$$

and

$$\int_{\overline{\mathcal{A}}_{noN}^L} \lambda_N(r)[1 - \Upsilon_N(r, y_L)] r \, d\varphi \, dr$$

$$= \lambda \int_{r_N^L}^{r_e + r_M} \varphi_1(r) P_N(r)[1 - \Upsilon_N(r, y_L)] r \, dr$$

$$+ \lambda \int_{r_N^L}^{r_e + r_M} [\pi - \varphi_2(r)] \mathcal{P}_N(r)[1 - \Upsilon_N(r, y_L)] r \, dr$$

$$= \lambda \int_{r_N^L}^{r_e + r_M} r \mathcal{P}_N(r)[1 - \Upsilon_N(r, y_L)][\varphi_1(r) + \pi - \varphi_2(r)] \, dr$$

$$\triangleq \lambda \, I_{2N}^L.$$

Accordingly,

$$\mathcal{L}_{I|R_S}(y_L) = e^{-2\lambda[I_{2L}^L + I_{2N}^L]}. \tag{4.42}$$

Similarly,

$$\mathcal{L}_{I|R_S}(y_N) = e^{-2\lambda[I_{2L}^N + I_{2N}^N]}, \tag{4.43}$$

where related parameters are specified in Theorem 4.1.

Appendix B
Proof of Lemma 1

The first moment of the aggregate interference I can be obtained using its Laplacian transform as

$$\mu_{I|R_S} = -\left. \frac{d}{dy_L} \mathcal{L}_{I|R_S}(y_L) \right|_{y_L=0}. \tag{4.44}$$

From Eq. (4.23), Eq. (4.44) can be written as

$$\mu_{I|R_S} = -2\lambda \left. \frac{d}{dy_L} I_{2L}^L \right|_{y_L=0} \mathcal{L}_{I|R_S}(0)$$

$$= 2\lambda \int_{r_S}^{r_e + r_M} \left. \frac{d}{dy_L} \Upsilon_L(r, y_L) \right|_{y_L=0} r \mathcal{P}_L(r)[\varphi_1(r) + \pi - \varphi_2(r)] \, dr$$

$$= 2\lambda P_{tx} G_{tot} \int_{r_S}^{r_e + r_M} r \mathcal{P}_L(r) \zeta_L(r)[\pi + \varphi_1(r) - \varphi_2(r)] \, dr. \tag{4.45}$$

As for the variance of I we can write

$$\sigma_{I|R_S}^2 = \left. \frac{d^2}{dy_L^2} \mathcal{L}_{I|R_S}(y_L) \right|_{y_L=0} - \mu_{I|R_S}^2. \tag{4.46}$$

By using Eq. (4.23) and a similar approach as in the computation of the mean value, the desired result can be obtained.

Appendix C
Proof of Corollary 1

Let us set $\phi_t = 0$, which results in $\varphi_1(r) = \varphi_2(r) = \pi/2$. Therefore, the integral in Eq. (4.20) can be expressed as

$$\mu_{I|R_S} = 2\pi \lambda P_{tx} G_{tot} \sum_{k=i}^{j} p_k \int_{r_k}^{r_{k+1}} r \zeta_L(r) \, dr$$

$$= 2\pi \lambda P_{tx} G_{tot} A_L \sum_{k=i}^{j} p_k \int_{r_k}^{r_{k+1}} r(r^2 + \Delta_h^2)^{-\alpha_L/2} \, dr.$$

Therefore,

$$
\mu_{I|R_\mathrm{s}} =
\begin{cases}
2\pi\lambda \mathrm{P_{tx}G_{tot}} A_\mathrm{L} \displaystyle\sum_{k=i}^{j} p_k \left[\dfrac{(r^2 + \Delta_\mathrm{h}^2)^{1-0.5\alpha_\mathrm{L}}}{2(1 - 0.5\alpha_\mathrm{L})} \right]_{r=r_k}^{r_{k+1}}, & \text{if } \alpha_\mathrm{L} > 2, \\[2em]
2\pi\lambda \mathrm{P_{tx}G_{tot}} A_\mathrm{L} \displaystyle\sum_{k=i}^{j} p_k \left[\dfrac{\ln(r^2 + \Delta_\mathrm{h}^2)}{2} \right]_{r=r_k}^{r_{k+1}}, & \text{if } \alpha_\mathrm{L} = 2,
\end{cases}
$$

which is the desired equation.

A similar method yields the expression of $\sigma_{I|R_\mathrm{s}}^2$.

References

1 Y. Zeng, R. Zhang, and T. J. Lim (2016). Wireless communications with unmanned aerial vehicles: opportunities and challenges. *IEEE Commun. Mag.* 54 (5): 36–42.

2 D. Joshi (2017). Commercial unmanned aerial vehicle (UAV) market analysis industry trends, companies and what you should know. *Business Insider*.

3 S. D. Intelligence (2017). The global UAV payload market 2017–2027.

4 L. Sundqvist (2015). *Cellular controlled drone experiment: evaluation of network requirements*. Master's Thesis, Aalto University, December.

5 GSMA, Mobile-enabled unmanned aircraft. Tech. Rep., Feb. 2018. https://www.gsma .com/iot/mobile-enabled-unmanned-aircraft.

6 M. M. Azari, F. Rosas, and S. Pollin, Reshaping cellular networks for the sky: major factors and feasibility. *Proceedings of the 2018 IEEE International Conference on Communications (ICC)*. IEEE, 2018, pp. 1–7.

7 3GPP (2010). 3rd Generation Partnership Project; technical specification group radio access network; evolved universal terrestrial radio access (E-UTRA); further advancements for E-UTRA physical layer aspects (release 9). Tech. Rep., March 2010. www.qtc .jp/3GPP/Specs/36814-900.pdf

8 M. M. Azari, F. Rosas, K.-C. Chen, and S. Pollin, Ultra reliable UAV communication using altitude and cooperation diversity. *IEEE Trans. Commun.*, vol. 66, no. 1, pp. 330–344, 2018.

9 M. M. Azari, Y. Murillo, O. Amin et al. (2017). Coverage maximization for a Poisson field of drone cells. *Proceedings of the 28th IEEE Annual International Symposium on Personal, Indoor, and Mobile Radio Communications (PIMRC)*. IEEE, 2017, pp. 1–6.

10 W. Khawaja, I. Guvenc, D. Matolak et al. (2018). A survey of air-to-ground propagation channel modeling for unmanned aerial vehicles. arXiv preprint arXiv:1801.01656.

11 M. M. Azari, F. Rosas, A. Chiumento, and S. Pollin, Coexistence of terrestrial and aerial users in cellular networks. *Proceedings of the IEEE Global Communications (GLOBECOM) Workshops*, Dec. 2017.

12 M. M. Azari, F. Rosas, and S. Pollin, Cellular connectivity for UAVs: network modeling, performance analysis and design guidelines. *IEEE Trans. Wireless Commun.*, 2019.

13 M. M. Azari, et al. 2020. UAV-to-UAV communications in cellular networks. *IEEE Trans. Wireless Commun.*, 19 (9): 6130–6144.

14 W. Mei, O. Wu, and R. Zhang (2018). Cellular-connected UAV: uplink association, power control and interference coordination. *Proceedings of the IEEE Global Communications (GLOBECOM) Conference*, pp. 206–212.

15 Qualcomm Technologies, Inc. (2017). LTE unmanned aircraft systems. Tech. Rep., May.

16 G. Geraci, A. Garcia-Rodriguez, L. G. Giordano et al. (2018). Understanding UAV cellular communications: from existing networks to massive MIMO. *IEEE Access* 6, 67 853–67 865.

17 Ericsson, Drones and networks: ensuring safe and secure operations. White Paper, Nov. 2018.

18 M. Mozaffari, W. Saad, M. Bennis, and M. Debbah (2016). 'Efficient deployment of multiple unmanned aerial vehicles for optimal wireless coverage. *IEEE Commun. Lett.* 20 (8): 1647–1650.

19 M. M. Azari, H. Sallouha, A. Chiumento et al. (2018). Key technologies and system trade-offs for detection and localization of amateur drones. *IEEE Commun. Mag.* 56 (1): 51–57.

20 M. Mozaffari, W. Saad, M. Bennis, and M. Debbah, Wireless communication using unmanned aerial vehicles (UAVs): optimal transport theory for hover time optimization. *IEEE Trans. Wireless Commun.*, vol. 16, no. 12, pp. 8052–8066, 2017.

21 Y. Chen, W. Feng, and G. Zheng, Optimum placement of UAV as relays. *IEEE Commun. Lett.*, 2017.

22 D. Yang, Q. Wu, Y. Zeng, and R. Zhang (2017). Energy trade-off in ground-to-UAV communication via trajectory design. arXiv:1709.02975.

23 M. Alzenad, A. El-Keyi, and H. Yanikomeroglu, 3D placement of an unmanned aerial vehicle base station for maximum coverage of users with different QOS requirements. *IEEE Wireless Commun. Lett.*, 2017.

24 R. I. Bor-Yaliniz, A. El-Keyi, and H. Yanikomeroglu (2016). Efficient 3-D placement of an aerial base station in next generation cellular networks. *Proceedings of the IEEE International Conference on Communications (ICC)*, pp. 1–5.

25 E. Vinogradov, H. Sallouha, S. De Bast et al. (2019). Tutorial on UAV: a blue sky view on wireless communication. arXiv:1901.02306.

26 H. Sallouha, M. M. Azari, A. Chiumento, and S. Pollin, Aerial anchors positioning for reliable RSS-based outdoor localization in urban environments. *IEEE Wireless Commun. Lett.*, vol. 7, no. 3, pp. 376–379, 2017.

27 H. Sallouha, M. M. Azari, and S. Pollin (2018). Energy-constrained UAV trajectory design for ground node localization. *Proceedings of the IEEE Global Communications (GLOBECOM) Conference*, pp. 1–7.

28 M. M. Azari, F. Rosas, A. Chiumento et al. (2017). Uplink performance analysis of a drone cell in a random field of ground interferers. *Proceedings of the IEEE Global Communications (GLOBECOM) Conference*, submitted.

29 M. Haenggi (2012). *Stochastic Geometry for Wireless Networks*. Cambridge University Press.

30 L. Kleinrock and J. Silvester, Optimum transmission radii for packet radio networks or why six is a magic number. *Proceedings of the IEEE National Telecommunications Conference*, vol. 4, 1978, pp. 1–4.

31 J. Silvester and L. Kleinrock, On the capacity of multihop slotted ALOHA networks with regular structure. *IEEE Trans. Commun.*, vol. 31, no. 8, pp. 974–982, 1983.

32 M. Di Renzo, W. Lu, and P. Guan, The intensity matching approach: a tractable stochastic geometry approximation to system-level analysis of cellular networks. *IEEE Trans. Wireless Commun.*, vol. 15, no. 9, pp. 5963–5983, 2016.

33 C. A. Balanis (2016). *Antenna Theory: Analysis and Design*, 4th ed. Wiley.

34 F. Rosas and C. Oberli, Nakagami-m approximations for multiple-input multiple-output singular value decomposition transmissions. *IET Commun.*, vol. 7, no. 6, pp. 554–561, 2013.

35 ITU-R (2012). Recommendation P.1410-5: Propagation data and prediction methods required for the design of terrestrial broadband radio access systems operating in a frequency range from 3 to 60 GHz. Tech. Rep.

36 C. Forbes, M. Evans, N. Hastings, and B. Peacock (2011). *Statistical Distributions*. Wiley.

37 F. Yilmaz and M.-S. Alouini, A unified MGF-based capacity analysis of diversity combiners over generalized fading channels. *IEEE Trans. Commun.*, vol. 60, no. 3, pp. 862–875, 2012.

38 3GPP (2018). 3rd Generation Partnership Project: Technical specification group radio access network; study on enhanced LTE support for aerial vehicles (release 15). Tech. Rep., Jan. www.3gpp.org/dynareport/36777.htm

39 3GPP (2013). Small cell enhancements for E-UTRA and E-UTRAN – physical layer aspects. Tech. Rep., December 2013. www.3gpp.org/dynareport/36872.htm

40 H. C. Rajpoot, Analysis of oblique frustum of a right circular cone. *Int. J. Math. Phys. Sci. Res.*, 2015. www.researchpublish.com

5

Performance Enhancements for LTE-Connected UAVs: Experiments and Simulations

Rafhael Medeiros de Amorim[1], Jeroen Wigard[1], István Z. Kovács[1], and Troels B. Sørensen[2]

[1]*Nokia Bell Labs, Denmark*
[2]*Aalborg University, Denmark*

5.1 Introduction

According to [1], the total number of unmanned aerial vehicles (UAVs), also referred to as drones, will reach 86.5 million by 2025. In [2], the commercial applications of drone technology are allowing companies ranging from agriculture to the film-making industry to create new business and operating models, which in turn creates large global market value. This increasing interest in drones is one of the best signs of how lower pricing in hardware can drive the Internet of Things (IoT). Currently, regulations in most countries only allow the operation of drones with visual line of sight (VLOS) between drone pilot and the drone, but it is expected that beyond-visual-line-of-sight (BVLOS) operations will be allowed, provided there is a reliable command and control (C2 or C&C in some literature) link to the drone. This link is important to ensure safe drone operations.

In the uplink, i.e., from a drone to a base station, the control link is used to update the unmanned aircraft system traffic management or flight control unit with status messages, including the drone location, and information from, e.g., sensors, which can be utilized to make its decisions on the flight control. In the downlink (towards the drone), it allows the flight control function to change the flight plan of drones to avoid potential collisions, to enable dynamic geofencing, or to command a range of sensor/actuator functions on-board the drone. One example of downlink usage of the C2 link is when, on the route of a drone, suddenly a helicopter or another drone needs to land for an emergency. In that case, the downlink C2 link can be used to notify the drone of a new ad-hoc no-fly zone and redirect the drone by providing new directions.

One attractive means to provide this C2 link is to utilize existing cellular networks, in particular the existing LTE (long-term evolution) and future 5G (fifth generation) systems, as the infrastructures are in place and investments can therefore be minimized. However, such networks are not designed for aerial coverage, as they are optimized for ground users, typically using, for instance, down-tilted antennas at the base stations. Nevertheless, as we will explain in this chapter, existing LTE networks and future 5G networks are able to ensure

UAV Communications for 5G and Beyond, First Edition.
Edited by Yong Zeng, Ismail Guvenc, Rui Zhang, Giovanni Geraci, and David W. Matolak.
© 2021 John Wiley & Sons Ltd. Published 2021 by John Wiley & Sons Ltd.

Table 5.1 Technical comparison on aggregation capabilities.

	Downlink throughput	Uplink throughput	Reliability	Delay
C2 link	100 kbps	100 kbps	99.9%	50 ms
Application	≤ 1 Mbps	Up to 20 Mbps	Similar to mobile broadband	

reliable C2 communication to drones and play an important role in the provisioning of end-to-end (E2E) reliability for drone communications, similar to other use cases requiring high reliability. In this way, cellular networks can enable BVLOS drone flights. For the network operators, the increasing number of drones is an attractive group of potential customers to address.

Not surprisingly, various regulation committees are striving to specify the rules to which drone operations must conform, to ensure a robust and well-organized transition towards the "aerial vehicles era." Among those organizations attempting to address drone use cases, one can find also the 3rd generation partnership project (3GPP), responsible for standardizing worldwide cellular technologies, such as UMTS (universal mobile telecommunications service) and LTE. In this chapter we use the reliability requirements set by 3GPP in its studies on aerial connectivity, which means 99.9% reliability within a one-way delay budget of 50 ms. At the same time, the required throughput is rather low, as the 3GPP [3] assumes 100 kbps (kilobits per second). On top of the C2 link, there may be other applications running on-board the drone which require radio communication. These applications may have very different requirements, varying from case to case; e.g., an autonomous transport drone will not have much extra information exchanged, except maybe a note when a packet is delivered, whereas a drone providing high-quality live streaming from a certain event will require high uplink throughput. These applications typically require heavier uplink load than the downlink. The C2 control link is more symmetric. The key drone traffic characteristics, considered in this chapter, are summarized in Table 5.1.

In the rest of this chapter we focus on the C2 link and the possibility of providing reliable communication over cellular networks, while also having a look at the ability of cellular networks to provide high uplink throughputs. To understand how existing cellular networks work when users are in the air, we first study the propagation characteristics for drones.

5.2 LTE Live Network Measurements

Experimental investigations of the radio channel, capturing real-life effects of the radio environment, are a necessity when analyzing complex propagation conditions and scenarios. In UAV scenarios, especially for C2 link performance in cellular networks, it is extremely valuable to determine how the cellular radio modem on the UAV would perceive the radio channel in the existing cellular networks. With this aim, measurements have been conducted to characterize the large- and medium-scale propagation, as well as interference, that a UAV will experience if communication takes place over the typical cellular LTE/LTE-A network deployments. The characterization has been done for operation

Table 5.2 Summary of network configuration for urban area; antenna height is above ground level.

Frequency (MHz)	ISD (m)	Average antenna height (m)	Average downtilt (degrees)
800	850	26	5.5
1800	580	30	5.5
2600	690	25	5.8

heights up to 120 m above average ground level, i.e., within the very low-level (VLL) airspace. Both rural and urban environments have been characterized over a number of downlink and uplink experiments, in each case applying a specific measurement methodology for characterizing one or more desired characteristics. For downlink, large-scale path loss, shadowing, and interference are the main interest, whereas for uplink experiments, it has primarily been interference.

Measurements have been conducted in two urban and two rural environments. The urban environments are small-sized city environments, with population density ranging from around 1000 to 2900 inhabitants per km^2 (population size of 110 000 to 260 000 people, covering the larger city area). The measurement areas are characterized by irregular street grids and a mix of new and old buildings. The building height varies over the measurement area, from as low as 4 m in residential areas on the outskirts, to 15–20 m in the center of the city; in the densest parts, average building height is in the 20–25 m range.

Most of the sites have tri-sector configurations and a typical angular sector separation of 120 degrees. The carrier frequency plays a significant role in the path-loss characterization, and therefore different frequency bands have been measured. However, our main focus was on the 800 MHz and 1800 MHz bands. A summary of the average inter-site distance (ISD), antenna height, and down-tilt angle are given in Table 5.2. The typical sector antenna half-power beamwidth is 65 degrees for the H-plane and 7 degrees for the E-plane, with a maximum antenna gain of 18 dBi.

The two rural environments differed primarily in the land topography, and hence also by the ISD. One is primarily flat with a cellular network ISD of 5000 m, whereas the other has terrain variations between 40 and 120 m and about half the ISD. Correspondingly, antenna heights and down-tilts were also higher, ranging from 19 to 54 m and from 0 to 9 degrees, respectively, over the two environments, but with the same typical tri-sector configuration and antenna types as in the urban environments. For rural environments, only the 800 MHz band was used for the measurements.

The details on the cell identity, location, operating frequency, antenna height, antenna type, antenna tilt, cell bearing, and transmit power have been available in the planning and processing of the measurements, provided by courtesy of the Danish mobile operators.

5.2.1 Downlink Experiments

Different measurement setups were applied to accommodate local regulations and practical execution of the measurements. Therefore, in some cases, measurements were conducted

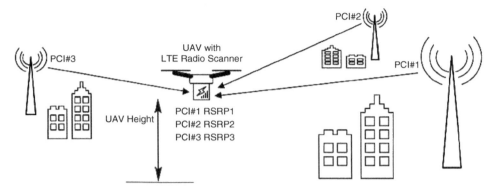

UAV: Unmanned Aerial Vehicle
PCI#x: LTE Physical Cell ID #x
RSRPx: Reference Signal Received Power from LTE radio cell with PCI#x

Figure 5.1 Experimental measurement methodology for radio channel investigations in live LTE networks.

with a drone, and in other cases with the measurement equipment mounted on a mobile crane lift. In the cases where a drone was used, measurements were in some cases collected while flying over street segments, and in the remaining, by hovering at a fixed position. All cases, including that with fixed position, achieved spatial averaging by collecting samples while moving the measurement antenna over linear or circular paths.

The basic experimental methodology followed the setup depicted in Figure 5.1, and has been described in [4–7]. For the characterization of path loss and cell environment, measurements were performed with the R&S TSME© mobile network scanner. The scanner actively detects LTE carriers over a configurable set of frequency bands, subject to a maximum number of detected cells and minimum possible detection level for demodulation and synchronization reference signals. Active detection means that measurements are only available subject to active decoding of the cell identification. In all measurements, the scanner has been configured to allow for detection of the maximum possible number of cells (32) at a rate between 5 and 9 Hz. Detection performance for cell identification depends on the absolute signal level of the reference signals, as well as the signal quality. The primary limitation lies in the signal quality, signal-to-interference-plus-noise ratio (SINR), of the synchronization channel, due to the interference faced on the central 1.08 MHz of the LTE carrier in the considered measurement environments. Detection therefore depends on the specific network configuration and propagation conditions at a particular measurement location, to achieve a synchronization SINR of minimum −20 dB.

The TSME was connected to a (cellular) multiband omnidirectional antenna. The antenna is seen mounted on top of the drone in Figure 5.2, about 50 cm above the drone fuselage and rotor plane. The antenna is a so-called paddle type, ground-plane-independent, antenna, which supports all major cellular bands in the frequency ranges 698–960, 1710–2170, and 2396–2700 MHz, with a nominal gain of 2 dBi.

Figure 5.3 shows the path loss versus distance results obtained in the rural environment, along with the corresponding average PL models fitted to a log-distance alpha–beta model

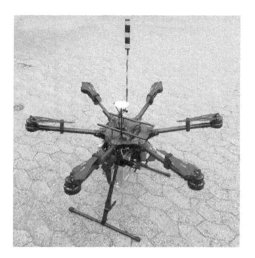

Figure 5.2 (left) Omnidirectional antenna mounted above the hexacopter fuselage. (right) Measurement equipment mounted on a custom-made carbon-fiber chassis underneath the drone (scanner is underneath the small form-factor computer). Source: Jeroen Wigard, Istvan Kovacs et al.

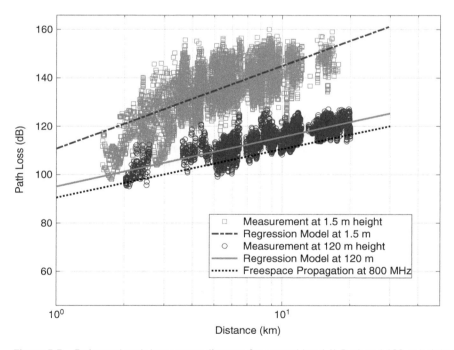

Figure 5.3 Estimated path loss versus distance for ground level (1.5 m) and 120 m height in rural environment (800 MHz).

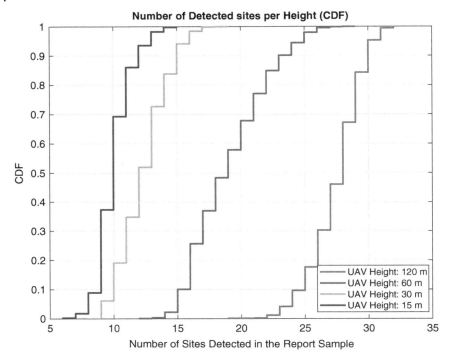

Figure 5.4 Number of neighbor cells detected per report sample for different heights in the rural scenario of measurements.

(see Section 5.2.2). From Figure 5.3 there are two main observations: the slope of the line of path loss versus distance decreases with height, and, similarly, the residuals of the path-loss variation relative to the average also reduces. In fact, at 120 m height, the average path loss behavior is approximately free-space propagation. Similar results have been obtained for urban environments, and for other heights, as will be summarized in Section 5.2.2, with the primary exception of slightly higher residual path-loss variation for the urban case.

Figure 5.4 shows another significant effect revealed by the measurements in both rural and urban environments. Specifically, the figure shows the case for the rural scenario, illustrating how the interference environment changes with increasing height. From the measurements, there is a clear trend that the average number of detected cells increases with height: for urban, from around five (depending on the exact environment) up to around 12 cells when above rooftop [5], and for rural, from around five to a few more than 20. But what Figure 5.4 also shows is that it is very likely to have one cell within 3 dB of the strongest, and also even two cells. This is particularly remarkable when the drone is well above the average rooftop level, i.e., in the 35–40 m range, where also the number of cases with three and four cells within 3 dB is significant.

A few of the experiments in urban and rural environments have investigated heights up to 120 m, confirming that the trend continues; at 120 m, the average number of detected cells for the urban case increased to 22 cells. The impact of frequency is less pronounced, particularly at elevated heights, since the improved propagation over long distances is already compensated by the (close to) free-space propagation. With regards to the increased interference, there is an increase in the average cell range with height, so that more and more distant cells play a role in the interference. The LTE reference signal received quality (RSRQ), which was also recorded by the scanner during the measurements, confirms the deterioration of the signal-to-interference ratio (SIR) with increasing height.

5.2.2 Path-Loss Model Characterization

From the estimated PL values and direct line of sight (LOS) 3D distances between the drone and the relevant cells, a log-distance alpha–beta model [8] was fitted to the data using least-squares regression. It is worth noting that some authors prefer to describe the path-loss model for airborne users as a function of the elevation angle between the UAV and base station and their distance apart [9]. It is possible to convert the parameters of one model to the other. The log-distance model is given by

$$\mathrm{PL}_{\mathrm{est}}(d) = 10\alpha\log_{10}(d) + \beta + X_{\sigma}. \tag{5.1}$$

The model was already exemplified in connection with Figure 5.3, and is furthermore used as the basis for the system-level performance evaluations presented in Section 5.3. In Eq. (5.1), d is the 3D distance in meters, α represents the path-loss exponent, and β is the intercept point at $d = 1$ m. The term X_{σ} is a random variable that accounts for the path-loss variation relative to the average path loss (shadowing), and modeled as a normally distributed random variable with standard deviation σ equal to the standard deviation of the regression residuals. In the regression, data were excluded based on criteria set for the maximum distance and bearing relative to the main beam direction of the cell, in order to minimize the cropping from detection capability and antenna side-lobe effects, respectively [4].

For the parameters in Eq. (5.1), height-dependent models were derived, valid for the rural environment. The corresponding behavior expressed by the models, is illustrated in Figure 5.5.

A similar analysis was carried out for the urban environment, confirming that the same trends and modeling framework apply, with the same or smaller PL slope as in Figure 5.5, whereas the standard deviation is about double, going from above 12 dB at ground level to around 6 dB at 40 m. A higher standard deviation is expected for urban, due to the increased clutter in this environment.

5.2.3 Uplink Experiments

The lower propagation losses observed by flying UAVs represent less attenuation of the radio signal radiated from the UAV toward the serving cell, but also to several neighbor cells. Especially because of the clearance in the UAV-to-neighbor-cell radio path, the UAV will cause potentially higher interference to the neighbor base stations, compared to terrestrial users.

The increase in uplink interference caused by airborne UAVs, when compared to the interference caused by terrestrial user equipment (UEs), is demonstrated by field measurements. The experimental setup is presented in Figure 5.6. For the airborne transmissions,

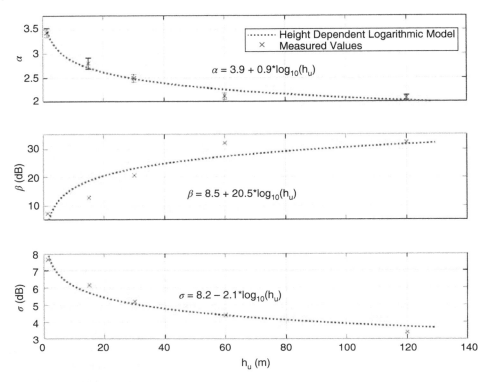

Figure 5.5 Height-dependent regression parameters for the model in Eq. (5.1) for the rural environment. The equations in the charts represent the model in the range 1.5 m $\leq h_u \leq$ 120 m.

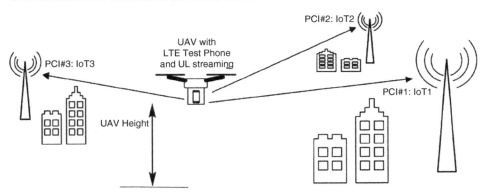

UAV: Unmanned Aerial Vehicle
PCI#x: LTE Physical Cell ID #x
IoTx: Interference Over Thermal noise at LTE radio cell with PCI#x

Figure 5.6 Experimental UAV measurement setup for uplink interference investigations in live LTE networks.

the UE is attached to a UAV, flying at 100 m height, in a circumference with 5 m radius. The terrestrial transmissions are made with the UE at approximately 1.5 m height, in a static position. In both cases, an LTE Category 6 UE device is used.

A full-buffer traffic type is emulated in the experiment, by means of a cyclic upload of a large file to a remote server through an Internet connection provided by a public LTE operator. The tests were performed between 2 and 5 a.m., a time suitably chosen, with the assistance of the network operator, to mitigate the effect of background interference caused by other active users as a confounding factor. The baseline for the experiments is constituted by seven days of interference over thermal noise rise data collected in a region of interest, represented by a 30 km radius circumference around the test location.

A battery of three tests with 15 minutes duration was repeated for both terrestrial and airborne UEs, at different times for each. Thereafter, the noise rise reported by each base station in the region of interest is collected and compared to the baseline. If the noise rise observed during the experiments for a given cell is above the 99% percentile of the baseline for the average of the three batteries, that cell is considered as an interference victim cell (IVC) from the test transmissions, motivated on the grounds that it is more likely that it was caused by the experiment than by statistical variation. The aforementioned battery of tests was repeated at three different scenarios: urban, suburban, and rural. At the rural location, the information of neighboring base stations was limited to a radius of 15 km.

Figure 5.7 shows the results measured at the rural location. The IVC are shown on the *x*-axis, and are ranked in descending order of interference over thermal noise for the aerial transmissions. It is worth observing that the airborne device has impacted on many more

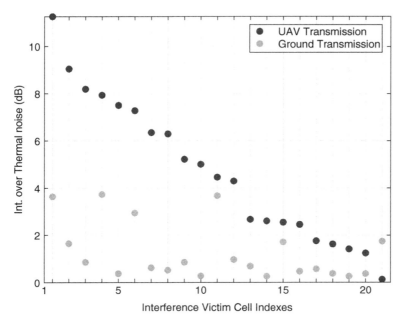

Figure 5.7 Interference over thermal noise observed in the interference victim cells for the UAV transmission (rural scenario). The UAV is flying at 100 m height.

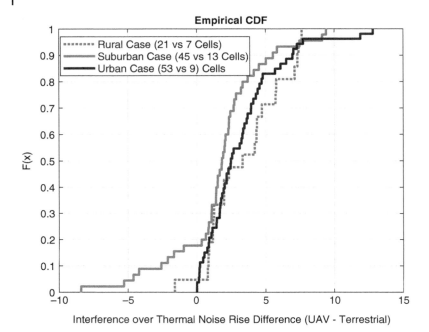

Figure 5.8 CDF plots of the difference between the interference over thermal noise caused by the UAV transmission at 100 m compared to a terrestrial transmission in the interference victim cells for the three scenarios. In the legend, "*X* vs *Y* cells" represents the number of interference victim cells for the UAV transmission (*X*) compared to the victim cells originated by terrestrial transmission (*Y*).

(20) cells than the terrestrial transmission (7), and, on average, with a much higher interference over thermal noise (5.0 dB versus 2.7 dB) in those cells.

Figure 5.8 shows the differences of the cumulative distribution function (CDF) of the interference over thermal noise measured in all airborne victim cells. It is observed that there is a significant increase in interference caused by the aerial transmissions, even though the receiving antennas in the base stations are usually down-tilted, optimized for ground coverage. Moreover, the legend in the figure shows the comparison between the number of victim cells perceived from an aerial transmission versus the number of victim cells perceived from a ground transmission. The larger number of cells impacted from the UAV transmission at 100 m is caused by the lower signal attenuation for large distances, compared to terrestrial UE.

However, the initial density of UAVs is much smaller than terrestrial users, and they are expected to transmit mostly low-data-rate C2 information, requiring few frequency resources. Because the load generated by UAVs is not expected to be that high in the first years, the effect measured in these experiments tends not to cause a major impact. On the other hand, if the density of UAVs increases significantly, or, especially, when some of them are used to transmit high-data-rate uplink traffic (e.g., by live video streaming from the UAV), actions may be needed to mitigate the potential performance degradation to all network users caused by UAV transmissions.

Controlling UAV transmit power is one way of adjusting for different propagation conditions, by limiting the UAV transmit power to lower levels. However, in this case the interference mitigation may come with a cost to the UAV throughput. Another option is to use highly directional transmissions from the UAV side, benefiting from the LOS sight likelihood and limiting the area affected by the UAV transmission.

5.3 Performance in LTE Networks

To show the impact of different propagation conditions explained in the previous section, we investigated the downlink connectivity for drones in both rural and urban areas by simulations. For the rural area, we consider a 70 km × 70 km rural space in Denmark. For the urban area, we study one of the major cities in Denmark, Aalborg. In both cases, we use a real-world LTE network as network layout, including actual base station locations, heights, antenna patterns, bearings, and down-tilting. In the rural area, the base station heights range from 19 to 50 m above the ground, and down-tilting angles are from 0 to 9 degrees; while in the urban area, the base station heights are around 30 m, with a bit larger down-tilt, ranging from 0 to 11 degrees. The terrain profile is taken into account and drones are assumed always to fly above the terrain at a constant height. The LTE system considered in our simulation is 2 × 2 multiple input–multiple output (MIMO). For the rural area, we use the propagation model deduced from our measurements mentioned in the previous section; while for the urban area, the model from 3GPP [3] has been used, with an additional height-dependent path-loss slope determined by the LOS probability.

Figure 5.9 shows the average downlink SINR for both the terrestrial UEs (TUEs) and drones (UAVs) at different heights for the urban and rural scenarios at medium and high downlink traffic loads. These load levels correspond to 30% and 55–65% average resource or physical resource block (PRB) utilization in the downlink. It should be noted that the load in today's real LTE networks is typically lower than the values we use for medium load. The reason for using these high values is to see whether higher reliability can be achieved for drones when the load will increase in the future. The following observations can be made on these results:

- The average downlink SINR drops with the increase of drone flying height. The reason for this is that interference increases with increasing height, as more interfering sources becomes visible and the propagation becomes more advantageous when the height increases.
- The average downlink SINR of drones is more sensitive to the load in the network than that of the terrestrial users, since aerial users see more potential interferers.

To enable a UE to connect to the network and stay connected, the downlink SINR needs to be above −6 dB [10]. While the downlink mean SINR values are all above this threshold, there are variations around these mean values that cause that UE at different heights to experience different levels of outage, i.e., the likelihood of being unable to connect to the network. The outage levels for different heights and loads for both rural and urban scenarios can be seen in Table 5.3. From the table it can be seen that the outage probabilities increase

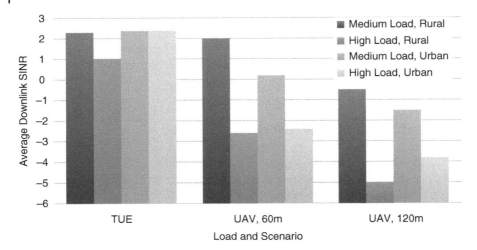

Figure 5.9 Average downlink SINR for the terrestrial UE (TUE) and for drones (UAV) at two different heights for rural and urban scenarios at medium and high load.

Table 5.3 Outage probabilities for terrestrial users and drones at different heights for different scenarios.

	Rural area (%)		Urban area (%)	
	Medium load	High load	Medium load	High load
Terrestrial users	0.3	1.5	1.1	2.8
Drones at 60 m	1.7	10.9	6.9	11.2
Drones at 120 m	3.5	23	14.5	21.4

with height and network load. Whereas, on the ground, in an urban area, the terrestrial users in a highly loaded network have 1.5% probability of experiencing outage, for drones at 120 m height in the same scenario, the outage probability reaches a soaring 23%. In general, it can be concluded that these outage probabilities show that the current cellular network at high load cannot provide a highly reliable connection to drones, whereas at medium load the outage probabilities may be acceptable. In other words, to ensure high reliability at all load conditions, interference mitigation is needed. This will be looked at in the next section.

5.4 Reliability Enhancements

As shown in the previous section, reliability-enhancing techniques may be needed to ensure a reliable C2 link. We can split the reliability-improving techniques into two categories: terminal-side enhancements, which corresponds to enhancing the receiver and/or transmitter on the drone, and network enhancements, which corresponds to adding features to the network to lower the interference effects. The next subsections give a description of a number of these techniques and show their potential for addressing the reliability issue in the worst case where a drone is at 120 m height and the network load is high. Furthermore, the same modeling as in the previous section is used.

5.4.1 Interference Cancellation

A traditional terminal-side interference mitigation solution for downlink transmissions in cellular systems is the use of receiver interference cancellation (IC) or interference rejection combining (IRC) algorithms. When two or more antenna elements are available at the UE in the receiver chain, the interference cancellation mechanism can suppress one or more downlink dominant interfering signals. In practice, the interference mitigation performance of such receivers is determined not only by the number of antenna elements, but also by the degree of antenna correlation, the level of interfering signals, and the spatio-temporal characteristics of the signal fading.

In our early studies we have assessed the potential performance of an ideal IC algorithm using measured signal traces in various rural network conditions and locations [11]. The measurements have been carried out as described in Section 5.2.1. These results have shown that the potential gains from suppressing three or four interfering signals change versus UAV flight height with a typical maximum height at 30–60 m, and at the same time the gains strongly depend on the network inter-site distance (ISD) and UAV location within the network: larger ISDs and lower SIR or SINR conditions typically lead to higher IC gains.

The last two observations lead to further investigations of the performance of interference mitigation algorithms using large-scale system-level simulations in urban and rural scenarios, with the same setup as described in Section 5.3. We have assumed that each UAV has a UE with a four-antenna receiver capable of ideally suppressing three downlink interfering signals [12]. The IC gains were evaluated in terms of the outage probability indicator, and the average achieved values for UAVs at 120 m height are presented in Figure 5.10. These results indicate that, even with an ideal IC and with three interferers rejected, the gains are insufficient to reach the target outage of 0.1% (99.9% reliability).

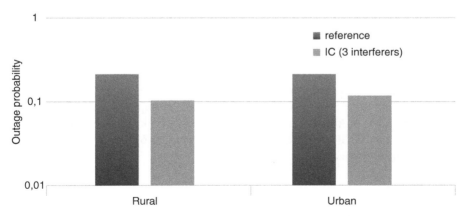

Figure 5.10 Average outage probabilities for UAV flights at 120 m in urban and rural radio network deployments when no IC (reference) and IC (three interferers rejected) is used at the UAV UE.

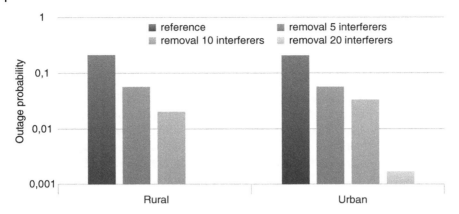

Figure 5.11 Drone outage probability for a drone at 120 m in urban and rural environments under high load with removal of a different number of interferers.

5.4.2 Inter-Cell Interference Control

A network-based solution is the downlink inter-cell interference coordination (ICIC). There are several standardized solutions for this. The simplest downlink ICIC scheme was introduced in 3GPP Release 8, and is purely based on inter-cell signaling and does not require any UE-side functionality. The general idea is to coordinate the usage of radio resources between cells to optimize the cell edge SINRs. Further possibilities were introduced with the enhanced and further enhanced ICIC (eICIC and feICIC) solutions, which were included in 3GPP Releases 10 and 11. These solutions can be considered as candidate solutions in the drone scenarios as well. The outage can be decreased by coordinating the downlink transmission from the right number of cells. Figure 5.11 shows the gain in outage versus the number of interfering sources removed, i.e., assuming an ideal ICIC capable of suppressing entirely their interference, from the receiving signal for the urban and rural scenario.

From the results in Figure 5.11, it can be concluded that several cells need to be muted, during the transmission intervals dedicated to a UAV, to reach a reliability of 99,9% in the case of high load. However, one needs to keep in mind that this reliability is for the C2 link, which only requires 100 kbps. This means that the data transmission to and from the drone can be concentrated in time, such that, for instance, every 10th transmission time interval (TTI) is used for drone-related transmissions. That would mean that only in those TTIs do the interfering cells need to be muted. An example of such a setup could be that a drone at 120 m in the air with a 100 kbps C2 link uses every 10th TTI. In the high-load scenario this will require muting the eight strongest cells every 10th TTI [13], corresponding to removing 10% of the capacity of those cells.

5.4.3 CoMP

To mitigate the impact of radio interference in downlink and uplink in cellular systems, a special class of techniques has been developed and standardized for LTE-Advanced: Joint Coordinated Multi-Point Transmission and Reception (JT/JR-CoMP) schemes [14]. In the transmission, the JT-CoMP is a simultaneous transmission of the same signal by more than one base stations toward a UE, aiming at enhancing the reliability/quality of the received signal, by the accumulation of received power. Likewise, in the uplink, the JR-CoMP works similarly, with the UE signal being received by more than one base station that combine their versions to obtain an output version with enhanced reliability/quality.

Both JT-CoMP and JR-CoMP require radio network mechanisms which can coordinate and combine the transmission/reception signals from several radio cells (the CoMP set). This, in turn, puts certain requirements on the performance of the inter-cell communication links and limits also the number of cells which can be included in the CoMP set.

In typical LTE-Advanced network deployments, the UEs are located on (or close to) the ground, and due to the propagation conditions the number of radio cells which need to be included in the CoMP set is relatively low. However, for the UAVs that fly high above the ground, and potentially even above the radio cells, the nearly non-obstructed propagation conditions lead to the need to use larger CoMP sets (in a larger geographical area) when the target is to achieve similar gains compared to the ground UE scenarios. As already discussed in Section 5.4.2, coordinating a high number of cells is not a realistic assumption, for the rural and sub-rural networks.

Taking into consideration the practicalities discussed above, in UAV scenarios we have investigated the potential gain from using JT/JR-CoMP only for a reduced CoMP set of up to five cells [11]. Furthermore, we used the experimental radio measurement data collected in the rural scenario (see Section 5.2.1). The benefit of JT/JR-CoMP is estimated using the metric of the ideal average SINR gain, without considering performance losses due to imperfect signalling or the possible limited communication between the cells in the CoMP set. The probability of achieving at least 3 dB downlink SIR, or 3 dB uplink received signal level, improvement is used as an evaluation metric [11].

Figures 5.12 and 5.13 summarize the JT- and JR-CoMP evaluation results for the rural scenario with ISD = 2.8 km. Two sub-cases are presented, corresponding to UAV flights in the areas with low downlink SIR conditions. The downlink JT-CoMP results indicate that UAV scenarios with flight heights above 60 m can benefit most from CoMP techniques with only three cooperating cells (one serving plus two additional), while at lower altitude the gains provided by JT-CoMP might not always be possible to harness. In the case of uplink JR-CoMP, our results show that the cooperation of five cells (one serving plus four additional) would be needed to benefit from JR-CoMP techniques, especially for UAV flights at low height, below 60 m; the gains are highest for UAV scenarios with 60 to 120 m flight heights.

5.4.4 Antenna Beam Selection

Contemporary wireless communication systems can use multiple antenna techniques to improve system performance by exploiting spatial and temporal diversity in the radio communication channel. One of the simplest multi-antenna configurations is when the available antenna elements are used selectively depending on the signal level (or signal quality) estimated at each antenna element. For example, by processing the reference signals received by each antenna element, the receiving terminal may infer the antenna elements that provide better signal reception, and modulate the power and phase of their antenna

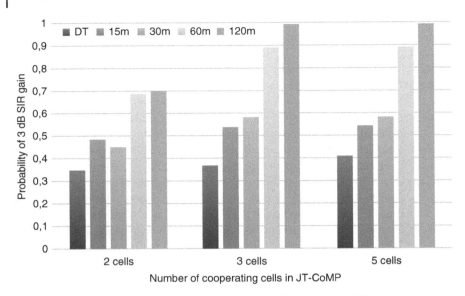

Figure 5.12 Summary of the downlink JT-CoMP 3 dB SIR gain probability results in a rural scenario with ISD = 2.8 km.

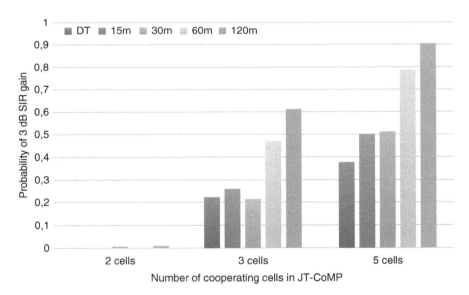

Figure 5.13 Summary of the uplink JR-CoMP 3 dB gain probability results in a rural scenario with ISD = 2.8 km.

array to favor that beam. This antenna beam selection diversity technique can be easily adapted to the physical geometry of the terminal, receiver, and/or transmitter.

In the case of medium-large UAVs, which have a considerably larger physical size compared to a mobile phone, the placement with the right spacing of directional antenna elements around the fuselage enables the implementation of a so-called grid of fixed beams

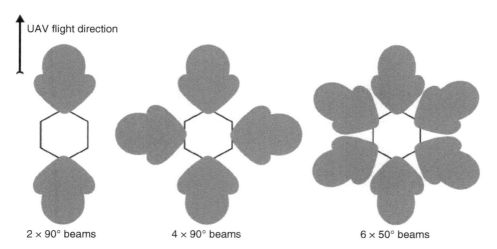

Figure 5.14 UAV GoB configurations with two, four and six beams (beamwidth not at scale).

(GoB). We have investigated the performance of such a GoB on the UAV in a large-scale system in urban and rural scenarios, with the same setup as described in Section 5.3. The tested GoB geometries are depicted in Figure 5.14. The modeled antenna beam patterns provide +6.6 dBi gain in the main direction and 13 dB front-to-side-lobe attenuation, which can be considered to account for the non-ideal shape of the beams. The radio network was simulated with the existing three-sector configuration and without any beamforming capabilities.

The receiver on the UAV selects the antenna beam direction with the best downlink signal quality (RSRQ) without adjusting the orientation of the UAV during flight. This antenna selection leads to the reduction of the amount of interference power received in the downlink, improves the receive SINR, and thus reduces the overall radio outage level experienced by the UAV. The GoB gains were evaluated in terms of the outage probability indicator, and the average achieved values for UAVs at 120 m height are presented in Figure 5.15. These results indicate that with a six-beam GoB configuration at the UAV, the target outage of 0.1% can be met in all scenarios [12, 13].

In the uplink, the use of a GoB configuration, with the antenna beam selection based on the downlink signal quality, results in both signal gain for the UAV transmissions and reduced uplink interference in the network. As a consequence, an improved overall network performance is achieved with more than 50% uplink throughput improvement for both terrestrial UEs and UAV UEs [12, 13].

5.4.5 Dual LTE Access

A complementary approach to system-level simulations such as those presented in Section 5.3 is to use a measurement setup able to record the typical LTE radio connection performance indicators, while actively performing data transfer to/from the UAV. Further, to obtain a more accurate view on the communication link performance, one needs to emulate the data traffic on the radio communication link (uplink and downlink data

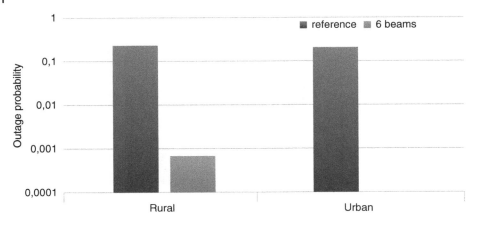

Figure 5.15 Average outage probabilities for UAV flights at 120 m in urban and rural radio network deployments when an omnidirectional antenna (reference) and GoB with six beams is used at the UAV UE.

packets) to reflect a specific UAV use case, e.g., the C2 (or C&C) link remote-piloted UAV via an LTE connection. An experimental setup for this purpose, would need at minimum to include the following: (i) on-board the UAV, an LTE radio modem capable of recording LTE radio connection performance indicators, and a client application, which generates the uplink packets and records the downlink packets; and (ii) a remote server application, connected to the Internet, which generates the downlink packets and records the uplink packets. This type of experimental setup has been used previously to investigate packet delays in the context of hybrid access transmission schemes [15].

For our UAV investigations, we have adopted the setup from [15] and modified it for dual-LTE network connectivity studies. For this, we have used two LTE radio modems – two mobile phones running special firmware – each connected to a different LTE network operator, and the client application generates, and records, data packets to/from both modems simultaneously. Similarly, the server application handles both data connections simultaneously. Figure 5.16 shows schematically this measurement setup. An urban radio environment was investigated, with the UAV flying along aerial paths approximately 2 km long at three different heights, 15 m, 40 m, and 100 m; additional drive tests, at ground level, have been also performed [16].

All data packets transmitted (uplink and downlink) are time-stamped, and the client and server applications are time-synchronized via a Global Navigation Satellite System (GNSS) [15]. This allows for delay measurements on both LTE connections, with precision in the order of hundreds of microseconds. Because the two LTE connections are measured simultaneously, the obtained results can be used to investigate the potential performance of realistic "hybrid" access scheme(s) for improving the reliability of the C2 link. In addition, the LTE radio connection performance indicators recorded by each modem can be further used to gain insights into the achieved C2 performance, and potential improvements [16]. It must be noted that the results obtained with such a setup are indicative of the total achievable end-to-end delays, and not only of the delays in the radio access networks. Furthermore, in order to achieve realistic results, the packet generation rates in both

UAV: Unmanned Aerial Vehicle
Dual-LTE: Simultaneous data connectivity to two LTE networks

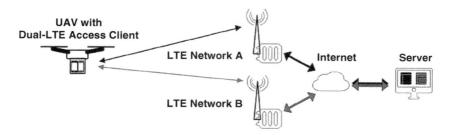

Figure 5.16 Experimental UAV measurement setup for dual network connectivity in live LTE networks.

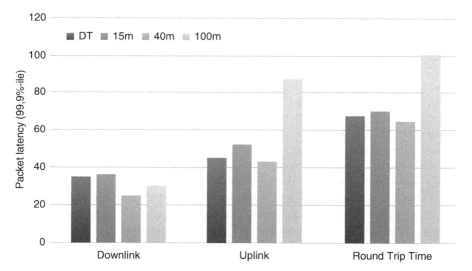

Figure 5.17 UAV measurement results in terms of achieved packet delays for different flight heights when hybrid access is assumed.

the client and server have to be configured according to the discontinuous reception (DRX) settings in the access networks, such that the UEs do not have to switch between connected and idle modes, which would lead to additional connection setup delays for each transmitted/received packet.

A representative set of results is presented in Figure 5.17, where we show the peak packet delay performance at the 99.9 percentile probability when dual-LTE hybrid access is assumed, for downlink, uplink, and total packet round-trip time. The hybrid solution we assume here is that packets are simply bi-casted over both network connections, and only the earliest packets (minimum delay) are retained at the receiver side.

These results indicate that the overall latency performance of the C2 traffic is determined by the latency in uplink (drone-to-network) transmissions. A dual-LTE hybrid access solution can reduce the peak E2E latencies to 40 ms and 90 ms in downlink and uplink, respectively, while achieving a close to 100 ms round-trip time. The UAV flight height can have significant impact on the achieved delay, with delays up to 90 ms for heights above 40 m.

5.4.6 Dedicated Spectrum

The usage of a dedicated spectrum allocated for UAVs, in a separate network deployment, seems particularly appealing to isolate the C2 link from external sources of interference, such as users on the ground. However, the infrastructure required for deploying such a network, for example, masts and backhauling connection, may entail unattractive costs. In particular, considering the current density of flying UAVs and the amount of time they spend in the air, it can be rather challenging to make the investment worthwhile from the business perspective.

To mitigate such costs, the infrastructure may be shared with current cellular networks, but still using a dedicated spectrum bandwidth reserved for UAVs. In [17], a forecast of the UAV fleet size and the consequent demand for C2-reserved bandwidth is proposed for the next 20 years, based on a public database of commercial non-hobbyist UAV registrations in the USA. The simulation results show that, in the majority of scenarios analyzed, 5 MHz of bandwidth are predicted to be enough to guarantee a reliable C2 link connectivity using currently available technical solutions for LTE networks. The simulations assumed that all UAVs were flying at 120 m height. In highly dense areas, such as Manhattan, though, the total required bandwidth can go up to more than 10 MHz.

Even though the numbers show that a relatively narrow bandwidth is required, and that the site density can be kept low, there are some caveats. First, the required bandwidth was forecast for peak usage, which can be significantly larger than the average use in some locations. Therefore, we should consider the amount of time the spectrum may be idle in the entire network. Moreover, because small commercial drones may land from and take off to anywhere on the ground, as the authors in [17] suggest, a higher density of base stations may be involved to enhance the ground coverage to enable such operations.

It is possible to minimize the waste of resources by adopting a smart spectrum management policy, where the amount of resources reserved for C2 increases and decreases following the fluctuations of instantaneous demand. The research community must investigate allocation strategies behind this policy, in order to obtain a good trade-off between protection of C2 link and the waste of resources in the dedicated bandwidth.

5.4.7 Discussion

In the previous section, four different interference mitigation techniques for drones are discussed. These interference mitigation techniques are needed when the same spectrum as for terrestrial users is used for drone communication. Obviously, when dedicated spectrum is available for drones, the situation changes, as the load in the network only depends on the load caused by drones. In that case, especially when the number of drones is small, no interference mitigation will be needed. However, coverage needs to be built for the dedicated spectrum. To make this economically viable, this can be done by using existing cell towers.

Table 5.4 Comparison of the interference mitigation techniques for drones (UL = uplink).

Technique	Gain potential	Complexity Drone	Network
Grid of fixed beams	High	Medium	Low
Interference cancellation	Low	Medium	Low
Power control	Medium (UL)	Low	Low–medium
Interference coordination	Medium–high	Low	High
Hybrid access	High	Low–medium	Low
Dedicated spectrum	High	Low	Low

In Table 5.4 we compare different interference mitigation techniques and the dedicated spectrum option in terms of gain potential and complexity for the drone and network side.

The dedicated spectrum option is a very attractive solution, but it requires that sufficient spectrum is reserved for this. From the four studied interference mitigation techniques, the grid of fixed beams is the most attractive, giving good gains at a slightly higher complexity at the drone. Interference cancellation alone does not have promising potential to solve the interference issues, while the gains from power control alone are limited to the uplink. Interference coordination can give good gains if enough cells are coordinated, but this will lead to a higher network signaling and synchronization complexity. The latter can be lowered by taking fewer cells into account, but this means lower gains. Furthermore, it needs to be noted that different techniques can be combined to reduce complexity of the different features and/or to get even better gains.

5.5 Summary and Outlook

The number of drones is increasing rapidly. It is well understood that, in order to allow for BVLOS, which opens up for further use cases, a reliable C2 link is needed. This C2 link requires a high reliability and wide area coverage to guarantee safety. A natural candidate for providing this C2 link are cellular networks, as they have already provided almost ubiquitous coverage, making it economically attractive.

Cellular networks are, however, designed to provide coverage for terrestrial users and not in the sky. Antennas are, for example, typically down-tilted. Furthermore, signals propagate further in the sky as compared to the ground, as there is less obstruction from buildings, vegetation, etc. This leads to enhanced signals, but, at the same time, results in stronger interference. On top of this, the number of visible interferers increases with UAV flying height. For relatively low-load networks, this is not a problem, but for medium- and high-load networks, the required reliability for the C2 service may not be reached without some additional measures. Additionally, when the drone is running an application with high uplink throughput towards the base station, e.g., video streaming, this causes significantly more interference to the other users in the network than a user device with similar service at ground level. This means that interference mitigation may be needed in the case of heavy load. From the studied interference mitigation techniques, the grid of fixed beams implemented on the drone and the hybrid access scheme are the most attractive solutions, giving good gains at a modest complexity increase on the drone. A completely different approach to ensure reliable C2 link communication is to operate in dedicated spectrum, but this may not be an option in all regions.

In general, today's cellular networks can provide coverage for drones in many application scenarios, and with small enhancements they will be able to provide high reliability for more challenging conditions with heavy network load. In highly challenging conditions, like Manhattan environments, 5G techniques are needed to provide a reliable C2 link by exploiting high density of drones, with operation above the VLL airspace, massive MIMO and beam space enhancements, improved reference signal design, etc.

References

1 GSMA (2019). A look into the future of mobile-enabled drones. https://www.gsma .com/iot/wp-content/uploads/2019/02/22166-Connected-Drones-Infographic-v3-002.pdf (accessed June 2019).

2 M. Mazur, A. Wiśniewski, J. McMillan et al. (2016). Clarity from above. PwC global report on the commercial applications of drone technology. PwC Tech. Rep., May.

3 3GPP (2018). Enhanced LTE support for aerial vehicles. 3rd Generation Partnership Project (3GPP), Technical Specification (TS) 36.777, version 15.0.0, January.

4 R. Amorim, H. Nguyen, P. Mogensen et al. (2017). Radio channel modeling for UAV communication over cellular networks. *IEEE Wireless Commun. Lett.* 6 (4): 514–517.

5 R. Amorim, H. Nguyen, J. Wigard et al. (2018). LTE radio measurements above urban rooftops for aerial communications. *Proceedings of the IEEE Wireless Communications and Networking Conference (WCNC)*, pp. 1–6.

6 T. B. Sørensen and R. Amorim (2018). DroC2om-763601. Preliminary report on first drone flight campaign. SESAR Joint Undertaking, Deliverable 5.1, March.

7 T. B. Sørensen, R. Amorim, and M. López (2018). DroC2om-763601. Report of first drone flight campaign. SESAR Joint Undetaking, Deliverable 5.2, October.

8 T. Rappaport (2002). *Wireless Communications: Principles and Practice*. Prentice-Hall Communications Engineering and Emerging Technologies Series. Prentice-Hall.

9 A. Al-Hourani and K. Gomez (2018). Modeling cellular-to-UAV path-loss for suburban environments. *IEEE Wireless Commun. Lett.* 7 (1): 82–85. February 2018.

10 3GPP (2012). Evolved Universal Terrestrial Radio Access (E-UTRA); Mobility enhancements in heterogeneous networks. 3GPP Tech. Rep. TS 36.839, version 2.0.0, September.

11 I. Kovacs, R. Amorim, H. C. Nguyen et al. (2017). Interference analysis for UAV connectivity over LTE using aerial radio measurements. *Proceedings of the 86th IEEE Vehicular Technology Conference (VTC)*, Fall, September, pp. 1–6.

12 Nokia, Reliable 3D connectivity for drones over LTE networks. Nokia, White Paper, May 2018.

13 H. C. Nguyen, R. Amorim, J. Wigard et al. (2018). How to ensure reliable connectivity for aerial vehicles over cellular networks. *IEEE Access* 6, 12304–12317.

14 D. Lee, H. Seo, B. Clerckx et al., E. Hardouin, D. Mazzarese, S. Nagata, and K. Sayana (2012). Coordinated multipoint transmission and reception in LTE-advanced: Deployment scenarios and operational challenges. *IEEE Communications Magazine* 50 (2): 148–155. February 2012.

15 G. Pocovi, T. Kolding, M. Lauridsen, R. Mogensen, C. Markmller, and R. Jess-Williams (2018). Measurement framework for assessing reliable real-time capabilities of wireless networks. *IEEE Communications Magazine*. 56 (12): 156–163. December 2018.

16 R. Amorim, J. Wigard, I. Z. Kovacs, T. Sorensen, P. Mogensen, and G. Pocovi (2019). Improving drone's command and control link reliability through dual-network connectivity. *2019 IEEE 89th Vehicular Technology Conference (VTC Spring)*, May 2019, pp. 1–5.

17 R. Amorim, I. Z. Kovács, J. Wigard, T. B. Sorensen, and P. Mogensen (2019). Forecasting spectrum demand for uavs served by dedicated allocation in cellular networks. *2019 IEEE Wireless Communications and Networking Conference (WCNC)*, pp. 1–6.

6

3GPP Standardization for Cellular-Supported UAVs

Helka-Liina Määttänen

Ericsson Research, Finland

Use cases for drones, or more officially unmanned aerial vehicles (UAVs) or unmanned aerial systems (UASs), requiring beyond-radio-line-of-sight (BRLOS) connectivity vary from package delivery to infrastructure monitoring and rescue services [9, 12, 15]. Using cellular networks for drone connectivity is seen as an interesting way to enable BRLOS connectivity to UAVs, as important functionality such as security, integrity, latency, capacity, and coverage are readily in place. Further, the mobile networks are already widely deployed, and the coverage and capacity are being enhanced continuously.

The 3rd generation partnership project (3GPP) has been the dominating standardization development body of several global mobile technologies. The international standardization effort helps to ensure compatibility among vendors and reduce network operation and device costs. The 3GPP completed the first global 5th generation (5G) new radio (NR) standard in its Release 15. In parallel, also the long-term evolution (LTE), 4th generation (4G), development continues.

In this chapter, we give a short introduction to LTE and NR from the UAV operation perspective. We discuss the characteristic when UAVs are served by cellular networks, and present in detail the specified LTE support for UAV operation in mobile networks, comparing it to NR. Finally, we discuss that drone flight mode detection is regarded as one of the critical aspects when drones are served by mobile networks.

6.1 Short Introduction to LTE and NR

The first phase of the NR standard was finalized in December 2017. It is called a non-standalone NR, as it is based on multi-radio dual connectivity (MR-DC) between an LTE radio access technology (RAT) and the NR RAT. The dual connectivity means that the user equipment (UE) is connected to two nodes simultaneously, and it is a concept introduced already in LTE Release 12. In the non-standalone NR operation, the NR connectivity may be added to the UE after the UE has initiated connection to LTE. The second phase of the NR standard is called the standalone NR and it enables UEs to access

UAV Communications for 5G and Beyond, First Edition.
Edited by Yong Zeng, Ismail Guvenc, Rui Zhang, Giovanni Geraci, and David W. Matolak.

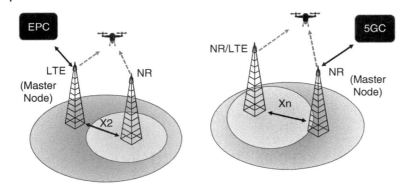

Figure 6.1 MR-DC with eNB as master node on the left-hand side and gNB as master node on the right-hand side.

directly the NR RAT and to have only NR connectivity. The specification for the standalone NR was finished in June 2018.

From the core network (CN) perspective, both Evolved Packet Core (EPC) and 5G Core Network (5GC, 5GCN) can be used with the MR-DC solution. The EPC is the existing CN for LTE, and 5GC is the new CN specified for NR. If the LTE node (eNB) is a master node in MR-DC, the CN connection from the radio access network (RAN) is to the EPC. If the NR node (gNB), or an enhanced LTE node called NG-eNB, is the master node, then connectivity is to 5GC [4]. Figure 6.1 depicts these options. In order to enable the MR-DC, there needs to be an X2 or Xn interface set up between the network nodes, as shown in the figure. This interface is used for exchanging control information between the nodes as well as for enabling the connectivity for the UE via the two nodes.

In both LTE and NR, the connection between UE and eNB/NG-eNB/gNB as well as eNB/NG-eNB/gNB and CN consists of control plane (CP) and user plane (UP). The CP connection as well as CP radio protocol between eNB/NG-eNB/gNB and UE is called Radio Resource Control (RRC) which is specified for LTE in [2] and for NR in [5]. With RRC, the eNB, NG-eNB, or gNB can control the LTE/NR connectivity of the drone. All data for the UE, like video streaming by the UAV, belongs to UP. It should be noted that the command and control, or traffic between UAV and the UAS traffic management (UTM) used for the drone flight or other control, is also user plane traffic from the 3GPP perspective.

Given that the first product roll-outs of NR will be MR-DC with LTE, it can be viewed that NR already has initial support for UAVs. This is because the UE is connected to LTE and if the eNB is master node and UE supports the LTE Release 15 aerial features like height reporting, there is no difference from the aerial functionality support point of view. However, it assumes that the UAV supports MR-DC.

In the rest of this section, we give a short comparison of LTE and NR physical layer. We concentrate on the differences in multi-antenna, multiple-input multiple-output (MIMO), aspects. The multiple antennas at the transmitter can be used for transmit beamforming, which means essentially that signal coverage can be steered. It also gives the possibility to control interference. For UAV operation, the interest in MIMO schemes stems from the theoretical potential to steer the beams towards the sky, especially if combined with active antennas.

6.1.1 LTE Physical Layer and MIMO

In LTE, the cell-specific reference signal (CRS) are transmitted "cell wide," which means that the bandwidth of the CRS is the same as the downlink bandwidth of the cell. The CRS is also always-on and is sent in every downlink subframe, that is, every 1 ms. The other always-on reference signals are the primary synchronization signal (PSS) and secondary synchronization signal (SSS) which are transmitted at the center of the cell downlink bandwidth spanning 72 subcarriers as a pair of PSS/SSS and the pair is repeated every fifth subframe. UE uses the PSS/SSS to detect the cell and CRS to measure the received signal strength from the cell. Spatially, the CRS as well as the PSS/SSS cover the whole cell. In LTE, no additional beamforming weights, that is, no precoding, is applied to PSS/SSS/CRS. From Release 10 onwards, LTE also supports a UE-specific reference signal called the channel state information reference signal (CSI-RS). The CSI-RS cannot be searched and found by the UE but has to be separately configured for each UE, and can thus be used only when the UE has active connection to the network. Further details on LTE physical layer can be found in [7].

LTE supports MIMO from the first release (Release 8), and since Release 10 up to eight transmit antennas are possible at the eNB. Even though the PSS/SSS/CRS are always non-precoded, it is possible to precode the user-specific transmission with selected antenna weights. In order to know which precoding weight should be applied to a given UE, feedback is needed from the UE. Until Release 13, also the CSI-RS is non-precoded. The possibility to configure CSI-RS in a user-specific manner makes it an enabler of more advanced MIMO schemes as well as Coordinated Multi-Point (CoMP) schemes introduced in Release 11.

The most advanced LTE MIMO scheme, called full-dimension MIMO (FD-MIMO), was introduced in Release 13 and was slightly improved in Release 14. There actually exist two different FD-MIMO schemes, Class A and Class B. The Class A FD-MIMO enables the UE to be configured with a CSI-RS that represents up to 32 antenna ports. The UE would then measure that CSI-RS and find a suitable precoding weight for a selected number of spatially multiplexed streams that the downlink transmission would best support. In Class B, the UE is configured with a set of CSI-RSs with maximum number of eight CSI-RS resources in the set. Each CSI-RS in the set represents up to eight antenna ports, which means there are potentially eight distinct beams in the set, as each CSI-RS can be differently precoded. For Class B, the UE measures the channel experienced from each of the CSI-RS beams and selects the strongest beam. The UE selects also an additional precoder weight as well as the best transmission rank assuming the selected beam. The UE feeds back these together with the selected beam index to the network.

The Class A and Class B seem to provide the capability to split resources between aerial and terrestrial UEs for LTE deployments that attempt to provide specialized UAV support. However, it should be kept in mind that the LTE MIMO schemes are for connected mode UEs and rely on UE-specific configurations given by the eNB. This means that, when the UE is mobile, it needs to receive a new configuration in each handover that applies to the new serving cell. There is a certain delay involved before the UE is able to feed back the first MIMO feedback to the network after receiving the MIMO configuration.

6.1.2 NR Physical Layer and MIMO

The NR counterpart for the LTE cell-specific always-on reference signal CRS is called Synchronization Signal and PBCH block (SSB). While CRS is always cell-bandwidth wide and is sent in every subframe, the SSB transmission is hugely reduced in both frequency and time. The width of the SSB is 240 subcarriers and the length in time is four OFDM (orthogonal frequency-division multiplexing) symbols. One SSB consists of PSS/SSS and the Physical Broadcast Channel (PBCH) carrying a Master Information Block (MIB) and demodulation reference signal (DMRS). The PSS/SSS of NR have similar structure as in LTE and enable the detection of the cell. The DMRS is used for measuring the SSB and for the UE to receive the MIB. The SSBs are transmitted in bursts, and the periodicity of a burst may vary from 5 ms to even 160 ms. For a cell that enables initial access, the maximum periodicity of the burst is 20 ms. Each SSB in a burst may be transmitted with different beamforming weights and is often referred to as SSB beam. A cell is thus covered by one or more of these SSB beams as depicted in Figure 6.2. This makes the biggest difference to LTE, as in NR the SSB (PSS/SSS/DMRS) is beamformed and one cell coverage is formed out of multiple such beams. As these are the always-on reference signals of NR, the initial access to the serving cell by the UE is via an SSB beam whereas in LTE it is via the "cell wide" PSS/SSS/CRS.

Depending on the frequency range, a different maximum number L_{max} of SSBs per burst are allowed: below 3 GHz, $L_{max} = 4$; between 3 GHz and 6 GHz, $L_{max} = 8$; and above 6 GHz, $L_{max} = 64$, as at higher frequency more beamforming gain is needed to compensate the higher signal loss. The maximum number refers to the number of possible nominal locations of an SSB in time. A network may choose how many SSBs are present in the bursts at a given frequency position where the bursts of SSBs are transmitted. All cells that have SSB transmissions on that frequency location need to have SSBs on those nominal locations present within the burst. In the example in Figure 6.2, $L_{max} = 4$ and the cell is formed of three SSB beams. If the cell supports initial access, that is, it is a standalone NR cell, a UE may access the cell via any of the SSB beams, as the initial system information giving

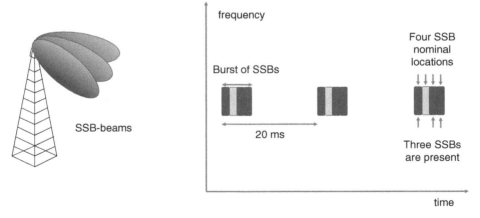

Figure 6.2 SSB basic principles in NR. Three consecutive bursts of SSBs with a periodicity of 20 ms are shown. Each burst has four nominal locations for an SSB ($L_{max} = 4$), and three SSBs are present in each burst.

information how the cell can be accessed is broadcast in all of the beams. Information on whether the initial system information is present and how to find it is given in the MIB within PBCH. Moreover, paging messages are sent via all of the beams.

NR also supports configuring the UE with UE-specific CSI-RS resources. The NR MIMO framework allows a lot of flexibility on how and with what kind of assumptions the different CSI-RS resources can be configured. As these NR MIMO schemes rely on UE-specific configurations, the same trade-offs as in LTE apply. That is, these are valid only in connected mode, and a new configuration is applied in every handover. However, as SSB beams are the always-on signals that enable initial access, some further advantages can be seen. In an NR network that is planned to support drones specifically, one or two SSB beams could be steered upwards and aerial UEs could access the cell via these beams. As an NR UE directs its uplink beam towards the selected SSB beam, the interference to other cells could be reduced. More information on the NR system can be found in [8].

6.2 Drones Served by Mobile Networks

Due to the significant interest in the possibility to use cellular networks for low-altitude UAV connectivity, the ability of existing LTE networks to provide connectivity services for UAVs has been studied by 3GPP. The 3GPP study item (SI) on enhanced LTE support for aerial vehicles was initiated in 2017 with objectives to study the performance of terrestrial mobile networks, identify challenges, and highlight possible solutions [13].

At the beginning of the study, 3GPP defined the performance requirements for UAV operation, such as data rate, reliability, and latency, as listed in Table 6.1. For UAV operation, the data traffic can be categorized into command and control data and application data. The command and control data includes critical information such as flight route for autonomous flights or real-time piloting, while application data includes, for example, video streaming and images. These data types have different requirements in terms of data rate, latency, and reliability. Reliability is defined as the success probability of transmitting X bytes within a latency bound from an eNB to UE. Latency is measured over the air interface and all radio access data protocol processing. That is, an eNB wrapping up an internet protocol (IP) packet containing the X bytes into a packet data convergence protocol (PDCP) packet, further to radio link control (RLC) and medium access control (MAC) packets, which are then delivered via the air interface by means of the physical layer, received

Table 6.1 3GPP SI performance requirements for UAV operation in LTE networks.

Data type	Command and control	Application data
Data example	Telemetry, flight route for autonomous flights, real-time piloting, flight authorization	Video streaming, image transfer, sensor data
Latency	50 ms from eNB to aerial UE	Similar to LTE terrestrial UEs
Data rate	60–100 kbps for both uplink and downlink	Up to 50 Mbps for uplink
Reliability	Up to 10^{-3} packet error loss rate	Not applicable

by the UE physical layer, processed by MAC, RLC, and PDCP to receive the IP packet. The definition of reliability was defined in the NR study item phase and described in the corresponding Technical Report [1]. The PDCP, RLC, and MAC are user plane protocols in both LTE and NR. Though those have separate specifications for LTE and NR, the protocols are similar to a large extent.

For drones, the command and control packet size was taken as $X = 1250$ bytes, one-way latency requirement was defined to be 50 ms, and reliability 10^{-3} packet error loss requirement. For application data type of traffic, the latency requirement was identified to be similar to LTE terrestrial users and the uplink data-rate requirement was identified to be up to 50 Mbps.

During the study item, both simulations and field studies were conducted in order to evaluate LTE performance for drones, and the results are reported in the Technical Report 36.777 [3]. In general, it is concluded that the existing LTE networks targeting terrestrial usage can support the initial deployments of low-altitude drones especially when the number of drones in the network is relatively low. The more detailed findings are that, although existing LTE networks can readily serve the drones, there are also challenges related to interference and mobility. Also the capability of the network to identify and authorize the aerial UE was identified to be essential.

The study item followed with a work item, during which further functionality to support UAV operation was specified in LTE Release 15. In the remainder of this section, we discuss certain study item findings in more detail and explain how the work item objectives were identified. In the next section we describe in detail LTE Release 15 standardization support.

6.2.1 Interference Detection and Mitigation

When flying above antenna boresight in a network that has been deployed for providing terrestrial coverage, the drones are served by side lobes of the base station (BS) antenna patterns. The drone has close to line-of-sight visibility to multiple BSs, which means that potentially both uplink (UL) and downlink (DL) interference increases. Indeed, one outcome of the study item was that, when the aerial UE is airborne, the close to line-of-sight propagation condition changes the overall interference situation. That is, an aerial UE above the base station height sees interference from multiple neighbor cells. Figure 6.3 depicts the situation.

Even if full reciprocity may not be assumed in a frequency-division duplexing (FDD) system, the link strengths in UL and DL are not totally independent. Consequently, the study item results show that an airborne aerial UE sees multiple reference signal reference power (RSRP) values from different cells which are close to equally strong [3]. The RSRP is measured from the DL reference signal and is the main measurement parameter used to indicate the signal strength of a given cell. As a result, detecting the event of a UE seeing multiple strong RSRP values was seen as an important interference detection method. This was also identified as one of the methods needing standardization changes; the details are described in Section 6.3.1.

As this happens when a UE is airborne, the reporting triggered based on number N of cells can also be used as input to flying mode detection. As the flying mode detection and drone identification are also relevant from the regulation perspective, we have dedicated

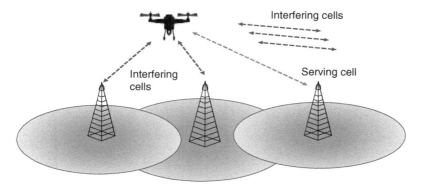

Figure 6.3 Due to close to line-of-sight connection to multiple BSs, the UL and DL interference potentially increases.

Section 6.4 at the end of this chapter in order to provide a summary with all the interrelations between implementation, standardization, and regulatory aspects. It should be noted that there is no explicit definition for the flying mode, as the drone may be flying at very low altitude and thus may not differ from a mobile phone used by a person, or the drone may be sitting on a shelf. From the network perspective, flying mode is linked to the changed interference conditions and may vary between network deployments.

In addition to RSRP reporting, feedback related to the MIMO techniques described in Section 6.1 was considered to be a valid input for interference detection. For example, the FD-MIMO feedback from the UE may indicate that the UE is seeing the beams directed towards the sky stronger than what a terrestrial UE would see. Also the existing MIMO techniques potentially work for interference mitigation, as the signal can be steered towards the drone, or from the drone towards the base station. However, no standardization impact to enhance the respective schemes was identified during the study item, and it was concluded that these can be readily used as per implementation choice.

Another identified method for interference detection and mitigation is cell coordination, also called CoMP techniques. Release 11 LTE has support for different CoMP techniques like joint transmission and dynamic point muting described in [11]. For example, eNBs may measure the uplink interference detected and exchange information such that the serving eNB knows whether the neighboring nodes are experiencing interference from a particular UE it is serving. This is possible if the eNBs have first exchanged information on how the uplink reference signals are configured for the UE. When this is known, the serving eNB is able to limit the uplink scheduling for that UE. However, it should be noted that in practice there may be too much delay involved in the configurations and setup for the information to be available for fast enough reaction for a flying drone. Also, the backhaul requirement and whether the backhaul even exists between the respective nodes affects the feasibility of the cell coordination techniques. The LTE study item conclusion was the same as for FD-MIMO. The standard already provides support for use of these techniques for drone interference detection and mitigation.

The LTE study item identified an enhancement related to uplink power control as an interference mitigation scheme needing LTE specification change. This enhancement is described in Section 6.3.3.

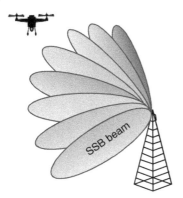

Figure 6.4 NR SSB beams forming coverage over the sky.

SSB beam

The advanced MIMO for both LTE and NR support relies on user-specific reference signals which need to be separately configured for each UE after RRC connection is established. However, in NR also the always-on reference signals are beamformed, thus potentially NR has better repertoire for MIMO-based interference management techniques for drones. In addition to connected mode operation, UEs use the always-on reference signals for initial access and idle mode mobility. Figure 6.4 depicts the NR SSB beams forming coverage over the sky. In NR, each of the SSB beams has a beam index that can be read from system information which is broadcast within each beam. When UE selects one beam, it also steers its own receive/transmit beam towards the selected beam. Thus, uplink interference is likely to be less in the NR case. However, 3GPP has not studied UAVs in the NR context so far.

In [16], field results from a commercial network were presented such that interference from one drone to a terrestrial UE was measured in a rural environment. The conclusion was that the interference caused by a drone is similar to the interference caused by another terrestrial UE. In [6], the coexistence is studied by simulations. In [10], further simulation results and field measurements are presented. The overall conclusion is that one or two drones do not cause considerable issues for cellular networks, but if the number of drones increases in an area, there is a risk of increased interference. Due to this, there is some relevance to know where the drones are flying or where the drones are planning to be flying in addition to single aerial UE flying mode detection.

6.2.2 Mobility for Drones

Another identified area for potential enhancement was mobility. When the flying altitude increases, the coverage area of the cells becomes more scattered and the signal strength fluctuates. The reason for this effect is that in existing LTE networks the antennas are usually tilted downwards so that the main lobe of the radiation pattern covers the ground around the base station where most of the UEs are normally located. A drone UE on the ground will be served as a regular LTE smartphone. Once the drone starts moving vertically upwards, the strength of this main lobe will decrease, and the side lobes of neighboring base stations will become stronger. These side lobes are narrower than the main lobe; thus naturally the coverage area of a side lobe is smaller.

The mobility evaluations as well as field trials conducted during the study item give slightly contradictory outcomes, as some deduce that mobility performance is not affected by flying state and some see increased handover failures and radio link failures. This is most likely due to the strong dependence on the deployment scenario applied for the study. For rural areas, the drone is more likely to see good mobility performance, whereas in more dense urban areas, due to signal fluctuations and interference, the mobility issues are more evident [3].

There is no straightforward specification-change-related improvement for drone mobility, as a similar situation for fast-changing coverage also applies to the high-speed terrestrial UEs. Thus, the mobility issues are similar, and there have been numerous past efforts to improve the mobility performance of LTE networks. Mostly what can be improved for drones is that, if the flying status of the drone is known, the RRC configuration can be adapted to the flying mode. For example, it is seen during the mobility simulations [3] that triggering measurement faster is beneficial when the UE is airborne. It can be concluded that the flying mode detection is important for drone mobility performance.

Similar to interference detection and mitigation, the NR specification enables better flexibility for deployments that could offer enhanced coverage towards the sky, as some SSB beams can be steered upwards (see Figure 6.4).

6.2.3 Need for Drone Identification and Authorization

During the study item, the need to identify and authorize an aerial UE was brought up. This relates to regulations that vary between different countries. In some countries, like in Japan, the UE cannot be flying while connected to an LTE network unless it is authorized. Depending on country-specific regulations, the aerial UE authorization may need to be verified by the network in order to allow the use of LTE networks for aerial UE connectivity. Another aspect is that there may be a drone-specific service or charging by the operator.

Even if the UE were to be authenticated by the Authentication, Authorization, and Accounting (AAA) server to the LTE network, this would not as such tell anything about authorization for aerial operation in the network. Similarly, the same device may have a connection to a UTM and receive aerial authorization from there. Nevertheless, as there is no standardized interface between the UTM and LTE network, there is no specification support for that kind of UE authorization.

One important distinction that needs to be taken into account is whether the discussion is about identifying aerial UEs that are based on Release 15 LTE specification, or identifying any potentially airborne UEs served by the mobile network. For the latter, the only means to identify drone operation is to try to detect the flying mode of any UE. This cannot be affected by specification work, as it is not possible to change earlier versions of the specification in a non-backward-compatible manner – or already manufactured UEs. Methods for detecting flying mode for any LTE UE are based on deducing potential flying state based on certain parameter values known by the network; this is discussed in Section 6.4.

To identify a Release 15 aerial UE, the UE capability signaling can be utilized. A complication here is that there is no such feature of "being an aerial UE" that could be indicated as one bit. Instead, the Release 15 specifies a set of features that may be considered as sub-features for an aerial UE. These features are described in Section 6.3 and each feature is associated with a capability bit that tells the network whether the UE supports a certain feature or not. When the UE reports such capabilities to the network, the network can in principle deduce that it is an aerial UE.

Yet another aspect is whether a third party, e.g., an authority inspecting a drone from the ground, needs to identify the drone. Identifying a flying drone by a third party might require a global identifier for the drone that can be recognized or read by the third party. This aspect is considered by different regulatory bodies like the Federal Aviation Administration (FAA) and the European Conference of Postal and Telecommunications Administrations (CEPT).

The LTE Release 15 solution for aerial UE identification and authorization is described in Section 6.3.5.

6.3 3GPP Standardization Support for UAVs

In this section, we provide a detailed review of specification support for UAVs in Release 15 LTE and describe the specification status and needs for NR. The key features are:

- interference detection based on measurement reporting that is triggered when a configured number of cells fulfills the triggering criteria;
- height and location reporting based on the event that the UE's altitude has crossed a network-configured threshold altitude;
- introduction of a UE-specific fractional path-loss compensation factor for UL power control;
- signalling of flight path information from the UE to the network; and
- subscription-based aerial UE identification and authorization.

6.3.1 Measurement Reporting Based on RSRP Level of Multiple Cells

A UE in connected mode is configured to measure and report serving and neighbor cells in order to know a strong enough neighbor cell as handover target when serving cell quality drops. The RSRP with event-based measurement report triggering is the main measurement configuration type for mobility purposes. In LTE, several different events are specified in [2]. The relevant events to be considered here are called A3, A4, and A5. Event A3 is defined as the event when a neighbor cell becomes a configured offset better than the serving cell. Event A4 is the event that a neighbor cell becomes better than an absolute threshold. Event A5 is the event where the serving cell becomes worse than one absolute threshold and the neighbor cell becomes better than another absolute threshold. Also a parameter called time-to-trigger (TTT) plays a role here. The legacy event-based reporting works such that an event entry condition needs to be fulfilled through a time window defined as TTT for the measurement to be triggered to be sent to the network. The TTT is a value that is given to the UE in a measurement configuration that also describes the events and corresponding thresholds the UE should consider when measuring other cells. There are also other parameters, but these are the main ones for this discussion.

As discussed in Section 6.2.1, detecting the case of a UE seeing multiple strong RSRP values was seen as an important interference detection method needing specification change.

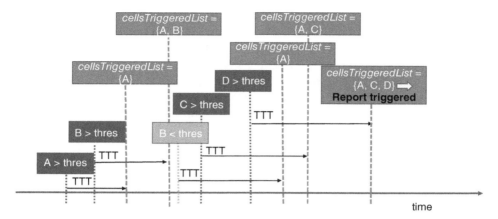

Figure 6.5 Triggering of measurements when an event is configured with $N = 3$, where N is the number of cells of which RSRP should fulfill the event.

The RSRP values from legacy reporting used for mobility can be used for flying mode detection, but it requires a machine learning algorithm implemented in the network, discussed further in Section 6.4. In order to directly identify the case where the UE sees multiple strong neighbor cells, an enhanced triggering condition was specified. The enhanced triggering would require, for example, three neighbor cell RSRP values to become higher than a threshold, or to fulfill a selected event entering condition for TTT. Figure 6.5 attempts to depict the triggering of measurement report when an event is configured with $N = 3$, where N is the number of cells of which RSRP should fulfill the event.

First, the RSRP of a cell A fulfills the entry condition of the event for TTT and is added to the *cellsTriggeredList* which is used by the UE to keep track of cells fulfilling the entry condition for TTT. The same happens for cells B and C, but before cell C is added to the list, the cell B fulfills a leaving condition of the event for a TTT and is thus removed from the *cellsTriggeredList*. Both entry and leaving conditions involve a configured hysteresis in order to avoid a ping-pong effect. Only when there are N cells that have all fulfilled the entry condition for TTT is the measurement report triggered. In this way, the report is triggered when the UE currently sees multiple strong neighbor cells and not when it has measured one cell while flying and another later.

After the report has been triggered, the specification does not allow another measurement report to be immediately triggered if a fourth cell fulfills the entry condition for TTT. This is to avoid unnecessary uplink data being sent by the UE, as the main purpose is to detect the case when N cells are above a threshold. Further, as the network may configure the UE with more than one reporting configuration, the UE can be configured to report for $N = 3$, $N = 5$, and so on.

Another remark to be made from Figure 6.5 is that, when $N > 1$, the reporting of RSRP is delayed compared to the reporting that would happen if $N = 1$. We discussed in Section 6.2.2 that the measurement report needs to be triggered earlier when the UE is flying than for terrestrial UEs, as the cell coverage area is smaller at higher altitude. Thus, the interference detection RSRP reporting does not compensate for measurement configuration for mobility purposes. That is, in order to support good mobility performance, which is enabled by timely RSRP reporting by the UE, the UE needs to be configured always with $N = 1$ in addition to $N > 1$, which is used for detecting certain interference situations.

The measurement framework for NR is based on the measurement framework of LTE. The biggest differences lie in how a cell quality is derived from RSRP values measured from SSB beams. The cell quality derivation is controlled by a separate configuration, and one RSRP value is derived per cell as a cell quality. Given this, the above-described interference detection reporting could be directly specified also to NR.

6.3.2 Height, Speed, and Location Reporting

An LTE network may configure a terrestrial UE to piggyback location coordinates and horizontal velocity to any RSRP measurement report. The UE would piggyback the location and speed if it has the location information and speed information available. The event-based triggering of when this report is sent is related, not to the location or velocity of the UE, but to the RSRP values according to, for example, event A3 as described in the previous subsection.

An enhancement identified and agreed to be specified for aerial UEs is that the triggering of the report would be height-dependent and that the report includes also height and vertical velocity in addition to horizontal velocity and location. That is, the event triggers the height, speed, and location reporting, where the event threshold is the height of the UE. When the UE is configured with this event, a report is triggered when the UE's altitude crosses the threshold altitude. Figure 6.6 depicts this situation.

3GPP is specifying a conformance test for UE for height reporting to test that height reporting works properly for UEs that are indicating it as their capability. The report needs to include the height of the UE, but location, vertical and horizontal speed are in the report if available at the UE.

The scope of the 3GPP work is for low-altitude UAVs and it was agreed to consider UAVs up to 300 m above ground. For this reason, the configured height threshold needs to cover heights up to 300 m above ground level. As ground level with respect to sea changes with location, the height threshold specified in [2] covers heights from −420 m to 8880 m from sea level.

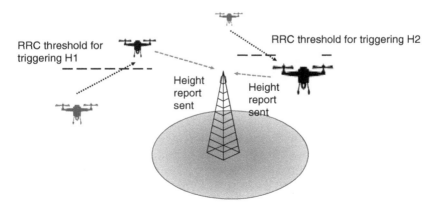

Figure 6.6 Drone height, speed, and location reporting based on height threshold.

The drone height, speed, and location reporting based on height threshold does not have an NR-specific component, which means also that this could be readily added to NR RRC. From the UE perspective, the only difference would be which RRC (LTE or NR) configures the height reporting.

6.3.3 Uplink Power Control Enhancement

As discussed in Section 6.2.1, the uplink interference mitigation discussion that led to specification change is related to UE power control. LTE has both open-loop and closed-loop power control mechanisms. The closed loop is by definition UE-specific, as the network gives power control command per UE. The open-loop power control was enhanced during the work item such that a UE-specific fractional path-loss compensation factor has been introduced. This enables to set this parameters differently for terrestrial and aerial UEs in a given cell, which was concluded to be beneficial for the cell's overall performance [3]. In addition, the range of the UE-specific P_0 parameter (which is an open-loop power control parameter) is extended to further help mitigate the UL interference for aerial UEs to multiple cells.

The introduction of a UE-specific power control parameter enables the network to reconfigure the power control parameter for the UE when the UE is detected to be in flying mode, or simply when above a configured height threshold. Flying mode detection is discussed in detail in Section 6.4.

UL interference mitigation for LTE is further discussed in [17]. NR specification already supports a UE-specific fractional path-loss compensation factor.

6.3.4 Flight Path Signalling

The Release 15 LTE introduces support for the network to request flight path information from the UE. An aerial UE may be connected to a UTM via an application layer or it may have a flight path from the user. The communication between the UTM and the UE is not visible to LTE or NR networks as per specification as there is no standardized interface between any LTE/NR logical node and the UTM. Thus, in order for the eNB to know the flight path information, it needs receive the information from the UE.

The simplest way to represent a flight path is the location information of the take-off point and the landing point as well as some mid-way points. In addition to the location coordinates of the planned waypoints, the UE may include also time stamps describing when the UE is planning to arrive on the waypoint. There are no requirements on how accurate the flight path information needs to be, so it has to be considered as best-effort additional information.

During the work item it was discussed whether the UE would simply send flight path information to the eNB when it becomes available, or whether it would be based on eNB requesting and UE providing the information if available. Typically all signalling from the UE to the network is controlled by the network, and in order to avoid unnecessary and redundant uplink transmissions, the request–response signalling was agreed upon. In addition, it was agreed that the UE may inform the eNB on the availability of the flight path

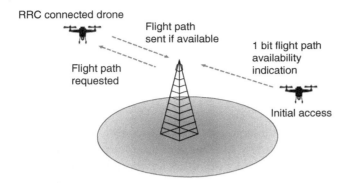

Figure 6.7 Flight path signalling from aerial UE to the network.

information, for example, when the RRC connection is established. The signalling of flight path information is depicted in Figure 6.7.

The use of the flight path information by the RAN can be debated. As discussed in Section 6.2.2, the cell coverage at higher altitudes is fragmented and it will be difficult to determine an accurate target cell based on planned waypoints. Thus, the flight path information cannot be directly used for a handover decision. However, the network can do coarse resource planning based on flight path information from drones. For example, if several drones are heading to a certain area, the network may prepare in advance for that.

The comparison between LTE and possible NR support for flight path reporting is the same as for height reporting. The only difference is which RRC (LTE or NR) would configure the reporting, as signalling the flight path is NR agnostic.

6.3.5 Drone Authorization and Identification

A drone may be connected to a UTM via the application layer and receive a flight authorization from there. This communication is not visible to LTE or NR networks, as there is no standardized interface between any LTE/NR logical node and UTM. Thus, even if the command and control data for the drone were to include drone authorization or drone identity, a network node would be unaware of that. As there are regulations, e.g., in the USA and in Japan, for the network to authorize drones that use mobile networks for connectivity, 3GPP specified a subscription-based authorization method.

The support of aerial UE function is stored in the user's subscription information in a home subscriber server (HSS), where also, for example, an International Mobile Subscriber Identity (IMSI) is stored as well as some security-related information. When a UE, terrestrial or aerial, accesses a cell and the RRC connection is about to be set up over the air interface, the eNB asks via an S1 AP interface from a Mobility Management Entity (MME) the subscription information for that UE. The MME receives the subscriber information from HSS via an S6 interface.

It should be noted that the subscription information is a user authorization, and in order to combine the user authorization to information about the device, the subscription information needs to be combined with certain radio capability information. Each of the specified aerial UE-specific features described in Section 6.3 are associated with capability bits.

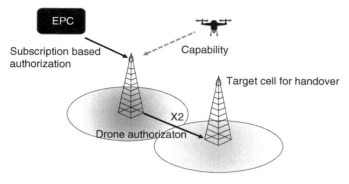

Figure 6.8 Forwarding subscription-based drone authorization to a target cell for handover.

Capability bits specified in current releases are typically optionally supported by UEs, which gives freedom for the UE manufacturer to select a combination of features to be implemented in a given UE.

However, here, due to the importance of drone identification, two of the aerial specific features have capabilities that are conditionally mandatory to the subscription that authorizes aerial usage. That is, if the device can have a subscription that has aerial UE authorization, it has to support both height reporting and measurement triggering based on *N* RSRP values. The eNB may then combine the subscription information with radio capability indication from the aerial UE in order to identify whether the aerial UE has been authorized to be connected to the E-UTRAN network while flying. The above-mentioned drone authorization does not yet describe when the UE is flying, but describes that the UE is allowed to fly while connected to the network.

Support to transfer the subscription authorization for connected mode UEs for mobility purposes was added to the specification. If there is an X2 interface between the source eNB and the target eNB to where the UE is handed over, the source eNB can include the subscription information in the X2–AP Handover Request message to the target eNB. This is depicted in Figure 6.8. If there is no X2 interface, the MME provides the subscription information to the target eNB after the handover procedure. Especially for drones, which may connect to a far-away cell and then be handed over to another far-away cell, an intra- or inter-MME S1-based handover may need to be performed.

6.4 Flying Mode Detection in Cellular Networks

As discussed, identifying a flying state of the drone by the network is fundamentally important, for several reasons. It is needed by regulations in some countries where a mobile phone cannot be in flying state without permission. Even without regulatory mandating it, flying mode detection is needed by the network in order for the network to control its resources and, for example, for interference detection and mitigation. In addition, adapting the measurement configuration may be beneficial to achieve better mobility performance for the airborne UE as discussed in Section 6.2.2.

Explicit flying mode determination is challenging, as a drone may be hovering at high altitude, or a terrestrial UE may just be, for example, on a balcony on a high floor.

A UE may be any regular pre- or post-Release 15 terrestrial UE that does not support the specified height reporting or RSRP reporting based on number N of cells. This kind of cellphone may be attached to a drone, or a drone may be manufactured without these aerial UE-specific features. Depending on a specific country's regulations, this kind of drone may be considered as a rogue drone or not. For example, currently, in Finland, if a pilot receives permission from authorities and from the operator, it is possible to attach a regular UE to a drone and fly while connected to the LTE network. Without these permissions, the same situation results in the device being a rogue drone. Regardless, detecting the flying status is important so the network can make appropriate actions, which may be limiting connectivity, controlling interference or optimizing the radio configuration. For example, the network may choose not to support many UEs flying in connected mode within a small area. In order to detect the flying mode of this type of UE, the network needs to rely on information that it is able to receive from these UEs. For example, the mobility RSRP reports can be used along with a machine learning algorithm to deduce the flying mode out of these individual RSRP reports. This has been studied in [14], where it was concluded that it is possible with good probability to determine the flying state of the drone based on individually triggered RSRP reports.

Another parameter that may in some scenarios reveal that UE is in flying mode is timing advance value. Timing advance is a value that each UE needs to apply in all uplink transmissions such that those arrive at the eNB during the correct uplink slot, which was indicated to the UE when the UL transmission was granted. The specified timing advance values cover cell ranges up to 100 km, and, if a UE needs a value that is considerably larger than the terrestrial coverage of the cells in a given area, it can be deduced that the UE is connecting to the cell from the sky. It should be noted that this is applicable only in scenarios where the terrestrial coverage of the cells is less than 100 km.

Yet another possible method also discussed during the study item relies on FD-MIMO feedback. If the deployment supports arranging for the Class B CSI-RS beams such that some beams are towards the sky, the flying mode can be deduced when UE reports these beams as the strongest beams. In order to be able to apply this method, in addition to the network, the UE also needs to support the FD-MIMO Class B feature.

All the above discussion applies to UEs that do not support Release 15 aerial features but also to UEs that support those. The Release 15 aerial features can be used on top of those already mentioned to improve and ease the flying mode detection. The most important features for flying mode detection are height reporting and RSRP reporting based on number N of cells. In real deployments, the base station antenna height may vary a lot. This means that the mere height of the UE does not necessarily mean that the UE is above all the base station antennas in the area. Or, the UE may already be considered to be airborne even if it is not above all base station antennas. Consequently, the height reporting or RSRP reporting based on number N of cells alone may not be sufficient for flying mode determination. Hence, when the information on these two reporting types is combined, the network has both height and interference situation knowledge, which makes flight mode detection more reliable. For this reason, and because detecting the flying mode was considered important,

Table 6.2 Functionality to support flying mode detection by a UE depending on whether the UE has subscription-based aerial UE authorization or not.

Functionality supported	Aerial authorization	Aerial authorization not supported
Height reporting	Yes	Depends on capability
RSRP for N cells	Yes	Depends on capability
Legacy RSRP	Yes	Yes
Timing advance	Yes	Yes
FD-MIMO	Depends on capability	Depends on capability

the support for these two features was specified to be conditionally mandatory for UEs that may have the subscription-based aerial authorization.

Table 6.2 summarizes the available methods that provide input for flying mode determination. The variety of methods depends on UE capabilities. Timing advance and RSRP measurements for mobility purposes, that is equivalent to $N = 1$, are supported by all LTE UEs. FD-MIMO related capabilities are all optional features. For example, UE might support Class A but not Class B. The height and RSRP reporting based on number N of cells is conditionally mandatory if an aerial UE has subscription-based authorization. For all other UEs, those are optional features.

References

1 3GPP (2017). Study on New Radio Access Technology Physical Layer Aspects, Release 15. *TR 38.802*, September. http://www.3gpp.org/

2 3GPP (2018). Radio Resource Control (RRC); Protocol Specification, Release 15. *TS 36.331*, September. http://www.3gpp.org/.

3 3GPP (2018). Enhanced LTE Support for Aerial Vehicles, Release 15. *TR 36.777*, June. http://www.3gpp.org/.

4 3GPP (2018). NR; Multi-Connectivity; Overall Description; Stage-2, Release 15. *TS 37.240*, September. http://www.3gpp.org/.

5 3GPP (2018). Radio Resource Control (RRC); Protocol Specification, Release 15. *TS 38.331*, September. http://www.3gpp.org/.

6 M. M. Azari, F. Rosas, A. Chiumento, and S. Pollin (2017). Coexistence of terrestrial and aerial users in cellular networks. *Proceedings of the IEEE Global Communications (GLOBECOM) Workshops, December, pages 1–6*.

7 J. Sköld E. Dahlman, S. Parkval. *LTE/LTE-Advanced for Mobile Broadband)*. Academic Press, 2011.

8 J. Sköld E. Dahlman, S. Parkval. *5G NR: The Next Generation Wireless Access Technology*. Academic Press, 2018.

9 M. Erdelj, E. Natalizio, K. R. Chowdhury, and I. F. Akyildiz (2017). Help from the sky: leveraging UAVs for disaster management. *IEEE Pervasive Comput.* 16: 24–32.

10 X. Lin, R. Wiren, S. Euler et al. (2019). Mobile networks connected drones: field trials, simulations, and design insights. *IEEE Vehic. Technol. Mag.* 14: 115–125.

11 H.-L. Määttänen, K. Hämäläinen, J. Venäläinen et al. (2012). System-level performance of LTE-Advanced with joint transmission and dynamic point selection schemes. *EURASIP J. Adv. Signal Process.* 54: 1–18. doi: 10.1186/1687-6180-2012-247y.

12 H. Menouar, I. Guvenc, K. Akkaya et al. (2017). UAV-enabled intelligent transportation systems for the smart city: applications and challenges. *IEEE Commun. Mag.* 55: 22–28.

13 Ericsson NTT Docomo (2017). Study on Enhanced LTE Support for Aerial Vehicles. *RP-170779.* http://www.3gpp.org/.

14 H. Ryden, S. B. Redhwan, and X. Lin (2019). Rogue drone detection: a machine learning approach. *Proceedings of the IEEE Wireless Communications and Networking Conference (WCNC)*, May, pages 1–6.

15 Goldman Sachs (2016). Drones: reporting for work. https://www.goldmansachs.com/insights/technology-driving-innovation/drones/.

16 J. Säe, R. Wiren, J. Kauppi et al. (2018). Public LTE network measurements with drones in rural environment. *Proceedings of the IEEE Vehicular Technology Conference (VTC) Workshops*, April, pages 1–5.

17 V. Yajnanarayana, Y.-P. E. Wang, S. Gao et al. (2018). Interference mitigation methods for unmanned aerial vehicles served by cellular networks. *Proceedings of the IEEE 5G World Forum*, July, pages 1–5.

7

Enhanced Cellular Support for UAVs with Massive MIMO

Giovanni Geraci[1], Adrian Garcia-Rodriguez[2], Lorenzo Galati Giordano[2], and David López-Pérez[2]

[1]*Universitat Pompeu Fabra, Barcelona, Spain*
[2]*Nokia Bell Labs, Dublin, Ireland*

7.1 Introduction

In this chapter, we adopt the view of a mobile network operator that, by rolling out its massive multiple input–multiple output (MIMO) based 5G network, aims to offer communication services to both ground users (GUEs) and unmanned aerial vehicles (UAVs) simultaneously, reusing its cellular spectrum and infrastructure. The chapter is thus devoted to answering the following question: Will a present-day network infrastructure suffice to meet the UAVs' link requirements – 100 kbps (kilobits per second) command and control (C&C) channel and uplink payloads demanding several megabits per second (Mbps) – set forth by the 3GPP standardization forum [3]? Or should the network, primarily catering to GUEs, undergo substantial upgrades to also accommodate UAVs?

In order to provide a well-founded answer, we evaluate the network performance and capture the propagation environment between ground base stations (BSs) and both GUEs and UAVs through the latest 3GPP 3D channel model [3]. In this model, parameters such as path loss, shadowing, probability of line-of-sight (LoS), and small-scale fading explicitly account for the users' height. In what follows, we provide a seminal evaluation of solutions that enable 5G-connected UAVs. The results of our extensive simulation campaigns [7, 16, 17] are overviewed, explained, and finally distilled into essential takeaways.

7.2 System Model

In this section, we introduce the network topology and channel model employed in this chapter. Further details on the parameters used are given in Table 7.1 (see subsequent text for acronyms).

UAV Communications for 5G and Beyond, First Edition.
Edited by Yong Zeng, Ismail Guvenc, Rui Zhang, Giovanni Geraci, and David W. Matolak.
© 2021 John Wiley & Sons Ltd. Published 2021 by John Wiley & Sons Ltd.

Table 7.1 System parameters.

Deployment		Ref.
BS distribution	Three-tier wrapped-around hexagonal grid, 37 sites, three sectors each, one BS per sector, 500 m inter-site distance	[3]
User distribution	15 users per sector on average	[3]
GUE distribution	80% indoor; horizontal: uniform; vertical: uniform in buildings of 4–8 floors	
	20% outdoor; horizontal: uniform; vertical: 1.5 m	
UAV distribution	100% outdoor; horizontal: uniform; vertical: uniform between 1.5 m and 300 m	[3]
UAVs/GUEs ratio	3GPP Case 3: 7.1%; Case 4: 25%; Case 5: 50%	[3]
User association	Based on RSRP (large-scale fading)	
Channel model		
Path loss, probability of LoS, shadowing, small-scale fading	Urban macro	[3, 5]
Channel estimation	Single-user: perfect channel estimation	
	Massive MIMO: UL sounding reference signals with Reuse 3	
Thermal noise	-174 dBm Hz^{-1} spectral density	[3]
PHY		
Carrier frequency	2 GHz	[3]
System bandwidth	10 MHz with 50 PRBs	[3]
BS transmit power	46 dBm	[3]
BS antenna elements	Horizontal and vertical half-power beamwidth: 65°; max. gain: 8 dBi	[3]
BS array	Height: 25 m; electrical down-tilt: 12°; element spacing: 0.5λ	[3]
BS array size	Single-user: 8×1 X-POL $\pm45°$, 1 RF chain	
	Massive MIMO: 8×8 X-POL $\pm45°$, 128 RF chains	
BS precoder	Single-user: none	
	Massive MIMO: zero-forcing	
Power control	DL: equal power allocation	
	UL: fractional with $\alpha = 0.5$, $P_0 = -58$ dBm, and $P_{\max} = 23$ dBm	[10]
User antenna	Omnidirectional with vertical polarization; gain: 0 dBi	[3]
Noise figure	BS: 7 dB; user: 9 dB	[3, 4]
MAC		
Traffic model	Full buffer	
Scheduler	Round robin with one/eight users per PRB for SU/mMIMO	[11]

7.2.1 Cellular Network Topology

We consider the downlink of a traditional cellular network (designed for GUEs) as depicted in Figure 7.1, where BSs are deployed on a hexagonal layout and communicate with their respective sets of connected users. Each deployment site is comprised of three co-located BSs, each covering one sector spanning an angular interval of 120°. The cellular network under consideration serves both GUEs and UAVs in downlink (DL) and uplink (UL), e.g., providing GUEs with DL/UL data streams and UAVs with UL data streams and DL/UL C&C information. In what follows, *users* denotes both GUEs and UAVs. GUEs are located both outdoors (at a height of 1.5 m) and indoors in buildings that consist of several floors. UAVs are located outdoors at variable heights between 1.5 m, which represents their height during take-off and landing, and 300 m, which is regarded as their maximum cruising altitude with cellular service. All deployment features comply with the ones specified by the 3GPP in [3].

The set of cellular BSs is denoted by $\mathcal{B}$, and we assume that all BSs employ a transmission power P_b per time–frequency physical resource block (PRB), given by the total

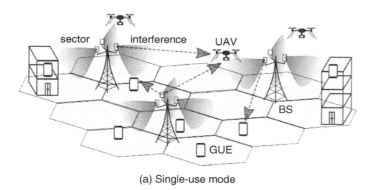

(a) Single-use mode

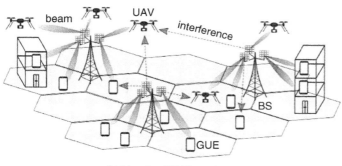

(b) Massive MIMO mode

Figure 7.1 Two examples of cellular infrastructure for supporting both ground and UAV users. In (a), similarly to many existing deployments, BSs cover a cellular sector with a vertical antenna panel and serve a single user on each PRB, potentially generating strong interference towards nearby users. In (b), which exemplifies next-generation deployments, BSs serve multiple users on each PRB through massive MIMO arrays and beamforming; this increases the useful signal power at each served user, and mitigates the interference towards nearby users.

BS transmit power divided by the number of PRBs. Users associate to the BS providing the largest reference signal received power (RSRP) across the whole communication band. Each BS is equipped with N_a antennas, and we assume all users to be equipped with a single antenna, unless otherwise specified. We denote as $\mathcal{K}_b$ the set of users served by BS b, and thus belonging to cell b, on a given PRB, and by K_b its cardinality. While the total number of associated users is determined by their density and distribution, the set $\mathcal{K}_b$ can be chosen adaptively by BS b through scheduling operations. In this regard, we identify two cases: the one where $K_b = 1$ (*single-user* mode operations) and the one where $K_b \geq 1$ (*massive MIMO* operations through spatial multiplexing). These two cases are illustrated in Figure 7.1(a) and 7.1(b), respectively, and described as follows.

- **Single-user mode (SU).** Each BS is equipped with an 8×1 antenna array of $\pm 45°$ cross-polarized (X-POL) radiating elements, electrically down-tilted by $12°$. The radio-frequency (RF) signals at the radiating elements are combined in the analog domain, and fed to a single RF chain. Hence, the BS can serve one device per PRB,[1] as illustrated in Figure 7.1(a).
- **Massive MIMO mode (mMIMO).** Each BS is still electrically down-tilted by $12°$, but equipped with an 8×8 antenna array of $\pm 45°$ cross-polarized radiating elements, as shown in Figure 7.1(b). We consider that each radiating element is connected to a separate RF chain; thus there are 128 single-element antennas connected to 128 RF chains. With this overhaul, each BS employs digital precoding and combining to spatially multiplex eight devices and perform 3D beamforming. To acquire the channel state information (CSI) essential to implement the above-mentioned capabilities, GUEs and UAVs transmit device-specific pilot sequences, which are reused every three cells, and massive MIMO BSs perform a conventional least-squares channel estimation [9].

A common feature of the SU and mMIMO paradigms is an equal time splitting between DL and UL data transmissions.

7.2.2 System Model

We adopt the latest 3GPP channel model for evaluating cellular support for UAVs [3]. In this model, all radio links are affected by large-scale fading (comprising antenna gain, path loss, and shadow fading) and small-scale fading. Among other real-world phenomena, the model accounts for 3D channel directionality, spatially correlated shadowing, and time-and-frequency correlated small-scale fading. Moreover, all propagation parameters for UAVs in the model – such as path loss, probability of LoS, shadow fading, and small-scale fading – have been derived as a result of numerous measurement campaigns, and explicitly account for the transmitter and receiver heights.

On a given PRB, $\boldsymbol{h}_{bjk} \in \mathbb{C}^{N_a \times 1}$ denotes the channel vector between BS b and user k in cell j. In the DL, the signal $y_{bk}^{\mathrm{DL}} \in \mathbb{C}$ received by user k in cell b can be expressed as

$$
y_{bk}^{\mathrm{DL}} = \sqrt{P_b}\, \boldsymbol{h}_{bbk}^{\mathrm{H}} \boldsymbol{w}_{bk} s_{bk}^{\mathrm{DL}} + \sqrt{P_b} \sum_{i \in \mathcal{K}_b \setminus k} \boldsymbol{h}_{bbk}^{\mathrm{H}} \boldsymbol{w}_{bi} s_{bi}^{\mathrm{DL}}
$$

$$
+ \sqrt{P_b} \sum_{j \in \mathcal{B} \setminus b} \sum_{i \in \mathcal{K}_j} \boldsymbol{h}_{jbk}^{\mathrm{H}} \boldsymbol{w}_{ji} s_{ji}^{\mathrm{DL}} + \epsilon_{bk}, \tag{7.1}
$$

---

1 In our study, a PRB occupies a bandwidth of 180 kHz and has a duration of 1 ms. This configuration is supported by existing LTE and upcoming 5G-NR systems.

where P_b is the power transmitted by BS b, assumed equal for all BSs, $s_{bk}^{DL} \in \mathbb{C}$ is the unit-variance DL signal intended for user k in cell b, $\epsilon_{bk} \sim \mathcal{CN}(0, \sigma_\epsilon^2)$ is the thermal noise, and $\boldsymbol{w}_{bk} \in \mathbb{C}^{N_a \times 1}$ is the transmit precoding employed by BS b to serve user k in cell b, normalized to satisfy the total power constraint. The four terms on the right-hand side of Eq. (7.1) represent, respectively: the useful signal, the intra-cell interference from the serving BS (only present for massive MIMO operations), the inter-cell interference from other BSs, and the thermal noise.

Assuming that the users have perfect CSI, the resulting instantaneous DL signal-to-interference-plus-noise ratio (SINR) γ_{bk}^{DL} at user k in cell b on a given PRB is obtained as an expectation over all symbols, and it is given by

$$\gamma_{bk}^{DL} = \frac{P_b \, |\boldsymbol{h}_{bbk}^H \boldsymbol{w}_{bk}|^2}{P_b \displaystyle\sum_{i \in \mathcal{K}_b \setminus k} |\boldsymbol{h}_{bbk}^H \boldsymbol{w}_{bi}|^2 + P_b \displaystyle\sum_{j \in \mathcal{B} \setminus b} \sum_{i \in \mathcal{K}_j} |\boldsymbol{h}_{jbk}^H \boldsymbol{w}_{ji}|^2 + \sigma_\epsilon^2}. \tag{7.2}$$

Similarly, in the UL, the vector $\boldsymbol{y}_b^{UL} \in \mathbb{C}^{N_a \times 1}$ of received signals at BS b can be expressed as

$$\boldsymbol{y}_b^{UL} = \sqrt{P_{bk}} \, \boldsymbol{h}_{bk} s_{bk}^{UL} + \sum_{i \in \mathcal{K}_b \setminus k} \sqrt{P_{bi}} \, \boldsymbol{h}_{bi} s_{bi}^{UL} + \sum_{j \in \mathcal{B} \setminus b} \sum_{i \in \mathcal{K}_j} \sqrt{P_{ji}} \, \boldsymbol{h}_{bji} s_{bji}^{UL} + \boldsymbol{e}_b, \tag{7.3}$$

where $s_{jk}^{UL} \in \mathbb{C}$ is the unit-variance UL signal transmitted by user k in cell j. The four terms on the right-hand side of Eq. (7.3) represent, respectively: the useful UL signal from user k in cell b, the intra-cell interference from other users in the same cell (only present for massive MIMO operations), the inter-cell interference from users in other cells, and the thermal noise vector at BS b with independent and identically distributed entries following $\mathcal{CN}(0, \sigma_e^2)$. As for the power P_{jk} transmitted by user k in cell j, we assume fractional UL power control as follows [8, 10]

$$P_{jk} = \min \{P_{\max}, P_0 \overline{h}_{jjk}^{-\alpha}\}, \tag{7.4}$$

where $P_{\max}$ is the maximum user transmit power, P_0 is a cell-specific parameter, α is a path-loss compensation factor, and $\overline{h}_{jjk}$ is the average channel gain measured at UE k in cell j based on the RSRP [2, 6]. The aim of Eq. (7.4) is to compensate only for a fraction α of the path loss, up to a limit specified by $P_{\max}$.

The resulting instantaneous UL SINR γ_{bk}^{UL} for user k in cell b on a given PRB is given by

$$\gamma_{bk}^{UL} = \frac{P_{bk} \, |\boldsymbol{w}_{bk}^H \boldsymbol{h}_{bbk}|^2}{\displaystyle\sum_{i \in \mathcal{K}_b \setminus k} P_{bi} |\boldsymbol{w}_{bk}^H \boldsymbol{h}_{bbi}|^2 + \sum_{j \in \mathcal{B} \setminus b} \sum_{i \in \mathcal{K}_j} P_{ji} |\boldsymbol{w}_{bk}^H \boldsymbol{h}_{bji}|^2 + \sigma_e^2}, \tag{7.5}$$

where $\boldsymbol{w}_{bk}$ is the receive filter employed by BS b for user k in cell b, and is assumed equal to the precoder employed in the DL.

For both DL and UL, each value of SINR is mapped to the rate achievable on a given PRB by assuming ideal link adaptation, i.e., choosing the maximum modulation and coding scheme (MCS) that yields a desired block error rate (BLER) [2].[2] When computing the achievable rates, we also account for the overhead due to control signaling [2].

---

2 We set the BLER to 10^{-1}, which we regard as a sufficiently low value considering that retransmissions further reduce the number of errors. This yields minimum and maximum spectral efficiencies of 0.22 b/s/Hz (bits per second per hertz) and 7.44 b/s/Hz, for SINRs in the range [−5.02 dB, −4.12 dB] and [25.87 dB, +∞], respectively.

7.2.3 Massive MIMO Channel Estimation

In the massive MIMO scenario, the network operates in a time-division duplexing (TDD) fashion, where the precoder and receive filters are calculated from the estimated channels. The latter are obtained at the BS via UL sounding reference signals (SRSs) – commonly known as *pilots* – sent by the users under the assumption of channel reciprocity [12, 21].

Let the pilot signals span M_p symbols. The pilot transmitted by user k in cell b is denoted by $v_{i_{bk}} \in \mathbb{C}^{M_p}$, where i_{bk} is the index in the pilot codebook, and all pilots in the codebook form an orthonormal basis [21]. Each pilot signal received at the BS undergoes *contamination* due to pilot reuse across cells. We assume pilot Reuse 3, i.e., the set of pilot signals is orthogonal among the three 120° BS sectors of the same site, but it is reused among all BS sites, creating contamination.[3] Each BS sector randomly allocates its pool of pilots to its served users. The collective received signal at BS b is denoted as $Y_b \in \mathbb{C}^{N_a \times M_p}$, and given by

$$Y_b = \sum_{j \in \mathcal{B}} \sum_{k \in \mathcal{K}_j} \sqrt{P_{jk}}\, h_{bjk} v_{i_{jk}}^{\mathrm{T}} + N_b, \tag{7.6}$$

where N_b contains the additive noise at BS b during pilot signaling with independent and identically distributed entries following $\mathcal{CN}(0, \sigma_\epsilon^2)$, and P_{jk} is the power transmitted by user k in cell j, assuming fractional power control as in Eq. (7.4).

The received signal Y_b in Eq. (7.6) is processed at BS b by correlating it with the known pilot signal $v_{i_{bk}}$, thus rejecting interference from other orthogonal pilots. BS b hence obtains the following least-squares channel estimate for user k in cell b [20]:

$$\hat{h}_{bbk} = \frac{1}{\sqrt{P_{bk}}} Y_b v_{i_{bk}}^* = h_{bbk} + \frac{1}{\sqrt{P_{bk}}} \left(\sum_{j \in \mathcal{B} \setminus b} \sum_{k' \in \mathcal{K}_j} \sqrt{P_{jk'}}\, h_{ijk'} v_{i_{jk'}}^{\mathrm{T}} + N_i \right) v_{i_{bk}}^* \tag{7.7}$$

where intra-cell pilot contamination is not present since BS b allocates orthogonal pilots for the users in its own cell.

7.2.4 Massive MIMO Spatial Multiplexing

In the DL, each BS simultaneously serves multiple users on each PRB via zero forcing (ZF) precoding, attempting to suppress all intra-cell interference.[4] Let us define the estimated channel matrix $\hat{H}_b \in \mathbb{C}^{N_a \times K_b}$ as

$$\hat{H}_b = [\hat{h}_{bb1}, \dots, \hat{h}_{bbK_b}]. \tag{7.8}$$

The ZF precoder

$$W_b = [w_{b1}, \dots, w_{bK_b}] \tag{7.9}$$

at BS b can be calculated as [14, 22]

$$W_b = \hat{H}_b (\hat{H}_b^{\mathrm{H}} \hat{H}_b)^{-1} (D_b)^{-1/2}, \tag{7.10}$$

---

3 This solution is particularly practical from an implementation perspective, since it involves coordination only between the three co-located BSs of the same site.
4 We consider inter-cell interference suppression techniques in Section 7.5.

where the diagonal matrix $\boldsymbol{D}_b$ is chosen to meet the transmit power constraint with equal user power allocation, i.e., $\|\boldsymbol{w}_{bk}\|^2 = P_b/K_b$ for all k, b. The same filter $\boldsymbol{w}_{bk}$ is employed in the UL to receive the signal transmitted by user k in cell b. The DL and UL SINR on a given PRB for user k can be obtained from (7.2) and (7.5), respectively, with the vectors $\boldsymbol{w}_{bk}$ as in Eq. (7.9).

7.3 Single-User Downlink Performance

In this section, we consider the cellular network depicted in Figure 7.1(a), where each BS is equipped with $N_a = 16$ antennas arranged in a vertical array of eight X-POL elements, each with 65° half-power beamwidth, electrically down-tilted by 12° and supported by a single RF chain. Such a configuration yields the BS antenna pattern depicted in Figure 7.2. In this setup, each BS serves at most one user on each PRB, without employing digital precoding [1, 4, 19]. This setup embodies many current cellular networks, and we refer to it as single-user mode. In this mode, Eqs. (7.1)–(7.5) simplify as follows: all vectors $\boldsymbol{w}$ consist of identical scalars, the second terms on the right-hand sides of Eqs. (7.1) and (7.3) vanish, and so do the first ones in the denominator of Eqs. (7.2) and (7.5). For this setup, we will examine how a UAV's height affect its DL C&C channel performance.

7.3.1 UAV Downlink C&C Channel

Six curves are plotted in Figure 7.3 in order to show the coupling loss and the SINR per PRB experienced by a UAV as a function of its height. The coupling loss (right y-axis)

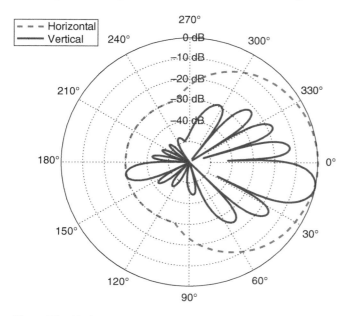

Figure 7.2 Horizontal and vertical antenna pattern (normalized to maximum gain) of a BS consisting of a vertical array of eight X-POL elements, each with 65° half-power beamwidth, electrically down-tilted by 12°.

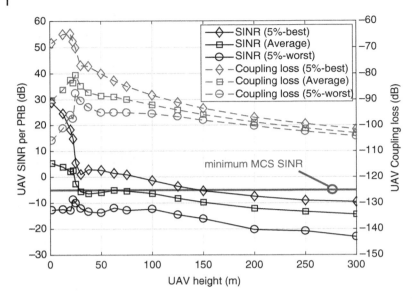

Figure 7.3 Coupling loss (right *y*-axis) and SINR per PRB (left *y*-axis) experienced by a UAV versus its height in a single-user scenario. The minimum MCS SINR threshold of −5.02 dB, required for nonzero rate, is also shown as a benchmark.

expresses the carrier signal attenuation between the serving BS and a UAV due to antenna gain, path loss, and shadow fading. On the other hand, the SINR per PRB (left *y*-axis) also accounts for small-scale fading and for the interference perceived at the UAV. For both metrics, Figure 7.3 shows the 5% best (i.e., 95 percentile), average, and 5% worst values, triggering the following observations:

(1) As UAVs rise from the ground up to a height of around 25 m, their average coupling loss improves due to closer proximity to the serving BS and increased probability of experiencing a LoS link with the latter. Instead, the 5% best UAVs, which were those located in the direction of the main lobe of the BSs, experience a degraded coupling loss as a consequence of a diminished antenna gain. As the UAV height keeps increasing, so does the BS-to-UAV distance, causing the coupling loss to decay.

(2) While the average UAV coupling loss is moderately improved, a UAV flying at around 25 m generally sees a degraded SINR per PRB. This is caused by the fact that more neighboring BSs become visible to the UAV, acting as strong LoS interferers. The opposite occurs for the 5% worst UAVs flying at 25 m, which experience a significant improvement in their coupling loss, and, as a result, also enhance their SINR. As the UAV height keeps increasing, the SINR keeps decreasing, though more slowly than the coupling loss. This trend is due to a simultaneous slight reduction of the interference as the UAV moves further away from neighboring interfering BSs.

(3) Overall, UAVs flying at heights of 25 m and above experience low values of SINR per PRB. In particular, for heights beyond 100 m the average SINR falls below the minimum

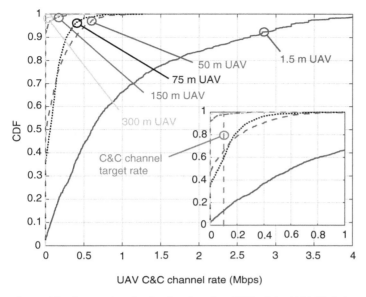

Figure 7.4 Cumulative distribution function (CDF) of the UAV C&C channel rates versus the UAV height in single-user mode. The target rate of 100 kbps is also shown in the enlargement.

MCS SINR threshold of −5.02 dB; for heights beyond 150 m, even the 5% best SINR per PRB falls below said minimum threshold.[5]

The measured values of SINR per PRB can be translated into the data-rate performance of the UAV C&C channel over a 10 MHz bandwidth. Figure 7.4 shows said performance for several UAV heights, motivating the following conclusions:

- UAVs at 1.5 m meet the target rate of 100 kbps 87% of the time, and 34% of the time their rates even exceed 1 Mbps;
- UAVs at around 50 m and 75 m, respectively, achieve the target rate 35% and 40% of the time only, and the achievable rates for this range of heights almost never reach 1 Mbps (0.3% of the time); and
- at higher heights, the UAV target rate of 100 kbps can only be achieved for small fractions of time, amounting to just 2% and 1% for heights of 150 m and 300 m, respectively.

The above results prompt us to conclude that, in cellular networks with heavy data traffic, simply relying on BS sectorization and single-user mode operations may not be sufficient to support the much-needed C&C channel for UAVs flying at reasonable heights.

---

5 In practice, opportunistic proportional-fair schedulers could be employed that outperform the round-robin scheduler considered in this chapter. However, this measure alone would not suffice to bring the UAV C&C channel performance to an acceptable level in single-user mode networks. Indeed, Figure 7.3 shows that even the 5% best UAVs, which corresponds to those with large channel fading gains, may experience very low values of SINR.

7.4 Massive MIMO Downlink Performance

In this section, we consider a network as depicted in Figure 7.1(b), where cellular BSs are equipped with massive MIMO antenna arrays and avail of beamforming and spatial multiplexing capabilities. In particular, we consider $N_a = 128$ antennas, arranged in an 8×8 planar array of $\pm 45°$ cross-polarized elements, fed by 128 RF chains. We allow each BS b to serve at most $K_b = 8$ users per PRB via digital ZF precoding.[6]

In the remainder of this section, we will first evaluate the UAV C&C channel performance improvement achieved through massive MIMO, and then study what the UAV presence entails for the GUEs' performance.

7.4.1 UAV Downlink C&C Channel

Similarly to Figures 7.3 and 7.4 for the single-user mode case, we now show the coupling loss (Figure 7.5, right y-axis), SINR per PRB (Figure 7.5, left y-axis), and C&C data rate (Figure 7.6) achieved by a UAV as a function of its height in a massive MIMO setup. Both Figures 7.5 and 7.6 consider the 3GPP Case 3, i.e., one UAV and 14 GUEs per sector [3]. In order to evaluate the gains theoretically achievable with massive MIMO, in these figures the

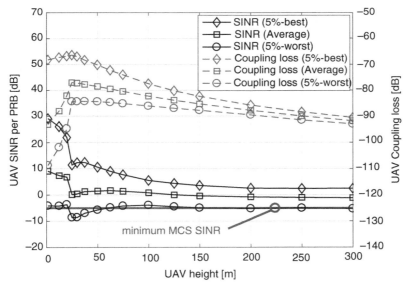

Figure 7.5 Coupling loss (right y-axis) and SINR per PRB (left y-axis) experienced by a UAV versus its height in a massive MIMO setup with perfect CSI (Case 3). The minimum MCS SINR threshold of -5.02 dB is also shown as a benchmark.

---

6 Scheduling a larger number of users may yield higher cell spectral efficiency. However, it may prevent achieving a minimum guaranteed rate for all users [9], which is the primary goal for DL UAV communications.

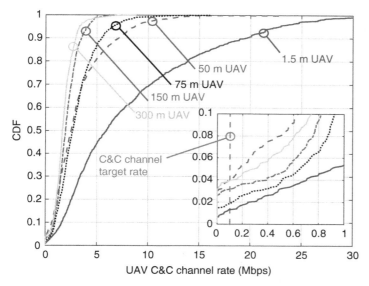

Figure 7.6 CDF of the UAV C&C channel rates for various UAV heights in a massive MIMO setup with perfect CSI (Case 3). The enlargement shows the target rate of 100 kbps as a benchmark.

BS is assumed to avail of perfect CSI, i.e., no pilot contamination is considered. A more realistic channel estimation through SRSs as in Eq. (7.7) and the effect of pilot contamination on the performance of both UAVs and GUEs will be discussed in the sequel.

Comparing Figure 7.5 to Figure 7.3 provides the following insights:

(1) Consistently with Figure 7.3, a UAV flying at around 25 m generally sees an improved coupling loss but a degraded SINR per PRB. This is due to the fact that more neighboring BSs become visible to the UAV, acting as strong LoS interferers.

(2) Employing massive MIMO at the BSs improves the UAVs' coupling loss, which is measured at the output of the first RF chain [5], thanks to an increased antenna gain towards the sky.

(3) The SINR per PRB experienced by a UAV is largely improved in a massive MIMO system, owing to two phenomena. First, UAVs benefit from a beamforming gain from the serving BS, which can now send beams into the sky as well. Second, since most users are GUEs, neighboring BSs tend to point most of their beams downwards, greatly mitigating the interference generated at the UAVs.

(4) Overall, most UAVs experience values of SINR per PRB above the minimum MCS threshold. In particular, the average SINR per PRB is well above said threshold for any UAV height. Moreover, even the 5% worst UAVs meet the minimum SINR threshold for most UAV heights.

Figure 7.6 shows the data-rate performance of the UAV C&C channel in a massive MIMO setup for various UAV heights. Comparing this figure to Figure 7.4 provides the reader with a key takeaway: compared to single-user mode cellular networks, massive MIMO networks have the potential to support a 100 kbps UAV C&C channel with a substantially higher reliability, namely, in at least 96% of the cases for all UAV heights under consideration.

Indeed, the data rates in a massive MIMO network are largely improved due to both an SINR gain (as per Figure 7.5) and a spatial multiplexing gain, owing to the fact that eight users, between UAVs and GUEs, are allocated the same PRB simultaneously.

7.4.2 UAV–GUE Downlink Interplay

We now study how supporting the UAV C&C channel through cellular networks may affect the performance of GUEs. In particular, we assess the impact of UAVs in both single-user and massive MIMO settings with the user height distributions specified in Table 7.1. For the latter, we discuss the impact of SRS reuse and contamination.

Figure 7.7 shows the SINR per PRB for both UAVs and GUEs in the presence of realistic CSI acquisition with SRS Reuse 3 and UL fractional power control. The figure considers the 3GPP Cases 3, 4, and 5, corresponding to one UAV and 14 GUEs, three UAVs and 12 GUEs, and five UAVs and 10 GUEs per sector, respectively. Figure 7.7 carries multiple consequential messages:

(1) In spite of an imperfect CSI available at the BSs, the UAV SINR per PRB greatly improves when moving from a single-user to a massive MIMO scenario. This is due to a beam-forming gain paired with a reduced interference from nearby BSs that focus most of their energy downwards.

(2) In line with the above, the UAV SINR per PRB in massive MIMO mode scenarios is reduced when moving from Case 3 to Cases 4 and 5, mainly because (i) a larger number of UAVs leads to an increased CSI pilot contamination due to their strong LoS channel to many BSs, and (ii) neighboring cells point more beams upwards, thus generating more inter-cell interference at the UAVs. On the other hand – although not explicitly

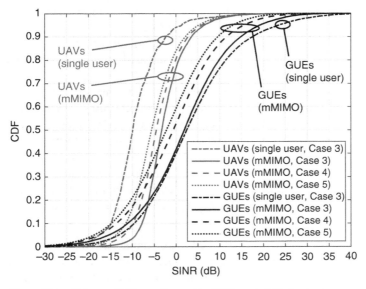

Figure 7.7 SINR per PRB experienced by UAVs and GUEs in single-user and massive MIMO scenarios with pilot Reuse 3 and UL fractional power control (various 3GPP cases).

shown in the figure – the number of UAVs does not affect the SINR in single-user mode scenarios.

(3) Unlike the UAV SINR, the GUE SINR does not improve when moving from a single-user to a massive MIMO scenario. This is mainly due to the severe pilot contamination incurred by GUEs, which outweighs any beamforming gains. Indeed, each GUE's SRS is likely to collide with the SRS of at least one UAV in a neighboring cell in the scenario considered, with said UAV being likely to experience a strong LoS link with the GUE's serving BS.

(4) Accordingly, the GUE SINR per PRB further degrades when moving from Case 3 to Cases 4 and 5, since the presence of more UAVs in neighboring cells causes the pilot contamination effect to increase its severity.

The ultimate DL rate performance achievable by UAVs and GUEs is shown in Figure 7.8 for the 3GPP Case 3, i.e., one UAV and 14 GUEs per sector. This figure not only illustrates the gains provided by massive MIMO networks, but it also highlights the crucial role played by CSI acquisition through a comparison of two scenarios: (i) perfect CSI ("Perfect") and (ii) imperfect CSI obtained through pilot Reuse 3 and fractional UL power control ("R3 PC"). Figure 7.8 motivates us to conclude this section with the following key takeaways:

- Pilot contamination can severely degrade the rate performance of both UAVs and GUEs. Indeed, the median UAV rates attained with imperfect CSI acquisition are reduced to 40% of those achievable without channel estimation errors.

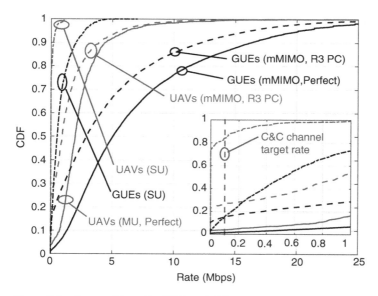

Figure 7.8 Rates achieved by UAVs and GUEs in mMIMO setups under: (i) perfect CSI, "mMIMO, Perfect" (solid); and (ii) SRS Reuse 3 and UL fractional power control, "mMIMO, R3 PC" (dashed). Also shown are the rates in a single-user scenario, "SU" (dash-dotted); and the UAV C&C target rate of 100 kbps (in the enlargement).

- Massive MIMO boosts the GUEs' data rates. This is due to the multiplexing gain rather than to SINR gain, as illustrated by the non-improving SINRs in Figure 7.7. As for the UAV C&C channel, massive MIMO is a key enabler, meeting the target rate of 100 kbps in 74% of the cases even under pilot contamination ("MU, R3 PC").
- Availing of massive MIMO with perfect CSI would allow one to achieve said C&C channel target rate in 96% of the cases, as opposed to a mere 16% in single-user scenarios. In order to close the performance gap caused by pilot contamination, one may resort to UAV-side and network-side enhancements as discussed in the next section.

7.5 Enhanced Downlink Performance

In this section, we explore the following two enhancements [7], each of which will be independently evaluated on top of the SU and mMIMO modes described in the previous sections:

- **UAVs with adaptive arrays (aaUAV).** UAVs integrate a 2×2 adaptive array comprised of omnidirectional antenna elements and a single RF chain. This seems a natural choice given that flying UAVs generally experience LoS propagation conditions with their serving BS, which generally prevents them from performing spatial multiplexing and makes the performance of analog and digital signal processing comparable. As illustrated in Figure 7.9, this hardware upgrade enables aerial devices with knowledge of the azimuth and elevation angles to their serving BS to perform a precise analog beam steering.
- **Massive MIMO BSs with null steering (mMIMOnulls).** BSs incorporate additional signal processing features that enable them to perform a twofold task. First, by leveraging channel directionality, which invariably occurs in UAV-to-BS links, BSs can spatially separate non-orthogonal pilots transmitted by different aerial devices [9]. Second, by placing spatial radiation nulls, BSs can mitigate interference to vulnerable users in

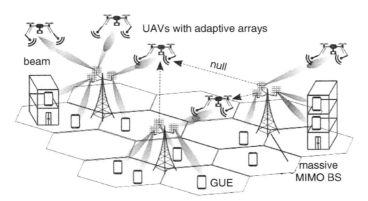

Figure 7.9 Illustration of enhanced cellular support for UAVs. Massive MIMO BSs serve multiple GUEs and UAVs on each PRB via digital precoding, also mitigating pilot contamination and inter-cell interference through radiation nulls, and UAVs point precise beams towards their serving BS.

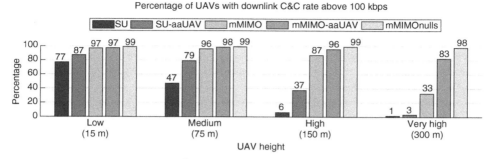

Figure 7.10 Percentage of UAVs with downlink C&C channel rates larger than 100 kbps as a function of their flying altitude.

other cells [13, 14, 18].[7] Intuitively, BSs will tend to steer their radiation nulls – 16 in our setup – towards the locations of the closest high-altitude UAVs connected to other BSs, since they undergo LoS propagation conditions [3].

7.5.1 UAV Downlink C&C Channel

Figure 7.10 illustrates the percentage of UAVs that achieve a downlink C&C channel rate larger than the minimum requirement of 100 kbps [3]. This percentage is shown by considering one UAV per cell, and by varying the height of all UAVs (15, 75, 150, and 300 m) to exemplify the crucial role of this parameter on the ground-to-air link performance.

Once again, Figure 7.10 demonstrates that, irrespective of the network and device capabilities, the performance of the downlink C&C channel diminishes as UAVs increase their height from 15 m to a maximum flying altitude of 300 m. The latter is due to an increased interference, and its effect is particularly noticeable for SU setups (with poorer interference coordination), where the percentage of UAVs with downlink rates larger than 100 kbps goes from 77%, for the low-altitude UAVs flying at 15 m, to a mere 1%, when UAVs fly at 300 m. The degradation is particularly severe at 150 m and above – where the interference does not decay at the same rate as the signal power does – and it also occurs in spite of equipping UAVs with an adaptive array (SU-aaUAV).

The trends of Figure 7.10 stress the need to employ more sophisticated hardware and signal processing when serving aerial users. For instance, it can be observed that complementing massive MIMO BS processing with explicit inter-cell interference suppression techniques (mMIMOnulls) is essential when catering for very-high-altitude UAVs. Indeed, these additional capabilities dramatically increase the percentage of UAVs that meet the 100 kbps requirement when these are flying at 300 m, from 33% (mMIMO) to a whopping 98% (mMIMOnulls). This is because the number of radiation nulls placed by cellular BSs suffices to effectively suppress the interference generated (received) towards (from) non-associated UAVs. Even though each BS requires channel knowledge from cell-edge devices associated to other BSs to perform this inter-cell interference suppression, its acquisition may

---

7 An inherent trade-off exists between allocating more spatial degrees of freedom for interference suppression and employing them to augment beamforming gain [15].

be facilitated for higher UAVs. In fact, due to the strong channel directionality, the problem boils down to estimating the UAV angle of arrival.

Figure 7.10 also tells us that, as far as UAV downlink performance is concerned, it may not be necessary to rely on devices with adaptive arrays (mMIMO-aaUAV), since the gain compared to mMIMO is minor at lower altitude and mMIMOnulls performs substantially better at very high altitudes.

Overall, Figure 7.10 corroborates the effectiveness of massive-MIMO-based networks to serve UAVs in downlink. Massive MIMO will thus take center stage in the remainder of this section.

7.5.2 UAV–GUE Downlink Interplay

To illustrate the impact of the presence of UAVs on the network performance, Figure 7.11 shows the CDF of the downlink SINR per PRB experienced by terrestrial users. Both cellular networks with and without UAVs are considered, with one UAV per cell flying at a height of 150 m in the former case. Figure 7.11 confirms that UAV-generated pilot contamination causes an overall degradation of the downlink SINRs attainable by GUEs. The impact is significant for cell-edge GUEs, i.e., those located in the lower tail of the CDF and shown in the inset, which lose around 5 dB when a single UAV per cell is deployed (mMIMO). Instead, the performance of cell-center GUEs is not severely affected because their serving BSs receive their uplink pilot signals with a large power, therefore making them less vulnerable to the uplink pilot interference from aerial UEs.

Interestingly, this performance loss is partially compensated when aerial devices are equipped with adaptive arrays (mMIMO-aaUAV), and focus their beam towards their serving BS. In fact, the increased link budget allows UAVs that perform (analog) beamforming

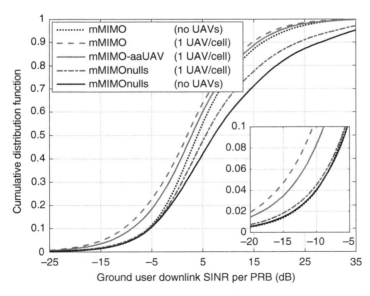

Figure 7.11 CDF of the downlink SINRs per PRB for the ground users. When present, one UAV per cell is flying at an altitude of 150 m.

to reduce their transmit power, in turn reducing the interference generated to other cells during the uplink pilot transmission phase. However, the performance of cell-edge GUEs is still reduced when compared to the mMIMO scenario without UAVs.

On the other hand, inter-cell interference suppression capabilities (mMIMOnulls) yield a shortened gap between scenarios with and without UAVs. Figure 7.11 reveals that such capabilities approximately preserve the cell-edge GUE performance even in the presence of one UAV per cell. This remarkable result demonstrates the benefits of explicitly accounting for the presence of UAVs in the network and mitigating UAV-to-GUE pilot contamination. The remaining performance gap can be explained as follows. In the absence of UAVs, BSs point their radiation nulls towards out-of-cell GUEs, including those located in the cell center. Instead, when UAVs are present, most of the nulls are targeted towards them and towards the vulnerable cell-edge GUEs.

7.6 Uplink Performance

After evaluating the downlink performance of networks with UAV users, we would like to assess whether cellular networks can provide high-speed uplink aerial links for video streaming purposes. An equally important concern is related to the influence that UAV transmissions have on the uplink performance of terrestrial users. These two facets of uplink transmission are treated through this section.

7.6.1 UAV Uplink C&C Channel and Data Streaming

Figure 7.12 shows both (a) the average and (b) the 95% likely UAV uplink data rates for a varying number of UAVs per cell. Intuitively, the average uplink data rates are indicative of the network support to real-time streaming applications, whereas the 95 percentile identifies the uplink C&C channel reliability. In the considered networks, UAVs are uniformly distributed between 1.5 m, to capture their performance during the critical take-off and landing operations, and 300 m [3]. Moreover, we remark that the average number of active devices per sector remains fixed to 15 as per the 3GPP specifications [3] and, consequently, the number of GUEs per cell diminishes as the number of UAVs per cell increases.

Figure 7.12 carries the consequential message that increasing the number of UAVs has a detrimental impact on their own performance, as close-by cells schedule UAVs more often. The reduction in the uplink data rates is especially significant for old-fashioned SU setups, where the minimum requirement of 100 kbps cannot be satisfied for the 5% worst UAVs when more than one aerial user is present per cell. Therefore, once again, we shift the focal point to massive MIMO systems.

Figure 7.12 confirms how networks with massive MIMO BSs are capable of substantially boosting the average uplink data rates attainable by SU architectures. However, massive MIMO is not immune to severe inter-UAV interference generated both during the CSI acquisition and uplink data transmission phases. It can be observed how, with massive MIMO, the average UAV uplink rates in Figure 7.12(a) drop from 8.9 Mbps to 2.7 Mbps when the number of UAVs is increased from one to five per cell. These results hint that UAV-agnostic signal processing techniques might not be enough to guarantee the 100 kbps

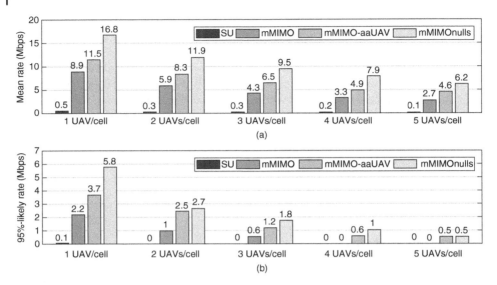

Figure 7.12 (a) Average and (b) 95% likely uplink UAV rates (Mbps) for cellular networks with one, two, three, four, or five UAVs per cell.

required by the uplink C&C channel, when a large number of single-antenna UAVs are present in the network.

In these challenging circumstances, one could get away without null steering if all cellular-connected UAVs were equipped with adaptive arrays (mMIMO-aaUAV). However, this being beyond the control of a network operator, it may be desirable to complement the network with inter-cell interference suppression capabilities (mMIMOnulls) to enhance the performance of the UAV uplink payload and C&C channels. Indeed, Figure 7.12 shows that the latter approach guarantees uplink rates of 0.5 Mbps to 95% of the UAVs, even in the presence of five active UAVs per cell.

7.6.2 UAV–GUE Uplink Interplay

Figure 7.13 puts a spotlight on the average uplink rates achieved by GUEs under the presence of a varying number of UAVs, from one to five per cell. Figure 7.13 illustrates that the average GUE uplink rates are dramatically curtailed when more UAVs are active.

Figure 7.13 also carries another fundamental message: GUEs may not be satisfied by the more aggressive spatial reuse offered by massive-MIMO-capable BSs, unless explicit inter-cell interference suppression mechanisms during the data reception and CSI acquisition phases are implemented (mMIMOnulls). Still, even this approach suffers when there are three or more UAVs per cell in a fully loaded network, with the mean GUE uplink rates dropping below those achieved in a UAV-free setup.

Finally, Figure 7.13 shows that equipping UAVs with adaptive arrays offers a limited performance improvement, in spite of allowing them to reduce the interference they generate thanks to both (i) their reduced radiated power as per the fractional power control logic applied, and (ii) the increased directionality of their transmissions. Altogether, Figure 7.13

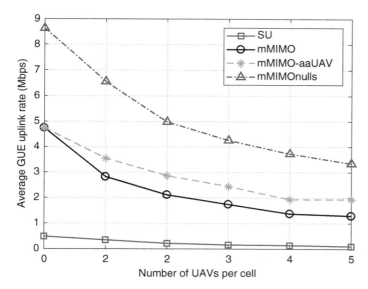

Figure 7.13 Average uplink data rates (Mbps) for the ground users as a function of the number of UAVs per cell.

highlights the need to account for the presence of UAVs throughout the network design stage, if the performance of the existing GUEs is to be preserved.

7.7 Conclusions

In this chapter, we made an effort to understand what it will take to realize cellular-connected UAVs. We showed that, owing to its capability of focusing multiple signals towards multiple users, massive MIMO is an important tool to achieve reliable UAV communications. Indeed, its adoption is critical in fully loaded networks to restrict the impact that UAV-generated interference has on legacy ground communications.

While being a viable solution, the efficacy of mMIMO fades in the presence of many high-altitude UAVs. Moreover, mMIMO requires updated and accurate CSI, whose acquisition could be facilitated by UAV location- and trajectory-related information. Finally, pilot contamination may pose a severe threat that, if not properly addressed, can jeopardize the performance of existing terrestrial users.

As a result, it seems desirable to complement mMIMO with appropriate infrastructure and signal-processing upgrades by both operators and UAV manufacturers. The latter, when demanding cellular service at high altitudes, may need to improve the hardware characteristics of their devices, e.g., by equipping UAVs with beamforming capabilities (mMIMO-aaUAV). The former may want to resort to inter-cell interference suppression (mMIMOnulls), exploiting BS–UAV channel directionality, to serve both UAVs and GUEs satisfactorily.

References

1 3GPP (2006). 3GPP Technical Report 25.814. Physical layer aspects for evolved Universal Terrestrial Radio Access (UTRA) (Release 7), September.

2 3GPP (2010). 3GPP Technical Report 36.213. Evolved Universal Terrestrial Radio Access (E-UTRA); physical layer procedures (Release 9), June.

3 3GPP (2017). 3GPP Technical Report 36.777. Technical Specification Group Radio Access Network; study on enhanced LTE support for aerial vehicles (Release 15), December.

4 3GPP (2013). 3GPP Technical Report 36.814. Further advancements for E-UTRA physical layer aspects (Release 9), March.

5 3GPP (2017). 3GPP Technical Report 38.901. Study on channel model for frequencies from 0.5 to 100 GHz (Release 14), May.

6 3GPP (2011). 3GPP Technical Specification 36.201. LTE; Evolved Universal Terrestrial Radio Access (E-UTRA); LTE physical layer (Release 10), June.

7 A. Garcia-Rodriguez, G. Geraci, D. López-Pérez, L. Galati Giordano, M. Ding, and E. Björnson (2019). The essential guide to realizing 5G-connected UAVs with massive MIMO. *IEEE Commun. Mag.* 57 (12): 84–90.

8 P. Baracca, L. Galati Giordano, A. Garcia-Rodriguez et al. (2018). Downlink performance of uplink fractional power control in 5G massive MIMO systems. *Proceedings of the IEEE Global Communications (GLOBECOM) Conference*, December, pages 1–7.

9 E. Björnson, J. Hoydis, and L. Sanguinetti (2017). Massive MIMO networks: spectral, energy, and hardware efficiency. *Found. Trends Signal Process.* 11 (3–4): 154–655.

10 C. Ubeda Castellanos, D. L. Villa, C. Rosa et al. (2008). Performance of uplink fractional power control in UTRAN LTE. *Proceedings of the Vehicular Technology Conference (VTC)*, May, pages 2517–2521.

11 H. Fattah and C. Leung (2002). An overview of scheduling algorithms in wireless multimedia networks. *IEEE Wireless Commun. Mag.* 9 (5): 76–83.

12 L. Galati Giordano, L. Campanalonga, D. López-Pérez et al. (2018). Uplink sounding reference signal coordination to combat pilot contamination in 5G massive MIMO. *Proceedings of the IEEE Wireless Communications and Networking Conference (WCNC)*, April, pages 1–6.

13 A. Garcia Rodriguez, G. Geraci, L. Galati Giordano et al. (2018). Massive MIMO unlicensed: a new approach to dynamic spectrum access. *IEEE Commun. Mag.* 56 (6): 186–192.

14 G. Geraci, A. Garcia-Rodriguez, D. López-Pérez et al. (2018). Indoor massive MIMO deployments for uniformly high wireless capacity. *Proceedings of the IEEE Wireless Communications and Networking Conference (WCNC)*, April.

15 G. Geraci, A. Garcia Rodriguez, D. López-Pérez et al. (2017). Operating massive MIMO in unlicensed bands for enhanced coexistence and spatial reuse. *IEEE J. Sel. Areas Commun.* 35 (6): 1282–1293.

16 G. Geraci, A. Garcia Rodriguez, L. Galati Giordano et al. (2018). Understanding UAV cellular communications: from existing networks to massive MIMO. *IEEE Access* 6: 67 853–67 865.

17 G. Geraci, A. Garcia Rodriguez, L. Galati Giordano et al. (2018). Supporting UAV cellular communications through massive MIMO. *Proceedings of the IEEE International Conference on Communications (ICC) Workshops*, May, pages 1–6.

18 H. H. Yang, G. Geraci, T. Q. S. Quek, and J. G. Andrews (2017). Cell-edge-aware precoding for downlink massive MIMO cellular networks. *IEEE Trans. Signal Process.* 65 (13): 3344–3358.

19 A. Kammoun, H. Khanfir, Z. Altman et al. (2014). Preliminary results on 3D channel modeling: from theory to standardization. *IEEE J. Sel. Areas Commun.* 32 (6): 1219–1229.

20 M. S. Kay (1998). *Fundamentals of Statistical Signal Processing: Detection Theory.* Prentice-Hall.

21 T. L. Marzetta, E. G. Larsson, H. Yang, and H. Q. Ngo (eds.) (2016). *Fundamentals of Massive MIMO.* Cambridge University Press.

22 Q. H. Spencer, A. L. Swindlehurst, and M. Haardt (2004). Zero-forcing methods for downlink spatial multiplexing in multiuser MIMO channels. *IEEE Trans. Signal Process.* 52 (2): 461–471.

8

High-Capacity Millimeter Wave UAV Communications

Nuria González-Prelcic[1,], Robert W. Heath[1], Cristian Rusu[2], and Aldebaro Klautau[3]*

[1]*Electrical and Computer Engineering Department, University of Texas at Austin, 2501 Speedway, Austin TX 78712, USA*
[2]*LCSL, Istituto Italiano di Tecnologia (IIT), via Morego, 16163 Genova, Liguria, Italy*
[3]*Computer and Telecommunication Engineering Department, Universidade Federal do Pará, Augusto Correa, Belem, 66075-110, Pará, Brazil*

8.1 Motivation

High-data-rate communication enables many applications of aerial vehicles. Millimeter wave (mmWave) carrier frequencies, a key feature of 5G cellular systems, are ideally suited to provide high data rates thanks to higher-bandwidth channels. In this chapter, we explain the potential of mmWave for unmanned aerial vehicles (UAVs). We motivate the study of mmWave for aerial vehicles by explaining how it enables important use cases that depend on high data rates, low latencies, or combinations of the two. Then we summarize important considerations for propagation at millimeter wave frequencies and review appropriate channel models for different scenarios. We describe the fundamental challenges related to UAV multiple input–multiple output (MIMO) communication at mmWave frequencies, including the use of large arrays, high mobility, and less structured mobility. We conclude the chapter by describing additional research opportunities that are pertinent for UAVs operating at mmWave frequencies.

UAVs have tremendous disruptive potential in diverse applications, as already outlined in the previous chapters of this book. Reaching this potential, however, requires a commensurate wireless communications technology like mmWave communication. Because of the higher bandwidths available at mmWave carrier frequencies, along with careful system design, mmWave cellular communication has the potential to offer high data rates (gigabits per second, Gbps) and low end-to-end latencies (sub-millisecond). This has several advantages in important applications of aerial vehicles, including sensing, wireless access, and transportation as illustrated in Figure 8.1.

In sensing, high data rates and low latency permit the delivery of high-fidelity sensor information from the UAV to the infrastructure to facilitate remote monitoring and control. The end destination for UAV data depends on the application: it may be a human user who is controlling the UAV, an algorithm that is extracting information from that sensor

---

* Corresponding Author: Nuria González-Prelcic; ngprelcic@utexas.edu

UAV Communications for 5G and Beyond, First Edition.
Edited by Yong Zeng, Ismail Guvenc, Rui Zhang, Giovanni Geraci, and David W. Matolak.

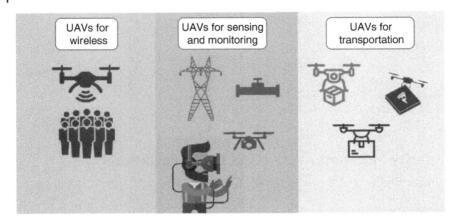

Figure 8.1 Key applications of aerial vehicles empowered by mmWave communication.

data, a centralized controller that is controlling the UAV mobility patterns, or even other UAVs that are using artificial intelligence to coordinate their own activities. For example, first-person-view drone racing (human or autonomous) [17], cinematography [43], agriculture [49], and search and rescue [18, 68] benefit from the ability to stream high-quality video, including multiple video streams and multiple camera views. Video and other sensor data is also useful as a means to perform remote control of UAVs, including positioning, navigation, obstacle avoidance, and cooperation [74]. High-data-rate connections permit offloading of computation, so that machine learning algorithms used to process the data may be run on the ground. This reduces the weight and power consumption required in aerial computation.

In wireless access, mmWave facilitates high-data-rate and low-latency communication between the UAV, the tower, and other users. For example, mmWave may be used as a means for high-data-rate backhaul so that a UAV may act as a hotspot during temporary events [42, 84] or for disaster relief [41]. Note that the wireless access technology from the UAV to the user need not be at mmWave; it could be delivered using a legacy 4G or wireless local area network (WLAN) technology. UAVs may also be a means to improve resilience in 5G mmWave cellular networks. For example, mmWave at the UAV may provide an alternative means to deal with blockage of the mmWave line-of-sight link for backhaul [21] or the access link [39]. mmWave has been used in recent aerial applications, like Facebook's Aquila system to support rural connectivity [31, 45]. High data rates ensure that the UAV has sufficient capacity to serve its intended application, while low latency makes applications like wireless Internet access operate seamlessly over those links.

In urban air mobility – an emerging application for UAVs – mmWave may also play an important role. mmWave is already being considered for ground vehicle applications to support raw sensor data sharing and infotainment [12, 70]. For example, with mmWave, vehicles can share their lightly processed sensor data for improved situational awareness, cooperative navigation, and remote operation through intersections. Similar benefits may be realized in the use of mmWave for urban air mobility, especially the delivery of packages and people. For example, mmWave provides a natural means for achieving ultra-precise positioning [75], which is challenging in urban areas where the Global Positioning System (GPS) suffers due to limited visibility of satellite or multipath effects [51]. The use of mmWave also facilitates sensor data sharing and remote operation, which are useful for traffic management and collision avoidance just as for ground vehicles. Thus far, though, the potential of mmWave for urban area mobility is still under investigation.

mmWave is now acknowledged as an important new feature of 5G New Radio (NR) cellular systems [7]. The major reason is that channel bandwidths can be larger, up to 400 MHz at carriers around 28 GHz and 40 GHz in 5G as defined in 3GPP Release 15 [16] compared to the 20 MHz available without carrier aggregation at carrier frequencies below 2 GHz. For a given operating signal-to-noise ratio (SNR), according to the classic Shannon capacity expression in an additive Gaussian noise channel, the communication rate increases in proportion to bandwidth B as $B\log(1 + \text{SNR})$. Of course, the rate achieved in real networks is much more complicated to define, but the linear increase of rate with bandwidth is common in many operating regimes of interest.

There are many challenges associated with communicating at mmWave frequencies [30]. It is fair to say that the entire beam-based design found in 5G NR was suggested in response to many of these challenges. One of the main differentiating features is the use of antenna arrays with a large number of radiators to achieve array and MIMO multiplexing gains. Further, these antenna arrays are normally controlled through networks of analog beamforming components like phase shifters – creating what is called the hybrid architecture. This makes mmWave MIMO communication different from the one at low frequencies [30]. For example, basic signal processing tasks like precoding or channel estimation become vastly more complicated. Despite these challenges, it is expected that mmWave arrays will be small and lightweight, making them well suited to small aerial vehicles. In addition, the hybrid arrays likely used in mmWave MIMO processing may offer important advantages of mmWave UAVs. For example, narrow beams may be beneficial in improving spatial reuse in general, avoiding interference with ground users, and enhancing security by making the signals more difficult to intercept.

The applications of mmWave to UAVs introduce additional complications to the usual mmWave MIMO signal processing steps. Mobility considerations are more significant compared with the fixed wireless and handheld applications targeted in the first release of 5G NR. This means that the communication channels (or beams) need to be configured quickly. There are also significant differences in the underlying channel assumptions that depend on the use case. For example, we would expect the channel from a high-altitude UAV acting as hotspot to the tower to be line-of-sight (LoS), with few significant multipath components. Alternatively, the channel from a street-level UAV delivering a package may likely be non-line-of-sight (NLoS), with many more multipath components. This means that the application of mmWave to UAVs is not a "one-size-fits-all" application; rather, each use case will entail a different set of operating conditions and requirements. As a result, there is still great potential to further develop mmWave in many UAV applications.

In this chapter, we explain the potential of mmWave for UAVs and its importance as a use case for 5G and beyond cellular systems. We begin with a more detailed assessment of the roles and use cases that become possible with mmWave communication, and then review the fundamentals of mmWave propagation and relevant aerial channel models for mmWave frequencies. Next, we describe the key aspects of communicating at millimeter wave, setting the stage for a more detailed description of mmWave MIMO communication. We describe one of the main challenges for aerial communication: establishing and maintaining communication with highly mobile aerial vehicles. We conclude with promising directions for future research in mmWave UAVs.

8.2   UAV Roles and Use Cases Enabled by Millimeter Wave Communication

mmWave communication provides a means of obtaining high-data-rate communication link to and from a UAV. This becomes a key ingredient for enabling several important use cases of UAVs, which cannot be adequately served using lower data rates provided by 4G, WiFi, or non-millimeter wave technologies. In this section, we describe first the ways that a UAV may be connected in a cellular network like 5G. Then we describe several applications enabled by UAVs taking these roles with high-data-rate communications. While the emphasis is on the benefits related to high data rates, most applications are assisted by low latency as well.

8.2.1   UAV Roles in Cellular Networks

UAVs may be connected to the cellular networks in different ways, acting as a piece of cellular infrastructure or as user equipment. Here we summarize these applications. While they do not necessarily require mmWave and are also discussed in other chapters of this book, they can potentially work better when mmWave is used. The different scenarios also have important implications on the propagation channel models, described later.

A cellular network operator may employ UAVs as aerial base stations. In this application, the UAV is backhauled to the cellular network over a wireless link. The link may be served from a base station, other ground infrastructure, or a satellite link, as may be useful in rural applications. The size of the backhaul needs to be commensurate with the peak data rates offered by the UAV on its access link, e.g., gigabits per second to support 5G data rates. Naturally, this is a great application for mmWave. The access link provided by the UAV does not have to be based on 5G technology. Conceivably, it could use something less advanced like 4G or even a WiFi hotspot. It should be noted that the UAV does not need to fly all the time to act as an aerial base station. For example, it could fly and then perch like a bird on top of a building, thus saving flight energy [66].

UAVs may also play the role of the user equipment in a cellular network. One of the distinguishing features of 5G was the development of the standard around several important industry vertical applications including transportation. 5G supports low end-to-end latencies of 1 ms, which means that 5G may be used for remote control of the UAV by humans. Note that recent (2019) consumer products advertise "ultra-low latencies" of 28 ms [33] for responsive human operator control; thus 1 ms is a substantial improvement over the state of the art. The high data rates also provide a latency advantage. For example, [67, table 1] gives an example of frame-by-frame latency for video capture and display. It includes 25 ms for video encoding and 27 ms for video decoding. Using high data rates as supported by mmWave in 5G, uncompressed video could be sent, saving substantial encoding and decoding latency.

There are different ways in which UAVs may operate as a relay in the cellular network. In this mode of operation, the UAV is playing the role of forwarding signals, and does not implement all the higher-layer functions as if it were acting as a base station. For example, a UAV may serve as a relay between other user equipment (potentially on the ground) and a base station [5]. This may be especially useful in mmWave cellular networks where users with an NLoS connection to the base station may be afforded a two-hop LoS link through the UAV with much better performance. A UAV may also be used as a means to facilitate backhaul [65]. In this case, the UAV is deployed to avoid costly wired/fiber backhaul costs to small cells. Again, the main advantage occurs in the case when the small cell has an NLoS connection to a service hub, but has LoS to the UAV and LoS to infrastructure connecting it to the network core. These use cases are well suited for mmWave communication, as it solves a key problem (blockage of the ground-to-base-station link) and leverages the high-data-rate capability to give the relay high capacity.

While we have emphasized the connection of a single UAV to the base station, most of the applications naturally generalize to UAV swarms [11, 19], which refers to a collection of UAVs working together. The aerial vehicles in the swarm may be coordinating as base stations, for example, to maximize coverage over the ground or to fill in dead zones. The swarm may also be more tightly coupled and engaged in a joint mission, e.g., accomplishing a surveillance activity. In this example, multi-hop or mesh communication directly between the UAVs has some advantages. Note that a framework for such communication is already supported under the term device-to-device communication in 5G. Millimeter wave offers the potential for high data rates in these networks.

8.2.2 UAV Use Cases Enabled by High-Capacity Cellular Networks

mmWave networks can circumvent some of the bottlenecks that the UAV industry is experiencing to make the more disruptive applications a reality [78]. In particular, mmWave technology can guarantee real-time transmission of high-rate sensing data, while jointly providing high-accuracy position information that facilitates accurate tracking of the UAVs [75]. In this section, we review some of the specific use cases that are enabled by UAVs with high-data-rate mmWave cellular communication.

UAVs have many applications to solve key problems in cellular networks. For example, they could act as dynamic hotspots in temporary events [84] with ultra-dense traffic demands. Alternatively, they can fill in coverage gaps due to loss of infrastructure, for example, after a natural disaster [41]. They can also be used to support that infrastructure, for example, backhauling base stations when the communication network is down (many base stations already have emergency power backup). UAVs help to solve the problems of blockage between the base station and ground terminal in cellular systems [39], for example, dynamic backhaul connections when the conventional mmWave backhaul link is blocked by moving vehicles, humans, or other dynamic objects [21]. These applications all benefit from the fundamental advantages of mmWave communication, including narrow beams (to reduce interference) and high bandwidth (to provide high data rates), while solving some of the key challenges found in deployment of mmWave, like sensitivity to blockage.

Using a cellular wireless connection like 5G affords high-data-rate transmission of sensor data, for example, for surveillance and rescue operations, which make use of a large variety of sensors, such as 4K, hyperspectral, or infrared cameras, that need to generate a high data rate in real time [18, 28]. Such data may be backhauled to a central control station and processed by human operators or machine learning algorithms to extract information. Precision agriculture is another potential application [49]. It can benefit from ultra-precise positioning and also the ability to communicate large sets of hyperspectral images [22]. Virtual reality (VR) is another application that benefits from 5G mmWave connections. First-person-view for remotely controlled UAVs is incredibly popular for drone racing [55]. In this setup, a user wears VR goggles to see the environment from the perspective of a pilot (admittedly a very small one) inside the UAV. This must be contrasted with the usual perspective of the operator controlling the UAV from its own perspective from the ground. A good experience with VR in this application requires high-quality video from the UAV to the user and a low-latency control link to the UAV.

There are several applications of UAVs in the transportation industry. One simple example is to monitor ground vehicle traffic, for example, monitoring vehicle flow rates at an intersection [46]. There are many new emerging applications of UAVs, which have the potential for great disruption in the transportation industry. Urban air mobility is one area that is starting to grow [53]. In this application, a UAV is employed to pick up and drop off passengers using an electric "vertical take-off and landing" aircraft. Essentially, the UAV plays the role filled now by the helicopter as a means of transport in urban areas, but avoiding the associated noise and reducing costs. While most urban air mobility projects now involve human pilots, having remotely controlled aerial vehicles has benefits in improving the payload (no human pilot needed) and also reducing cost (as more tasks are offloaded to an automated computer). A related much-hyped urban application of UAVs is for package delivery, including fast food and small packages. The benefits of a 5G mmWave connection are numerous: low-latency remote control, high-accuracy positioning in urban canyons, and improved situational awareness for coordinated operation of the UAVs. Cellular networks have many advantages as well, whether or not they use mmWave, as part of the coordination of multiple aircraft [73] in terms of coordinating deliveries, managing congestion in the sky, and avoiding collisions [13].

8.3 Aerial Channel Models at Millimeter Wave Frequencies

8.3.1 Propagation Considerations for Aerial Channels

In this section, we highlight some notable features of mmWave aerial propagation. We begin by addressing the impact of the atmosphere, which becomes more important for longer links and higher frequencies. Then we explain the sources of mmWave link blockage in aerial networks. Both atmospheric effects and blockage depend on the physical environment and the height above ground level (or just height for short).

8.3.1.1 Atmospheric Considerations

mmWave transmissions experience additional attenuation beyond the typical free-space path loss due to atmospheric effects [52]. The impact of the atmosphere is relatively more important for aerial vehicles than for ground applications of mmWave. Aerial links may be relatively longer than those used in terrestrial 5G networks. For longer distances, atmospheric attenuation becomes an important consideration in maintaining the communication link.

There are three primary sources of atmospheric attenuation. The first is gaseous molecular absorption, in which the mmWave signal excites the molecules that make up the gases in the atmosphere. The second is suspended liquid found, for example, in fog or clouds. The third is a byproduct of the scattering due to rain or other types of precipitation. All these sources create a frequency-dependent attenuation. In general terms, the effects increase with increasing frequency (decreasing wavelength), with peaks of higher attenuation due to interactions with specific molecules at certain frequencies.

The two atmospheric gases responsible for attenuation of mmWave signals below 100 GHz, which is of interest for 5G, are water vapor and oxygen. Detailed calculations may be found in International Telecommunication Union (ITU) documents like [63] up to 1000 GHz, or by using software at [14]. The peak due to water vapor absorption is centered around 23 GHz. Using data computed assuming a temperature of 15°C, total barometric pressure of 1 atm, and relative humidity of 58%, the peak at 23 GHz adds about 0.3 dB km^{-1} extra attenuation. This is only a significant consideration for longer links. The peak due to oxygen occurs around 60 GHz. Using the same parameters, the attenuation is more than 10 dB km^{-1}. This is a substantial additional factor. It is one of the reasons that 60 GHz was selected very early as an unlicensed band (spectrum viewed as only useful for very short-range links) and also that 60 GHz is used for satellite-to-satellite links (interference to the ground is heavily attenuated). Nonetheless, it is likely that 60 GHz will continue to be used in future releases of 5G following trends on license-assisted access at lower frequencies.

Suspended water vapor creates another source of frequency-dependent attenuation, which also increases with frequency (decreases with wavelength) [6]. The attenuation increases with the density of water (measured in grams per meter cubed) and may increase or decrease with temperature depending on the frequency. Based on [6, figure 1], for a dense fog of 0.1 g m^{-3}, the attenuation is 0.5 dB km^{-1} at 100 GHz. The attenuation is much lower at lower frequencies. As a result, we conclude that suspended water vapor is not a significant issue for aerial mmWave communication. Of course, suspended water vapor can obstruct optical sensors, and so it still has an impact on overall UAV operations.

Scattering due to precipitation like rain, sleet, and snow is complicated due to the many parameters involved with precipitation. There are a number of different models for attenuation as a function of rainfall rate in millimeters per hour and the signal polarization [62, 69]. As an example given in [52], for a rainfall rate of 50 mm h^{-1} (moderately heavy rain), the attenuation may be between 8 and 18 dB km^{-1}. As a result, it is important to consider the impact of rain on the link budget. Rain can also add additional penetration losses on foliage and additional scattering near the ground. Of course, heavy rain has the obvious disadvantage of poor flying conditions. In a nutshell, rain is significant and should be included, but the target maximum rain rate should be determined based on the UAV mission and application.

Snow, sleet, and icing rain also introduce additional attenuation in mmWave signals. As these conditions, though, introduce additional challenges in operating the UAV, they are of lower priority for most of the anticipated 5G UAV use cases.

Computing the combined impact of the various sources of attenuation is complicated. In general, it requires a detailed understanding of how the link connecting transmitter and receiver transects the body of the atmosphere. Clouds and fog, for example, are all localized phenomena whose impact varies with the UAV's position and height above ground level. Temperature and relative humidity also vary with altitude and weather conditions. A good understanding of the use cases and operating environment is important to determine the impact of atmospheric effects on the link budget.

A more detailed survey of aerial propagation considerations and modeling may be found in [15]. Code for reproducing the results in [15], and for estimating atmospheric losses according to the ITU, may be found in [14].

8.3.1.2 Blockages

The significance of blockages is an important differentiating feature of mmWave relative to lower-frequency communication systems [30]. The primary sources of blockage in mmWave cellular systems are fixed objects like buildings and foliage, and mobile objects such as other vehicles, as illustrated in Figure 8.2. Blockage considerations apply to both signal and interfering links. In the aerial setting, the altitude of the aerial vehicle with respect to the environment partly determines the likelihood that a link is blocked. The extent of the impact of the environment depends on the scenario and the relative position of the other pair in the communication.

The likelihood of blockage is a function of the height above ground level. For example, at high elevations, unblocked links are more likely for air-to-tower links but less so for air-to-ground links. At low altitudes, the proximity of buildings, people, and foliage is higher, making blockage a more significant consideration. The significance of airframe blockages from other aerial vehicles does not depend on the altitude, but more strongly on the use case, and is dependent on the density of other vehicles. For example, the risk of such blockage may be significant for swarms. Self-body-blockage is also possible in the aerial case, but this can be overcome through judicious antenna placement, e.g., both

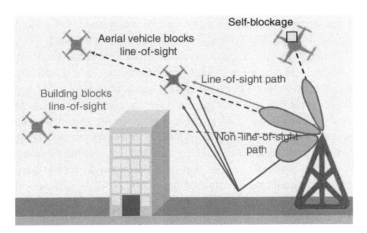

Figure 8.2 Different blocking scenarios in a mmWave aerial communication system.

above and below the vehicle. The direct impact of blockage on the link budget is in terms of an additional path-loss contribution, though this neglects aspects related to mobility.

Accounting for the impact of blockage in a system-level simulation often entails the use of a two-state path-loss model with a free-space exponent for LoS and a higher exponent for the NLoS or blocked case. In this situation, the impact of blockage is dependent on the length of the link in terms of both a distant-dependent blockage probability function and a different distant-dependent loss when blocked. Such models have been used to analyze mmWave cellular networks [9] and extended to analyze mmWave aerial networks [80]. A widely used model for the LoS probability based on random shape theory with parameters related to the fraction of land occupied, the mean number of buildings per unit area, and a scale parameter related to the height distribution, depending on the transmitter and receiver elevations, is available in [[1], eqn. (4)]. Other generalizations are possible to include multiple types of blockages [21].

The relationship between height above ground level and blockage probability leads to some interesting conclusions about mmWave UAV networks. For example, consider the case of air-to-ground communication. If the UAV is close to the ground, the link is more likely to be obstructed by a building than if the UAV is high in the air. When the UAV is high, though, the link distances become longer and the impacts of attenuation and atmosphere become more significant. As a result, there is usually an optimum operating point above ground level in these cases that balances blockage probability and the path loss [27].

The self-blockage by the UAV propellers, which is an issue when the UAV is required to communicate with both ground and aerial devices, was addressed in [10]. The authors used a hardware testbed to characterize relevant parameters of the blockage due to propellers and proposed schemes to improve the throughput by tracking the consequent periodic fading.

The effects of human body blockage are discussed in [20], which also presented a methodology for optimizing capacity via mmWave communication provided by UAVs. The proposed methodology takes into account human blockage, informing the optimum position of the UAV (3D coordinates) and also the corresponding coverage area. The numerical results in this work indicate how the coverage radius decreases with the density of human blockers.

8.3.2 Air-to-Air Millimeter Wave Channel Model

LoS channel models are appropriate for free-space communication, as will be found often in air-to-air (A2A) and air-to-ground (A2G) communication with high vehicle altitudes. The A2A MIMO channel is parameterized by azimuth and elevation angles of arrival (AoA) and angles of departure (AoD), the atmospheric path loss, array geometry, and small-scale effects resulting from phase changes and local scattering. This channel model assumes that there is only a single propagation path between transmitter and receiver, which leads to a rank-one MIMO channel. A rank-two model is possible for A2A scenarios, though, with polarized transmit and receive antennas.

Simulating a LoS channel requires characterization of both large-scale and small-scale propagation effects. These include atmospheric attenuation effects, phase changes driven by the relative geometry between transmitter and receiver, and knowledge of azimuth and elevation AoD and AoA. Denoting the number of antennas at the transmitter as N_t, the number of antennas at the receiver as N_r, the joint effect of atmospheric attenuation and phase changes as a complex number α_1, the array steering vectors at the transmit and receive side as $\boldsymbol{a}_\mathrm{T}$ and $\boldsymbol{a}_\mathrm{R}$, the azimuth AoA and AoD as $\theta_{\mathrm{AoA},1}$ and $\theta_{\mathrm{AoD},1}$, the elevation AoA and AoD as $\phi_{\mathrm{AoA},1}$ and $\phi_{\mathrm{AoD},1}$, the propagation delay of the LoS path as $\tau_{1,1}$, the joint effect of transmit and receive pulse shapes and analog filtering evaluated at τ as $p(\tau)$, and the sampling time as T_s, the MIMO matrix corresponding to the LoS channel can be geometrically modeled as

$$\boldsymbol{H} = \sqrt{N_\mathrm{t} N_\mathrm{r}} \; \alpha_1 p(dT_\mathrm{s} - \tau_{1,1}) \; \boldsymbol{a}_\mathrm{R}(\theta_{\mathrm{AoA},1}, \phi_{\mathrm{AoA},1}) \; \boldsymbol{a}_\mathrm{T}^*(\theta_{\mathrm{AoD},1}, \phi_{\mathrm{AoD},1}). \tag{8.1}$$

Besides the channel modeling between antenna arrays at both transmitter and receiver, it is important to consider phase variations due to non-ideal radio-frequency (RF) oscillators used at both ends [61].

8.3.3 Air-to-Ground Millimeter Wave Channel Model

NLoS channels have multipaths coming from reflections and scattering in addition to a possible LoS component. This is the type of channel that better characterizes the altitude-dependent A2G channel. Sources of reflection common in aerial and non-aerial mmWave applications include buildings, foliage, and street fixtures. Sources unique to the aerial case are the ground and the local aircraft body. In general, communication in NLoS without the LoS component is the most challenging, and consequently the most important to capture correctly in an NLoS model. Differences between A2A and A2G channel models at mmWave frequencies are illustrated in Figure 8.3.

NLoS channels can be generated by using a clustered geometric channel model [30]. Uniform planar arrays (UPAs) are a common choice of array geometry. Let C be the number of

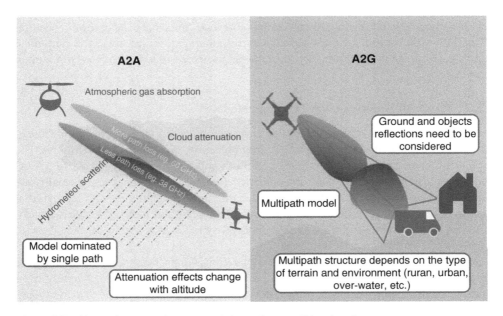

Figure 8.3 Air-to-air versus air-to-ground channels at mmWave bands.

multipath clusters and R_c be the number of rays for the cth cluster, and let $\alpha_{c,r}$ be the joint effect of atmospheric attenuation and phase changes for the rth ray within cluster c, and $\tau_{c,r}$ be the time of arrival of the rth ray within cluster c. With these definitions, the complex baseband discrete-time channel model may be written as

$$
\boldsymbol{H}_d = \sqrt{\frac{N_t N_r}{\sum_{c=1}^C R_c}} \sum_{c=1}^C \sum_{r=1}^{R_c} \alpha_{c,r} p_{r,c}(dT_s - \tau_{c,r})
$$
$$
\times \boldsymbol{a}_R(\theta_{\text{AoA},c,r}, \phi_{\text{AoA},c,r}) \, \boldsymbol{a}_T^*(\theta_{\text{AoD},c,r}, \phi_{\text{AoD},c,r}), \tag{8.2}
$$

for $d = 1, \ldots, D$, with D being the number of channel taps to be considered. Including the pulse shaping effect is important for frequency selective channels, as the impact of bandwidth and temporal leakage (one continuous-time ray creating multiple discrete-time samples) is exactly captured in that response. This is important even for LoS, as imperfect synchronization means that there will still be a few discrete-time coefficients even if there is only one ray.

Given the difficulty of flying prototypes of millimeter wave communication systems, ray tracing simulations have become a valid tool to obtain experimental channel measurements. In [34], the results of ray tracing simulations with Remcom's Wireless InSite are presented. Four scenarios (3D models and corresponding electromagnetic parameters in InSite) are studied: rural, urban, suburban, and over sea. The simulations used 28 or 60 GHz as carrier frequencies. The received signal strength and delay spread were obtained, for UAV heights of 2, 50, 100, and 150 m. The simulations indicated that the impact of the UAV height on the delay spread depends on the scenario. For urban, the delay spread increases as a function of the UAV height, while it decreases for rural. The main conclusion of these ray tracing simulations is that a two-ray model may hold at high flying altitudes, while more multipath components are needed at low altitudes.

In [35], the authors extended their work in [34], and presented results of ray tracing simulations for A2G LoS channels at 28 GHz in the previously mentioned four scenarios. Parameters such as direction and time of arrival (DOF and TOA, respectively) were analyzed for two distinct categories of multipath components (MPCs): persistent and non-persistent. The MPCs in the former category depend mainly on the scenario geometry, while those in the non-persistent category depend on properties of the scatterers and are modeled by birth/death processes.

In [79], the authors used machine learning methods to predict path loss and delay spread in A2G mmWave channels in which the UAV has the receiver. There, K nearest neighbors (KNN) and random forests are the learning algorithms adopted for training the regression models. A feature selection stage is incorporated, in order to pre-select a subset of sensible features among an initial set of eight features for path-loss prediction (the prediction of delay spread had the correct path-loss value as the ninth feature). For instance, some of the features are the 2D distance from the transmitter to the UAV and the number of buildings located in the line between the transmitter and the UAV. The features are highly dependent on the specific site, and transfer learning methods are evaluated in order to predict performance in a new site by leveraging data from another site. Ray tracing simulations with Remcom's Wireless InSite used two distinct urban areas (in Ottawa and Helsinki) and distinct carrier frequencies (2.4, 5.8, 28, and 37 GHz). The simulated data enabled the evaluation of two transfer learning strategies proposed by the authors: frequency-based and scene-based. The results showed that these two strategies can predict path loss with good accuracy. However, the experiments used only urban areas and the input features rely on detailed knowledge of the 3D scenarios along the UAV trajectories.

8.3.4 Ray Tracing as a Tool to Obtain Channel Measurements

Light equipment implementing a mmWave MIMO system is expensive and not generally available. As mentioned in previous sections, ray tracing simulation is an interesting approach to obtain channel data. A methodology to simulate realistic mmWave A2G channels was proposed in [37]. Some previous work using ray tracing to simulate A2G mmWave channels (e.g., [34, 35]) adopt well-defined trajectories for the UAV. For instance, the UAV is assumed to fly at a fixed height. The mmWave channels generated by this simplified approach may not suffice for applications such as channel tracking algorithms, in which the time evolution of the channel needs to reflect variations that occur in practice. In order to obtain UAV trajectories that incorporate, for instance, the effect of wind, Microsoft's AirSim [64] was incorporated into the methodology proposed in [37]. AirSim models some physical phenomena and is used to obtain the positions (x, y, z) at a rate $1/T_s$, where T_s is the sampling interval specified by the user. As depicted in Figure 8.4, while AirSim specifies the UAV trajectories, the open-source Simulation of Urban MObility (SUMO) [38] generates the positions of all other moving objects (vehicles and pedestrians). Given a 3D scenario of interest, the transmitter antenna(s) is(are) manually positioned. As depicted in Figure 8.4, a Python orchestrator code collects the positions of all mobile objects (UAVs, vehicles, and pedestrians), composes a scene incorporating all information required by a ray tracing simulation, and finally invokes the ray tracing simulator. For each scene, Remcom's Wireless InSite ray tracing simulator is executed and information about the L strongest rays (gain, phase, angles, etc.) to each receiver is stored in a database.

As an example of the methodology in [37] applied to UAVs, we used ray tracing simulations to obtain worst-case values for the Ricean K-factor in [59]. We generated the data with the scenario depicted in Figure 8.5, which is part of Wireless InSite's examples

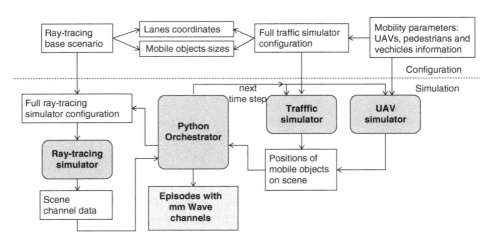

Figure 8.4 Methodology that integrates ray tracing, traffic, and UAV simulators.

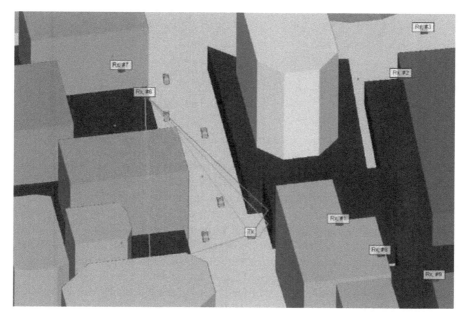

Figure 8.5 Urban canyon 3D scenario with one transmitter, seven receivers at UAVs, and vehicles composing a scene for ray tracing simulation. The four strongest rays to UAV 6 (receiver Rx, #6) are shown.

and represents a 337×202 m^2 area of Rosslyn, Virginia. The adopted carrier frequency was 60 GHz, $T_s = 1$ s, the transmitter antenna height was 5 m, and "diffuse scattering" was enabled. The channels estimated with these ray tracing simulations are then used to guide the design of mmWave communications for UAVs while measurements are not plentifully available.

8.4 Key Aspects of UAV MIMO Communication at mmWave Frequencies

Multiple antennas play an important role in mmWave communication systems [30]. The main reason derives from the role of antenna aperture gain as it contributes to path loss in the channel. To illustrate, consider free-space propagation. With transmit power P_t, the far-field receive power is

$$P_r = G_r G_t \left(\frac{\lambda}{4\pi d} \right)^2 P_t, \tag{8.3}$$

where the powers are in linear scale, d is the transmit–receive separation distance, λ is the wavelength, and G_t and G_r are the transmit and receive antenna gains. If the antenna gains are fixed, then received power decreases in proportion to the square of the wavelength, leading to the notion that mmWave signals are very lossy. If the physical antenna aperture, however, is fixed, then multiple antenna elements can be fitted into the same space. The resulting array gain G from a simple directional beamforming generally scales $\propto \lambda^{-2}$, which

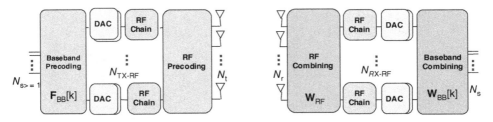

Figure 8.6 Typical hybrid precoding architecture for a single mmWave MIMO link.

compensates for the λ term and actually leads to improved received power. The importance of multiple antennas for path loss, not to mention many other secondary benefits, is the main reason that mmWave is empowered by MIMO communication.

The multiple antennas in a mmWave MIMO communication link may be used for other purposes besides obtaining array gain. The most prominent example is the spatial multiplexing of multiple data streams, which is the core reason for the enthusiasm in MIMO for the past 20 years [29]. The multiple antennas may also be used to support multiple users, to suppress interference, to enhance security [72], or for diversity against airframe blockage [10]. The types of signal processing performed may differ in each application.

Power consumption leads to different RF architectures in mmWave MIMO, compared with lower-frequency MIMO communication systems. Among the most power-consuming devices are the mixed signal components: the analog-to-digital converter (ADC) and the digital-to-analog converter (DAC). The hybrid architecture, illustrated in Figure 8.6, is one approach to reduce the number of DACs and ADCs. The key idea is to combine the signals to/from the antennas using analog phase shifters and (possibly) attenuators, which are analog components. The drawback is increased analog complexity, requiring a more complicated network connecting the mixed signal devices to the RF components.

To understand the challenges associated with mmWave MIMO, it is useful to consider the received signal model for a hybrid precoding link, contrasting it with an unconstrained architecture. To start, consider a MIMO communication link with N_t transmit antennas, N_r receive antennas, and N_s data streams. We will consider the received signal model assuming a MIMO orthogonal frequency-division multiplexing (OFDM) setup, with k indicating the subcarrier index and n indicating the OFDM symbol index. Denote $\mathbf{H}[k]$ as the $N_r \times N_t$ discrete-time frequency response of the MIMO channel defined in Eq. (8.2), $\mathbf{F}^{(n)}$ as the $N_t \times N_s$ precoding matrix, and $\mathbf{W}^{(n)}$ as the $N_r \times N_s$ combining matrix. Let $\mathbf{y}^{(n)}[k]$ denote the $N_r \times 1$ vector of samples at the receiver, assuming perfect synchronization. Let $\mathbf{s}^{(n)}[k]$ denote the vector of transmitted symbols, e.g., from a quadrature amplitude modulation (QAM) constellation. Finally, let $\mathbf{n}^{(n)}[k]$ denote the additive noise term including thermal noise and residual interference. Then the received signal may be written as

$$\mathbf{y}^{(n)}[k] = \mathbf{W}^{(n)*}[k]\mathbf{H}[k]\mathbf{F}^{(n)}\mathbf{s}^{(n)}[k] + \mathbf{W}^{(n)*}[k]\mathbf{n}^{(n)}[k]. \qquad (8.4)$$

If channel state information were available at the transmitter, then the precoder and combiner would be chosen based on the singular value decomposition of the channel, possibly with a water filling power allocation [29].

Now consider the case of hybrid precoding. We use a similar notation as in Eq. (8.4) with additional subscripts of BB to denote baseband quantities and RF to denote passband quantities (the effective model is the same whether the analog operations are at an intermediate frequency or the carrier frequency). Then the received signal is

$$\mathbf{y}^{(n)}[k] = \mathbf{W}_{\mathrm{BB}}^{(n)*}[k]\mathbf{W}_{\mathrm{RF}}^{(n)*}\mathbf{H}[k]\mathbf{F}_{\mathrm{RF}}^{(n)}\mathbf{F}_{\mathrm{BB}}^{(n)}[k]\mathbf{s}^{(n)}[k] + \mathbf{W}_{\mathrm{BB}}^{(n)*}[k]\mathbf{W}_{\mathrm{RF}}^{(n)*}\mathbf{n}^{(n)}[k]. \tag{8.5}$$

Note that the precoding operation has been split into the matrices $\mathbf{F}_{\mathrm{RF}}^{(n)}$ and $\mathbf{F}_{\mathrm{BB}}^{(n)}[k]$, producing an equivalent $N_t \times N_s$ matrix, but with the product of an $N_t \times N_{\mathrm{TX-RF}}$ and $N_{\mathrm{TX-RF}} \times N_s$ matrix. As $N_{\mathrm{TX-RF}}$ is the number of mixed signal chains, it is normally much smaller than N_t, but is bigger than N_s. Because $\mathbf{F}_{\mathrm{BB}}^{(n)}[k]$ is applied in the digital domain, it is possible to implement frequency-selective precoding; whereas because $\mathbf{F}_{\mathrm{RF}}^{(n)}$ is applied in the analog domain, the components act the same (ideally) for all frequencies. Further, the RF precoding may be subject to additional constraints such as quantized phase and no amplitude control. The same considerations for precoding also apply for the combining at the receiver with $N_{\mathrm{RX-RF}}$ pairs of ADCs.

The main challenges introduced by the hybrid architecture are summarized as follows:

(1) The precoding and combining requires the optimization of additional precoding and combining matrices. One challenge lies in the rank constraint due to the dimensions $N_s \leq N_{\mathrm{TX-RF}} \ll N_t$. Another is the frequency flatness of the RF operations and the constraints on the gain and phase values. This makes calculating the perfect decomposition into the hybrid precoders of either the optimal $\mathbf{F}$ or $\mathbf{W}$ not possible in general. That said, a common algorithmic approach is to find an optimal all-digital precoder and then to approximate it with a hybrid pair.

(2) The channel, as measured at the receiver, is contaminated by the RF precoding and combining. Sending training in the usual way, as a non-hybrid MIMO system, leads to the measurement of $\mathbf{W}_{\mathrm{RF}}^{(n)*}[k]\mathbf{H}[k]\mathbf{F}_{\mathrm{RF}}^{(n)}\mathbf{F}_{\mathrm{BB}}^{(n)}[k]$, whose dimensions $N_{\mathrm{RX-RF}} \times N_{\mathrm{TX-RF}}$ are much smaller than the underlying $N_r \times N_t$ channel. This means that new approaches are needed either to compressively estimate the channel [58] or to configure the analog precoders and combiners using beam training [47]. In the case of estimation, training data is sent combined with smart compressive sensing algorithms that make use of pseudo-random beam patterns. For beam training, different beam-based precoders and combiners are successively tried. An iterative process is used to select and refine the best option. Additional complications occur in the channel tracking task, which is important in UAV applications [59].

Though described for a single-user MIMO link, hybrid architectures may be employed in multiuser mmWave MIMO settings. For example, one could envision a hybrid array used at the base station and an analog array (special case with $N_{\mathrm{TX-RF}} = N_{\mathrm{RX-RF}} = 1$) on the device to reduce power consumption [30]. A general received signal model for the uplink, for example, is an extension of that in Eq. (8.5) as for user u:

$$\mathbf{y}_u^{(n)}[k] = \mathbf{W}_{\mathrm{BB},u}^{(n)*}[k]\mathbf{W}_{\mathrm{RF},u}^{(n)*}\sum_{m=1}^{U}\mathbf{H}_m[k]\mathbf{F}_{\mathrm{RF},m}^{(n)}\mathbf{F}_{\mathrm{BB},m}^{(n)}[k]\mathbf{s}_m^{(n)}[k]$$
$$+ \mathbf{W}_{\mathrm{BB},u}^{(n)*}[k]\mathbf{W}_{\mathrm{RF},u}^{(n)*}\mathbf{n}^{(n)}[k]. \tag{8.6}$$

The base station would apply a different set of combiners in an effort to extract the signals from the individual users. Alternatively, for the downlink, the received signal at user u is

$$\mathbf{y}_u^{(n)}[k] = \mathbf{W}_{\mathrm{BB},u}^{(n)*}[k]\mathbf{W}_{\mathrm{RF},u}^{(n)*}\mathbf{H}_u[k]\sum_{m=1}^{U}\mathbf{F}_{\mathrm{RF},m}^{(n)}\mathbf{F}_{\mathrm{BB},m}^{(n)}[k]\mathbf{s}_m^{(n)}[k]$$

$$+ \mathbf{W}_{\mathrm{BB},u}^{(n)*}[k]\mathbf{W}_{\mathrm{RF},u}^{(n)*}\mathbf{n}^{(n)}[k]. \qquad (8.7)$$

In the downlink, the base station combines the contributions of all users in the transmitted signal. Each user must work to extract its intended signal while suppressing the impact of other users.

Configuring the arrays in the multiuser setting is even more challenging compared with the single-user case. One reason is that, even in the unconstrained case, it is difficult to compute exactly the optimal precoders and combiners [29]. Another explanation is that interference considerations may trump noise concerns. This means that imprecision in configuring the precoders can lead to a significant performance disadvantage in terms of a low signal-to-interference ratio. The greedy approach is one practical strategy for multiuser mmWave MIMO communication. On the downlink, for example, the analog precoder may be configured for the user with the best channel, that contribution removed, and the analog precoder for the next user chosen, and so on. The digital precoding is then used to eliminate residual multiuser interference.

There are many challenges associated with implementing mmWave MIMO communication on UAV platforms [76] beyond considerations that apply to all mmWave MIMO applications. One consideration is that the channel for many use cases is LoS with just a single dominant propagation path. This means that there is a significant disadvantage faced when transmitting on beams that do not align with the dominant channel directions. As a result, acquiring the channel may be more challenging. Alternatively, channel tracking is potentially simpler, as there can be fewer channels to track. Another issue is that UAVs experience a high level of mobility in three dimensions. Unlike ground vehicles, they have more degrees of freedom in their movement. This makes it harder to exploit machine learning approaches for mmWave MIMO beam training that leverage items like the cars moving on the same streets [71]. Further, important tasks like tracking must be done more frequently [59]. Antenna placement also becomes relatively more important. To avoid airframe blockage, a UAV may need arrays on the top and bottom, possibly multiple arrays in different places. Additional smart algorithms would be needed to adaptively change the arrays.

UAVs offer an interesting interplay between sensing and communication. This can be leveraged to aid mmWave communication. For example, prior work has shown how radar may assist in the beam training problem [3, 25]; LIDAR (light detection and ranging), a sensor on high-end automated vehicles, has also been suggested as a means of determining whether a link is blocked [36]. Other communication signals may also be used to help configure the mmWave link [2]. In the UAV case, sensors on the aerial vehicle used for navigation may also be employed to assist mmWave link configuration. The navigational sensors could feed their outputs to make a smarter location-based channel tracking algorithm [59]. Sensors may also be deployed in UAV networks to further assist with navigation and traffic management. Sensors have already been suggested as a way to support ground vehicles [12].

8.5 Establishing Aerial mmWave MIMO Links

Links between UAVs are challenging to establish and maintain because both nodes are highly mobile, although the channel is usually LoS and a single path is enough to accurately model the propagation environment. When the goal is to establish a mmWave communication link between a ground base station and a UAV, the channel is usually less dynamic, but several multipath components need to be detected and tracked.

8.5.1 Beam Training and Tracking for UAV Millimeter Wave Communication

The general beam training/tracking approaches described in Section 8.4 can also be applied to air-to-everything (A2X) aerial communications. There are, however, some specific designs that leverage particular information usually available in a UAV to reduce beam training overhead.

For example, beam training/tracking strategies for an analog-only MIMO architecture in an A2A scenario try to find and track a single pair AoA/AoD. Most UAVs are equipped with sensors such as GPS and inertial measurement units (IMUs), so that it can be assumed that a UAV knows its own absolute position with a given error. Alternatively, in the A2A setup, one transmit UAV could estimate and predict the position of the receive UAV using different signal processing approaches. Position information can then be used to reduce training overhead, since estimates of the positions can be translated into estimates of the AoA/AoD by using a transformation in spherical coordinates, which can easily be refined later.

Following this idea, a beam tracking algorithm has been described [83] that leverages 3D position information predicted by a Gaussian process (GP) based learning algorithm. The transmitter UAV exploits the estimated position for the receive UAV to track AoA/AoD. Simulation results show that the method performs well even when the UAVs are highly mobile. One potential problem is the choice of the machine learning algorithm, which does not scale well with data and needs significant computational resources. A similar approach, also based on a GP learning strategy, was proposed in [32] to predict the position of multiple UAVs and restrict them within a spatial region where a beam search is then performed. To address the potential computational problem, a more efficient (online) variant of GP is used to reduce overhead – not just symbols used for tracking, but the actual time needed to calculate the predictions of the UAV positions.

Other works on beam training for UAV communication have exploited motion information to speed up the beam search process. For example, an A2G scenario is considered in [85] where aerial base stations serve ground users. The maximum speed of the users is used to predict the directions of two beams around the currently active one. The beam search is then restricted to these two possible beams, which requires only two tracking pilots.

8.5.2 Channel Estimation and Tracking in Aerial Environments

In the initial compressive channel estimation problem [58], a sequence of training symbols are sent to the receiver, each through a different pseudo-random beam pattern. The symbols are generally OFDM symbols or single-carrier frequency-domain equalization (SC-FDE) frames, with the beam switching occurring during a zero prefix separating the symbols. The measurements are collected together at the receiver and used to reconstruct an estimate of the channel. For example, in [58] we developed an algorithm that exploits noise statistics and spatially common support among the subcarriers to estimate the wideband channel with around 80 symbols. Though our algorithm works in a channel with many clusters and rays, 80 symbols may still be too long in highly mobile aerial scenarios. As a result, adaptive approaches are attractive that make use of past measurements to track the channel evolution with lower overheads. For the simpler case of an LoS A2G communication system, it is possible to exploit the high sparsity of the channel in both angle and delay domains. An ESPRIT algorithm is proposed in [40] for this particular case to recover the AoA and delay, while the path gains are obtained through a straightforward least squares (LS) estimator.

In channel tracking, new measurements are combined with past measurements or a channel estimate to update the channel estimate. This approach is sensible during the periods where the channel parameters are spatially consistent and change smoothly with aerial vehicle location, as illustrated in Figure 8.7. To deal with cluster arrival or departures, the algorithm could be periodically reinitialized or change-point detection could be applied to estimate such events.

One interesting approach in aerial vehicles is to incorporate knowledge about the position or the trajectory into the adaptive algorithm that tracks the channel. This is a valuable form of side information, which is particular to the vehicular setting [26]. For example, in [57] a channel tracking algorithm for a multiuser A2A scenario is proposed, where all UAVs communicate with a single aerial access point, that exploits location information and a uniform error location model. This allows the definition of a reduced space search around the angles corresponding to the estimated position, as illustrated in Figure 8.8. The compressive approach is able to track the channel with as few as eight OFDM symbols. In a later

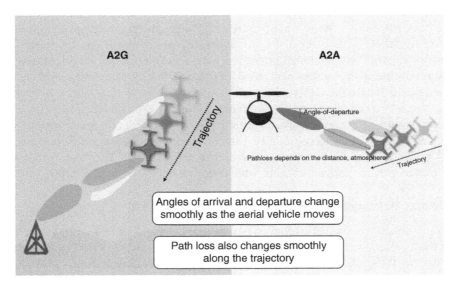

Figure 8.7 Concept of spatial consistency in aerial mmWave channels.

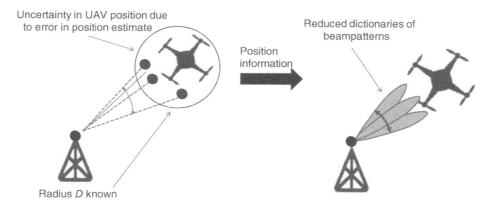

Figure 8.8 Leveraging position information to reduce the space for beam pattern search or channel estimate in aerial mmWave MIMO.

work [59], a scenario where multiple UAVs communicate with the ground station is considered. The difficulty lies again in the high mobility of the aerial units. A channel tracking algorithm is proposed based on the sparse nature of the mmWave channel and maximum likelihood (ML) techniques, which are augmented with the prior information about the position and trajectory of the UAVs. The experimental results show that only a few OFDM training symbols are needed to track three UAVs, given 256 subcarriers. Furthermore, it is interesting to note that the proposed method is able to maintain spectral efficiency rates over time, with little overhead, in the absence of blockage.

8.5.3 Design of Hybrid Precoders and Combiners

The goal of hybrid precoding is to approximate the performance of an optimal, fully digital precoder, while taking into account the hardware and energy constraints. Most of the algorithms in prior work factorize the optimal all-digital solution into the analog and digital counterparts. The first works on the design of hybrid precoders and combiners for a general mmWave system [4, 8, 44, 60, 77] considered a narrowband channel model, which may be accurate for A2A configurations when reflections on the aircraft itself are disregarded. The solution in [77] does not seem feasible, however, for UAV communication, since it requires a large number of RF chains to obtain perfect factorization. The design in [60], implementing a successive interference cancellation (SIC) based hybrid precoding, is interesting for the hybrid equipment at the UAV side, which avoids costly numerical operations such as the singular value decomposition (SVD) and matrix inversions that were necessary in the initial papers.

More recent works consider the frequency selectivity of the mmWave channel, a feature of A2G channels, proposing solutions that either (i) assume a near-optimum combiner [48, 50, 81], or (ii) consider a more general setting where this assumption is avoided [23, 24, 56, 82]. The solution in [23] makes use of alternative uplink and downlink transmissions to iteratively update the precoder and combiner. The main limitations of this work are the computational complexity and the high training overhead, which make this approach unsuitable for aerial communication. The algorithm proposed in [24] is interesting for A2G scenarios, since it provides a good performance when the number of multipath components is small. The greedy SVD-based algorithm proposed in [56] is interesting for aerial scenarios because of its extremely low complexity. In this case, the design is based on an iterative technique that minimizes the Euclidean distances between the optimum all-digital precoders and combiners and the hybrid filters.

Hybrid precoding and combining has been specifically considered also in the context of mmWave systems of UAVs, where the power constraints on the hardware are even more pressing. One of the first considerations of using hybrid processing with UAVs was made in [76], which discussed the challenges and opportunities of using mmWave for UAV cellular networks and identified hybrid processing as a mandatory step towards making UAVs cost and energy sustainable. A recent paper [54] considers a scenario where aerial devices with a lens antenna behave as base stations for ground users, establishing only LoS connections, which is a limitation for A2G channels. The authors recognize the potential energy consumption issues of the aerial base stations and propose an energy-saving hybrid precoding technique where the digital part has only a few RF chains and the phase shifters are replaced by simpler inverters and switches. The proposed algorithm uses robust supervised machine learning techniques to design the analog/digital precoders. The results show that the energy efficiency gap against the fully digital precoding closes with the increasing number of users.

8.6 Research Opportunities

In this section, we conclude by describing some research ideas that are in part enabled by the use of mmWave communication.

8.6.1 Sensing at the Tower

Sensing is useful for providing situational awareness and preventing collisions. UAVs typically have small payloads and may not be able to support long-range sensing technologies. An alternative is to combine sensing with communication infrastructure on the tower. In this way, the tower may use cameras, radar, and acoustic sensors to collect and rebroadcast information about aerial vehicle positions. Such an architecture would aid in other applications, including detection of unauthorized UAV flights and managing airspace amount aerial vehicles in general. Sensors may also be used to help facilitate communication, by providing out-of-band side information about position or the environment, which is relevant especially for A2G mmWave communication [26].

8.6.2 Joint Communication and Radar

mmWave frequencies are used both for 5G and in automotive radar. It is conceivable that the spectrum may be shared in time or frequency between communication and radar signals. Using the same waveform, with full duplex cancellation, the communication waveform may even be used for both purposes. A main advantage of joint radar and communication on aerial vehicles is that the same hardware can be shared for two purposes. For example, an aerial vehicle may use such capability to both communicate with and monitor the speed and velocity of ground vehicles. Such capability may also be valuable for longer-range radar imaging, and for avoiding larger manned aerial vehicles.

8.6.3 Positioning and Mapping

5G and aerial vehicles may work together in other synergistic ways. For example, the higher bandwidths of both low- and high-frequency 5G waveforms provide another means of positioning. This may be used to aid GPS in denied environments, or as a separate mechanism for positioning. With accurate positioning, aerial vehicles may also aid the creation of 3D maps, from a birds-eye view, which support activities like automated driving. In this case, the mmWave communication is used to transport sensor data from the vehicle to an edge or cloud processing device, which in turn updates the maps. Aerial vehicles may also leverage knowledge of their location to aid in the creation of radio maps. For example, dead zones in A2G mmWave communication could be identified, and then avoided by the navigation algorithms in the aerial vehicles.

8.7 Conclusions

mmWave is a sensible way to provide a high-bandwidth connection to aerial vehicles. The higher mobility of aerial platforms changes lower-mobility design considerations found in more conventional applications of mmWave. The propagation channel has sparsity, which is altitude-dependent. Atmospheric effects become more significant at higher altitudes and for longer links. Establishing mmWave communication links with aerial vehicles is expected to be a major challenge, especially with hybrid architectures and NLoS links in which cluster blockage is expected. Nonetheless, mmWave is well positioned to be a key technology for many critical aerial vehicle missions, including communications and sensing.

References

1 A. Al-Hourani, S. Kandeepan, and S. Lardner (2014). Optimal LAP altitude for maximum coverage. *IEEE Wireless Commun. Lett.* 3 (6): 569–572. doi: 10.1109/LWC.2014.2342736.

2 A. Ali, N. González-Prelcic, and R. W. Heath (2018). Millimeter wave beam-selection using out-of-band spatial information. *IEEE Trans. Wireless Commun.* 17 (2): 1038–1052. doi: 10.1109/TWC.2017.2773532.

3 A. Ali, N. González-Prelcic, and A. Ghosh (2019). Millimeter wave V2I beam-training using base-station mounted radar. *IEEE Radar Conference (RadarConf)*, April, pp. 1–5. doi: 10.1109/RADAR.2019.8835615.

4 A. Alkhateeb, G. Leus, and R. W. Heath. Limited feedback hybrid precoding for multi-user millimeter wave systems. *IEEE Trans. Wireless Commun.*, 14(11):6481–6494, Nov 2015. doi: 10.1109/TWC.2015.2455980.

5 A. Almohamad, M. O. Hasna, T. Khattab, and M. Haouari. Maximizing dense network flow through wireless multihop backhauling using UAVs. *Proceedings of the International Conference on Information and Communication Technology Convergence (ICTC)*, pages 526–531, Oct 2018. doi: 10.1109/ICTC.2018.8539573.

6 E. Altshuler. A simple expression for estimating attenuation by fog at millimeter wavelengths. *IEEE Trans. Antennas Propag.*, 32(7):757–758, July 1984. doi: 10.1109/TAP.1984.1143395.

7 J. G. Andrews, S. Buzzi, W. Choi et al. What will 5G be? *IEEE J. Sel. Areas Commun.*, 32(6):1065–1082, June 2014. doi: 10.1109/JSAC.2014.2328098.

8 O. E. Ayach, S. Rajagopal, S. Abu-Surra et al. Spatially sparse precoding in millimeter wave MIMO systems. *IEEE Trans. Wireless Commun.*, 13 (3):1499–1513, Mar. 2014.

9 T. Bai, A. Alkhateeb, and R. W. Heath Jr. Coverage and capacity of millimeter-wave cellular networks. *IEEE Commun. Mag.*, 52(9):70–77, Sep. 2014.

10 J. Bao, D. Sprinz, and H. Li. Blockage of millimeter wave communications on rotor UAVs: demonstration and mitigation. *Proceedings of the 2017 IEEE Military Communications Conference (MILCOM)*, pages 768–774, Oct 2017. doi: 10.1109/MILCOM.2017.8170850.

11 M. Campion, P. Ranganathan, and S. Faruque. A review and future directions of UAV swarm communication architectures. *Proceedings of the 2018 IEEE International Conference on Electro/Information Technology (EIT)*, pages 0903–0908, May 2018. doi: 10.1109/EIT.2018.8500274.

12 J. Choi, N. G. Prelcic, R. Daniels et al. Millimeter wave vehicular communication to support massive automotive sensing. *IEEE Commun. Mag.*, 54(12):160–167, Dec. 2016.

13 W. B. Cotton. Adaptive autonomous separation for UAM in mixed operations. *Proceedings of the 2019 Integrated Communications, Navigation and Surveillance Conference (ICNS)*, pages 1–11, April 2019. doi: 10.1109/ICNSURV.2019.8735196.

14 T. Cuvelier. mmWaveAerialNetworks. https://github.com/travisCuvelier/ mmWaveAerialNetworks, May 2019.

15 T. Cuvelier and R. W. Heath Jr. (2018). MmWave MU-MIMO for aerial networks. Preprint. arXiv:1804.03295v2.

16 E. Dahlman, S. Parkvall, and J. Sköld (2018). *5G NR The Next Generation Wireless Access Technology.* Academic Press.

17 J. Delmerico, T. Cieslewski, H. Rebecq et al. (2019). Are we ready for autonomous drone racing? The UZH-FPV drone racing dataset. *Proceedings of the 2019 International Conference on Robotics and Automation (ICRA)*, May, pp. 6713–6719. doi: 10.1109/ICRA.2019.8793887.

18 M. Erdelj, E. Natalizio, K. R. Chowdhury, and I. F. Akyildiz. Help from the sky: leveraging UAVs for disaster management. *IEEE Pervasive Comput.*, 16(1):24–32, Jan 2017. doi: 10.1109/MPRV.2017.11.

19 Z. Feng, L. Ji, Q. Zhang, and W. Li. Spectrum management for mmwave enabled UAV swarm networks: challenges and opportunities. *IEEE Commun. Mag.*, 57(1):146–153, January 2019. doi: 10.1109/MCOM.2018.1800087.

20 M. Gapeyenko, I. Bor-Yaliniz, S. Andreev et al. (2018). Effects of blockage in deploying mmWave drone base stations for 5G networks and beyond. *Proceedings of the IEEE International Conference on Communications (ICC) Workshops*, pp. 1–6.

21 M. Gapeyenko, V. Petrov, D. Moltchanov et al. Flexible and Reliable UAV-Assisted Backhaul Operation in 5G mmWave Cellular Networks. *IEEE J. Sel. Areas Commun.*, 36(11):2486–2496, Nov 2018. doi: 10.1109/JSAC.2018.2874145.

22 C. M. Gevaert, J. Suomalainen, J. Tang, and L. Kooistra (2015). Generation of spectral–temporal response surfaces by combining multispectral satellite and hyperspectral UAV imagery for precision agriculture applications. *IEEE J. Sel. Topics Appl. Earth Observ. Remote Sens.* 8 (6): 3140–3146. doi: 10.1109/JSTARS.2015.2406339.

23 J. P. González-Coma, N. González-Prelcic, L. Castedo, and R. W. Heath. Frequency selective multiuser hybrid precoding for mmWave systems with imperfect channel knowledge. *Proceedings of the 50th Asilomar Conference on Signals, Systems and Computers*, pages 291–295, Nov 2016. doi: 10.1109/ACSSC.2016.7869044.

24 J. P. González-Coma, J. Rodríguez-Fernández, N. González-Prelcic et al. (2018). Channel estimation and hybrid precoding for frequency selective multiuser mmWave MIMO systems. *IEEE J. Sel. Topics Signal Process.* 12 (2): 353–367. doi: 10.1109/JSTSP.2018.2819130.

25 N. González-Prelcic, R. Méndez-Rial, and R. W. Heath. Radar aided beam alignment in mmWave V2I communications supporting antenna diversity. *Proceedings of the 2016 Information Theory and Applications Workshop (ITA)*, pages 1–7, Jan 2016. doi: 10.1109/ITA.2016.7888145.

26 N. Gonzalez-Prelcic, A. Ali, V. Va, and R. W. Heath. Millimeter-wave communication with out-of-band information. *IEEE Commun. Mag.*, 55 (12):140–146, December 2017. doi: 10.1109/MCOM.2017.1700207.

27 K. Han, K. Huang, and R. W. Heath. Connectivity and blockage effects in millimeter-wave air-to-everything networks. *IEEE Wireless Commun. Lett.*, 8(2):388–391, April 2019. doi: 10.1109/LWC.2018.2873361.

28 S. Hayat, E. Yanmaz, and R. Muzaffar. Survey on unmanned aerial vehicle networks for civil applications: a communications viewpoint. *IEEE Commun. Surveys Tuts*, 18(4):2624–2661, 2016. doi: 10.1109/COMST.2016.2560343.

29 R. W. Heath Jr. and A. Lozano. *Foundations of MIMO Communication*. Cambridge University Press, 2018.

30 R. W. Heath Jr., N. Gonzalez-Prelcic, S. Rangan et al. An overview of signal processing techniques for millimeter wave MIMO systems. *IEEE J. Sel. Topics Signal Process.*, 10(3):436–453, 2016.

31 J. Hempel (2016). Inside FaceBook's ambitious plan to connect the whole world, January. https://www.wired.com/2016/01/facebook-zuckerberg-internet-org/.

32 Y. Ke, H. Gao, W. Xu et al. (2019). Position prediction based fast beam tracking scheme for multi-user UAV-mmWave communications. *Proceedings of the IEEE International Conference on Communications (ICC)*, May, pp. 1–7. doi: 10.1109/ICC.2019.8761775.

33 H. Kesteloo (2019). New DJI digital FPV transmission system with low latency and HD video for drone racing, July. https://dronedj.com/2019/07/31/dji-digital-fpv-transmission-system/.

34 W. Khawaja, O. Ozdemir, and I. Guvenc (2017). UAV air-to-ground channel characterization for mmWave systems. *Proceedings of the IEEE 86th Vehicular Technology Conference (VTC)*, Fall, pp. 1–5.

35 W. Khawaja, O. Ozdemir, and I. Guvenc (2018). Temporal and spatial characteristics of mmWave propagation channels for UAVs. *Proceedings of the 11th Global Symposium on Millimeter Waves (GSMM)*, pp. 1–6.

36 A. Klautau, N. González-Prelcic, and R. W. Heath. LIDAR data for deep learning-based mmWave beam-selection. *IEEE Wireless Commun. Lett.*, 8(3):909–912, June 2019. doi: 10.1109/LWC.2019.2899571.

37 A. Klautau, P. Batista, N. González-Prelcic et al. (2018). 5G MIMO data for machine learning: application to beam-selection using deep learning. *Proceedings of the 2018 Information Theory and Applications Workshop (ITA)*, pp. 1–9.

38 D. Krajzewicz, J. Erdmann, M. Behrisch, and L. Bieker (2012). Recent development and applications of SUMO – Simulation of Urban MObility. *Int. J. Adv. Syst. Measurem.* 5 (3-4): 128–138.

39 B. Li, Z. Fei, and Y. Zhang. UAV communications for 5G and beyond: recent advances and future trends. *IEEE Internet Things J.*, 6(2): 2241–2263, April 2019. doi: 10.1109/JIOT.2018.2887086.

40 A. Liao, Z. Gao, Y. Wu et al. Multi-user wideband sparse channel estimation for aerial BS with hybrid full-dimensional MIMO. *Proceedings of the IEEE International Conference on Communications (ICC) Workshops*, pages 1–6, May 2019. doi: 10.1109/ICCW.2019.8757125.

41 C. Luo, W. Miao, H. Ullah et al. Unmanned aerial vehicles for disaster management. *Geological Disaster Monitoring Based on Sensor Networks*, Springer, 2018.

42 J. Lyu, Y. Zeng, and R. Zhang (2017). UAV-aided offloading for cellular hotspot. arXiv:1705.09024.

43 I. Mademlis, V. Mygdalis, N. Nikolaidis, and I. Pitas. Challenges in autonomous UAV cinematography: an overview. *Proceedings of the 2018 IEEE International Conference on Multimedia and Expo (ICME)*, pages 1–6, July 2018. doi: 10.1109/ICME.2018.8486586.

44 R. Méndez-Rial, C. Rusu, N. González-Prelcic, and R. W. Heath. Dictionary-free hybrid precoders and combiners for mmWave MIMO systems. *Proceedings of the IEEE 16th International Workshop on Signal Processing Advances in Wireless Communications (SPAWC)*, pages 151–155, June 2015. doi: 10.1109/SPAWC.2015.7227018.

45 C. Metz (2016). Inside Facebook's first efforts to rain internet from the sky, November https://www.wired.com/2016/11/inside-facebooks-first-efforts-rain-internet-sky/.

46 H. Niu, N. Gonzalez-Prelcic, and R. W. Heath. A UAV-based traffic monitoring system – invited paper. *Proceedings of the IEEE 87th Vehicular Technology Conference (VTC)*, Spring, pages 1–5, June 2018. doi: 10.1109/VTCSpring.2018.8417546.

47 E. Onggosanusi, M. S. Rahman, L. Guo et al. Modular and high-resolution channel state information and beam management for 5G new radio. *IEEE Commun. Mag.*, 56(3):48–55, March 2018. doi: 10.1109/MCOM.2018.1700761.

48 S. Park, A. Alkhateeb, and R. W. Heath. Dynamic subarrays for hybrid precoding in wideband mmWave MIMO systems. *IEEE Trans. Wireless Commun.*, 16(5):2907–2920, May 2017. doi: 10.1109/TWC.2017.2671869.

49 Y. A. Pederi and H. S. Cheporniuk. Unmanned aerial vehicles and new technological methods of monitoring and crop protection in precision agriculture. *Proceedings of the IEEE International Conference on Actual Problems of Unmanned Aerial Vehicles Developments (APUAVD)*, pages 298–301, Oct 2015. doi: 10.1109/APUAVD.2015.7346625.

50 R. Peng and Y. Tian. Wideband hybrid precoder design in MU-MIMO based on channel angular information. *Proceedings of the IEEE 18th International Workshop on Signal Processing Advances in Wireless Communications (SPAWC)*, pages 1–5, July 2017. doi: 10.1109/SPAWC.2017.8227759.

51 K. M. Pesyna, R. W. Heath Jr., and T. E. Humphreys. Centimeter accurate positioning with a smartphone-grade antenna. *Proceedings of the ION GNSS+ Conference*, 2014.

52 T. S. Rappaport, R. W. Heath Jr., R. C. Daniels, and J. N. Murdock (2014). *Millimeter Wave Wireless Communications*. Prentice-Hall.

53 C. Reiche, C. McGillen, J. Siegel, and F. Brody. Are we ready to weather urban air mobility (UAM)? *Proceedings of the 2019 Integrated Communications, Navigation and Surveillance Conference (ICNS)*, pages 1–7, April 2019. doi: 10.1109/ICN-SURV.2019.8735297.

54 H. Ren, L. Li, W. Xu et al. (2019). Machine learning-based hybrid precoding with robust error for UAV mmWave massive MIMO. *Proceedings of the IEEE International Conference on Communications (ICC)*, May, pp. 1–6. doi: 10.1109/ICC.2019.8761112.

55 R. Ribeiro, J. Ramos, D. Safadinho, and A. M. de Jesus Pereira. UAV for everyone: an intuitive control alternative for drone racing competitions. *Proceedings of the 2nd International Conference on Technology and Innovation in Sports, Health and Wellbeing (TISHW)*, pages 1–8, June 2018. doi: 10.1109/TISHW.2018.8559538.

56 J. Rodríguez-Fernández and N. González-Prelcic (2018). Low-complexity multiuser hybrid precoding and combining for frequency selective millimeter wave systems. *Proceedings of the IEEE Signal Processing Advances in Wireless Communications (SPAWC) Conference*, June.

57 J. Rodríguez-Fernández, N. González-Prelcic, and R. W. Heath. Position-aided compressive channel estimation and tracking for millimeter wave multi-user MIMO air-to-air communications. *Proceedings of the IEEE International Conference on Communications (ICC) Workshops*, May 2018.

58 J. Rodríguez-Fernández, N. González-Prelcic, K. Venugopal, and R. W. Heath. Frequency-domain compressive channel estimation for frequency-selective hybrid millimeter wave MIMO systems. *IEEE Trans. Wireless Commun.*, 17(5):2946–2960, May 2018. doi: 10.1109/TWC.2018.2804943.

59 J. Rodríguez-Fernández, N. González-Prelcic, I. Pamplona-Trindade, and A. Klautau. Position-aided compressive channel estimation and tracking for millimeter wave multi-user MIMO air-to-ground communications. *Proceedings of the IEEE 20th International Workshop on Signal Processing Advances in Wireless Communications (SPAWC)*, pages 1–5, July 2019. doi: 10.1109/SPAWC.2019.8815594.

60 C. Rusu, R. Méndez-Rial, N. González-Prelcic, and R. W. Heath. Low complexity hybrid precoding strategies for millimeter wave communication systems. *IEEE Trans. Wireless Commun.*, 15(12):8380–8393, Dec 2016. doi: 10.1109/TWC.2016.2614495.

61 T. Schenk (2008). Phase noise. *RF Imperfections in High-Rate Wireless Systems*, Springer.

62 ITU (2005). ITU Radiocommunication Sector. Specific attenuation model for rain for use in prediction methods.

63 ITU (2016). ITU Radiocommunication Sector. Attenuation by atmospheric gases.

64 S. Shah, D. Dey, C. Lovett, and A. Kapoor (2018). AirSim: high-fidelity visual and physical simulation for autonomous vehicles. *Field and Service Robotics*, pp. 147–196. Springer. doi: https://doi.org/10.1007/978-3-319-67361-5_40.

65 W. Shi, J. Li, W. Xu et al. Multiple drone-cell deployment analyses and optimization in drone assisted radio access networks. *IEEE Access*, 6: 12518–12529, 2018. doi: 10.1109/ACCESS.2018.2803788.

66 R. Shinkuma and N. B. Mandayam. Design of ad hoc wireless mesh networks formed by unmanned aerial vehicles with advanced mechanical automation. https://arxiv.org/abs/1804.07428, 2018.

67 Texas Instruments (2016). Low-latency design considerations for video-enabled drones. http://www.ti.com/lit/wp/spry301/spry301.pdf.

68 T. Tomic, K. Schmid, P. Lutz et al. (2012). Toward a fully autonomous UAV: research platform for indoor and outdoor urban search and rescue. *IEEE Robot. Autom. Mag.* 19 (3): 46–56. doi: 10.1109/MRA.2012.2206473.

69 R. N. Trebits (1987). MMW propagation phenomena. *Principles and Applications of Millimeter-Wave Radar*, pp. 131–188. Artech House.

70 V. Va, T. Shimizu, G. Bansal, and R. W. Heath Jr. (2016). *Millimeter Wave Vehicular Communications: A Survey.* Now Publishers. https://www.nowpublishers.com/BookSeries.

71 V. Va, T. Shimizu, G. Bansal, and R. W. Heath (2019). Online learning for position-aided millimeter wave beam training. *IEEE Access* 7: 30 507–30 526. doi: 10.1109/ACCESS.2019.2902372.

72 N. Valliappan, A. Lozano, and R. W. Heath Jr. Antenna subset modulation for secure millimeter-wave wireless communication. *IEEE Trans. Commun.*, 61(8):3231–3245, Aug. 2013.

73 M. D. Villaluz, L. Gan, J. Sia et al. Preliminary 4.5G cellular network assessment with calibrated standard propagation model (SPM) for uTM-UAS operations in Singapore airspace. *Proceedings of the International Conference on Unmanned Aircraft Systems (ICUAS)*, pages 796–805, June 2018. doi: 10.1109/ICUAS.2018.8453326.

74 X. Wang, V. Yadav, and S. N. Balakrishnan (2007). Cooperative UAV formation flying with obstacle/collision avoidance. *IEEE Trans. Control Syst. Technol.* 15 (4): 672–679. doi: 10.1109/TCST.2007.899191.

75 H. Wymeersch, G. Seco-Granados, G. Destino et al. (2017). 5G mmWave positioning for vehicular networks. *IEEE Wireless Commun.* 24 (6): 80–86. doi: 10.1109/MWC.2017.1600374.

76 Z. Xiao, P. Xia, and X. Xia. Enabling UAV cellular with millimeter-wave communication: potentials and approaches. *IEEE Commun. Mag.*, 54(5): 66–73, May 2016. doi: 10.1109/MCOM.2016.7470937.

77 X. Zhang, A. F. Molisch, and S.-Y. Kung (2005). Variable-phase-shift-based RF-baseband codesign for MIMO antenna selection. *IEEE Trans. Signal Process.* 53 (11): 4091–4103. doi: 10.1109/TSP.2005.857024.

78 G. Yang, X. Lin, Y. Li et al. (2018). A telecom perspective on the internet of drones: from LTE-Advanced to 5G. arXiv:1803.11048, March.

79 G. Yang, Y. Zhang, Z. He et al. (2019). Machine-learning-based prediction methods for path loss and delay spread in air-to-ground millimetre-wave channels. *IET Microw. Antennas Propag.* 13 (8): 1113–1121.

80 W. Yi, Y. Liu, A. Nallanathan, and G. K. Karagiannidis. A unified spatial framework for clustered UAV networks based on stochastic geometry. *Proceedings of the IEEE Global Communications Conference (GLOBECOM)*, pages 1–6, Dec 2018. doi: 10.1109/GLO-COM.2018.8648138.

81 X. Yu, J. C. Shen, J. Zhang, and K. B. Letaief (2016). Alternating minimization algorithms for hybrid precoding in millimeter wave MIMO systems. *IEEE J. Sel. Topics Signal Process.* 10 (3): 485–500. doi: 10.1109/JSTSP.2016.2523903.

82 X. Yu, J. Zhang, and K. B. Letaief. A hardware-efficient analog network structure for hybrid precoding in millimeter wave systems. *IEEE J. Sel. Topics Signal Process.*, 12(2):282–297, May 2018. doi: 10.1109/JSTSP.2018.2814009.

83 J. Zhang, W. Xu, H. Gao et al. Position–attitude prediction based beam tracking for UAV mmWave communications. *Proceedings of the IEEE International Conference on Communications (ICC)*, pages 1–7, May 2019. doi: 10.1109/ICC.2019.8761981.

84 L. Zhang, H. Zhao, S. Hou et al. A survey on 5G millimeter wave communications for UAV-assisted wireless networks. *IEEE Access*, 7: 117460–117504, 2019. doi: 10.1109/AC-CESS.2019.2929241.

85 W. Zhang and W. Zhang. Beam training and tracking efficiency analysis for UAV mmWave communication. *Proceedings of the IEEE International Conference on Communication Systems (ICCS)*, pages 115–119, Dec 2018. doi: 10.1109/ICCS.2018.8689233.

Part III

UAV-Assisted Wireless Communications

9

Stochastic Geometry-Based Performance Analysis of Drone Cellular Networks

Morteza Banagar, Vishnu V. Chetlur, and Harpreet S. Dhillon

Wireless@VT, Department of ECE, Virginia Tech, Blacksburg, VA, USA

9.1   Introduction

The unmanned aerial vehicle (UAV) base stations (BSs), commonly known as drone base stations (DBSs), possess several advantages over conventional terrestrial BSs due to their mobility, ease of deployment, cost-effectiveness, and a high likelihood of line-of-sight (LoS) link to the ground user equipment (UE). Owing to these features, DBSs are considered a valuable addition to the current cellular networks. DBSs extend the coverage of cellular networks by providing connectivity to locations which are beyond the service zones of terrestrial cellular BSs. Moreover, DBSs play a crucial role in providing connectivity in times of natural disasters and emergencies, as the terrestrial networks may be completely incapacitated [22, 37]. Further, the ease of deployment of DBSs has made them an effective solution for providing network access in scenarios where there is a short-term need for network resources, such as sporting events or concerts [21]. To leverage the numerous benefits offered by drone cellular networks, the third generation partnership project (3GPP) has recently included the support for DBS communications as part of the fifth generation (5G) standard [1, 2].

Since DBSs are typically deployed at much higher altitudes than the terrestrial BSs, the propagation characteristics of the air-to-ground links of DBSs differ significantly from those of the terrestrial links. The large-scale and small-scale fading characteristics of the air-to-ground communications were studied through extensive measurement campaigns for a variety of environments (e.g., rural, semi-urban, and urban) in [35, 36, 43]. A probabilistic model for the occurrence of LoS and non-line-of-sight (NLoS) links in an aerial network as a function of the altitude of the drones, link distance, and the type of the environment was proposed in [5, 6] and further developed in [23]. Although the mobility of DBSs provides numerous advantages to the drone cellular networks, the power constraints and the nature of backhaul links limit the performance of DBSs in these networks. The trade-offs between the benefits and limitations of DBSs and their impact on the performance of the network need to be clearly understood for planning, design, and deployment of DBSs. This has stimulated a lot of research into diverse aspects of drone cellular networks using various mathematical tools. Motivated by the mobility of

UAV Communications for 5G and Beyond, First Edition.
Edited by Yong Zeng, Ismail Guvenc, Rui Zhang, Giovanni Geraci, and David W. Matolak.
© 2021 John Wiley & Sons Ltd. Published 2021 by John Wiley & Sons Ltd.

DBSs, several works have focused on determining optimal trajectories that maximize the utility of DBSs [30, 46, 48]. An algorithm to dynamically adjust the heading of the drones in order to maximize the sum rate of uplink communications is proposed in [30]. In [46], the minimum throughput of ground users for downlink communication was maximized by jointly optimizing user scheduling and drone trajectory. Using tools from game theory, an effective algorithm for bandwidth resource allocation in order to balance the service costs and the network performance has been proposed in [47]. The authors in [3] modeled drones as mobile relays to minimize the average peak age of information for a source–destination pair. Using machine learning, an energy-efficient drone control method for maximizing the coverage of UEs on the ground has been proposed in [34].

As is the case in any wireless setting, the system-level performance analysis of drone cellular networks can, in principle, be carried out by simulation-based approaches. However, since these approaches are not scalable by the number of network parameters, it is important to further develop analytical approaches to complement the simulations. Given the irregularity in the locations of DBSs and UEs, the use of tools from stochastic geometry is a natural choice for modeling and analysis of aerial communication networks. The main idea in this approach is to endow the locations of wireless nodes with appropriate distributions and then analytically characterize key performance metrics such as coverage and data rate by leveraging the properties of these distributions [8]. The downlink signal-to-interference ratio (SIR) based coverage probability of a reference receiver and the network spectral efficiency have been analyzed for a finite network of DBSs in [24, 25], where the locations of DBSs were modeled by a binomial point process (BPP) [4]. The results provided in that work offer useful insights into the network performance as a function of the altitude of DBSs, the number of DBSs, and the location of the user on the ground. Considering LoS and NLoS probabilities, the authors of [10] have derived the coverage probability in a homogeneous Poisson network of DBSs.

As DBSs often coexist with terrestrial cellular networks in practice, the coverage probability has been computed for a typical user in a multi-tier heterogeneous network in [11, 31], where the locations of DBSs and terrestrial BSs are modeled by independent homogeneous Poisson point processes (PPPs). The performance of a multi-tier drone network in terms of the downlink spectral efficiency has been investigated in [40]. Analysis of a UAV-to-UAV communication link has been analyzed in [13, 15], where the authors evaluated the coverage probability and rate for two spectrum sharing strategies. A user-centric deployment of DBSs was considered in [45], where the DBSs were assumed to be deployed at a certain height above the cluster centers of UEs modeled by a Poisson cluster process [39]. For this setup, the signal-to-interference-plus-noise ratio (SINR) based coverage probability of a typical UE and the area spectral efficiency (ASE) have been analyzed. Analytical expressions for the uplink coverage probability of a DBS are derived in [12, 50]. In all these works, the DBSs are assumed to hover at a fixed location and the analysis is carried out for one snapshot of the network. However, as we will discuss in this chapter, the mobility of the DBSs has a significant impact on the network performance.

In order to design efficient application-oriented protocols, it is important to understand the performance of drone cellular networks by accounting for the mobility of DBSs. Two key performance metrics that are directly related to the mobility of nodes in cellular networks are the handover probability and the handover rate. Furthermore, as the spatial distribution of DBSs is affected by their mobility, it is also of interest to investigate the temporal evolution of SINR-based performance metrics, such as coverage probability and achievable data rate. Building further on the finite network model of DBSs considered in [25], two classes of trajectory processes were proposed in [27] to reduce the average fade duration while ensuring the same coverage performance as that of static DBSs. The performance gains achieved in terms of spectral efficiency by dynamically deploying DBSs have been demonstrated in [28]. Using the random walk (RW) and random waypoint (RWP) mobility models for the horizontal and vertical displacements in a finite 3D network of DBSs, the authors in [41, 42] derived the coverage probability for a reference ground UE. Performance analysis of mobile drone cellular networks using various canonical mobility models, such as RW and RWP, has been studied [7, 16–18]. Handover probability and rate in drone networks have also been studied [14, 19].

In this chapter, we focus on a 3GPP-inspired drone mobility model in which the initial locations of DBSs are modeled as a homogeneous 2D PPP at a constant height from the ground and each DBS moves along a straight line in a random direction, independently of the other DBSs. Inspired by how cell association is done in single-tier cellular networks in practice, we assume that a typical ground UE is served by its nearest DBS and all the other DBSs act as interferers. We propose two service models for the mobility of the serving DBS, namely: (i) UE dependent model (UDM), in which the serving DBS moves at a fixed height above the ground in the direction of the typical UE and keeps hovering directly above the typical UE, and (ii) UE independent model (UIM), where the direction of motion of the serving DBS is completely random. For these two service models, we characterize the density of the interfering DBSs at every time t and analyze the downlink average rate as seen by the typical UE under Nakagami-m fading. We also present a comparison between our 3GPP-inspired mobility model and other nonlinear mobility models, and demonstrate that the rate performance of our proposed model is a lower bound on the rate performance of these more sophisticated nonlinear models. Further, we provide the analysis of the handover probability for both the UDM and the UIM, and show that the handover probability in the UIM is equivalent to that of a conventional terrestrial network where BSs are static and distributed as a homogeneous PPP, and UEs are moving in random directions along straight lines. More details about our system model are presented next.

9.2 Overview of the System Model

9.2.1 Spatial Model

We consider a network of mobile DBSs as illustrated in Figure 9.1, in which the DBSs are deployed at a fixed height h to serve UEs on the ground. We assume that the xy-plane of the Cartesian coordinate system is aligned with the ground and refer to the $z = h$ plane as the DBS plane throughout this chapter. We model the initial locations of DBSs as a homogeneous PPP $\Phi_D(0)$ with density λ_0 in the DBS plane. The locations of the UEs are modeled by an independent homogeneous PPP Φ_U. We denote the origin and the projection of the origin onto the DBS plane by $\boldsymbol{o} = (0, 0, 0)$ and $\boldsymbol{o}' = (0, 0, h)$, respectively. The focus of our analysis will be on the *typical* UE placed at $\boldsymbol{o}$. For a DBS located at $\boldsymbol{x}(t) \in \Phi_D(t)$ at time t, its distances from $\boldsymbol{o}'$ and $\boldsymbol{o}$ are denoted by $u_x(t) = \| \boldsymbol{x}(t) - \boldsymbol{o}' \|$ and $r_x(t) = \sqrt{u_x(t)^2 + h^2}$, respectively. In this chapter, we use subscript 0 for the terms corresponding to the closest DBS to $\boldsymbol{o}$.

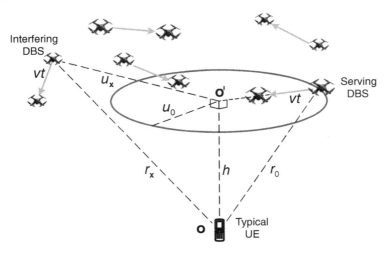

Figure 9.1 An illustration of the system model with the 3GPP-inspired straight-line mobility model.

So, we denote the location of the closest DBS to o at time t by $x_0(t)$ and its distance from o' and o by $u_0(t)$ and $r_0(t) = \sqrt{u_0(t)^2 + h^2}$, respectively. For notational simplicity, we drop the time index for $u_0(t)$ and $r_0(t)$ defined at $t = 0$, i.e., $u_0 \triangleq u_0(0)$ and $r_0 \triangleq r_0(0)$, and we drop t for $u_x(t)$ and $r_x(t)$ whenever the time index is clearly understood from the context.

9.2.2 3GPP-Inspired Mobility Model

Recently, the 3GPP has considered a simulation model for the placement and trajectories of DBSs [2], in which DBSs are initially placed at uniformly random locations at a constant height from the ground. They then start their motion at a constant velocity in uniformly random directions along straight lines. We refer to this model as the "straight-line mobility model" in this chapter. We consider a nearest-neighbor association policy in which the typical UE connects to its closest DBS, referred to as the serving DBS, and all the other DBSs are regarded as interfering DBSs for the typical UE. We model the mobility of interfering DBSs using the straight-line mobility model with constant velocity v. On the other hand, for the serving DBS, we consider the following two service models:

(1) UDM – the serving DBS moves with velocity v in the DBS plane towards o' and stops at this location.
(2) UIM – the serving DBS moves with velocity v in the DBS plane along a straight line in a random direction.

It is clear from our construction that the serving DBS in the UDM will not change over time for the typical UE, and thus no handover will occur in this service model. As the motion of the serving DBS in the UDM minimizes the link distance to the typical UE, this can be considered as the best-case service model. On the other hand, in the UIM, because of the random direction of movement of the serving DBS, the closest DBS to the typical UE may not remain the same over time, which will result in handovers. Furthermore, since the locations of the UEs are independent of each other, DBS trajectories will also be independent of each other in both the UDM and the UIM.

Remark 9.1 In this chapter, we represent DBSs as "points", which makes the occurrence of collision (defined as the event when two points arrive at the same location at the same time) a zero-measure event.

9.2.3 Channel Model

The SIR at the typical UE at time t is defined as

$$\text{SIR}(t) = \frac{h_0(t)r_0(t)^{-\alpha}}{\sum\limits_{x(t)\,\in\,\Phi'_{\mathrm{D}}(t)} h_x(t)r_x(t)^{-\alpha}}, \tag{9.1}$$

where $h_0(t)$ and $h_x(t)$ represent the small-scale channel fading gains corresponding to the serving and interfering links, respectively, α is the fixed path-loss exponent, and $\Phi'_{\mathrm{D}}(t) \equiv \Phi_{\mathrm{D}}(t)\backslash x_0(t)$ is the point process of the locations of interfering DBSs. For ease of notation, we will henceforth denote the denominator of Eq. (9.1) by $I(t)$. Since the air-to-ground channels may experience a variety of scattering scenarios, we invoke Nakagami-m fading to capture differences in the severity of fading in these environments. We denote the Nakagami-m fading parameter for the serving and interfering links by m_0 and m, respectively. Hence, the channel gains $h_0(t)$ and $h_x(t)$ follow Gamma distributions with the following probability density function (pdf)

$$f_{\mathrm{H}}(h) = \frac{\beta^{\beta}}{\Gamma(\beta)}h^{\beta-1}e^{-\beta h}, \tag{9.2}$$

where $\Gamma(x) = \int_0^\infty t^{x-1}e^{-t}\,\mathrm{d}t$ is the Gamma function and $\beta = m_0$ for the pdf of $h_0(t)$ and $\beta = m$ for the pdf of $h_x(t)$. In the interest of mathematical tractability, we consider integer values for m_0 and m.

9.2.4 Metrics of Interest

We characterize the network performance for both the UDM and the UIM using the following metrics.

- Average rate $(R(t))$, which is defined as $R(t) = \mathbb{E}[\log(1 + \text{SIR}(t))]$, where the expectation is taken over the PPP Φ_{D} and the trajectories. Note that this is the average rate as seen by the typical UE for different network and trajectory realizations at time t.
- Handover probability $(P_{\mathrm{H}}(t))$, which is the probability of the occurrence of a handover by time t.

In the next two sections of this chapter, we derive analytical expressions for these metrics under both service models.

9.3 Average Rate

In this section, we first characterize the point process of the interfering DBSs for both service models, using which we derive the average received rate at the typical UE. We start our discussion with the following lemma that directly results from the displacement theorem of a PPP, because of which we state it without proof [29].

Lemma 9.1 *Let Φ be a homogeneous PPP with density λ_0. If all the points of Φ are displaced independently of each other with identically distributed displacements, then the displaced points also form a homogeneous PPP with the same density λ_0.*

It can be inferred from Lemma 9.1 that, if all the DBSs move in random directions based on our 3GPP-inspired straight-line mobility model and independently from each other, then the spatial distribution of the DBSs does not change. Since this is essentially the mobility model for the UIM, the network of *interfering* DBSs in this service model follows an inhomogeneous PPP with density

$$\lambda(t; u_x, u_0) = \begin{cases} \lambda_0, & \text{for } u_x > u_0(t), \\ 0, & \text{for } u_x \leq u_0(t). \end{cases} \tag{9.3}$$

Although the serving distance $u_0(t)$ is a function of time, its distribution over time remains the same.

From our construction in the UDM, it is clear that initially there is no other DBS within an *exclusion zone* $\mathcal{X} = b(\mathbf{o}', u_0)$, where $b(\mathbf{o}', u_0)$ is a disk of radius u_0 centered at $\mathbf{o}'$. Consequently, the point process of the interfering DBSs in the UDM will be an inhomogeneous PPP with the initial density given by Eq. (9.3) at $t = 0$. Note that, as the serving DBS moves towards $\mathbf{o}'$, the interfering DBSs can enter $\mathcal{X}$, thereby altering the point process of the interfering DBSs. We characterize the point process of the interfering DBSs for the UDM in the next lemma.

Lemma 9.2 *In the UDM, as the interfering DBSs follow the straight-line mobility model with the same velocity as the serving DBS, the interfering DBSs are distributed as an inhomogeneous PPP at time t with density*

$$\lambda(t; u_x, u_0) = \lambda_0 \begin{cases} 1, & \text{for } u_0 + vt \leq u_x, \\ \dfrac{1}{\pi}\cos^{-1}\left(\dfrac{u_0^2 - u_x^2 - v^2t^2}{2u_x vt}\right), & \text{for } |u_0 - vt| \leq u_x \leq u_0 + vt, \\ \mathbf{1}(t > u_0/v), & \text{for } 0 \leq u_x \leq |u_0 - vt|, \end{cases} \tag{9.4}$$

where $\mathbf{1}(\,\cdot\,)$ is the indicator function.

Proof: The interfering DBSs in the UDM are initially distributed as an inhomogeneous PPP with density given in Eq. (9.3) at $t = 0$. Based on the displacement theorem [29] and due to the fact that the displacements of DBSs are independent of each other, the resulting network of interfering DBSs will also be an inhomogeneous PPP at any time t. In order to characterize the exact density of the resulting network, we need to find the distribution of the displaced locations of points.

Let x and y denote the initial location of a DBS and its displaced location after a displacement of vt in a uniformly random direction Θ, respectively. Further, let u_x and u_y denote the distances from the DBS to o' before and after the displacement, respectively. Writing the cosine law in triangle $\triangle o'xy$, we have $u_y^2 = u_x^2 + v^2t^2 - 2u_xvt\cos(\Theta)$. Now, using the basic transformation of random variables, we can write the distribution of the new locations of DBSs as

$$\rho(u_y; u_x) = \frac{2u_y}{\pi\sqrt{[u_y^2 - (u_x - vt)^2][(u_x + vt)^2 - u_y^2]}}, \tag{9.5}$$

when $|u_x - vt| \leq u_y \leq u_x + vt$, and zero otherwise. Note that these conditions are the triangle inequalities, which can also be written as $|u_y - vt| \leq u_x \leq u_y + vt$.

We also have $u_x \geq u_0$ by considering the exclusion zone $\mathcal{X}$. Thus, we obtain the condition $\max\{u_0, |u_y - vt|\} \leq u_x \leq u_y + vt$ as the support for Eq. (9.5). Now, we use the displacement theorem in polar coordinates to derive the density $\lambda(t; u_y, u_0)$ of the resulting network as follows:

$$2\pi u_y \lambda(t; u_y, u_0) = 2\pi \int_{u_0}^{\infty} \lambda_0 \rho(u_y; u_x) u_x \, du_x$$

$$= \int_{\max\{u_0, |u_y - vt|\}}^{u_y + vt} \frac{4\lambda_0 u_x u_y}{\sqrt{[u_y^2 - (u_x - vt)^2][(u_x + vt)^2 - u_y^2]}} \, du_x$$

$$= 2\lambda_0 u_y \cos^{-1}\left(\frac{(\max\{u_0, |u_y - vt|\})^2 - (u_y^2 + v^2t^2)}{2vtu_y}\right),$$

which gives Eq. (9.4) by some algebraic manipulations. Therefore, the proof is complete. ∎

Remark 9.2 The following observations directly stem from Eq. (9.4):

- Eq. (9.4) is continuous at the boundaries, i.e., at $u_x = |u_0 \pm vt|$.
- The point process of the interfering DBSs will become homogeneous as $u_0 \to 0$ or $t \to \infty$.
- Eqs. (9.4) and (9.3) will become identical as $u_0 \to 0$ or $t \to 0$.

The result of Lemma 9.2 could have also been derived using the following argument. From Lemma 9.1, it follows that the locations of all the DBSs are distributed as a homogeneous PPP with density λ_0 when there is no $\mathcal{X}$. As a result, when $\mathcal{X}$ is taken into account, the density of the DBSs can be viewed as a superposition of the following two parts: (i) the density of the DBSs initially inside $\mathcal{X}$, and (ii) the density of the DBSs initially outside $\mathcal{X}$, i.e., the density of the interfering DBSs. Thus, the density of the point process of the interfering DBSs is derived by subtracting the density of the DBSs initially inside $\mathcal{X}$ from the density of all the DBSs (λ_0). Using this argument and based on Figure 9.2, we can gain insights into the density of the interfering DBSs in different regions as follows.

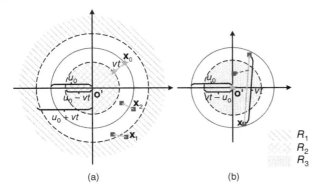

Figure 9.2 An illustrative explanation of the network density for the straight-line and nonlinear mobility models. Stars and squares represent serving and interfering DBSs, respectively. (a) Serving DBS is moving towards o' from x_0, and (b) serving DBS is hovering at o'.

(1) Region $R_1 = \{u_x \geq u_0 + vt\}$: No DBS can enter $\mathcal{X}$ until time t, and thus $\lambda(t; u_x, u_0) = \lambda_0$ in this region.

(2) Region $R_2 = \{0 \leq u_x \leq u_0 - vt \mid vt \leq u_0\}$: The serving DBS is moving towards o' and, since there is no interfering DBS in this region, we have $\lambda(t; u_x, u_0) = 0$.

(3) Region $R_3 = \{0 \leq u_x \leq vt - u_0 \mid vt \geq u_0\}$: In order to get the density of the interfering DBSs, we can first calculate the density of the DBSs initially inside $\mathcal{X}$ and then subtract it from λ_0. The density of the DBSs, which were initially inside $\mathcal{X}$, will be zero in R_3 after the displacement of vt (see Figure 9.2(b)), thereby yielding $\lambda(t; u_x, u_0) = \lambda_0$.

Although we have derived the network density for the straight-line mobility model, it would be instructive to give some intuitions on the density of the interfering DBSs when they follow more sophisticated nonlinear mobility models such as RW or RWP [20, 33]. In these mobility models, DBSs change their directions during their flights, which complicates the characterization of the resulting point process at a given time t. Nonetheless, using the insights that we have provided so far, we compare the performance of our straight-line mobility model with these nonlinear models as follows.

Let x be the location of an interfering DBS that is initially inside $\mathcal{X}$. As illustrated in Figure 9.2(b), after the interfering DBS at x travels a distance vt, where $t \geq u_0/v$, it will not lie inside R_3 in the straight-line mobility model. However, this is not the case when we employ other nonlinear mobility models. Specifically, as the interfering DBS at x is allowed to change its direction during the flight, it could possibly lie inside R_3 after traveling a distance of vt. Therefore, the density of the DBSs initially inside $\mathcal{X}$ will be nonzero for these mobility models, and hence the density of the interfering DBSs in R_3 will be less than λ_0. This means that fewer interfering DBSs are in the proximity of the typical UE in these nonlinear mobility models as compared to the straight-line mobility model. Consequently, the rate performance of the typical UE in nonlinear mobility models is lower-bounded by the rate performance of our straight-line mobility model.

We now analyze the distribution of $\mathrm{SIR}(t)$ and the average rate of the typical UE for both service models.

Theorem 9.1 *In the UDM, the average received rate at the typical UE at time t is given as*

$$R^{\text{UDM}}(t) = \int_0^\infty \int_0^\infty \frac{2\pi \lambda_0 u_0 \, e^{-\pi \lambda_0 u_0^2}}{1 + \gamma}$$

$$\times \left[\sum_{k=0}^{m_0-1} \frac{(-s)^k}{k!} \frac{\partial^k}{\partial s^k} \mathcal{L}_{I(t)}(s \mid x_0(t)) \right]_{s = m_0 \gamma r_0^\alpha(t)} du_0 \, d\gamma, \qquad (9.6)$$

where

$$\mathcal{L}_{I(t)}(s \mid x_0(t)) = \exp \left\{ -2\pi \int_0^\infty u_x(t) \lambda(t; u_x, u_0) \right.$$

$$\left. \times \left[1 - \left(1 + \frac{s(u_x^2(t) + h^2)^{-\alpha/2}}{m} \right)^{-m} \right] du_x(t) \right\}. \qquad (9.7)$$

Proof: We write the complementary cumulative distribution function (CCDF) of SIR(t) conditioned on the location of the serving DBS as

$$\mathbb{P}[\text{SIR}(t) \geq \gamma \mid x_0(t)] = \mathbb{E}[\mathbb{P}[h_0(t) \geq \gamma r_0^\alpha(t) I(t) \mid x_0(t), I(t)]]$$

$$\overset{(a)}{=} \mathbb{E} \left[\frac{\Gamma(m_0, m_0 \gamma r_0^\alpha(t) I(t))}{\Gamma(m_0)} \,\middle|\, x_0(t) \right]$$

$$\overset{(b)}{=} \mathbb{E} \left[\sum_{k=0}^{m_0-1} \frac{(m_0 \gamma r_0^\alpha(t) I(t))^k}{k!} \, e^{-m_0 \gamma r_0^\alpha(t) I(t)} \,\middle|\, x_0(t) \right]$$

$$= \sum_{k=0}^{m_0-1} \left[\frac{(-s)^k}{k!} \frac{\partial^k}{\partial s^k} \mathcal{L}_{I(t)}(s \mid x_0(t)) \right]_{s = m_0 \gamma r_0^\alpha(t)},$$

where (a) follows from the Nakagami-m fading assumption, and (b) follows from the definition of the incomplete Gamma function for integer values of m_0.

The conditional Laplace transform of interference $\mathcal{L}_{I(t)}(s \mid x_0(t))$ at time t is computed as

$$\mathcal{L}_{I(t)}(s \mid x_0(t))$$

$$= \mathbb{E}[e^{-sI(t)} \mid x_0(t)] = \mathbb{E} \left[\exp \left(-s \sum_{x(t) \in \Phi_D'(t)} h_x(t) r_x(t)^{-\alpha} \right) \,\middle|\, u_0(t) \right]$$

$$\overset{(a)}{=} \mathbb{E} \left[\prod_{x(t) \in \Phi_D'(t)} \left(1 + \frac{s r_x(t)^{-\alpha}}{m} \right)^{-m} \,\middle|\, u_0(t) \right]$$

$$\overset{(b)}{=} \exp \left\{ -2\pi \int_0^\infty u_x(t) \lambda(t; u_x, u_0) \left[1 - \left(1 + \frac{s r_x(t)^{-\alpha}}{m} \right)^{-m} \right] du_x(t) \right\},$$

where (a) results from the moment generating function (mgf) of the Gamma distribution, and (b) follows from the probability generating functional (pgfl) of a PPP.

Now, the average rate at time t can be written as

$$R^{\text{UDM}}(t) = \mathbb{E}[\log(1 + \text{SIR}(t))]$$

$$= \int_0^\infty \log(1 + \gamma) f_\Gamma(\gamma; t) \, d\gamma$$

$$= \int_0^\infty \int_0^\infty \frac{2\pi \lambda_0 u_0 e^{-\pi \lambda_0 u_0^2}}{1 + \gamma} \; \mathbb{P}[\mathrm{SIR}(t) \geq \gamma \mid \boldsymbol{x}_0(t)] \; \mathrm{d}u_0 \; \mathrm{d}\gamma,$$

where $f_\Gamma(\gamma; t)$ is the pdf of $\mathrm{SIR}(t)$. The last equation follows from integration by parts and deconditioning on $u_0(t)$. This completes the proof. ∎

As the point process of DBSs does not change with time in the UIM, the received rate at the typical UE can be computed by evaluating the expression for average rate in the UDM given in Eq. (9.6) at $t = 0$, i.e., $R^{\mathrm{UIM}} = R^{\mathrm{UDM}}(0)$.

9.4 Handover Probability

In this section, we characterize the handover probability for both service models. In terrestrial cellular networks, under the nearest-neighbor association policy, the coverage footprints are determined by Voronoi cells [9, 26]. Consequently, handover occurs in these networks when a reference UE crosses the boundary of a Voronoi cell. As mentioned earlier in Section 9.2, the handover probability as seen by the typical UE is zero for the UDM. However, derivation of the handover probability for the UIM is not straightforward and is the main focus of this section. We start our analysis by stating the following lemma.

Lemma 9.3 *Consider the UIM and let DBS D_0 be the serving DBS at time $t = t_0$. Assume that a handover occurs at time $t = t_1$, where $t_1 > t_0$, and DBS D_1 becomes the serving DBS. Then, the DBS D_0 cannot become the serving DBS again at any $t > t_1$.*

Proof: Figure 9.3 shows two sample trajectories for the DBSs D_0 and D_1. At time $t = t_0$, the DBSs D_0 and D_1 are located at points A_0 and B_0 with distances a_0 and b_0 from $\boldsymbol{o}'$, respectively. We assume that a handover occurs at time $t = t_1$, where the DBSs D_0 and D_1 are located at points A_1 and B_1 with distances a_1 and b_1 from $\boldsymbol{o}'$, respectively. At a later time $t = t_2$, the locations of the DBSs D_0 and D_1 and their distances from $\boldsymbol{o}'$ are denoted by A_2, B_2, and a_2, b_2, respectively. Assuming that $a_0 < b_0$ and $a_1 > b_1$, we need to show that $a_2 > b_2$.

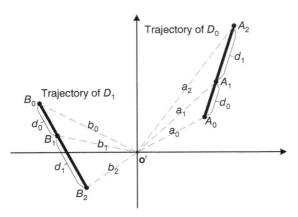

Figure 9.3 Trajectories of the DBSs D_0 and D_1 used in the proof of Lemma 9.3.

Figure 9.4 A different representation of Figure 9.3, where triangles $\triangle o'A_0A_2$ and $\triangle o'B_0B_2$ are drawn by placing bases A_0A_2 and B_0B_2 on top of each other.

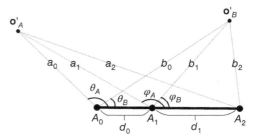

Since $|A_0A_1| = |B_0B_1| = d_0$ and $|A_1A_2| = |B_1B_2| = d_1$, we can draw triangles $\triangle o'A_0A_2$ and $\triangle o'B_0B_2$ by placing bases A_0A_2 and B_0B_2 on top of each other as shown in Figure 9.4. The points o'_A and o'_B in Figure 9.4 are virtual representations of the point o' corresponding to triangles $\triangle o'A_0A_2$ and $\triangle o'B_0B_2$, respectively. Define $\theta_A = \angle o'_A A_0 A_2$, $\theta_B = \angle o'_B A_0 A_2$, $\varphi_A = \angle o'_A A_1 A_2$, and $\varphi_B = \angle o'_B A_1 A_2$. Without loss of generality, we assume that $\theta_A > \theta_B$, otherwise o'_A lies on the right-hand side of o'_B and the same reasoning will follow due to symmetry (θ_A will be defined as $\angle o'_A A_2 A_0$ and the definitions of the other angles will change accordingly). Note that we also assume that $\pi > \theta_A$ due to symmetry.

We consider two cases:

(1) $\theta_B > \pi/2$.

In this case, we first show that $\varphi_A > \varphi_B > \pi/2$. First, observe that $\varphi_B > \theta_B > \pi/2$. Now writing the sine law in triangles $\triangle o'_A A_0 A_1$ and $\triangle o'_B A_0 A_1$, we have

$$\frac{a_0}{\sin(\pi - \varphi_A)} = \frac{a_1}{\sin(\theta_A)},$$

$$\frac{b_0}{\sin(\pi - \varphi_B)} = \frac{b_1}{\sin(\theta_B)},$$

and, since $\sin(\theta_A) < \sin(\theta_B)$ and $a_0/a_1 < b_0/b_1$, we conclude that $\sin(\pi - \varphi_A) < \sin(\pi - \varphi_B)$, which gives $\varphi_A > \varphi_B > \pi/2$. Writing the cosine law in triangles $\triangle o'_A A_1 A_2$ and $\triangle o'_B A_1 A_2$, we have

$$a_2^2 = a_1^2 + d_1^2 - 2a_1 d_1 \cos(\varphi_A),$$

$$b_2^2 = b_1^2 + d_1^2 - 2b_1 d_1 \cos(\varphi_B),$$

and since by assumption $a_1 > b_1$ and $\cos(\varphi_A) < \cos(\varphi_B) < 0$, we obtain $a_2 > b_2$.

(2) $\theta_B < \pi/2$.

In this case, we have $\cos(\theta_B) > \max\{0, \cos(\theta_A)\}$. Since $b_0 > a_0$, we have

$$b_0 \cos(\theta_B) > a_0 \cos(\theta_A). \tag{9.8}$$

Now we write the cosine law in the four triangles $\triangle o'_A A_0 A_1$, $\triangle o'_A A_0 A_2$, $\triangle o'_B A_0 A_1$, and $\triangle o'_B A_0 A_2$ as follows:

$$a_1^2 = a_0^2 + d_0^2 - 2a_0 d_0 \cos(\theta_A),$$

$$a_2^2 = a_0^2 + (d_0 + d_1)^2 - 2a_0(d_0 + d_1)\cos(\theta_A),$$

$$b_1^2 = b_0^2 + d_0^2 - 2b_0 d_0 \cos(\theta_B),$$

$$b_2^2 = b_0^2 + (d_0 + d_1)^2 - 2b_0(d_0 + d_1)\cos(\theta_B).$$

In order for $a_2 > b_2$ to hold, we must have

$$
\begin{aligned}
a_2^2 > b_2^2 \quad &\Longleftrightarrow \quad a_0^2 - 2a_0(d_0 + d_1)\cos(\theta_A) > b_0^2 - 2b_0(d_0 + d_1)\cos(\theta_B), \\
&\Longleftrightarrow \quad a_1^2 - 2a_0 d_1 \cos(\theta_A) > b_1^2 - 2b_0 d_1 \cos(\theta_B), \\
&\Longleftrightarrow \quad (a_1^2 - b_1^2) + 2d_1(b_0 \cos(\theta_B) - a_0 \cos(\theta_A)) > 0,
\end{aligned}
$$

and the last inequality is valid due to Eq. (9.8) and our assumption that $a_1 > b_1$.

Hence, the proof is complete. ∎

When a handover occurs in the UIM, a DBS that was the serving DBS before the occurrence of the handover will act as an interfering DBS after the occurrence of the handover. Lemma 9.3 states that this interfering DBS cannot become the serving DBS again under the UIM. This behavior is also observed in terrestrial single-tier cellular networks, where all the BSs are static and are modeled by a homogeneous 2D PPP and a reference UE is moving in a uniformly random direction along a straight line. Due to the convexity of the Voronoi cells in a single-tier network, a reference UE traveling along a straight line enters a cell only once.

Motivated by this behavior, we state the following theorem on the duality between the mobility of the DBSs in an aerial network with static UEs and the mobility of the UEs in a terrestrial network with static BSs.

Theorem 9.2 *Under the system setup of this chapter, the handover probabilities of the following two networks are equivalent:*

(1) Terrestrial model. *BSs are static and distributed as a homogeneous PPP, and the reference UE moves based on the straight-line mobility model.*
(2) Aerial model. *A typical UE is static and the DBSs, which are initially distributed as a homogeneous PPP, move based on the straight-line mobility model (UIM).*

Proof: Consider the terrestrial model mentioned in the theorem statement, where the reference UE moves in direction $\theta \sim U[0, 2\pi)$ along a straight line. We represent the resulting translated point process by $\Phi_B - x(t)$, where Φ_B and $x(t)$ denote the point process of the static BS locations and the trajectory of the UE, respectively. Now observe that the performance analysis of the reference UE in this network is equivalent to that of an aerial network where the typical UE is static and all the DBSs move in the *same* direction $\pi + \theta$ along straight lines. Denoting this point process by $\tilde{\Phi}_D(t)$, we can write $\tilde{\Phi}_D(t) \equiv \Phi_B - x(t)$. Since Φ_B is a homogeneous PPP, it is translation-invariant, and thus $\tilde{\Phi}_D(t)$ will also be a homogeneous PPP with density λ_0.

On the other hand, Lemma 9.1 states that the DBS locations under our straight-line mobility model follow a homogeneous PPP with density λ_0. Hence, as seen from the UE of interest at any time t, the terrestrial and aerial models mentioned in the theorem statement are both distributed based on a homogeneous PPP with density λ_0, and thus are equivalent in distribution. Consequently, the handover probability is the same for both models, and the proof is complete. ∎

Theorem 9.2 establishes duality in terms of handover probability between a conventional terrestrial network, where BSs are static and UEs are mobile with constant velocity v, and a drone network, where DBSs are mobile with the same velocity v and UEs are static. The handover probability for terrestrial cellular networks has been well investigated in the literature. Using the straight-line mobility model for the reference UE, the authors in [38] derived the handover probability of a network where the BSs are distributed as a homogeneous PPP. This result has also been derived as a part of the downlink joint coverage analysis of cellular networks [32], and more recently as a part of the tutorial on mobility-aware performance characterization of cellular networks [44]. The latter also explicitly discusses a correction to the original result [38]. In the following theorem, we state this result and provide a slightly simpler correction to the proof of [38].

Theorem 9.3 *Consider a network of BSs distributed as a homogeneous PPP with density λ_0 on the ground. Assume that a UE moves at a constant velocity v in a random direction along a straight line and connects to its nearest BS. Then the handover probability as seen by the UE is given as*

$$
P_{\mathrm{H}}(t) = 1 - \frac{1}{2\pi} \int_0^{2\pi} \int_0^{\infty} 2\pi \lambda_0 r
$$
$$
\times \exp\left\{ -\lambda_0[r^2(\pi - \varphi_1 + \frac{1}{2}\sin(2\varphi_1)) + R^2(\pi - \varphi_2 + \frac{1}{2}\sin(2\varphi_2))] \right\} \, dr \, d\theta,
$$
$$(9.9)$$

where

$$
R = \sqrt{r^2 + v^2 t^2 - 2rvt\cos(\theta)} \,,
\tag{9.10}
$$

$$
\varphi_1 = \cos^{-1}\left(\frac{v^2 t^2 + r^2 - R^2}{2vtr} \right),
\tag{9.11}
$$

$$
\varphi_2 = \cos^{-1}\left(\frac{v^2 t^2 + R^2 - r^2}{2vtR} \right).
\tag{9.12}
$$

Proof: Similar to the proof of theorem 1 in [38], assume that the serving BS is located at $\boldsymbol{o}$ and the UE is initially located at $\boldsymbol{x}_1$, which is at a distance r from $\boldsymbol{o}$ (see Figure 9.5). The UE then moves in a uniformly random direction θ with a constant velocity v to arrive at location $\boldsymbol{x}_2$ after time t. Let R be the distance from $\boldsymbol{x}_2$ to $\boldsymbol{o}$, which can be written as in Eq. (9.10) using the cosine law. Define $C_1 = b(\boldsymbol{x}_1, r)$ and $C_2 = b(\boldsymbol{x}_2, R)$.

By definition, handover will not occur if the serving BS is not changed until time t, i.e., if there is no BS other than the initial serving BS in C_2. Since we know that the serving BS was present in C_1 as well, handover will not occur if there is no BS in $C_2 \backslash C_1$ (shaded area in Figure 9.5). Hence, conditioning on r and θ, the handover probability can be written as

$$
P_{\mathrm{H}}(t \mid r, \theta) \overset{(a)}{=} 1 - \mathbb{P}[N(C_2 \backslash C_1) = 0] = 1 - \mathbb{P}[N(C_2 \backslash (C_1 \cap C_2)) = 0]
$$
$$
\overset{(b)}{=} 1 - e^{-\lambda_0 |C_2 \backslash (C_1 \cap C_2)|}
$$
$$
\overset{(c)}{=} 1 - e^{-\lambda_0(\pi R^2 - A_{C_1 \cap C_2})}.
\tag{9.13}
$$

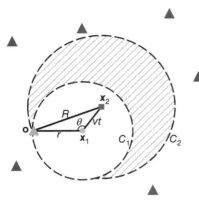

Figure 9.5 An illustration for the proof of Theorem 9.3. Triangles, small circle, and square denote BSs, initial location of the UE, and the location of the UE at time t, respectively.

Here $N(B)$ in (a) denotes the number of points in set B, in (b) we used the null probability of $\text{PPP}(\lambda_0)$, and $A_{C_1 \cap C_2}$ in (c) is the intersection area between C_1 and C_2, which is given from plane geometry as

$$A_{C_1 \cap C_2} = r^2(\varphi_1 - \frac{1}{2}\sin(2\varphi_1)) + R^2(\varphi_2 - \frac{1}{2}\sin(2\varphi_2)), \tag{9.14}$$

where φ_1 and φ_2 are as given in Eqs. (9.11) and (9.12), respectively.

Substituting Eq. (9.14) into Eq. (9.13), we get the handover probability conditioned on the serving distance r and angle θ. Since the serving distance in $\text{PPP}(\lambda_0)$ is distributed as a Rayleigh random variable with parameter $1/\sqrt{2\pi\lambda_0}$ and $\theta \sim U[0, 2\pi)$, deconditioning on these random variables gives the final result for the handover probability as Eq. (9.9), and the proof is complete. ∎

Remark 9.3 To the best of our knowledge, this is the first work to establish the fact that the handover probability in a drone cellular network with mobile DBSs and static UEs is equivalent to that of a terrestrial network with static BSs and mobile UEs. From Theorems 9.2 and 9.3, we conclude that the handover probability for the UIM is as given in Eq. (9.9).

9.5 Results and Discussion

In this section, we shed light on the effect of different network parameters on the system-level performance for both the UDM and the UIM through numerical results. We assume that DBSs are initially distributed as a homogeneous PPP with density $\lambda_0 = 10^{-6}$ at height h from the ground. We place the typical UE at the origin and assume that the closest DBS to the origin is the serving DBS, which can either move towards $\boldsymbol{o}'$ and stop at $\boldsymbol{o}'$ (the UDM), or follow the straight-line mobility model in a random direction, independent of the typical UE (the UIM). All the interfering DBSs follow the straight-line mobility model at a constant velocity of $v = 45$ km h^{-1} in the DBS plane and are transmitting at all times. The serving DBS also moves at the same velocity and transmits at all times. We assume that $h \in \{100,200\}$ m, which are the nominal heights for low-altitude DBSs [37]. The path-loss exponent take values $\alpha \in \{2.5, 3, 3.5\}$ depending on the network environment.

9.5.1 Density of Interfering DBSs

In the UDM, the evolution of the density of the interfering DBSs with time is plotted in Figure 9.6. As mentioned earlier, the density of the point process of the interfering DBSs for $t > u_0/v$ will have two homogeneous parts (where $\lambda = \lambda_0$) and one inhomogeneous part, which shrinks to λ_0 as $t \to \infty$. Hence, the point process of the interfering DBSs eventually becomes a homogeneous PPP with density λ_0. However, this is not the case for the UIM, and the point process of the interfering DBSs will remain inhomogeneous for all times (see Eq. (9.3)).

9.5.2 Average Rate

We plot the analytic and simulation results for the average rate of the network for both the UDM and the UIM in Figures 9.8, and 9.9. Figure 9.7 shows the effect of fading parameters m and m_0 on the average rate received by the typical UE. The selected parameters are $m = m_0 \in \{1, 2\}$, $h = 100$ m, and $\alpha = 3$. As is clear in this figure, increasing m and m_0 reduces the severity of fading, and thus the average received rate increases.

The comparison of the received rate when DBSs are at height $h = 100$ m with various path-loss exponents is presented in Figure 9.8. We can see from this figure that, as α increases, the average rate also increases. This fact can also be shown mathematically by rewriting the SIR(t) as

$$\text{SIR}(t) = \frac{h_0(t)}{\displaystyle\sum_{x(t) \in \Phi_D'(t)} h_x(t) \left(\frac{r_0(t)}{r_x(t)}\right)^{\alpha}}. \tag{9.15}$$

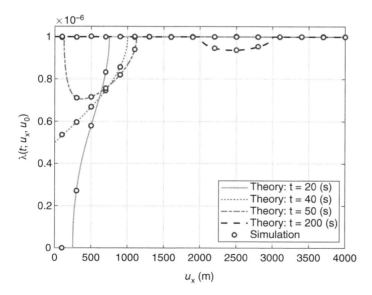

Figure 9.6 Density of the network of interfering DBSs in the UDM. The exclusion zone radius is $u_0 = 500$ m. The homogenization of the point process of the interfering DBSs with time is evident.

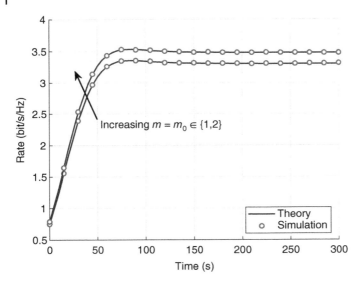

Figure 9.7 Average rate of the network for the UDM for different values of the fading parameter, i.e., $m = m_0 \in \{1, 2\}$. Other parameters are $h = 100$ m and $\alpha = 3$.

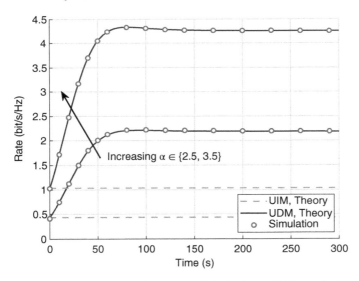

Figure 9.8 Average rate of the network for both the UDM and the UIM for $\alpha \in \{2.5, 3.5\}$, $h = 100$ m, and $m = m_0 = 1$.

Now since $r_0(t)/r_x(t) \le 1$, as α increases, the denominator of Eq. (9.15) decreases, and thus SIR(t) and thereby the rate increases.

To observe the effect of height, we plot the average received rate at various heights with the same path-loss exponent in Figure 9.9. Note that the received rate will decrease as height increases, which could have been understood directly from Eq. (9.6) as well. In Figures 9.8 and 9.9, we have also plotted the average rate for the UIM to highlight the advantage of the UDM over the UIM.

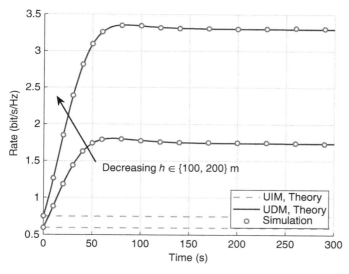

Figure 9.9 Average rate of the network for both the UDM and the UIM for $h \in \{100,200\}$m, $\alpha = 3$, and $m = m_0 = 1$.

9.5.3 Handover Probability

We plot the handover probability of the UIM as a function of time for different DBS velocities in Figure 9.10. Clearly, the handover probability will increase as we increase the velocity of the DBSs. We also provide a plot for the handover probability of the UIM for different values of the network density in Figure 9.11. By definition, as λ_0 increases, the

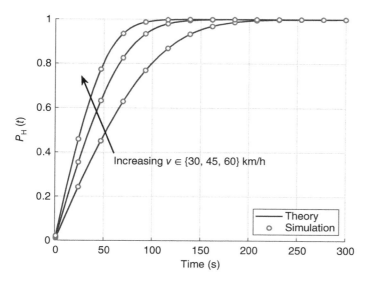

Figure 9.10 Handover probability of the network for the UIM for $v \in \{30, 45, 60\}$ km h^{-1} and $\lambda_0 = 10^{-6}$.

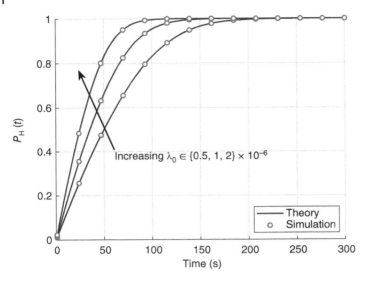

Figure 9.11 Handover probability of the network for the UIM for $\lambda_0 \in \{0.5, 1, 2\} \times 10^{-6}$ and $v = 45$ km h^{-1}.

average number of DBSs per unit area increases, and thus the probability of occurrence of a handover increases. Note that the handover probability for the UDM is zero for all times, and thus is not plotted in these two figures.

9.6 Conclusion

Inspired by the mobility model considered in the 3GPP simulations, we consider a straight-line mobility model for DBSs in a drone cellular network in which the DBSs are initially distributed based on a homogeneous PPP at a constant height serving UEs on the ground. We assumed that the serving DBS is selected based on the nearest-neighbor association policy and moves based on the following two service models: (i) UDM and (ii) UIM. In the UDM, the serving DBS moves towards the typical UE at a constant height and velocity, and keeps hovering above the location of the typical UE. In the UIM, the serving DBS follows the 3GPP-inspired straight-line mobility model in a random direction. All the other DBSs (referred to as interfering DBSs) also move based on the 3GPP-inspired straight-line mobility model in random directions, independently of each other and the serving DBS. For both service models, we characterized the point process of interfering DBSs as seen from the typical UE by applying the displacement theorem, using which we derived the average received rate under the Nakagami-m fading model.

We also compared our 3GPP-inspired straight-line mobility model with other more sophisticated nonlinear mobility models which allow direction changes as well. We demonstrated that our straight-line mobility model can be thought of as a lower bound on the system performance to these nonlinear mobility models. We then performed a comprehensive analysis on the handover probability for both service models and showed that, under the setup of this chapter, the handover probability of the UIM is equivalent to that of a conventional terrestrial scenario, where BSs are static and modeled as a homogeneous PPP and the reference UE moves based on the straight-line mobility model. The analysis of a mobile drone cellular networks under a multi-slope path-loss model [49] is a promising future work.

Acknowledgment

The support of the US NSF (Grants CNS-1617896 and CNS-1923807) is gratefully acknowledged.

References

1 3GPP (2018). Enhancement for unmanned aerial vehicles. 3rd Generation Partnership Project (3GPP), Tech. Rep. 22.829, Version 0.0.0.

2 3GPP (2018). Enhanced LTE support for aerial vehicles. 3rd Generation Partnership Project (3GPP), Tech. Rep. 36.777, Version 1.1.0.

3 M. A. Abd-Elmagid and H. S. Dhillon (2019). Average peak age-of-information minimization in UAV-assisted IoT networks. *IEEE Trans. Veh. Technol.* 68 (2): 2003–2008, Feb. 2019.

4 M. Afshang and H. S. Dhillon (2017). Fundamentals of modeling finite wireless networks using binomial point process. *IEEE Trans. Wireless Commun.* 16 (5): 3355–3370, May 2017.

5 A. Al-Hourani, S. Kandeepan, and A. Jamalipour (2014). Modeling air-to-ground path loss for low altitude platforms in urban environments. *Proceedings of the IEEE Global Communications (GLOBECOM) Conference*, pages 2898–2904, Dec. 2014.

6 A. Al-Hourani, S. Kandeepan, and S. Lardner (2014). Optimal LAP altitude for maximum coverage. *IEEE Wireless Commun. Lett.* 3 (6): 569–572, Dec. 2014.

7 R. Amer, W. Saad, and N. Marchettic (2020). Mobility in the sky: Performance and mobility analysis for cellular-connected UAVs. *IEEE Trans. Commun.*, 68(5): 3229–3246, May 2020.

8 J. G. Andrews, A. K. Gupta, and H. S. Dhillon (2016). A primer on cellular network analysis using stochastic geometry. http://arxiv.org/abs/1604.03183.

9 J. G. Andrews, F. Baccelli, and R. K. Ganti (2011). A tractable approach to coverage and rate in cellular networks. *IEEE Trans. Commun.*, 59(11): 3122–3134, Nov. 2011.

10 M. M. Azari, Y. Murillo, O. Amin et al. (2017). Coverage maximization for a Poisson field of drone cells. *Proceedings of the IEEE 28th Annual International Symposium on Personal, Indoor, and Mobile Radio Communications (PIMRC)*, pages 1–6, Oct. 2017.

11 M. M. Azari, F. Rosas, A. Chiumento, and S. Pollin (2017). Coexistence of terrestrial and aerial users in cellular networks. *Proceedings of the IEEE Global Communications (GLOBECOM) Workshops*, pages 1–6, Dec. 2017.

12 M. M. Azari, F. Rosas, A. Chiumento et al. (2018). Uplink performance analysis of a drone cell in a random field of ground interferers. *Proceedings of the IEEE Wireless Communications and Networking Conference (WCNC)*, pages 1–6, Apr. 2018.

13 M. M. Azari, G. Geraci, A. Garcia-Rodriguez, and S. Pollin (2019). Cellular UAV-to-UAV communications. *Proceedings of the IEEE 30th Annual International Symposium on Personal, Indoor, and Mobile Radio Communications (PIMRC)*, pages 1–7, Sep. 2019.

14 M. M. Azari, F. Rosas, and S. Pollin (2019). Cellular connectivity for UAVs: Network modeling, performance analysis, and design guidelines. *IEEE Trans. Wireless Commun.*, 18(7):3366–3381, July 2019.

15 M. M. Azari, G. Geraci, A. Garcia-Rodriguez, and S. Pollin (2020). UAV-to-UAV communications in cellular networks. *IEEE Trans. Wireless Commun.*, 19(9): 6130–6144, Sep. 2020.

16 M. Banagar and H. S. Dhillon (2019). 3GPP-inspired stochastic geometry-based mobility model for a drone cellular network. *Proceedings of the IEEE Global Communications (GLOBECOM) Conference*, pages 1–6, Dec. 2019.

17 M. Banagar and H. S. Dhillon (2019). Fundamentals of drone cellular network analysis under random waypoint mobility model. *Proceedings of the IEEE Global Communications (GLOBECOM) Conference*, pages 1–6, Dec. 2019.

18 M. Banagar and H. S. Dhillon (2020). Performance characterization of canonical mobility models in drone cellular networks. *IEEE Trans. Wireless Commun.*, 19(7): 4994–5009, July 2020.

19 M. Banagar, V. V. Chetlur, and H. S. Dhillon (2020). Handover probability in drone cellular networks. *IEEE Wireless Commun. Lett.* 9 (7): 933–937, July 2020.

20 C. Bettstetter, G. Resta, and P. Santi (2003). The node distribution of the random waypoint mobility model for wireless ad hoc networks. *IEEE Trans. Mobile Comput.*, 2(3):257–269, July 2003.

21 I. Bor-Yaliniz and H. Yanikomeroglu (2016). The new frontier in RAN heterogeneity: Multi-tier drone-cells. *IEEE Commun. Mag.*, 54(11):48–55, Nov. 2016.

22 S. Chandrasekharan, K. Gomez, A. Al-Hourani et al. (2016). Designing and implementing future aerial communication networks. *IEEE Commun. Mag.* 54 (5): 26–34, May 2016.

23 N. Cherif, M. Alzenad, H. Yanikomeroglu, and A. Yongacoglu (2020). Downlink coverage and rate analysis of an aerial user in vertical heterogeneous networks (VHetNets). https://arxiv.org/abs/1905.11934.

24 V. V. Chetlur and H. S. Dhillon (2016). Downlink coverage probability in a finite network of unmanned aerial vehicle (UAV) base stations. *Proceedings of the IEEE 17th International Workshop on Signal Processing Advances in Wireless Communications (SPAWC)*, pages 1–5, July 2016.

25 V. V. Chetlur and H. S. Dhillon (2017). Downlink coverage analysis for a finite 3-D wireless network of unmanned aerial vehicles. *IEEE Trans. Commun.*, 65(10):4543–4558, Oct. 2017.

26 H. S. Dhillon, R. K. Ganti, F. Baccelli, and J. G. Andrews (2012). Modeling and analysis of K-tier downlink heterogeneous cellular networks. *IEEE J. Sel. Areas Commun.*, 30(3):550–560, Apr. 2012.

27 S. Enayati, H. Saeedi, H. Pishro-Nik, and H. Yanikomeroglu (2019). Moving aerial base station networks: A stochastic geometry analysis and design perspective. *IEEE Trans. Wireless Commun.*, 18(6):2977–2988, June 2019.

28 A. Fotouhi, M. Ding, and M. Hassan (2016). Dynamic base station repositioning to improve performance of drone small cells. *Proceedings of the IEEE Global Communications (GLOBECOM) Workshops*, pages 1–6, Dec. 2016.

29 M. Haenggi (2012). *Stochastic Geometry for Wireless Networks*. Cambridge University Press.

30 F. Jiang and A. L. Swindlehurst (2012). Optimization of UAV heading for the ground-to-air uplink. *IEEE J. Sel. Areas Commun.*, 30(5):993–1005, June 2012.

31 M. G. Khoshkholgh, K. Navaie, H. Yanikomeroglu et al. (2019). Coverage performance of aerial–terrestrial HetNets. *Proceedings of the IEEE 89th Vehicular Technology Conference (VTC)*, Spring, pages 1–5, Apr. 2019.

32 S. Krishnan and H. S. Dhillon (2017). Spatio-temporal interference correlation and joint coverage in cellular networks. *IEEE Trans. Wireless Commun.*, 16(9):5659–5672, Sep. 2017.

33 X. Lin, R. K. Ganti, P. J. Fleming, and J. G. Andrews (2013). Towards understanding the fundamentals of mobility in cellular networks. *IEEE Trans. Wireless Commun.*, 12(4):1686–1698, Apr. 2013.

34 C. H. Liu, Z. Chen, J. Tang et al. (2018). Energy-efficient UAV control for effective and fair communication coverage: A deep reinforcement learning approach. *IEEE J. Sel. Areas Commun.*, 36(9):2059–2070, Sep. 2018.

35 D. W. Matolak and R. Sun (2017). Air–ground channel characterization for unmanned aircraft systems – Part I: Methods, measurements, and models for over-water settings. *IEEE Trans. Veh. Technol.*, 66 (1):26–44, Jan. 2017.

36 D. W. Matolak and R. Sun (2017). Air–ground channel characterization for unmanned aircraft systems – Part III: The suburban and near-urban environments. *IEEE Trans. Veh. Technol.*, 66(8):6607–6618, Aug. 2017.

37 A. Merwaday, A. Tuncer, A. Kumbhar, and I. Guvenc (2016). Improved throughput coverage in natural disasters: Unmanned aerial base stations for public-safety communications. *IEEE Veh. Technol. Mag.*, 11 (4):53–60, Dec. 2016.

38 S. Sadr and R. S. Adve (2015). Handoff rate and coverage analysis in multi-tier heterogeneous networks. *IEEE Trans. Wireless Commun.*, 14 (5):2626–2638, May 2015.

39 C. Saha, M. Afshang, and H. S. Dhillon (2018). 3GPP-inspired HetNet model using Poisson cluster process: Sum–product functionals and downlink coverage. *IEEE Trans. Commun.*, 66(5):2219–2234, May 2018.

40 S. Sekander, H. Tabassum, and E. Hossain (2018). Multi-tier drone architecture for 5G/B5G cellular networks: Challenges, trends, and prospects. *IEEE Commun. Mag.*, 56(3):96–103, Mar. 2018.

41 P. K. Sharma and D. I. Kim (2019). Coverage probability of 3-D mobile UAV networks. *IEEE Wireless Commun. Lett.*, 8(1):97–100, Feb. 2019.

42 P. K. Sharma and D. I. Kim (2019). Random 3D mobile UAV networks: Mobility modeling and coverage probability. *IEEE Trans. Wireless Commun.*, 18 (5):2527–2538, May 2019.

43 R. Sun and D. W. Matolak (2017). Air–ground channel characterization for unmanned aircraft systems – Part II: Hilly and mountainous settings. *IEEE Trans. Veh. Technol.*, 66(3):1913–1925, Mar. 2017.

44 H. Tabassum, M. Salehi, and E. Hossain (2019). Fundamentals of mobility-aware performance characterization of cellular networks: A tutorial. *IEEE Commun. Surveys Tuts.*, 21 (3): 2288–2308, 3rd Quart. 2019.

45 E. Turgut and M. C. Gursoy (2018). Downlink analysis in unmanned aerial vehicle (UAV) assisted cellular networks with clustered users. *IEEE Access*, 6:36313–36324, May 2018.

46 Q. Wu, Y. Zeng, and R. Zhang (2018). Joint trajectory and communication design for multi-UAV enabled wireless networks. *IEEE Trans. Wireless Commun.*, 17(3):2109–2121, Mar. 2018.

47 S. Yan, M. Peng, and X. Cao (2019). A game theory approach for joint access selection and resource allocation in UAV assisted IoT communication networks. *IEEE Internet Things J.*, 6 (2):1663–1674, Apr. 2019.

48 Y. Zeng, R. Zhang, and T. J. Lim (2016). Throughput maximization for UAV-enabled mobile relaying systems. *IEEE Trans. Commun.*, 64(12): 4983–4996, Dec. 2016.

49 X. Zhang and J. G. Andrews (2015). Downlink cellular network analysis with multi-slope path loss models. *IEEE Trans. Commun.*, 63(5):1881–1894, May 2015.

50 X. Zhou, J. Guo, S. Durrani, and H. Yanikomeroglu (2018). Uplink coverage performance of an underlay drone cell for temporary events. *Proceedings of the IEEE International Conference on Communications (ICC) Workshops*, pages 1–6, May 2018.

10

UAV Placement and Aerial–Ground Interference Coordination

Abhaykumar Kumbhar[1] and Ismail Guvenc[2]

[1]*Department of Electrical and Computer Engineering, Florida International University, Miami, FL 33199, USA*
[2]*Department of Electrical and Computer Engineering, North Carolina State University, Rayleigh NC 27695, USA*

10.1   Introduction

An air–ground heterogeneous cellular network (AG-HetNet) is regarded as one of the key components of future 5G wireless networks and beyond. They consist of ground-based fixed macro base stations (MBSs) and small cells such as ground-based pico base stations (PBSs) and unmanned aerial vehicles (UAVs) [18, 34]. This concept of an AG-HetNet is depicted in Figure 10.1, which illustrates the deployment of unmanned aerial base stations (UABSs) such as balloons, quadcopters, and gliders equipped with (Long Term Evolution) LTE-Advanced and 5G New Radio (NR) capabilities. These nodes complement the existing terrestrial infrastructure by serving ground user equipment (GUE) that is in either inadequate coverage or complete outage. Such UABSs can be deployed with minimum interdependencies at low cost, and they provide virtually omnipresent coverage. Several telecommunications service providers are already considering the integration of UABSs into an existing LTE-Advanced HetNet to enhance wireless connectivity, restore damaged infrastructure, and enable various new services and applications [4, 8, 11, 37].

There are several recent studies in the literature on LTE-Advanced AG-HetNets, which explore challenges such as the computation of optimal UABS deployment position and height in the geographical area of interest, path planning, mitigating inter-cell interference, developing suitable channel models, and optimizing network performance metrics such as spectral efficiency (SE) and coverage probability. The main goal of this chapter is to jointly study and optimize UABS placement and interference coordination in AG-HetNets. To this end, a genetic algorithm (GA) is considered for solving the optimization problem and studying the effect of different propagation conditions on the cell-edge SE using extensive computer simulations.

The rest of this chapter is organized as follows. Section 10.2 provides a review of UAV placement and interference coordination techniques in the literature, while Section 10.3 discusses several use cases of UABSs for cellular networks. Section 10.4 introduces the optimal placement problem of UABS in an AG-HetNet using the GA. Section 10.5 describes the UABS-based AG-HetNet model, different path-loss models, and the definition of 5th percentile spectral efficiency (5pSE) as a function of network parameters. In Section 10.7, we

UAV Communications for 5G and Beyond, First Edition.
Edited by Yong Zeng, Ismail Guvenc, Rui Zhang, Giovanni Geraci, and David W. Matolak.
© 2021 John Wiley & Sons Ltd. Published 2021 by John Wiley & Sons Ltd.

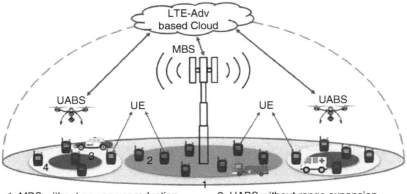

1: MBS without any power reduction. 3: UABS without range expansion.
2: MBS with reduced power. 4: UABS with range expansion.

Figure 10.1 An illustration of AG-HetNet, with MBS, GUE, and UABSs. MBS can use inter-cell interference coordination techniques defined in LTE-Advanced. The UABSs can dynamically change their position to maintain good wireless coverage and can utilize range expansion bias to take over MBS user equipment.

analyze and compare the 5pSE of the HetNet using extensive computer simulations for various inter-cell interference coordination (ICIC) techniques. Table 10.1 lists the notation and symbols used throughout the chapter.

10.2 Literature Review

There have been extensive studies in the literature that recently explored 3GPP Release 10 enhanced inter-cell interference coordination (eICIC) and 3GPP Release 11 further-enhanced inter-cell interference coordination (FeICIC) techniques for HetNets [9, 21, 27]. For example, [9] proposes algorithms which jointly optimize the eICIC parameters, user equipment (UE) cell association rules, and the spectrum resources shared between the macro and fixed small cells. However, in [9], the 3GPP Release 11 FeICIC technique is not considered, which provides better radio resource utilization and can offload a larger number of UEs to small cells through cell range expansion (CRE); whereas the effectiveness of 3GPP Release 10 and Release 11 ICIC techniques with ICIC parameter optimization has been studied in [21], without considering any mobility for small cells.

Recent advances in UAV technology has enabled the possibility of deploying small cells as UABSs mounted with a communication system. UABSs such as balloons, quadcopters, and gliders equipped with LTE-Advanced capabilities can be utilized to further enhance the capabilities of HetNets. The ability of UABSs to dynamically reposition in a HetNet environment can improve the overall SE of the network by filling the coverage gaps and offloading UEs in high-traffic regions. Hence, it is critical to optimize the locations of UABSs in a UAV-based HetNet to maximize SE gains.

Recent studies [1, 5, 7, 26, 28, 33] are mainly focused on finding the optimal locations of UAVs in the geographical area of interest to meet traffic demands. In [1, 5, 26, 33],

Table 10.1 Notation and symbols used in the system model (see text for full forms of acronyms).

Symbol	Description
$\lambda_{mbs}, \lambda_{ue}$	Density of the MBS and GUE nodes
X_{mbs}, X_{ue}	Locations of MBS and GUE
P_{mbs}, P_{uabs}	Maximum transmit power of MBS and UABS
P'_{mbs}, P'_{uabs}	Effective transmit power of MBS and UABS
K, K'	Attenuation factors due to geometrical parameters of antennas for both MBS and UABS
H	Exponentially distributed random variables that account for Rayleigh fading
δ	Path-loss exponent (PLE)
f_c	Carrier frequency (LTE band class 14)
h_{bs}	Height of the base station in Okumura–Hata model
h_{ue}	Height of a UE in Okumura–Hata model
d_{mn}, d_{mu}	Distance of a UE from its MOI and UOI, respectively
$S_{mbs}(d_{mn})$	RSRP from the MOI
$S_{uabs}(d_{mu})$	RSRP from the UOI
Z	Total interference at a UE from USF and CSF, respectively
γ, γ'	SIR from MOI and UOI, respectively, during USF
$\gamma_{csf}, \gamma'_{csf}$	SIR from MOI and UOI, respectively, during CSF
α	Power reduction factor for MBS during the transmission of CSFs
β	Duty cycle for the transmission of USF
τ	Cell range expansion bias
ρ, ρ'	Scheduling threshold for MUE and UUE, respectively
$N_{usf}^{mbs}, N_{csf}^{mbs}$	Number of USF MUEs and CSF MUEs, respectively, in a cell
$N_{usf}^{uabs}, N_{csf}^{uabs}$	Number of USF UUEs and CSF UUEs, respectively, in a cell
$C_{usf}^{mbs}, C_{csf}^{mbs}$	Aggregate SEs for USF MUEs and CSF MUEs, respectively, in a cell
$C_{usf}^{uabs}, C_{csf}^{uabs}$	Aggregate SEs for USF UUEs and CSF UUEs, respectively, in a cell
$\hat{X}_{uabs}^{(hex)}$	Fixed hexagonal locations of deployed UABS
$\hat{X}_{uabs}$	Optimized UABS locations computed using GA
S_{mbs}^{ICIC}	Matrix of ICIC parameters for MBSs
S_{uabs}^{ICIC}	Matrix of ICIC parameters for UABSs

UAV location optimization has been explored; however, ICIC techniques are not explicitly taken into account. The authors of [20, 22] explore UABS-assisted LTE-Advanced HetNets, where the UABSs employ CRE for offloading users from a macro-cell; however, they do not consider any ICIC in the cell expanded region. To maximize the 5pSE of the HetNet, a brute-force method is used to find the optimal UAV locations in [20], while the GA is used for optimizing UAV locations in [22].

Table 10.2 Literature review for UAV location placement and AG-HetNet interference coordination.

Ref.	Wireless nodes	Path-loss model	Optimization techniques	Optimization goal
[22]	MBS, UABS, GUE	Log-distance	Brute force, genetic algorithm	Location, 5pSE, coverage
[33]	MBS, UABS, GUE	Log-distance	Neural model	Location
[34]	MBS, UABS, GUE	Log-distance	Q-learning, deep Q-learning, brute force, sequential algorithm	Location, 5pSE, energy efficiency, interference
[17]	MBS, UABS, GUE	Log-distance, Okumura–Hata	fixed hexagonal, brute force, genetic algorithm	Location, 5pSE, energy efficiency, interference
[18]	MBS, PBS, UABS, GUE, AUE	Okumura–Hata, ITU-R P.1410-2, 3GPP RP-170779	Fixed hexagonal, brute force	Location, 5pSE, coverage, energy efficiency, interference
[31]	UABS, GUE	Log-distance, Close-in mmWave model	—	5pSE, coverage
[36]	UABS, GUE	ITU-R P.1410-2	Region partition strategy, backtracking line search algorithm	Location, GUE load balancing
[12]	UABS swarm	MIMO channel	Brute force, gradient descent location optimization	Location, SE
[6]	MBS, GUE, UABS	ITU-R P.1410-2, 3GPP TR 25.942	Deep reinforcement learning	Location, energy efficiency, wireless latency, interference
[41]	UABS, GUE	ITU-R P.1410-2	Centralized machine learning	Location, energy efficiency
[42]	MBS, UABS, GUE	ITU-R P.1410-2	Wavelet transform machine learning	Location, GUE load balancing
[32]	MBS, GUE, UABS	ITU-R P.1410-2	Greedy approach	3D location, GUE load balancing
[19]	UABS, GUE	Free space	Alternating optimization, successive convex programming	Location, bandwidth allocation, energy efficiency

The effect of interference in a UAV-based network is investigated in [25]. By calculating the optimal distance between the two interfering UAVs, each UAV is positioned at a fixed height to maximize the coverage area. However, this UAV-based network is not designed for LTE-Advanced HetNets. A priority-based UE offloading and UE association with mobile small cells for public safety communications (PSC) is studied in [15]. To improve the overall system throughput, 3GPP Release 10 eICIC and CRE are taken into account. However, using almost blank subframes (ABSs) at an MBS results in under-utilization of radio resources when compared to the use of reduced-power FeICIC defined in 3GPP Release 11.

The authors of [17, 18, 34] study joint optimization of UABS deployment location, interference coordination parameters, and 5pSE of the network for an LTE-Advanced AG-HetNet. A brute-force technique and a heuristic algorithm are used in [17], whereas [34] uses Q-learning, deep Q-learning, brute-force, and sequential algorithms, while only a brute-force technique is used in [18] for optimization. A literature review of the related work on UAV placement optimization and interference coordination is provided in Table 10.2.

10.3   UABS Use Case for AG-HetNets

An LTE-Advanced AG-HetNet with UABSs is illustrated in Figure 10.2. Such a network has a great potential to revolutionize cellular networks, by handling network congestion and high traffic during any sizeable public gathering, and by providing much-needed high-speed real-time data, video, and multimedia services [16, 17, 40]. In recent times, UABSs were used as base stations practically for the first time in Puerto Rico in November 2017: after the ground base stations were destroyed by hurricane Maria in 2017, AT&T used UAVs to temporarily restore wireless voice, text, data, and multimedia services [11].

As an example, consider the PSC scenario in Figure 10.2, where only two of the seven macro-cell base stations (MBSs) with large coverage areas remain operational after a disaster. The figure also illustrates several small-cell base stations (SCBSs), which can be critical to maintaining connectivity in PSC scenarios. In such scenarios, range expansion techniques [22] can be used with the SCBSs to extend coverage and equitably distribute users among different cells. To sustain ubiquitous broadband connectivity, Figure 10.2 shows how different types of UABSs can be utilized. In particular, in hotspot regions with denser GUE population, quadcopters can hover at a fixed location, while gliders have to follow a circular trajectory. Relaying and multi-hop communication methods can also be used for extending the coverage through the incident scene, either through UAVs or other GUEs. Thus, by exploiting UAV mobility, broadband connectivity can be delivered to desired regions, including congested areas and indoor environments.

However, a significant challenge with LTE-Advanced AG-HetNet is to optimize dynamically changing UABS positions and address severe and highly dynamic interference patterns. In particular, these unique challenges are due to: (i) potentially damaged base-station infrastructure, yielding outage problems; (ii) dynamically changing locations of GUEs, potentially clustered into some hotspot areas; (iii) HetNet traffic with bursty data transmission, which may temporarily overload the network infrastructure; and

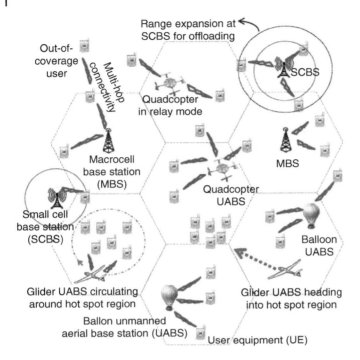

Figure 10.2 The MBSs, SCBSs, and UABSs constitute an AG-HetNet infrastructure, where the UABSs can dynamically change their positions for optimized coverage and seamless broadband connectivity.

(iv) the need to maintain a high quality of service. The next section will discuss the UABS placement problem in further detail.

10.4 UABS Placement in AG-HetNet

As a simulation study, this chapter considers a PSC scenario with wireless network having been destroyed after a disaster, as shown in Figure 10.3. In particular, Figure 10.3(a) shows that most of the geographical area in a typical PSC network is under SE coverage before a disaster. In the event of a disaster, the PSC network infrastructure is destroyed, and the first responders and victim users experience outages as illustrated by the white areas in Figure 10.3(b). In this scenario, the existing MBSs become overloaded with many UEs, and, as a result, these UEs begin to experience poor quality-of-service (QoS). Subsequently, at the site of the emergency, the first responders and victim users located in the outage regions will observe very low SE or possibly complete outage. To address the outage in a scenario such as in Figure 10.3(b), the design considers the optimization of UABS locations.

Consider individual locations (x_i, y_i) of each UABS $i \in \{1, 2, \ldots, N\}$ deployed over a geographical area of interest. In the design guidelines of the chapter, the UABSs are initially deployed on a fixed hexagonal grid as shown in Figure 10.4 and each UABS sends its location and the SE information of its users to a centralized server. The UABSs are placed within

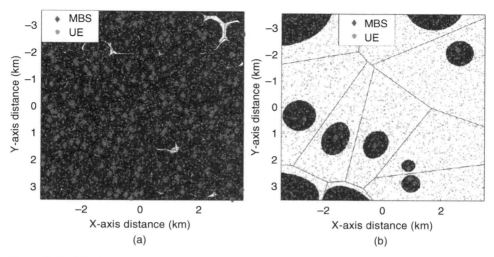

Figure 10.3 Wireless network spectral efficiency coverage before/after a disaster: (a) typical network before, and (b) wireless network after a disaster.

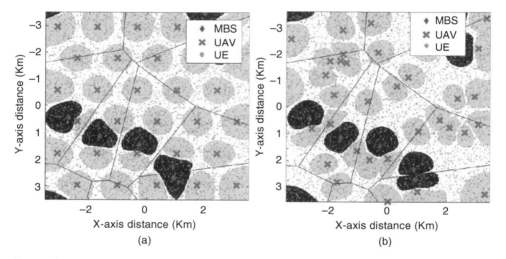

Figure 10.4 Wireless network after a disaster with UABS deployed at the height of 100 m: (a) UABS deployed on a fixed hexagonal grid, and (b) UABS locations optimized and reorganized using the GA.

the rectangular simulation area regardless of the existing MBS locations. Subsequently, the 5pSE for this fixed AG-HetNet will be determined by using a brute-force technique described in Listing 10.1. In Listing 10.1, Calc5thPercentileSE(*) is the objective function that computes the 5pSE of the network. Finally, the optimized ICIC parameters are captured in the matrix S_{ICIC} and the ICIC parameters that maximize the 5pSE can then be calculated as:

$$[\hat{\alpha}, \hat{\rho}, \hat{\tau}, \hat{\rho}'] = \arg\max_{\alpha,\rho,\tau,\rho'} C_{5\text{th}}(X_{\text{uabs}}^{(\text{hex})}, \alpha, \rho, \tau, \rho'), \tag{10.1}$$

where $X_{\text{uabs}}^{(\text{hex})}$ are the fixed and known hexagonal locations of the deployed UABSs within the simulation area.

Listing 10.1: Steps for computing 5pSE for hexagonal grid deployment.

```
Input: set of UABS locations and ICIC parameters
Output: SE: 5pSE for the network
Method:
    StopCondition: Number of iterations = 100
    while(! StopCondition)
    {
      Generate UABS locations
      for t = 1 to ICICParms.tau[t] do
      {
        for a = 1 to ICICParms.alpha[a] do
        {
          for r = 1 to ICICParms.rho[r] do
          {
            for p = 1 to ICICParms.rhoprime[p] do
            {
              SE = Calc5thPercentileSE(nodal locations, nodal Tx powers,
                  path-loss, tau, beta, alpha, rho, rhoprime)
            }
          }
        }
      }
    }
```

Then, a server can run any appropriate heuristic algorithm and compute the optimum locations of the UABSs. This chapter considers a GA proposed by Holland [13], which is a global optimization technique based on genetic science. One of the advantages of GA is that it runs with a population of candidate solutions rather than a single solution. Due to its parallel search capability, it can search over the whole working environment simultaneously, so that the optimum solution can be obtained more quickly than the classical optimization techniques such as brute-force search. Recently, GA has been used to solve the deployment problem of wireless sensor networks, which we will also adopt in this chapter for UABS deployment optimization.

With the approach of GA, a candidate solution for the optimization problem is referred to as a *chromosome*, which is a set of all the UABSs' location coordinates and inter-cell interference coordinates, as illustrated in Figure 10.5. A real-coded chromosome is used for representing UABSs' location coordinates and inter-cell interference coordinates. The optimization process begins with an initial population of chromosomes that are generated randomly and runs for a certain number of iterations until the optimum solution is reached. In each iteration of the GA process, the following steps are performed:

(1) All chromosomes are evaluated according to a *fitness function*. In this AG-HetNet model, the 5pSE of the network is the fitness function. This fitness function is referenced as Calc5thPercentileSE(*) in Listing 10.2 and calculated in Eq. (10.2).

(2) The *selection* process is used for determining the best chromosomes in the population which provide higher 5pSE results. There are many different selection strategies used

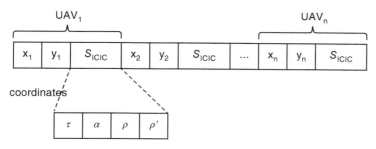

Figure 10.5 An example of a chromosome for FeICIC simulation, where the UABS locations, ICIC parameter τ, α, ρ, and ρ' are optimized. The ICIC parameter β is not optimized and is fixed at 50% duty cycle.

in GA, such as the roulette wheel selection, tournament selection, and rank selection. In this chapter, the roulette wheel selection method is used in which the probability of a chromosome k being selected is $P_k = f_k / \sum_{k=1}^{K} f_k$, where f_k is the fitness value of chromosome k, and K is the number of chromosomes in the population. In roulette wheel selection, the selection probability of each chromosome is proportional to its fitness value. After the chromosomes are selected, they are put into a mating pool to produce new chromosomes.

(3) A *crossover* process is used which combines the features of two parent chromosomes to generate two offspring. In this process, the genes of two chromosomes are swapped after the crossover point generated randomly. The single-point crossover operator is used in this study.

(4) After a new generation is created through the crossover operator, new chromosomes are randomly chosen for the *mutation* process. The mutation process is used in order to avoid the solution converging to local optima. For mutation, each member in a chromosome is randomly selected and replaced by another randomly chosen member under mutation probability.

Listing 10.2: Steps for optimizing population using GA.

```
Input:
    Population: set of UABS locations and ICIC parameters
    FITNESS function: Calc5thPercentileSE(*)
Output:
    Args: Best individuals of ICIC parameters and highest 5pSE
Procedure:
    NewPopulation <- empty set
    StopCondition: Number of iterations = 6
    SELECTION: Roulette wheel selection method
    while(! StopCondition)
    {
      for i = 1 to Size do
      {
        Parent1 <- SELECTION(NewPopulation, FITNESS function)
        Parent2 <- SELECTION(NewPopulation, FITNESS function)
        Child <- Reproduce(Parent1, Parent2)
```

```
    if(small random probability)
    {
      child <- MUTATE(Child)
      add child to NewPopulation
    }
  }
}
EVALUATE(NewPopulation, FITNESS function);
Args <- GetBestSolution(NewPopulation)
Population <- Replace(Population, NewPopulation)
}
```

Listing 10.2 describes the main steps used in GA to optimize the UABS locations and inter-cell interference network parameters in order to maximize the 5pSE of the network over a given geographical area of interest. The UABS locations and the ICIC parameters that maximize the 5pSE objective function can be calculated as

$$[\hat{X}_{\text{uabs}}, \hat{\alpha}, \hat{\rho}, \hat{\tau}, \hat{\rho}'] = \arg \max_{X_{\text{uabs}}, \alpha, \rho, \tau, \rho'} C_{\text{5th}}(X_{\text{uabs}}, \alpha, \rho, \tau, \rho'). \tag{10.2}$$

Since searching for optimal X_{uabs} and ICIC parameters using a brute-force approach is computationally intensive, this chapter considers a GA to find optimum UABS locations and the best-fit ICIC parameters τ, α, ρ, and ρ'. For the MBS locations shown in Figure 10.4(a), an example outcome of UABS locations using the GA is shown in Figure 10.4(b). Given the mobility and agility of UABSs, using the GA, the UAV positions can be dynamically reorganized to optimized locations to achieve the best network performance in the geographical area of interest.

10.5 AG-HetNet Design Guidelines

To address the outages in a scenario illustrated in Figure 10.3(b), the design considers a two-tier AG-HetNet deployment with MBSs and UABSs as shown in Figure 10.1, where all the MBS and UABS locations (in three dimensions) are captured in matrices $X_{\text{mbs}} \in \mathbb{R}^{N_{\text{mbs}} \times 3}$ and $X_{\text{uabs}} \in \mathbb{R}^{N_{\text{uabs}} \times 3}$, respectively, where N_{mbs} and N_{uabs} denote the number of MBSs and UABSs within the simulation area. The MBS and UE locations are each modeled using a two-dimensional Poisson point process (PPP) with densities λ_{mbs} and λ_{ue}, respectively [10, 21]. The UABSs are deployed at a fixed height, and their locations are either optimized using the GA or deployed on a fixed hexagonal grid.

The design assumes that the MBSs and the UABSs share a common transmission bandwidth, round-robin scheduling is used in all downlink transmissions, and a full buffer traffic is used in every cell. The transmit power of the MBS and UABS are P_{mbs} and P_{uabs}, respectively, while K and K' are the attenuation factors due to the geometrical parameters of the antennas for the MBS and the UABS, respectively. Then, the effective transmit power of the MBS is $P'_{\text{mbs}} = K P_{\text{mbs}}$, while the effective transmit power of the UABS is $P'_{\text{uabs}} = K' P_{\text{uabs}}$.

An arbitrary UE n is always assumed to connect to the nearest MBS or UABS, where $n \in \{1, 2, \ldots, N_{\text{ue}}\}$. Then, for the nth UE the reference symbol received power (RSRP) from the macro-cell of interest (MOI), and the UAV cell of interest (UOI) are given by [21]

$$S_{\text{mbs}}(d_{mn}) = \frac{P'_{\text{mbs}} H}{10^{\varphi/10}}, \qquad S_{\text{uabs}}(d_{un}) = \frac{P'_{\text{uabs}} H}{10^{\varphi'/10}}, \tag{10.3}$$

where the random variable $H \sim \exp(1)$ accounts for Rayleigh fading, φ is the path loss observed from MBS in dB, φ' is the path loss observed from UABS in dB, d_{mn} is the distance from the nearest MOI, and d_{un} is the distance from the nearest UOI. The Rayleigh fading channel is considered while presenting AG-HetNet design guidelines.

10.5.1 Path-Loss Model

To measure the path loss observed by the nth UE, we consider a log-distance path-loss model (LDPLM), which is an approximation to the real propagation channel and free-space suburban Okumura–Hata path-loss model (OHPLM) with LTE band class 14 frequency [16, 29].

10.5.1.1 Log-Distance Path-Loss Model

The LDPLM gives a coarse analysis of signal propagation and is a function of path-loss exponent and the distance between the serving base station and the nth UE [38]. The LDPLM is a free-space model and does not consider any physical structures or other obstacles which might affect the coverage of UABSs in real-world deployments. Based on the LDPLM, the path loss (in dB) observed by the nth UE from mth MOI and uth UOI is given by

$$\varphi = 10 \log_{10}(d_{mn}^{\delta}), \qquad \varphi' = 10 \log_{10}(d_{un}^{\delta}), \tag{10.4}$$

where δ is the path-loss exponent, and d_{un} depends on the locations of the UABSs that will be dynamically optimized.

The cumulative distribution functions (CDFs) for the combined path loss are plotted in Figure 10.6, for the cases when 50% and 97.5% of the MBS are destroyed. Figure 10.6(a) shows the empirical path-loss CDF, calculated for all the distances between base stations (X_{mbs} and X_{uabs}) and UEs (X_{ue}) using Eq. (10.4). Inspection of Figure 10.6(b) reveals that

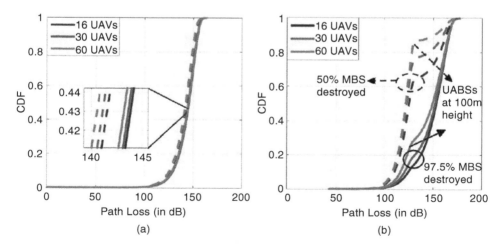

Figure 10.6 The CDFs of the combined path loss observed from all the base stations. Dashed lines correspond to the scenario with 50% of the MBS destroyed, while solid lines correspond to the scenario with 97.5% of the MBS destroyed. (a) CDF of the log-distance path-loss model, and (b) CDF of the Okumura–Hata model.

the variation in CDFs is minimum for the different number of UABSs deployed and for a different number of the MBSs destroyed. This is because the LDPLM does not consider external terrestrial factors. Nevertheless, the maximum allowable path loss for the system is 160 dB.

10.5.1.2   Okumura–Hata Path-Loss Model

The OHPLM is more suited for a terrestrial environment with manmade structures and an environment in which the base-station height does not vary significantly [30, 39]. This model is a function of the carrier frequency, the distance between the UE and the serving cell, base station height, and UE antenna height [24, 39]. Based on curve fitting of Okumura's original results, the path loss (in dB) observed by the nth UE from MOI and UOI is given by [3, 23]

$$\varphi = A + B\log(d_{mn}) + C, \tag{10.5}$$

$$\varphi' = A + B\log(d_{un}) + C, \tag{10.6}$$

where the distances d_{mn} and d_{un} are in km, and the factors A, B, and C depend on the carrier frequency and antenna height.

In a suburban environment, the factors A, B, and C are given by

$$A = 69.55 + 26.16\log(f_c) - 13.82\log(h_{bs}) - a(h_{ue}), \tag{10.7}$$

$$B = 44.9 - 6.55\log(h_{bs}), \tag{10.8}$$

$$C = -2\log(f_c/28)^2 - 5.4, \tag{10.9}$$

where f_c is the carrier frequency in MHz, h_{bs} is the height of the base station in meters, and $a(h_{ue})$ is a correction factor for the UE antenna height h_{ue} in meters, which is defined as

$$a(h_{ue}) = 1.1\log(f_c) - 0.7h_{ue} - 1.56\log(f_c) - 0.8. \tag{10.10}$$

Furthermore, OHPLM assumes the carrier frequency (f_c) to be between 150 MHz and 1500 MHz, the height of the base station (h_{bs}) between 30 m and 200 m, UE antenna height (h_{ue}) between 1 m and 10 m, and the distances d_{mn} and d_{un} between 1 km and 10 km [3, 23].

In Figure 10.6(b), we plot the empirical path-loss CDFs using Eqs. (10.5) and (10.10) and the OHPLM parameters in Table 10.3. Moreover, we plot the OHPLM path-loss CDFs for the cases when 50% and 97.5% of the MBSs are destroyed. Inspection of Figure 10.6(b) reveals a stepwise distribution of path loss in the CDFs. This behavior is due to the variation in the height of base stations, i.e., UABSs are deployed at the height of 100 m (larger path loss) while the height of MBSs is 30 m (smaller path loss). With 50% of the MBSs destroyed, it can be seen that most UEs are connected to the MBSs; while with 97.5% of the MBSs destroyed, the UABSs serve most of the UEs. Regardless, the maximum allowable path loss when 50% and 97.5% of the MBSs are destroyed is 225 dB, as shown in Figure 10.6(b).

10.6   Inter-Cell Interference Coordination

Due to their low transmission power, the UABSs are unable to associate a substantial number of UEs compared to that of MBSs. However, by using the cell range expansion

Table 10.3 Simulation parameters.

Parameter	Value
MBS and UE intensity	4 per km^2 and 100 per km^2
MBS and UABS transmit powers	46 dBm and 30 dBm
Path-loss exponent	4
Altitude of MBSs	30 m
Altitude of UABSs	100 m
Height of UE	3 m
LTE Band 14 center frequency	763 MHz for downlink. 793 MHz for uplink
$d_{mn}^{\min}, d_{mu}^{\min}$	30 m, 10 m
Simulation area	10×10 km^2
GA population size and generation number	60 and 100
GA crossover and mutation probabilities	0.7 and 0.1
Cell range expansion (τ)	0 to 15 dB
Power reduction factor for MBS during (α)	0 to 1
Duty cycle for the transmission of USF (β)	0.5 or 50%
Scheduling threshold for MUEs (ρ)	20 dB to 40 dB
Scheduling threshold for UUEs ($\rho\prime$)	−20 dB to −5 dB
MBS destroyed sequence	50% and 97.5%

(CRE) technique defined in 3GPP Release 8, UABSs can associate a large number of UEs by offloading traffic from the MBSs. A negative side effect of CRE includes increased interference in the downlink of cell-edge UEs or the UEs in the CRE region of the UABS, which is addressed by using ICIC techniques in LTE and LTE-Advanced [2, 14, 35].

3GPP Release 10 introduced a time-domain-based eICIC technique to address interference problems. In particular, it uses ABSs which require the MBS to completely blank the transmit power on the physical downlink shared channel (PDSCH) resource elements as shown in Figure 10.7(a). This separates the radio frames into coordinated subframes (CSFs) and uncoordinated subframes (USFs). On the other hand, 3GPP Release 11 defines FeICIC, where the data on PDSCH is still transmitted but at a reduced power level, as shown in Figure 10.7(b).

The MBSs can schedule their UEs either in USF or in CSF based on the scheduling threshold ρ. Similarly, the UABSs can schedule their UEs either in USF or in CSF based on the scheduling threshold ρ'. Let β denote the USF duty cycle, defined as the ratio of the number of USF subframes to the total number of subframes in a radio frame. Then, the duty cycle of CSFs is $(1 - \beta)$. For ease of simulation, the design considers a fixed USF duty cycle β of 0.5 for all the MBSs, which is shown in [21] to have limited effect on system performance when ρ and ρ' are optimized. Finally, let $0 \leq \alpha \leq 1$ denote the power reduction factor in coordinated subframes of the MBS for the FeICIC technique; $\alpha = 0$ corresponds to Release 10

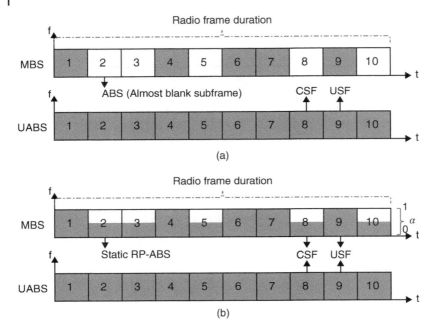

Figure 10.7 LTE-Advanced frame structures for time-domain ICIC: (a) 3GPP Release 10 eICIC with ABS, and (b) 3GPP Release 11 FeICIC with reduced power ABS (RP-ABS).

eICIC, while $\alpha = 1$ corresponds to no ICIC (e.g., as in 3GPP Release 8). The design assumes that the ABS and reduced power pattern are shared via the X2 interface, which is a logical interface between the base stations.

Given the eICIC and FeICIC framework in 3GPP LTE-Advanced as in Figure 10.7, and following an approach similar to that in [21] for a HetNet scenario, the signal-to-interference ratio (SIR) experienced by an arbitrary UE can be defined for CSFs and USFs for the MOI and the UOI as follows:

$$\Gamma = \frac{S_{\mathrm{mbs}}(d_{mn})}{S_{\mathrm{uabs}}(d_{un}) + Z} \quad \rightarrow \quad \text{USF SIR from MOI,} \tag{10.11}$$

$$\Gamma_{\mathrm{csf}} = \frac{\alpha S_{\mathrm{mbs}}(d_{mn})}{S_{\mathrm{uabs}}(d_{un}) + Z} \quad \rightarrow \quad \text{CSF SIR from MOI,} \tag{10.12}$$

$$\Gamma' = \frac{S_{\mathrm{uabs}}(d_{un})}{S_{\mathrm{mbs}}(d_{mn}) + Z} \quad \rightarrow \quad \text{USF SIR from UOI,} \tag{10.13}$$

$$\Gamma'_{\mathrm{csf}} = \frac{S_{\mathrm{uabs}}(d_{un})}{\alpha S_{\mathrm{mbs}}(d_{mn}) + Z} \quad \rightarrow \quad \text{CSF SIR from UOI,} \tag{10.14}$$

where Z is the total interference power at a UE during USF or CSF from all the MBSs and UABSs, excluding the MOI and the UOI. In the hexagonal grid UABS deployment model (and in [21]), the locations of the UABSs (and small cells) are fixed. To maximize the 5pSE of the network, the proposed design actively considers the SIRs in Eqs. (10.11)–(10.14) while optimizing the locations of the UABSs using the GA.

10.6.1   UE Association and Scheduling

The cell selection process relies on Γ and Γ' in Eqs. (10.11) and (10.13), respectively, for the MOI and UOI SIRs, as well as the CRE τ. If $\tau\Gamma'$ is less than Γ, then the UE is associated with the MOI; otherwise, it is associated with the UOI. After cell selection, the MBS-UE (MUE) and UABS-UE (UUE) can be scheduled either in USF or in CSF radio subframes as:

$$\text{if } \Gamma > \tau\Gamma' \text{ and } \Gamma \leq \rho \quad \rightarrow \quad \text{USF-MUE,} \tag{10.15}$$

$$\text{if } \Gamma > \tau\Gamma' \text{ and } \Gamma > \rho \quad \rightarrow \quad \text{CSF-MUE,} \tag{10.16}$$

$$\text{if } \Gamma \leq \tau\Gamma' \text{ and } \Gamma' > \rho' \quad \rightarrow \quad \text{USF-UUE,} \tag{10.17}$$

$$\text{if } \Gamma \leq \tau\Gamma' \text{ and } \Gamma' \leq \rho' \quad \rightarrow \quad \text{CSF-UUE.} \tag{10.18}$$

Once a UE is assigned to an MOI/UOI and is scheduled within the USF/CSF radio frames, then the SE for this scheduled UE can be expressed for the four different scenarios defined in Eqs. (10.15)–(10.18) as follows:

$$C_{\text{usf}}^{\text{mbs}} = \frac{\beta \log_2(1 + \Gamma)}{N_{\text{usf}}^{\text{mbs}}}, \tag{10.19}$$

$$C_{\text{csf}}^{\text{mbs}} = \frac{(1 - \beta)\log_2(1 + \Gamma_{\text{csf}})}{N_{\text{csf}}^{\text{mbs}}}, \tag{10.20}$$

$$C_{\text{usf}}^{\text{uabs}} = \frac{\beta \log_2(1 + \Gamma')}{N_{\text{usf}}^{\text{uabs}}}, \tag{10.21}$$

$$C_{\text{csf}}^{\text{uabs}} = \frac{(1 - \beta)\log_2(1 + \Gamma'_{\text{csf}})}{N_{\text{csf}}^{\text{uabs}}}, \tag{10.22}$$

where $N_{\text{usf}}^{\text{mbs}}$, $N_{\text{csf}}^{\text{mbs}}$, $N_{\text{usf}}^{\text{uabs}}$, and $N_{\text{csf}}^{\text{uabs}}$ are the number of MUEs and UUEs scheduled in USF and CSF radio subframes, and Γ, Γ_{csf}, Γ', and Γ'_{csf} are as in Eqs. (10.11)–(10.14).

The proposed design considers the use of 5pSE, which corresponds to the worst fifth percentile UE capacity among the capacities of all the N_{ue} UEs (calculated based on Eqs. (10.19)–(10.22)) within the simulation area. The 5pSE is a critical metric, particularly for PSC scenarios, to maintain a minimum QoS level at all the UEs in the environment. Moreover, the dependence of the 5pSE on UABS locations and ICIC parameters is defined as

$$C_{\text{5th}}(\boldsymbol{X}_{\text{uabs}}, \boldsymbol{S}_{\text{mbs}}^{\text{ICIC}}, \boldsymbol{S}_{\text{uabs}}^{\text{ICIC}}), \tag{10.23}$$

where $\boldsymbol{X}_{\text{uabs}} \in \mathbb{R}^{N_{\text{uabs}} \times 3}$ captures the UABS locations as defined earlier, $\boldsymbol{S}_{\text{mbs}}^{\text{ICIC}} = [\alpha, \rho] \in \mathbb{R}^{N_{\text{mbs}} \times 2}$ is a matrix that captures individual ICIC parameters for each MBS, while $\boldsymbol{S}_{\text{uabs}}^{\text{ICIC}} = [\tau, \rho'] \in \mathbb{R}^{N_{\text{uabs}} \times 2}$ is a matrix that captures individual ICIC parameters for each UABS. In particular,

$$\alpha = [\alpha_1, \ldots, \alpha_{N_{\text{mbs}}}]^{\text{T}}, \qquad \rho = [\rho_1, \ldots, \rho_{N_{\text{mbs}}}]^{\text{T}} \tag{10.24}$$

are $N_{\text{mbs}} \times 1$ vectors that include the power reduction factor and MUE scheduling threshold parameters for each MBS. On the other hand,

$$\tau = [\tau_1, \ldots, \tau_{N_{\text{uabs}}}]^{\text{T}}, \qquad \rho' = [\rho'_1, \ldots, \rho'_{N_{\text{uabs}}}]^{\text{T}} \tag{10.25}$$

are $N_{\text{uabs}} \times 1$ vectors that involve the CRE bias and UUE scheduling threshold at each UABS. As noted in Section 10.6, the duty cycle β of ABS and reduced power subframes is assumed to be set to 0.5 at all MBSs to reduce the search space and complexity.

Considering that the optimum values of the vectors $\boldsymbol{\alpha}$, $\boldsymbol{\rho}$, $\boldsymbol{\rho}'$, and $\boldsymbol{\tau}$ are to be searched over a multi-dimensional space, the computational complexity of finding the optimum parameters is prohibitively high. Hence, to reduce the system complexity (and simulation runtime) significantly, we consider that the same ICIC parameters are used for all MBSs and all UABSs. In particular, we consider that for $i = 1, \ldots, N_{\text{mbs}}$ we have $\alpha_i = \alpha$ and $\rho_i = \rho$, while for $j = 1, \ldots, N_{\text{uabs}}$ we have $\tau_j = \tau$ and $\rho'_j = \rho'$. Therefore, the dependence of the 5pSE on the UABS locations and ICIC parameters can be simplified as

$$C_{\text{5th}}(\boldsymbol{X}_{\text{uabs}}, \alpha, \rho, \tau, \rho'), \tag{10.26}$$

which we will seek ways to maximize through simulations.

10.7 Simulation Results

In this section, using extensive computer simulations, we compare the 5pSE of an AG-HetNet with and without ICIC techniques while considering different UABS deployment strategies and path-loss models for all the UEs covered by the base stations. Unless otherwise specified, the system parameters for the simulations are set to the values given in Table 10.3.

10.7.1 5pSE with UABSs Deployed on Hexagonal Grid

In the following, the key 5pSE observations are discussed when the UABSs are deployed on a hexagonal grid and utilizing optimized ICIC parameters (see Eq. (10.1) and Listing 10.1). In Figures 10.8 and 10.9, we plot the variations in 5pSE with respect to CRE while using LDPLM and OHPLM, respectively.

10.7.1.1 5pSE with Log-Normal Path-Loss Model

Figure 10.8(a) plots the 5pSE variation with respect to CRE for no-ICIC mechanism (NIM). In the case of NIM, all the base stations (MBSs and UABSs) always transmit at full power (P'_{mbs} and P'_{uabs}). The close evaluation of Figure 10.8(a) shows that the peak value of 5pSE for NIM is observed at around 0 dB CRE. This is because, with no CRE, the number of UEs associated with the UABSs and the interference experienced by these UEs is minimal. Moreover, as the CRE increases, the number of UEs associated with the UABSs increases, and so does the interference experienced by these UEs. Hence, with NIM, the 5pSE decreases with increasing CRE as seen in Figure 10.8(a).

The performance of 3GPP Release 10 and Release 11 ICIC techniques in terms of 5pSE and the variation in CRE are plotted in Figure 10.8(b) and (c). As noted in Section 10.6, the transmission power during blank subframes at the MBSs for eICIC is 0, and power reduction of the CSFs at the MBSs for FeICIC is $\alpha P'_{\text{mbs}}$. Using this understanding, the analysis Figure 10.8(b) and (c) shows that the 5pSE for ICIC techniques at 0 dB CRE are relatively lower. On the other hand, the ICIC techniques observe improvement in 5pSE performance

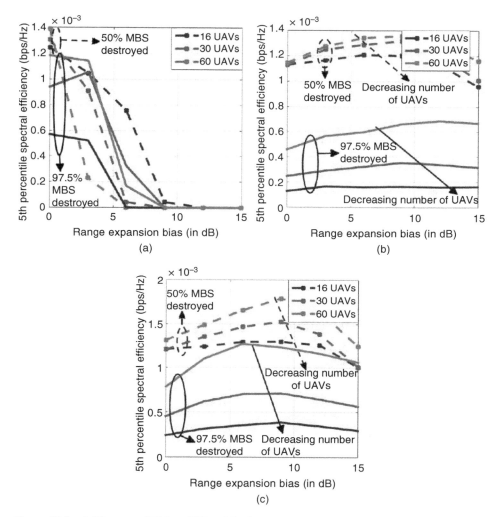

Figure 10.8 5pSE versus CRE for eICIC and FeICIC techniques with LDPLM (UABSs deployed on a hexagonal grid): 5pSE (a) without ICIC, (b) with eICIC, and (c) with FeICIC.

with increasing CRE, and the peak value of the 5pSE for the ICIC techniques is observed when the CRE is between 6 and 9 dB. This influence of CRE on the 5pSE for NIM and 3GPP Release 10 and Release 11 ICIC techniques is summarized in Figure 10.10(a).

10.7.1.2 5pSE with Okumura–Hata Path-Loss Model

Figure 10.9 plots the 5pSE variation with respect to CRE for NIM. In the case of NIM, all the base stations (MBSs and UABSs) always transmit at full power (P'_{mbs} and P'_{uabs}). The peak value of 5pSE for NIM is observed at around 3 dB CRE when 50% of the MBSs are destroyed. On the other hand, when 97.5% of the MBSs are destroyed, even though the number of existing MBSs are small and the interference is minimum, the higher path-loss presents higher probability for a cell-edge UE to fall out of the coverage area. Moreover,

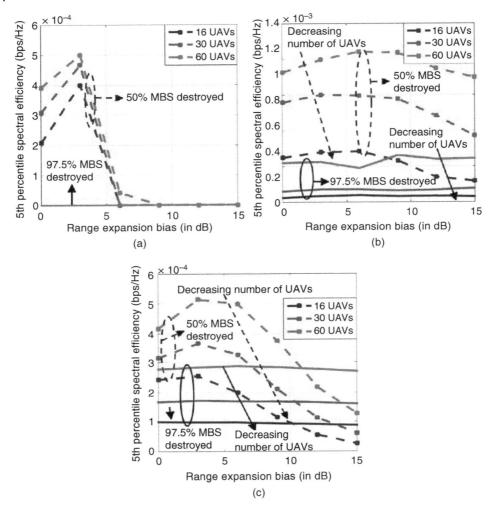

Figure 10.9 5pSE versus CRE for eICIC and FeICIC techniques with OHPLM (UABSs deployed on a hexagonal grid): 5pSE (a) without ICIC, (b) with eICIC, and (c) with FeICIC.

in the absence of any ICIC, using CRE can magnify the impact of interference. Hence, the 5pSE gains with NIM are close to zero.

Figure 10.9(b) and (c) plot the variation of 5pSE with respect to CRE for 3GPP Release 10/11 ICIC techniques. The analysis of Figure 10.9(b) and (c) shows that the ICIC techniques observe improvement in 5pSE performance with increasing CRE. When 50% of the MBSs are destroyed, the 5pSE peak values for the eICIC and FeICIC are observed when the CRE is between 6 and 9 dB and between 3 and 6 dB, respectively. When 97.5% of the MBSs are destroyed, even though the cell-edge UEs observe higher path loss, using the 3GPP Release 10/11 ICIC techniques along with CRE can decrease the probability of cell-edge UE going out of coverage, thus sustaining the 5pSE of the network as seen in

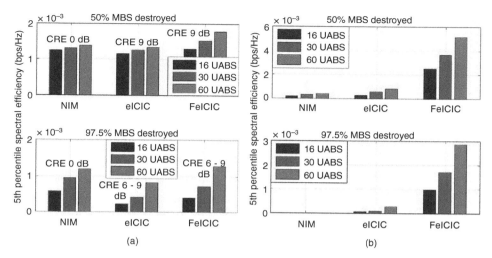

Figure 10.10 Peak observations for the 5pSE with UABSs deployed on a hexagonal grid: (a) LDPLM and (b) OHPLM.

Figure 10.9(b) and (c). Further analysis shows that the FeICIC technique observes significant improvement in SE performance when compared to NIM and eICIC. This influence of CRE on the 5pSE for NIM and 3GPP Release 10/11 ICIC techniques is summarized in Figure 10.10(b).

On comparison of Figure 10.10(a) and (b), we observe modest deviation in peak values of 5pSE between NIM, eICIC, and FeICIC with LDPLM. This is because UEs experience better SIR with lower path losses. With OHPLM, on the other hand, a significant deviation in the peak values of 5pSE is observed due to higher path losses. However, the higher path losses in OHPLM can be compensated by using modest/higher CRE values and 3GPP Release 10/11 ICIC techniques.

10.7.2 5pSE with GA-Based UABS Deployment Optimization

In the following, critical 5pSE observations are presented when UABS locations and ICIC parameters are optimized through the GA, defined in Eq. (10.2) and Listing 10.2. Figures 10.11 and 10.12 plot the peak 5pSE for the network using the GA, versus the optimized CRE value while using LDPLM and OHPLM, respectively. In the GA-based simulations, the optimum CRE value is directly related to the locations of the UABSs with respect to the MBSs, the number of UEs offloaded to the UABSs, and the amount of interference observed by the UEs.

10.7.2.1 5pSE with Log-Normal Path-Loss Model
Figure 10.11(a) and (b) plot the peak 5pSE with respect to the optimized CRE value for eICIC and FeICIC, respectively, for SPLM. Inspection of Figure 10.11(a) and (b) reveals higher values of CRE when 50% of the MBSs are destroyed and implies the presence of substantial interference from this large number of MBSs. Hence, offloading a large number

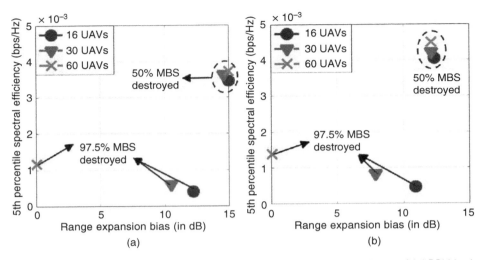

Figure 10.11 Peak 5pSE versus optimized CRE for eICIC and FeICIC techniques with LDPLM, when the UABS locations and ICIC parameters are optimized using the GA: (a) 5pSE with eICIC, and (b) 5pSE with FeICIC.

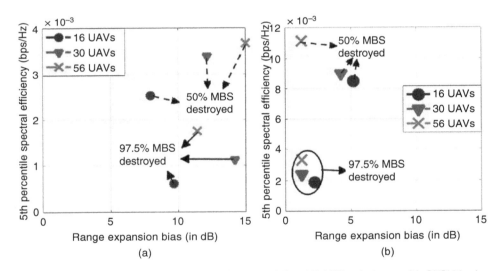

Figure 10.12 Peak 5pSE versus optimized CRE for eICIC and FeICIC techniques with OHPLM, when the UABS locations and ICIC parameters are optimized using the GA: (a) 5pSE with eICIC, and (b) 5pSE with FeICIC.

of UEs from MBSs to UABSs with higher values of CRE is necessary for achieving better 5pSE gains.

On the other hand, when most of the infrastructure is destroyed (i.e., when 97.5% of the MBSs are destroyed), the interference observed from the MBSs is limited, and a larger number of UEs need to be served by the UABSs. Therefore, with fewer UABSs deployed, higher CRE is required to serve a larger number of UEs and achieve better 5pSE. On the other hand, when a larger number of UABSs are deployed, smaller values of CRE will result in better 5pSE gains. This behavior is recorded in Figure 10.11(a) and (b) for eICIC and FeICIC, respectively.

10.7.2.2 5pSE with Okumura–Hata Path-Loss model

Using Eqs. (10.11)–(10.14), the SIR observations are plotted in Figure 10.13 for 3GPP Release 10/11 ICIC techniques. As illustrated in Figure 10.6(a), the higher path loss results in lower SIR values as seen in Figure 10.13. Using this understanding, we inspect the peak 5pSE with respect to the optimized CRE for the eICIC and FeICIC as shown in Figure 10.12(a) and (b), respectively.

With 3GPP Release 10 ABS, higher values of CRE are required for UABSs to compensate for the high path loss and under-utilization of radio resources by the MBSs in CSF radio subframes. When 50% and 97.5% of the MBSs are destroyed, the peak 5pSE values for eICIC are achieved with minimal SIR values, and by offloading a large number of UEs from MBSs to UABSs as seen in Figure 10.12(a).

On the other hand, with 3GPP Release 11 reduced power subframes (FeICIC), MBSs can establish and maintain connectivity with a sufficient number of cell-edge MUEs, while offloading the out-of-coverage UEs to UABSs for better QoS. When 50% and 97.5% of the MBSs are destroyed, the peak 5pSE values for FeICIC are achieved with minimal SIR and moderate CRE values as shown in Figure 10.12(b).

To summarize, using GA for both path-loss models, FeICIC in Release 11 is seen to outperform Release 10 eICIC in terms of the overall 5pSE of the network. When larger number of UABSs are deployed and when fewer MBSs are destroyed, the 5pSE of the network is higher. On the other hand, the 5pSE decreases with the increasing number of destroyed MBSs, as seen in Figure 10.12.

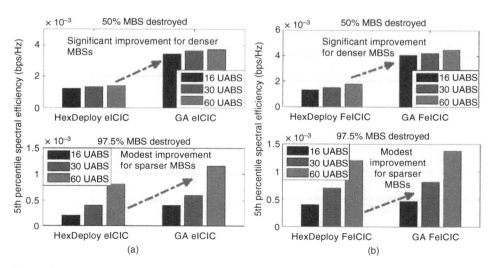

Figure 10.13 SIR observations for eICIC and FeICIC with OHPLM, when UABS locations are optimized using the GA: (a) SIR observations for eICIC, and (b) SIR observations for FeICIC.

10.7.3 Performance Comparison Between Fixed (Hexagonal) and Optimized UABS Deployment with eICIC and FeICIC

We summarize the key results from earlier simulations in Figures 10.14 and 10.15 for both path-loss models and compare the key trade-offs between fixed (hexagonal) deployment and GA-based deployment of UABSs.

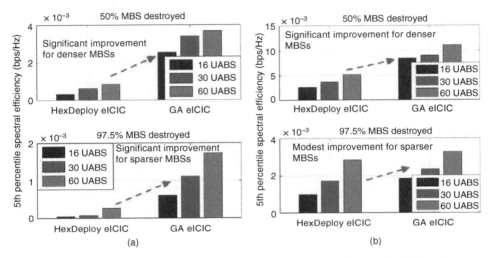

Figure 10.14 5pSE comparisons for eICIC and FeICIC with LDPLM, when the UABS locations are optimized using the GA and when the UABSs are deployed in a fixed hexagonal grid: (a) 5pSE with eICIC, and (b) 5pSE with FeICIC.

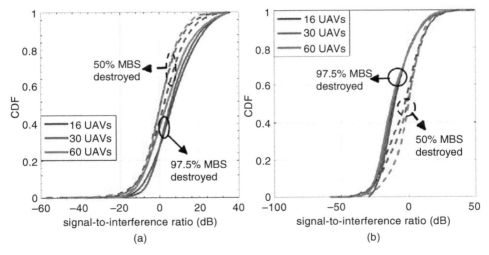

Figure 10.15 5pSE comparisons for eICIC and FeICIC with OHPLM, when the UABS locations are optimized using the GA and when the UABSs are deployed in a fixed hexagonal grid: (a) 5pSE with eICIC, and (b) 5pSE with FeICIC.

10.7.3.1   Influence of LDPLM on 5pSE

In Figure 10.14, the 5pSE observations are compared with LDPLM shown in Figures 10.8 and 10.11. The comparative analysis reveals that UABS deployment with optimized CRE and optimized location provides a better 5pSE than the UABSs deployed on a fixed hexagonal grid. Furthermore, Figure 10.14 shows that the 5pSE gains from the optimization of UABS locations are more significant when 50% of the MBSs are destroyed and less significant when 97.5% of the MBSs are destroyed.

When 50% MBSs are destroyed, there are still a large number of MBSs present, which causes substantial interference. Hence, in such an interference-driven scenario, it is important to optimize the locations of the UABSs, and use of a larger number of UABSs to provide significant gains in the 5pSE. On the other hand, with 97.5% of the MBSs destroyed, the interference from the MBSs is small, and deploying the UABSs on a hexagonal grid will perform close-to-optimum UABS deployment.

10.7.3.2   Influence of OHPLM on 5pSE

In Figure 10.15, we compare the 5pSE observations with OHPLM shown in Figures 10.9 and 10.12. The comparative analysis reveals that UABS deployment with optimized CRE and optimized locations provides a better 5pSE than the UABSs that are deployed on a fixed hexagonal grid.

With eICIC in Release 10, when 50% and 97.5% of the MBSs are destroyed, the 5pSE gains from the optimized UABS locations are significant, as shown in Figure 10.15(a). On the other hand, with FeICIC in Release 11, the 5pSE gains from the optimized UABS locations are more significant when 50% of the MBSs are destroyed, as seen in Figure 10.15(b). However, the difference between the hexagonal deployment and optimized deployment is especially small since the power reduction factor α in the MBS CSFs provides an additional optimization dimension for improving the 5pSE. Use of a larger number of UABSs when 97.5% of the MBSs are destroyed is shown to provide modest gains in the 5pSE, in contrast to significant gains in the 5pSE when 50% of the MBSs are destroyed.

10.7.4   Comparison of Computation Time for Different UABS Deployment Algorithms

In this subsection, the comparison of computational time between the GA and hexagonal grid deployment techniques with ICIC optimization is presented. Using an Intel Core i7-4810 central processing unit operating at 2.8 GHz, 24 GB of random access memory, and a Monte Carlo experimental approach, the mean runtime is calculated for the Matlab simulations. Figure 10.16(a) plots the mean runtime required for calculating the optimal ICIC network parameters and optimized UABS locations using Eq. (10.2), Listing 10.2, and the simulation values defined in Table 10.3. Inspection of Figure 10.16(a) reveals that the FeICIC technique requires similar computational time when compared to the eICIC technique for GA-based optimization. The main reason for this is that the large search space for UABS locations, which are common in both FeICIC- and eICIC-based approaches, dominates the computation time when compared with the optimization of ICIC parameters.

On the other hand, Figure 10.16(b) plots the mean runtime required for UABS deployment on a hexagonal grid using (10.1), the simulation values defined in Table 10.3, and with

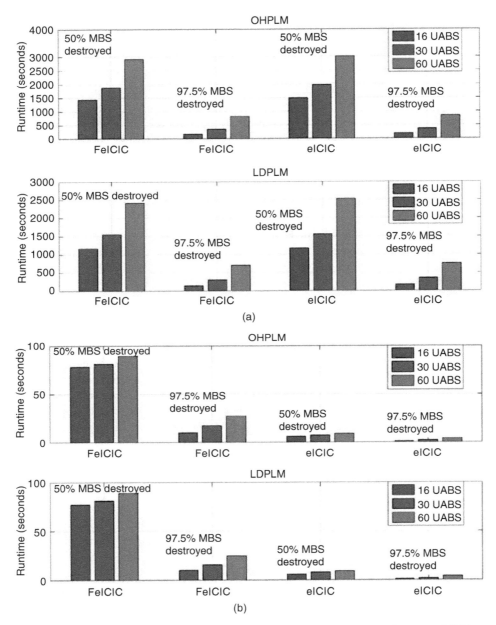

Figure 10.16 Simulation runtime using FeICIC and eICIC techniques with OHPLM and LDPLM: (a) fixed hexagonal grid, and (b) genetic algorithm.

fixed step size for the ICIC parameters. Inspection of Figure 10.16(b) reveals that the FeICIC technique requires significantly higher computational time when compared to the eICIC technique. The main reason for this behavior is due to the additional computation required for optimizing the power reduction factor α for the FeICIC approach. In general, with the GA and the hexagonal grid deployment, when a larger number of UABSs are deployed, and a larger number of the MBSs are present, the mean runtime is the largest. On the other hand, the mean runtime decreases with a smaller number of UABSs deployed, and when a smaller number of MBSs are present. Moreover, the comparative analysis of Figure 10.16(a) and (b) reveals that optimization (ICIC parameters and UABS locations) using GA requires significantly more computational time when compared to UABSs deployment on a hexagonal grid.

10.8 Concluding remarks

This chapter explores the suitability of a meta-heuristic GA for finding optimal or close-to-optimal locations for UAV placement. However, the computational complexity required to optimize the considered UAV deployment optimization problems in the real world using the GA techniques requires further investigations. Our simulation shows that optimizing UABS locations and ICIC parameters using GA yields significant improvement when compared to the deployment of UABSs on a fixed hexagonal grid.

Using the proposed design, the AG-HetNet with reduced power subframes (FeICIC) yields better 5pSE than that with almost blank subframes (eICIC). In a simulated network with SPLM and when 60 UABS locations are optimized using the GA, the FeICIC observes a modest improvement over eICIC: approximately 17% and 15% when 50% and 97.5% of the MBSs are destroyed, respectively. On the other hand, with OHPLM and when 60 UABS locations are optimized using the GA, the FeICIC yields a significant improvement over eICIC: approximately 66% and 51% when 50% and 97.5% of the MBSs are destroyed, respectively.

Finally, through simulations, the 5pSE of the network is compared and analyzed for different path-loss models and different UABS deployment strategies. With SPLM, the analysis shows that deployment of the UABSs on a hexagonal grid is close to optimal when the observed interference is limited. In the presence of substantial interference, the GA approach is more effective for deploying UABSs. On the other hand, with OHPLM, the network observes high path loss when compared to SPLM. To subdue the effects of high path loss, the GA approach is shown to be more effective.

References

1 Akram Al-Hourani, Sithamparanathan Kandeepan, and Simon Lardner. Optimal LAP altitude for maximum coverage. *IEEE Wireless Commun. Lett.*, 3(6):569–572, 2014.

2 Md Shipon Ali. An Overview on Interference Management in 3GPP LTE-Advanced Heterogeneous Networks. *Int. J. Future Generation Commun. Netw.*, 8(1):55–68, June 2015.

3 Yazan A Alqudah, Belal Sababha, Ayman Elnashar, and Sohaib H Sababha. On the validation of path loss models based on field measurements using 800 MHz LTE network. In *Proc. IEEE Annual Sys. Conf. (SysCon)*, pages 1–5, Orlando,FL, 2016.

4 AT&T. Flying COW Connects Puerto Rico. Nov. 2017. URL https://about.att.com/inside_connections_blog/flying_cow_puertori.

5 R Irem Bor-Yaliniz, Amr El-Keyi, and Halim Yanikomeroglu. Efficient 3-D placement of an aerial base station in next generation cellular networks. In *Proc. IEEE Intl. Conf. Commun. (ICC)*, pages 1–5, Kuala Lumpur, Malaysia, 2016.

6 Ursula Challita, Walid Saad, and Christian Bettstetter. Interference Management for Cellular-Connected UAVs: A Deep Reinforcement Learning Approach. *IEEE Trans.Wireless Commun.*, 2019.

7 Evander Christy, Rina Pudji Astuti, Budi Syihabuddin, Bhaskara Narottama, Obed Rhesa, and Furry Rachmawati. Optimum UAV flying path for Device-to-Device communications in disaster area. In *Proc. IEEE Int. Conf. Sig Sys. (ICSigSys)*, pages 318–322, Bali, Indonesia, 2017.

8 CNBC. AT&T and Verizon drones provide cell service in natural disasters. Aug. 2018. URL https://www.cnbc.com/2018/06/22/att-and-verizon-drones-provide-cell-service-in-natural-disasters.html.

9 Supratim Deb, Pantelis Monogioudis, Jerzy Miernik, and James P Seymour. Algorithms for enhanced inter-cell interference coordination (eICIC) in LTE hetnets. *IEEE/ACM Trans. Nwk.*, 22(1):137–150, 2014.

10 Radha Krishna Ganti, François Baccelli, and Jeffrey G Andrews. A new way of computing rate in cellular networks. In *Proc. IEEE Int. Conf. Commun. (ICC)*, pages 1–5, Kyoto, Japan, June 2011.

11 Claudia Geib. An ATT Drone Is Connecting Puerto Ricans to Wireless Service. Nov. 2017. URL https://futurism.com/att-drone-connecting-puerto-ricans-wireless-service/.

12 Samer Hanna, Han Yan, and Danijela Cabric. Distributed UAV Placement Optimization for Cooperative Line-of-sight MIMO Communications. In *Proc. IEEE Intl. Conf. Acoustics, Speech, Signal Process. (ICASSP)*, pages 4619–4623, 2019.

13 John Holland. Holland, adaptation in natural and artificial systems, 1992.

14 Harri Holma, Antti Toskala, and Jussi Reunanen. *LTE Small Cell Optimization: 3GPP Evolution to Release 13*. John Wiley & Sons, Jan. 2016.

15 Zeeshan Kaleem and Kyunghi Chang. Public Safety Priority-Based User Association for Load Balancing and Interference Reduction in PS-LTE Systems. *IEEE Access*, 4:9775–9785, 2016.

16 Abhaykumar Kumbhar, Farshad Koohifar, Ismail Guvenc, and Bruce Mueller. A survey on legacy and emerging technologies for public safety communications. *IEEE Commun. Survery Tuts.*, 18:97–124, Sep. 2016.

17 Abhaykumar Kumbhar, Ismail Güvenç, Simran Singh, and Adem Tuncer. Exploiting LTE-Advanced HetNets and FeICIC for UAV-assisted public safety communications. *IEEE Access*, 6:783–796, 2018.

18 Abhaykumar Kumbhar, Hamidullah Binol, Ismail Guvenc, and Kemal Akkaya. Interference Coordination for Aerial and Terrestrial Nodes in Three-Tier LTE-Advanced HetNet. In *Proc. IEEE Radio and Wireless Symposium (RWS)*, pages 1–4, 2019.

19 Peiming Li and Jie Xu. Placement Optimization for UAV-Enabled Wireless Networks with Multi-Hop Backhauls. *J. Commun. Information Netw.*, 3(4):64–73, 2018.

20 Arvind Merwaday and Ismail Guvenc. UAV assisted heterogeneous networks for public safety communications. In *Proc. IEEE Wireless Commun. Nwk. Conf. Workshops (WCNCW)*, pages 329–334, New Orleans, LA, 2015.

21 Arvind Merwaday, Sayandev Mukherjee, and Ismail Güvenç. Capacity analysis of LTE-Advanced HetNets with reduced power subframes and range expansion. *EURASIP J. Wireless Commun. Netw.*, (1):1–19, Nov. 2014.

22 Arvind Merwaday, Adem Tuncer, Abhaykumar Kumbhar, and Ismail Guvenc. Improved Throughput Coverage in Natural Disasters: Unmanned Aerial Base Stations for Public-Safety Communications. *IEEE Vehic. Technol. Mag.*, 11(4):53–60, Dec. 2016.

23 Andreas F Molisch. *Wireless communications*, volume 34. John Wiley & Sons, 2012.

24 Michael S Mollel and Michael Kisangiri. Comparison of Empirical Propagation Path Loss Models for Mobile Communication. *Comp. Eng. Intelligent Sys.*, 5:1–10, 2014.

25 Mohammad Mozaffari, Walid Saad, Mehdi Bennis, and Merouane Debbah. Drone small cells in the clouds: Design, deployment and performance analysis. In *Proc. IEEE Global Commun. Conf. (GLOBECOM)*, pages 1–6, San Diego, CA, 2015.

26 Mohammad Mozaffari, Walid Saad, Mehdi Bennis, and Mérouane Debbah. Optimal transport theory for power-efficient deployment of unmanned aerial vehicles. In *Proc. IEEE Intl. Conf. Commun. (ICC)*, pages 1–6, Kuala Lumpur, Malaysia, 2016.

27 Sayandev Mukherjee and Ismail Güvenç. Effects of range expansion and interference coordination on capacity and fairness in heterogeneous networks. In *Proc. IEEE Asilomar Conf. on Signals, Systems and Computers*, pages 1855–1859, 2011.

28 Mamta Narang, Simon Xiang, William Liu, Jairo Gutierrez, Luca Chiaraviglio, Arjuna Sathiaseelan, and Arvind Merwaday. UAV-assisted edge infrastructure for challenged networks. In *Proc. IEEE Conf. Computer Commun. Workshops (INFOCOM WKSHPS)*, pages 60–65, 2017.

29 Stagg Newman, Jon M Peha, and Jennifer A Manner. The FCC plan for a public safety broadband wireless network. 2010.

30 Sylvain Ranvier. Path loss models. Technical report, Helsinki University of Technology, Nov. 2004.

31 Nadisanka Rupasinghe, Yavuz Yapıcı, Ismail Güvenç, and Yuichi Kakishima. Non-orthogonal multiple access for mmwave drone networks with limited feedback. *IEEE Trans. Commun.*, 67(1):762–777, 2019.

32 Sanaa Sharafeddine and Rania Islambouli. On-demand deployment of multiple aerial base stations for traffic offloading and network recovery. *Computer Networks*, 2019.

33 Vishal Sharma, Mehdi Bennis, and Rajesh Kumar. UAV-assisted heterogeneous networks for capacity enhancement. *IEEE Commun. Lett.*, 20(6):1207–1210, 2016.

34 Simran Singh, Abhaykumar Kumbhar, Ismail Güvenç, and Mihail L Sichitiu. Distributed Approaches for Inter-cell Interference Coordination in UAV-based LTE-Advanced HetNets. In *Proc. IEEE 88th Vehic. Technol. Conf. (VTC-Fall)*, pages 1–6, 2019.

35 Beatriz Soret and Klaus I Pedersen. Macro transmission power reduction for hetnet co-channel deployments. In *Proc. IEEE Global Commun. Conf. (GLOBECOM)*, pages 4126–4130, Anaheim, CA, Dec. 2012.

36 Yang Sun, Tianyu Wang, and Shaowei Wang. Location Optimization for Unmanned Aerial Vehicles Assisted Mobile Networks. In *Proc. IEEE Intl. Conf. Commun. (ICC)*, pages 1–6, 2018.

37 The Drive. AT&T and Verizon Test 4G LTE Drones in New Jersey. Jun. 2018. URL https://www.thedrive.com/tech/21756/att-and-verizon-test-4g-lte-drones-in-new-jersey.

38 Michael Tsai. Path-loss and Shadowing (Large-scale Fading). Technical report, National Taiwan University, Oct. 2011.

39 Xiro Online. Okumura-Hata, 2017. URL https://www.xirio-online.com/help/en/okumurahata.html.

40 Murat Yuksel, Ismail Guvenc, Walid Saad, and Naim Kapucu. Pervasive spectrum sharing for public safety communications. *IEEE Commun. Mag.*, 54(3):22–29, 2016.

41 Qianqian Zhang, Mohammad Mozaffari, Walid Saad, Mehdi Bennis, and Merouane Debbah. Machine learning for predictive on-demand deployment of UAVs for wireless communications. In *Proc. IEEE Global Commun. Conf. (GLOBECOM)*, pages 1–6, 2018.

42 Qianqian Zhang, Walid Saad, Mehdi Bennis, Xing Lu, Merouane Debbah, and Wangda Zuo. Predictive Deployment of UAV Base Stations in Wireless Networks: Machine Learning Meets Contract Theory. *arXiv preprint arXiv:1811.01149*, 2018.

11

Joint Trajectory and Resource Optimization

Yong Zeng[1,2], Qingqing Wu[3], and Rui Zhang[4]

[1] *National Mobile Communications Research Laboratory, Southeast University, China*
[2] *Purple Mountain Laboratories, Jiangsu, China*
[3] *State Key Laboratory of Internet of Things for Smart City, University of Macau, China*
[4] *Department of Electrical and Computer Engineering, National University of Singapore, Singapore*

11.1 General Problem Formulation

Different from the conventional terrestrial communication infrastructure, such as ground base stations (BSs), access points (APs) and relays, an unmanned aerial vehicle (UAV) enabled communication platform is able to move dynamically to best suit the communication requirement. This offers a new design degree of freedom (DoF) apart from the conventional communication resource allocation, namely, UAV trajectory optimization, to enhance the communication performance. This chapter aims to discuss the main techniques of exploiting the fully controllable UAV mobility for UAV-assisted wireless communications.

We consider a generic UAV-assisted wireless communication system as shown in Figure 11.1, where a UAV is dispatched to serve multiple ground users. Denote the UAV trajectory by $q(t) \in \mathbb{R}^{3 \times 1}$, which is in general a three-dimensional (3D) vector containing the time-varying coordinate of the UAV locations. Let $r(t)$ denote the relevant communication resource allocations, which might be transmit power, bandwidth, channel allocation, beamforming, etc. Then a generic mathematical problem for joint UAV trajectory and communication resource optimization can be formulated as follows [1]:

$$\max_{\{q(t)\},\{r(t)\}} U(\{q(t)\}, \{r(t)\}) \tag{11.1}$$

$$\text{s.t. } f_i(\{q(t)\}) \geq 0, \quad i = 1, \ldots, I_1, \tag{11.2}$$

$$g_i(\{r(t)\}) \geq 0, \quad i = 1, \ldots, I_2, \tag{11.3}$$

$$h_i(\{q(t)\}, \{r(t)\}) \geq 0, \quad i = 1, \ldots, I_3, \tag{11.4}$$

where $U(\cdot, \cdot)$ denotes the utility function to be maximized, which could be the communication throughput, energy efficiency, coverage probability, negative of energy consumption, etc., the $f_i(\cdot)$ represent the UAV mobility constraints, some of which are discussed in the following, the $g_i(\cdot)$ denote the communication resource constraints, such as the maximum

UAV Communications for 5G and Beyond, First Edition.
Edited by Yong Zeng, Ismail Guvenc, Rui Zhang, Giovanni Geraci, and David W. Matolak.
© 2021 John Wiley & Sons Ltd. Published 2021 by John Wiley & Sons Ltd.

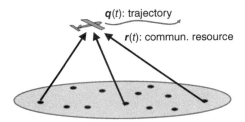

q(t): trajectory

r(t): commun. resource

Figure 11.1 UAV-assisted wireless communication with joint trajectory and communication resource optimization.

transmit power, total available bandwidth, etc., and finally the $h_i(\cdot, \cdot)$ specify the coupled constraints involving both UAV trajectories and communication resource allocation. One typical example of such coupled constraints is the co-channel interference constraint, which limits the transmit power and trajectory of each UAV such that its imposed interference to any other co-channel receiver is below a certain threshold.

Some typical examples for UAV trajectory constraints include the following.

- Minimum/maximum altitude:

$$H_{\min} \leq [q(t)]_3 \leq H_{\max}, \quad \forall\, t, \tag{11.5}$$

where $[q]_3$ denotes the third element for the vector q.
- Initial/final locations: In many scenarios, the UAV's initial and/or final locations for the time horizon of interest $[0, T]$ are predetermined when, e.g., the UAV can only be launched or landed at certain given locations, or its mission specifies the initial and final locations (e.g., for package delivery). Mathematically, we have

$$q(0) = q_I, \qquad q(T) = q_F, \tag{11.6}$$

where $q_I, q_F \in \mathbb{R}^{3\times1}$ are the given initial/final locations.
- Maximum/minimum UAV speed:

$$V_{\min} \leq \| v(t) \| \leq V_{\max}, \quad \forall\, t, \tag{11.7}$$

where $v(t) \triangleq \dot{q}(t)$ denotes the UAV velocity. Note that we usually have $V_{\min} = 0$ for rotary-wing UAVs, whereas $V_{\min} > 0$ for fixed-wing UAVs.
- Maximum acceleration constraint:

$$\| a(t) \| \leq a_{\max}, \quad \forall\, t, \tag{11.8}$$

where $a(t) \triangleq \ddot{q}(t)$ denotes the UAV acceleration. Note that, as shown in [2], for fixed-wing UAVs with banked level turn, the maximum acceleration constraint in Eq. (11.8) implies a constraint on the UAV's maximum turning angle.
- Obstacle avoidance: To ensure that the UAV avoids a given obstacle with known location $r \in \mathbb{R}^{3\times1}$, we could impose the constraint

$$\| q(t) - r \| \geq D_1, \quad \forall\, t, \tag{11.9}$$

where D_1 is the safety distance with the obstacle.
- Collision avoidance: For a multi-UAV system, the collision-avoidance constraint among the UAVs can be expressed as

$$\| q_m(t) - q_j(t) \| \geq \Delta_{\min}, \quad \forall\, m > j, \forall\, t, \tag{11.10}$$

where m and j represent the UAV indices, and $\Delta_{\min}$ is the safety distance to avoid collision.

- No-fly zone: The mathematical constraints of a given no-fly zone depend on its shape. For example, if the no-fly zone is of a ball shape, constraints in the form of Eq. (11.9) can be imposed. On the other hand, if it is a cubic volume, the following constraints need to be satisfied,

$$\bigcup_{i=1}^{6} \boldsymbol{a}_i^{\mathrm{T}} \boldsymbol{q}(t) \geq b_i, \quad \forall\, t, \tag{11.11}$$

where $\{\boldsymbol{a}_i, b_i\}_{i=1}^{6}$ specifies the six hyperplanes corresponding to the faces of the cubic volume, and for two conditions C_1 and C_2, $C_1 \cup C_2$ denotes that either C_1 or C_2 needs to be satisfied.

The optimization problem (11.1)–(11.4) is in general difficult to solve, for two main reasons. First of all, the optimization variables are functions of the continuous time t, and thus essentially involve infinite variables and are difficult to directly optimize. Second, the utility and constraint functions are usually non-concave with respect to $\boldsymbol{q}(t)$ and $\boldsymbol{r}(t)$, which renders the optimization problem non-convex. In the following, we introduce several useful techniques for solving problem (11.1)–(11.4) to obtain effective solutions for UAV trajectory and communication resource allocation.

11.2 Initial Path Planning via the Traveling Salesman and Pickup-and-Delivery Problems

Intuitively, an effective scheme for joint UAV trajectory design and communication resource allocation should fly the UAV sufficiently close to the ground user when the latter is scheduled for communication. This helps enhance the communication link quality due to not only the reduced link distance, but also the increased possibility of establishing line-of-sight (LoS) communication links with the ground users, especially in urban environments. From this perspective, the UAV trajectory optimization problem resembles the classic traveling salesman problem (TSP) [3–6]. The standard TSP is described as follows: A traveler wishes to visit K cities, for which the distances between each pair of cities are known. The objective is to find the traveling route or visiting order such that the total traveling distance is minimized, while each city is visited exactly once and the traveler will return to the original city where the journey was started. TSP is known to be NP-hard, but various efficient algorithms have been proposed to find high-quality solutions [4–6], e.g., via solving binary integer problems. An example of a closed path obtained by solving the TSP is shown in Figure 11.2(a).

Note that the standard TSP described above deals with the scenario that the traveler/UAV has to return to the initial city/location where it starts the tour. However, this may not be the case for UAV communications. For example, after completing the mission, the UAV may not necessarily return to the original location, and, in some scenarios, the initial and/or final locations might be pre-specified [2, 7], e.g., determined by those designated locations for UAV take-off/landing. To cater to such scenarios, we need to apply some variations to the standard TSP solution.

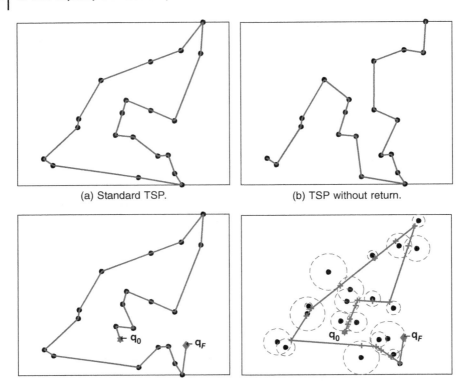

Figure 11.2 The UAV path obtained by solving TSP and its variants. Black dots denote the locations of ground users. (a) Standard TSP (b) TSP without return (c) TSP with given initial and final locations (d) TSPN with given initial and final locations

11.2.1 TSP without Return

In the first variation of TSP, we have the scenario where the traveler does not need to return to the original city after completing the mission. In this case, instead of a closed path, we need to obtain an open path connecting all the cities. One straightforward approach might be by first applying the standard TSP algorithms to the K given cities to obtain a closed path, and then removing the longest edge to get an open path. However, such a naive approach is strictly suboptimal for TSP without return. To obtain the optimal solution, we may apply the simple trick of adding a dummy city [8], as illustrated in Figure 11.3(a). Specifically, we may first add a dummy city by setting its distances to all the existing K cities to 0. Obviously, such a dummy city does not physically exist but it can be numerically represented. Then we solve the standard TSP problem for these $K + 1$ cities to obtain a closed path, and then remove the two edges associated with the dummy city. It can be shown by contradiction that such a solution is optimal for TSP without returning. An example of the UAV initial path obtained by TSP without return is shown in Figure 11.2(b). By comparing it with Figure 11.2(a), it is observed that the visiting order of the ground users is generally different from that in standard TSP, due to the relaxed constraint that the UAV does not have to return to the original location.

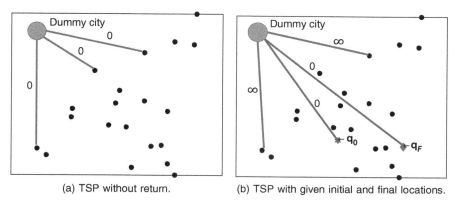

(a) TSP without return.   (b) TSP with given initial and final locations.

Figure 11.3 Solving TSP variations by adding a dummy city. (a) TSP without return (b) TSP with given initial and final locations. Source: From Zeng et al. [8].

11.2.2 TSP with Given Initial and Final Locations

In the second variation of TSP, the initial and final locations of the UAV flying path are predetermined, instead of being free design variables, as in the first variation. Denote by q_0 and q_F the given initial and final locations, respectively. To solve the TSP with given initial and final locations, as shown in Figure 11.3(b), we can similarly add a dummy city [8], with its distance to both q_0 and q_F set to 0, whereas those to all the given K cities are set to sufficiently large values (so as to avoid traveling along the edges from the dummy city to these K cities). By solving the standard TSP problem for these $K + 3$ cities and then removing the two edges associated with the dummy city, we obtain the optimal solution for TSP with pre-specified initial and final locations. An example of the UAV path with given initial and final locations is shown in Figure 11.2(c). Similar techniques can be applied to other variations of TSP, e.g., when only one of the initial or final locations (not both) is predetermined [8].

11.2.3 TSP with Neighborhood

The TSP-based problems discussed above require that the traveler/UAV reaches the exact location of each city/ground user. However, this is feasible only when the UAV's endurance is sufficiently long to visit all ground users. Besides, for UAV communications, it is usually unnecessary for the UAV to reach the location directly above each ground user, e.g., when only very little data needs to be collected from the ground user in UAV-enabled data collection. In this case, we may apply a generalization of the TSP technique, namely, *TSP with neighborhood (TSPN)*. As illustrated in Figure 11.4, with TSPN, each city/ground user is associated with a given neighborhood region, and the traveler only needs to visit each neighborhood to minimize the total traveling distance. TSPN is also NP-hard, while various algorithms have been proposed to obtain approximate solutions for it [8–10]. For the special case of disk-shaped neighborhood region, TSPN can be formulated as

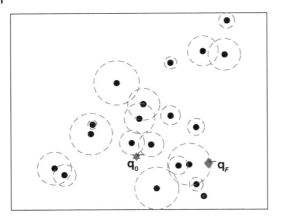

Figure 11.4 Traveling salesman problem with neighborhood (TSPN).

$$\min_{\{\boldsymbol{q}_k\},\{\pi(k)\}} \sum_k \| \boldsymbol{q}_{\pi(k+1)} - \boldsymbol{q}_{\pi(k)} \| \tag{11.12}$$

$$\text{s.t. } [\pi(1),\dots,\pi(K)] \in \mathcal{P}, \tag{11.13}$$

$$\| \boldsymbol{q}_k - \boldsymbol{w}_k \| \le r_k, \quad \forall\, k, \tag{11.14}$$

where $\boldsymbol{w}_k$ is the location of ground user k, $\boldsymbol{q}_k$ gives the waypoint of the UAV path located within the neighborhood of user k, and $\mathcal{P}$ is the set of all the possible permutations for the K users.

It is noted that the main difficulty for solving the optimization problem (11.12)–(11.14) lies in finding the optimal visiting order $\{\pi(k)\}$, for which an exhaustive search method has complexity of order $K!$. In fact, for UAV communications, the resultant problem is even more challenging to solve than TSPN, since the shape and/or size of each neighborhood area may also be a design variable that depends on the communication requirement. One useful method for solving TSPN is to decouple the optimizations of the waypoints $\{\boldsymbol{q}_k\}$ and ordering $\{\pi(k)\}$ [8]. Specifically, we can first solve the TSP based on the locations of the K users to obtain the visiting order, by ignoring the neighborhood regions. Then with the obtained order substituted in Eqs. (11.12)–(11.14), the remaining problem for waypoint optimization is convex, which can be efficiently solved by using standard convex optimization techniques or a software toolbox such as CVX [11]. The above process can be repeated to alternately update the waypoints and the visiting order till convergence. An example of the UAV path obtained by TSPN is shown in Figure 11.2(d).

11.2.4 Pickup-and-Delivery Problem

The TSP/TSPN and its variations discussed above do not impose any preference on the serving order among the users, as long as the corresponding UAV flying distance is minimized. This is only suitable for UAV-enabled uplink or downlink communications. In contrast, for UAV-enabled multipair relaying as shown in Figure 11.5, we have the additional *information-causality constraint* [7, 12], i.e., the UAV can only forward to the destination node those data that has already been received from the corresponding source node. Therefore, the TSP/TSPN techniques discussed above, which do not consider such restrictions, cannot be applied to determine the initial UAV flying path.

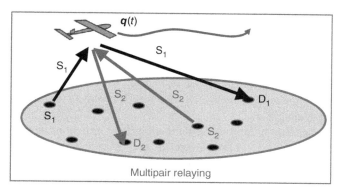

Figure 11.5   UAV-enabled multipair relaying.

Fortunately, we may apply another useful technique by solving the *pickup-and-delivery problem* (PDP). PDP can be regarded as another generalization of TSP, with the additional precedence constraints, i.e., for each pair of source–destination nodes, the UAV needs to first visit the source node before the destination node to meet the above information-causality constraint. PDP is also NP-hard, but various algorithms have been proposed to yield high-quality approximate solutions for it. Furthermore, in the general scenario where the UAV does not have to reach the exact location of each ground user, the more general PDP with neighborhood (PDPN) can be applied to obtain the serving order of the ground users. An example of the UAV path obtained by PDPN is shown in Figure 11.6, as compared to its counterpart TSPN.

Note that, although TSP and PDP are useful approaches to determine the initial UAV flying path or serving order of the ground users, they are in general strictly suboptimal or even infeasible for the generic optimization problem (11.1)–(11.4). On the one hand, the UAV flying trajectory needs to take into account the communication performance more

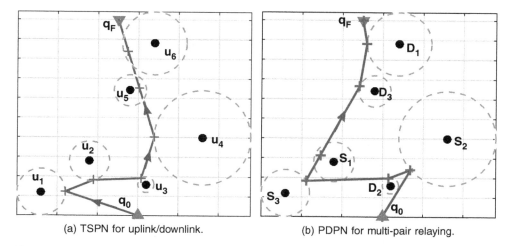

(a) TSPN for uplink/downlink.   (b) PDPN for multi-pair relaying.

Figure 11.6   UAV paths obtained by TSPN and PDPN [1]. (a) TSPN for uplink/downlink (b) PDPN for multipair relaying. Source: From Zeng et al. [1].

explicitly, which also depends on various factors, such as the communication environment, user scheduling design and resource allocation. On the other hand, in practical scenarios where UAVs are subject to various mobility constraints as exemplified in the previous section, the simple TSP and PDP solutions, which ignore such constraints, may lead to infeasible UAV path solutions. For example, the paths shown in Figure 11.2 mainly constitute line segments connecting the different waypoints, which sometimes require sharp turning and are thus infeasible for fixed-wing UAVs.

To tackle these issues, we need to address the joint trajectory and communication resource optimization problem (11.1)–(11.4). However, the optimization problem (11.1)–(11.4) is difficult to solve directly since the optimization variables are functions of the continuous time t and it is generally a non-convex problem with respect to $\boldsymbol{q}(t)$ and $\boldsymbol{r}(t)$. In the following, we first introduce two trajectory discretization techniques to convert the optimization problem (11.1)–(11.4) into more tractable forms with a finite number of optimization variables, and then introduce the block coordinate descent (BCD) and successive convex approximation (SCA) techniques to deal with the non-convexity.

11.3 Trajectory Discretization

The basic idea of trajectory discretization is to approximate the continuous UAV trajectory by a piecewise linear trajectory, which can be represented by a finite number of line segments and the duration that the UAV needs to spend on each line segment. In order to ensure sufficient discretization accuracy, the length of each line segment should not exceed a certain threshold, say $\Delta_{\max}$, whose value could be pre-specified based on practical requirements. For any given $\Delta_{\max}$, there are in general two discretization approaches, namely, *time discretization* [2, 7] and *path discretization* [13], visualized in Figure 11.7 and explained as follows.

11.3.1 Time Discretization

With time discretization, the given time horizon $[0, T]$ is partitioned into N equal time slots with sufficiently small slot length δ_t [2, 7], where $T = N\delta_t$. Let $V_{\max}$ denote the UAV's maximum flying speed. Then it is necessary to ensure that each segment length does not exceed $\Delta_{\max}$ even with the maximum flying speed, for which δ_t should be chosen as $\delta_t \leq \Delta_{\max}/V_{\max}$. Thus, the minimum number of segments required with time discretization is $N = \lceil TV_{\max}/\Delta_{\max} \rceil$. As a result, the continuous UAV trajectory $\boldsymbol{q}(t)$, $0 \leq t \leq T$, can be approximated by the N-length sequence $\{\boldsymbol{q}[n]\}_{n=1}^{N}$, which needs to satisfy the maximum UAV speed and acceleration constraints.

With time discretization, the UAV movement can be approximated by a linear state-space model with respect to the UAV location, velocity, and acceleration, by applying the simple first- and second-order Taylor expansions to the UAV trajectory [2]. As such, the UAV mobility constraints in terms of the UAV locations, velocity, and acceleration can be easily incorporated in the problem formulation (11.1)–(11.4).

While time discretization is very intuitive to understand and simple to apply, it may result in an unnecessarily large number of sampling points when the UAV flies slowly or even hovers for a significant portion of the time. For instance, consider the scenario in which

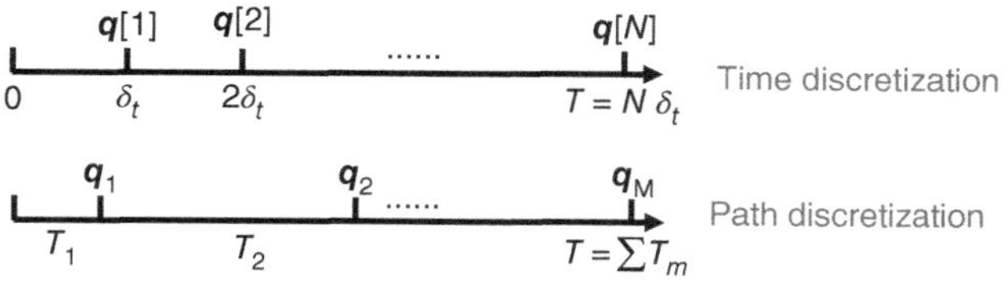

Figure 11.7 Time discretization versus path discretization. Source: From Zeng et al. [1].

the UAV needs to hover at a particular location for 1000 s. If time discretization is used (say, with a time interval of 1 s), then we need 1000 time slots or variables $q[1], \ldots, q[1000]$ (all are equal) to represent this simple status, which is undesirable. Such an issue can be addressed by using another discretization technique, namely, path discretization.

11.3.2 Path Discretization

The main idea of path discretization [13] is to divide the UAV path, instead of time, into M consecutive line segments of generally unequal lengths, as shown in Figure 11.7(b). In this case, the UAV trajectory is represented by a sequence of segments specified by the start/end locations $\{q_m\}$ along the path, together with the time sequence $\{T_m\}$ representing the duration that the UAV spends on each line segment. Path discretization can be interpreted as the more general form of time discretization, with flexibly chosen time slot lengths for different line segments.

Specifically, instead of fixing the slot length to $\delta_t = \Delta_{\max}/V_{\max}$ that is bottlenecked by the maximum flying speed, with path discretization, the time slot length T_m is dynamically determined by the actual flying speed V_m. In this case, we have $T_m V_m \leq \Delta_{\max}$ for all m. Note that, since $V_m \leq V_{\max}$, we have $T_m \geq \delta_t$ for all m. In other words, given the same value for the maximum segment length $\Delta_{\max}$, path discretization entails longer time slot length in general. As a result, given the same trajectory to be discretized with the total operation duration $T = N\delta_t = \sum_{m=1}^{M} T_m$, we have $M \leq N$ in general, i.e., fewer line segments are needed by path discretization than time discretization, especially when the UAV flies with a speed lower than the maximum speed for a significant portion of the operation duration. For the UAV hovering example discussed previously, only three variables are needed to represent the hovering status, namely, q_1 and q_2 (with $q_1 = q_2$) representing the hovering location and $T_1 = 1000$ s representing the hovering duration.

However, it is worth noting that, to represent each sampling segment, time discretization requires only one variable, namely, q_m, while path discretization requires two variables, q_m and T_m. Thus, if $N = M$, then path discretization needs twice as many variables as compared to time discretization. Therefore, the choice of which discretization technique to use depends on the practical scenario.

11.4 Block Coordinate Descent

With the above trajectory discretization techniques, the joint trajectory and communication resource optimization problem (11.1)–(11.4) can be transformed into the following more tractable form with a finite number of variables:

$$\max_{\{\boldsymbol{q}[n]\},\{\boldsymbol{r}[n]\}} U(\{\boldsymbol{q}[n]\}, \{\boldsymbol{r}[n]\}) \tag{11.15}$$

$$\text{s.t. } f_i(\{\boldsymbol{q}[n]\}) \geq 0, \quad i = 1, \dots, I_1, \tag{11.16}$$

$$g_i(\{\boldsymbol{r}[n]\}) \geq 0, \quad i = 1, \dots, I_2, \tag{11.17}$$

$$h_i(\{\boldsymbol{q}[n]\}, \{\boldsymbol{r}[n]\}) \geq 0, \quad i = 1, \dots, I_3, \tag{11.18}$$

where $\{\boldsymbol{q}[n]\}$ and $\{\boldsymbol{r}[n]\}$ denote the discretized UAV trajectory and communication resource allocation variables, respectively.

The above problem involves the joint optimization of UAV trajectory and communication resource allocation, which is usually non-convex and difficult to solve optimally. While there are various ways to tackle non-convex problems, one effective method to obtain a generally locally optimal solution for it is by alternately updating one block of variables with the other block fixed, which is known as the block coordinate descent (BCD) method [14, 15]. The main idea of BCD for solving problem (11.15)–(11.18) is illustrated in Figure 11.8. At the lth iteration, denote the currently obtained UAV trajectory as $\{\boldsymbol{q}^{(l)}[n]\}$. By fixing $\boldsymbol{q}[n]$ in problem (11.15)–(11.18) to $\{\boldsymbol{q}^{(l)}[n]\}$, we solve the subproblem to obtain the optimized resource allocation, which is denoted as $\{\boldsymbol{r}^{(l+1)}[n]\}$. With $\boldsymbol{r}[n]$ in problem (11.15–11.18) fixed to $\{\boldsymbol{r}^{(l+1)}[n]\}$, the UAV trajectory is optimized to obtain $\{\boldsymbol{q}^{(l+1)}[n]\}$. The iteration process continues till convergence.

Note that, for any fixed UAV trajectory, the resultant communication resource optimization problems have been extensively studied for conventional terrestrial communication systems, for which the existing techniques developed can be directly applied. However, for any fixed communication resource allocation, the UAV trajectory optimization problem is relatively new and also usually non-convex due to the generally non-concave objective functions and non-concave constraints with respect to UAV trajectory $\{\boldsymbol{q}[n]\}$. In the following, we discuss an effective technique for solving the non-convex UAV trajectory optimization problems, namely, successive convex approximation.

11.5 Successive Convex Approximation

For any fixed communication resource allocation in problem (11.15–11.18), the resultant UAV trajectory optimization subproblem can be compactly written as

$$\max_{\{\boldsymbol{q}[n]\}} f_0(\{\boldsymbol{q}[n]\}) \tag{11.19}$$

$$\text{s.t. } f_i(\{\boldsymbol{q}[n]\}) \geq 0, \quad i = 1, \dots, I, \tag{11.20}$$

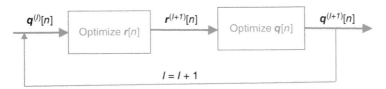

Figure 11.8 Block coordinate descent for UAV trajectory and communication resource optimization.

where $f_0(\cdot)$ represents the utility to be maximized, and the $f_i(\cdot)$ are the corresponding constraints in Eqs. (11.16) and (11.18) involving the UAV trajectory with $I = I_1 + I_3$. The above problem is non-convex if at least one of the functions $f_i(\cdot)$ is non-concave with respect to $\{q[n]\}$, $i = 0, 1, \ldots, I$. This is usually the case, since most utility and constraint functions are non-concave over $\{q[n]\}$, due to which standard convex optimization techniques cannot be directly applied. Fortunately, recent work has shown that SCA is a useful technique for transforming the non-convex optimization problem into a series of convex optimization problems, with guaranteed monotonic convergence to at least a Karush–Kuhn–Tucker (KKT) solution under some mild conditions [16, 17].

SCA is an iterative optimization technique. Specifically, at iteration l, denote the currently obtained UAV trajectory as $\{q^{(l)}[n]\}$, based on which we need to first find a global *concave* lower bound for those non-concave functions $f_i(\{q[n]\})$ in (11.19)–(11.20), such that

$$f_i(\{q[n]\}) \geq f_{i,\mathrm{lb}}^{(l)}(\{q[n]\}), \quad \forall\, q[n]. \tag{11.21}$$

Then by replacing those non-concave functions $f_i(\{q[n]\})$ in (11.19)–(11.20) with their corresponding concave lower bounds $f_{i,\mathrm{lb}}^{(l)}(\{q[n]\})$, we have the following optimization problem:[1]

$$\max_{\{q[n]\}}\ f_{0,\mathrm{lb}}^{(l)}(\{q[n]\}) \tag{11.22}$$

$$\text{s.t. } f_{i,\mathrm{lb}}^{(l)}(\{q[n]\}) \geq 0, \quad i = 1, \ldots, I. \tag{11.23}$$

Since all the functions in (11.22)–(11.23) are concave functions, the above problem is convex, which can be efficiently obtained by standard convex optimization techniques or a readily available software tool such as CVX [11]. Denote the optimal solution of (11.22)–(11.23) as $\{q^{(l)}[n]\}$. Thanks to the global lower bound of (11.21), it is not difficult to see that $\{q^{(l)}[n]\}$ is also feasible for the non-convex problem (11.19)–(11.20), and the corresponding optimal value provides at least a lower bound to that of problem (11.19)–(11.20).

Furthermore, if the lower bound (11.21) is tight at the local point $\{q^{(l-1)}[n]\}$ at the lth iteration, i.e.,

$$f_{i,\mathrm{lb}}^{(l)}(\{q^{(l-1)}[n]\}) = f_i(\{q^{(l-1)}[n]\}), \tag{11.24}$$

then the sequence $f_0(\{q^{(l)}[n]\})$ monotonically increases and converges to a finite limit [17]. With the additional condition that the gradient at the local point is also tight, i.e.,

$$\nabla f_{i,\mathrm{lb}}^{(l)}(\{q^{(l-1)}[n]\}) = \nabla f_i(\{q^{(l-1)}[n]\}), \tag{11.25}$$

then, under some mild constraint qualifications, $\{q^{(l)}[n]\}$ converges to a solution fulfilling the KKT conditions of problem (11.19)–(11.20) [17]. Thus, by iteratively updating the local point $\{q^{(l)}[n]\}$ and solving a sequence of convex optimization problems (11.22)–(11.23), a KKT solution of the non-convex trajectory optimization problem (11.19)–(11.20) can be obtained. The main idea of SCA for trajectory optimization is shown in Figure 11.9.

The remaining task is then to find the global concave lower bounds for those non-concave utility and constraint functions satisfying the above properties. Fortunately, such bounds can be found for the typical utility/constraints functions.

---

1 Note that those concave functions $f_i(\{q[n]\})$ in (11.19)–(11.20) can remain unchanged, but we also denote them as $f_{i,\mathrm{lb}}^{(l)}(\{q[n]\})$ for notational simplicity.

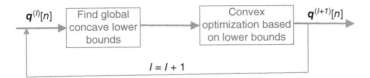

Figure 11.9 Successive convex approximation for trajectory optimization.

For example, for the collision-avoidance constraint in Eq. (11.10), as the left-hand side is a convex function with respect to the UAV trajectory, the resulting constraint (11.10) is non-convex. Fortunately, by using the fact that, for convex differentiable functions, the first-order Taylor approximation provides a global lower bound [18], for any given local UAV trajectory $\{\boldsymbol{q}_m^{(l)}[n]\}$ and $\{\boldsymbol{q}_j^{(l)}[n]\}$ at the lth iteration, the following lower bounds holds globally:

$$\| \boldsymbol{q}_m[n] - \boldsymbol{q}_j[n] \|^2 \geq - \| \boldsymbol{q}_m^{(l)}[n] - \boldsymbol{q}_j^{(l)}[n] \|^2$$
$$+ 2(\boldsymbol{q}_m^{(l)}[n] - \boldsymbol{q}_j^{(l)}[n])^{\mathrm{T}}(\boldsymbol{q}_m[n] - \boldsymbol{q}_j[n]), \quad \forall \boldsymbol{q}_m[n], \boldsymbol{q}_j[n]. \quad (11.26)$$

Note that, since $\{\boldsymbol{q}_m^{(l)}[n]\}$ and $\{\boldsymbol{q}_j^{(l)}[n]\}$ are the given local point, the right-hand side (RHS) of Eq. (11.26) is an affine function with respect to the optimization variables $\boldsymbol{q}_m[n]$ and $\boldsymbol{q}_j[n]$, and hence is concave. Thus, the trajectory-discretized correspondence of the collision-avoidance constraint (11.10) is satisfied as long as the RHS of Eq. (11.26) is no smaller than $\Delta_{\min}^2$, which is now a convex constraint. Similar bounds can be obtained for most other UAV mobility constraints given in Section 11.1.

Besides, for the average communication rate expression

$$R_k[n] = \log_2 \left(1 + \frac{\gamma_k}{\| \boldsymbol{q}[n] - \boldsymbol{w}_k \|^\alpha} \right),$$

where $\boldsymbol{w}_k$ is the location of the user k, α is the path-loss exponent, and γ_k is the received signal-to-noise ratio (SNR) at the reference distance of 1 m, the following concave lower bound can be obtained [1]:

$$\log_2 \left(1 + \frac{\gamma_k}{\| \boldsymbol{q}[n] - \boldsymbol{w}_k \|^\alpha} \right) \geq A_k[n] - B_k[n](\| \boldsymbol{q}[n] - \boldsymbol{w}_k \| - \| \boldsymbol{q}^{(l)}[n] - \boldsymbol{w}_k \|),$$
$$(11.27)$$

where

$$A_k[n] = \log_2 \left(1 + \frac{\gamma_k}{\| \boldsymbol{q}^{(l)}[n] - \boldsymbol{w}_k \|^\alpha} \right), \quad (11.28)$$

$$B_k[n] = \frac{\gamma_k \alpha (\log_2 e)}{\| \boldsymbol{q}^{(l)}[n] - \boldsymbol{w}_k \| (\| \boldsymbol{q}^{(l)}[n] - \boldsymbol{w}_k \|^\alpha + \gamma_k)}. \quad (11.29)$$

Note that the RHS of Eq. (11.27) is given by the negative of the UAV–user distance, which is concave. In fact, a global concave lower bound satisfying Eqs. (11.24) and (11.25) might not be unique. An alternative concave lower bound given in terms of the square of the UAV–user distance has been extensively used in the literature [2, 7, 19, 20]. A graphical illustration of the global concave lower bounds for the communication rate is shown in Figure 11.10.

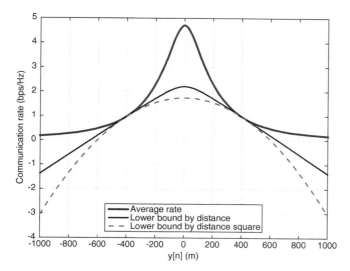

Figure 11.10 An illustration of the global concave lower bounds for communication rate. Source: From Zeng et al. [1].

11.6 Unified Algorithm

Based on the above discussions, a unified algorithm for joint UAV trajectory and communication resource optimization is summarized in Algorithm 1.

Algorithm 1 Joint trajectory and communication resource optimization for solving problem (11.15)–(11.18).

1: **Initialization**: obtain a feasible initial UAV trajectory $\{q^{(l)}[n]\}$.
2: **repeat**
3:   Fix the UAV trajectory to $\{q^{(l)}[n]\}$, optimize the communication resource allocation, and denote the solution as $r^{(l)}[n]$.
4:   Fix the communication resource allocation to $r^{(l)}[n]$.
5:   **repeat**
6:     With the current local trajectory $\{q^{(l)}[n]\}$, obtain the global concave lower bound for those non-concave functions in problem (11.19)–(11.20) and formulate the convex problem (11.22)–(11.23).
7:     Solve the convex problem (11.22)–(11.23), and denote the optimal solution as $\{q^{\star}[n]\}$.
8:     Update the local trajectory $q^{(l)}[n] \leftarrow q^{\star}[n]$.
9:   **until** convergence or reaches a prescribed number of inner iterations.
10: **until** convergence or reaches a prescribed number of outer iterations.

Note that finding a feasible initial UAV trajectory for Algorithm 1 is non-trivial due to the various UAV mobility constraints, as discussed in Section 11.1. One effective approach is that we first ignore the specific UAV mobility constraints, so as to determine the serving orders of the ground users by applying the TSP and PDP techniques as discussed in Section 11.2, and then with the fixed serving order, obtain a feasible path satisfying the mobility constraint by applying the global lower bounds as discussed above. It is also noted that Algorithm 1 constitutes both outer and inner iterations, with the outer iteration corresponding to BCD and inner iteration for SCA. In general, the inner iteration does not have to be continued till convergence, since the performance will be further optimized in the outer iterations.

11.7 Summary

In this chapter, we introduce the main techniques for joint trajectory design and communication resource optimization in UAV-assisted wireless communications. For multiuser UAV communication systems, the classic TSP/PDP and their variations are found to be effective techniques to determine the initial UAV flying path and serving order for ground users. To convert the joint optimization problem into a more tractable form with a finite number of optimization variables, time- and path-discretization techniques are introduced. To deal with the non-convexity of the associated problems, BCD can be applied to alternately update the communication resource allocation and UAV trajectory. In particular, for the non-convex trajectory optimization subproblem, SCA is found to be effective to obtain a KKT suboptimal solution in general. Note that, as the BCD- and SCA-based optimization requires iterative procedures, a feasible initial UAV trajectory needs to be specified. The TSP/PDP-based path planning offers an efficient means to obtain the initial UAV trajectory.

Note that one limitation of the BCD technique for alternately updating UAV trajectory and communication resource allocation is the possibility of trapping into undesirable local optima, if the initialization is not properly designed. Therefore, there have been recent efforts on investigating the simultaneous update of these two blocks of variables for certain setups via developing new concave lower bound functions [13, 21]. Finding efficient algorithms for joint UAV trajectory and communication resource allocation under various setups and more sophisticated channel models deserves further investigations.

References

1 Y. Zeng, Q. Wu, and R. Zhang, "Accessing from the sky: a tutorial on UAV communications for 5G and beyond," *Proc. of the IEEE*, vol. 107, no. 12, pp. 2327–2375, Dec. 2019.

2 Y. Zeng and R. Zhang, "Energy-efficient UAV communication with trajectory optimization," *IEEE Trans. Wireless Commun.*, vol. 16, no. 6, pp. 3747–3760, Jun. 2017.

3 E. L. Lawler, J. K. Lenstra, A. H. G. R. Kan, and D. B. Shmoys, *The Traveling Salesman Problem: A Guided Tour of Combinatorial Optimization*, 1st ed. Wiley, 1985.

4 G. Laporte, "The traveling salesman problem: an overview of exact and approximate algorithms," *EUR. J. Oper. Res.*, vol. 59, no. 2, pp. 231–247, Jun. 1992.

5 "Travelling Salesman Problem: Solver-Based". [Online]. Available https://www.mathworks.com/help/optim/ug/travelling-salesman-problem.html, accessed on May 18, 2019.

6 C. Rego, D. Gamboa, F. Glover, and C. Osterman, "Traveling salesman problem heuristics: leading methods, implementations and latest advances," *European Journal of Operational Research*, vol. 211, no. 3, pp. 427–441, 2011.

7 Y. Zeng, R. Zhang, and T. J. Lim, "Throughput maximization for UAV-enabled mobile relaying systems," *IEEE Trans. Commun.*, vol. 64, no. 12, pp. 4983–4996, Dec. 2016.

8 Y. Zeng, X. Xu, and R. Zhang, "Trajectory design for completion time minimization in UAV-enabled multicasting," *IEEE Trans. Wireless Commun.*, vol. 17, no. 4, pp. 2233–2246, Apr. 2018.

9 A. Dumitrscu and J. Mitchell, "Approximation algorithms for TSP with neighborhoods in the plane," *J. Algorithms*, vol. 48, no. 1, pp. 135–159, 2003.

10 B. Yuan, M. Orlowska, and S. Sadiq, "On the optimal robot routing problem in wireless sensor networks," *IEEE Trans. Knowledge and Data Eng.*, vol. 19, no. 9, pp. 1252–1261, Sep. 2007.

11 M. Grant and S. Boyd, CVX: Matlab software for disciplined convex programming, version 2.1, available online at http://cvxr.com/cvx.

12 J. Zhang, Y. Zeng, and R. Zhang, "UAV-enabled radio access network: multi-mode communication and trajectory design," *IEEE Trans. Signal Process.*, vol. 66, no. 20, pp. 5269–5284, Oct. 2018.

13 Y. Zeng, J. Xu, and R. Zhang, "Energy minimization for wireless communication with rotary-wing UAV," *IEEE Trans. Wireless Commun.*, vol. 18, no. 4, pp. 2329–2345, Apr. 2019.

14 M. Hong, M. Razaviyayn, Z.-Q. Luo, and J.-S. Pang, "A unified algorithmic framework for block-structured optimization involving big data: with applications in machine learning and signal processing," *IEEE Signal Process. Mag.*, vol. 33, no. 1, pp. 57–77, Jan. 2016.

15 Y. Xu and W. Yin, "A block coordinate descent method for regularized multiconvex optimization with applications to nonnegative tensor factorization and completion," *SIAM J. imaging Sci.*, vol. 6, no. 3, pp. 1758–1789, Sep. 2013.

16 B. R. Marks and G. P. Wright, "A general inner approximation algorithm for non-convex mathematical programs," *Oper. Res.*, vol. 26, no. 4, pp. 681–683, 1978.

17 A. Zappone, E. Bjornson, L. Sanguinetti, and E. Jorswieck, "Globally optimal energy-efficient power control and receiver design in wireless networks," *IEEE Trans. Signal Process.*, vol. 65, no. 11, pp. 2844–2859, Jun. 2017.

18 S. Boyd and L. Vandenberghe, *Convex Optimization.* Cambridge, U.K.: Cambridge Univ. Press, 2004.

19 C. Zhan, Y. Zeng, and R. Zhang, "Energy-efficient data collection in UAV enabled wireless sensor network," *IEEE Wireless Commun. Lett.*, vol. 7, no. 3, pp. 328–331, Jun. 2018.

20 Q. Wu, Y. Zeng, and R. Zhang, "Joint trajectory and communication design for multi-UAV enabled wireless networks," *IEEE Trans. Wireless Commun.*, vol. 17, no. 3, pp. 2109–2121, Mar. 2018.

21 C. Shen, T.-H. chang, J. Gong, Y. Zeng, and R. Zhang, "Multi-UAV interference coordination via joint trajectory and power control," *IEEE Trans. Signal Process.*, vol. 68, pp. 843–858, Jan. 2020.

12

Energy-Efficient UAV Communications

Yong Zeng[1,2] and Rui Zhang[3]

[1]*National Mobile Communications Research Laboratory, Southeast University, China*
[2]*Purple Mountain Laboratories, Jiangsu, China*
[3]*Department of Electrical and Computer Engineering, National University of Singapore, Singapore*

12.1   UAV Energy Consumption Model

The limited on-board energy of unmanned aerial vehicles (UAVs) brings a critical challenge to practically realize UAV-assisted wireless communications. While battery technology has been continuously advanced to facilitate the manufacturing of UAVs with much longer endurance than ever before, from the operation perspective, UAVs could also expend their energy more efficiently to improve the communication performance with whatever given energy budget. This leads to an important line of research for UAV systems, namely, *energy-efficient UAV communications*. Note that, besides the conventional communication-related energy consumption arising from signal processing, communication circuits, and signal radiation, UAVs are subject to the additional propulsion energy consumption for them to remain aloft and move freely. More importantly, the propulsion energy consumption of UAVs is usually much more significant than their communication-related energy. Thus, energy-efficient communication for UAV systems is significantly different from that in conventional terrestrial systems, which mainly concern communication-related energy consumption. This chapter aims to introduce the fundamental concepts and main design techniques for energy-efficient UAV communications.

The proper mathematical modeling of UAV energy consumption is the prerequisite for the performance evaluation and optimization of energy-efficient UAV communications. The energy consumption of UAVs mainly consists of two components: the communication-related energy and the propulsion energy. For the former, as there is no fundamental difference in transceivers between UAV and conventional terrestrial communication systems, the existing models developed from the extensively studied terrestrial systems are directly applicable for UAV communications. In contrast, the UAV propulsion energy consumption is unique for UAV communications, and its mathematical modeling has received little attention in the past.

Some empirical or heuristic energy consumption models have been used in early works on UAV systems. For example, in [1], experimental measurements were conducted to study the energy consumption of a specific quadrotor UAV at different speeds. However,

UAV Communications for 5G and Beyond, First Edition.
Edited by Yong Zeng, Ismail Guvenc, Rui Zhang, Giovanni Geraci, and David W. Matolak.
© 2021 John Wiley & Sons Ltd. Published 2021 by John Wiley & Sons Ltd.

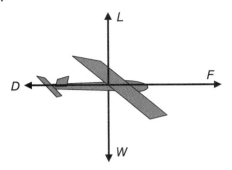

Figure 12.1 A schematic of the forces on a fixed-wing aircraft in straight and level flight. Source: From Zeng and Zhang [6].

no mathematical model on UAV energy consumption was suggested in [1], which makes the result difficult to generalize for other UAVs. In [2] and [3], the UAV energy (fuel) cost was modeled as the L_1 norm of the control force or acceleration vector, whereas in [4], it was modeled to be proportional to the square of the UAV speed. However, no rigorous mathematical derivation was provided for such heuristic models. In fact, although the power consumption of mobile robots moving on the ground can be modeled as a polynomial and monotonically increasing function with respect to its moving speed [5], such results are not applicable for UAVs due to their fundamentally different moving mechanisms from ground robots. Rigorous mathematical derivations were performed in [6] and [7] to obtain the theoretical closed-form propulsion energy consumption models for fixed-wing and rotary-wing UAVs, respectively.

12.1.1 Fixed-Wing Energy Model

12.1.1.1 Forces on a UAV

Figure 12.1 gives the schematic of the forces exerted on a fixed-wing aircraft, which includes four components: weight, drag, lift, and thrust [8].

- *Weight* (W): the force of gravity, $W = mg$, with m denoting the aircraft's mass including all its payload, and g being the gravitational acceleration in m s^{-2}.
- *Drag* (D): the aerodynamic force component parallel to the airflow direction. For zero wind speed, D is in the opposite direction to the aircraft's motion. Drag is essentially the air resistance force to the motion of the aircraft.
- *Lift* (L): the aerodynamic force component normal to the drag and pointing upward.
- *Thrust* (F): the force produced by the aircraft engine, which overcomes the drag to move the aircraft forward.

For a fixed-wing aircraft moving at a subsonic speed V, the drag D can be expressed in a simple form as [8]

$$D = c_1 V^2 + \frac{c_2 \kappa^2}{V^2}, \tag{12.1}$$

where c_1 and c_2 are two parameters related to the aircraft's weight, wing area, air density, wingspan efficiency, and aspect ratio (i.e., the ratio of the wingspan to its aerodynamic breadth), etc. [6], and $\kappa \triangleq L/W$ is known as the *load factor*, which is the ratio of the aircraft's lift to its weight. The first term in Eq. (12.1) is known as the *parasitic drag*, which is a combination of multiple drag components, such as form drag, skin friction drag, and interference drag, etc. The parasitic drag increases quadratically with the UAV speed V. The second term in Eq. (12.1) is called the *lift-induced drag*, which is the drag force due to wings redirecting air to generate lift L.

12.1.1.2 Straight and Level Flight

Straight and level flight refers to a flight that maintains a constant heading (or direction) and altitude. This implies that: (i) the horizontal acceleration, if any, must be in parallel with the aircraft's flying direction, so that no turning occurs; and (ii) the lift and weight are balanced, so that there is no vertical acceleration. Furthermore, if a constant level speed V is assumed, the net force on the UAV is then equal to zero. Based on Figure 12.1, we have

$$L = W, \qquad F = D. \tag{12.2}$$

Then by combining Eqs. (12.1) and (12.2), and using the equation for κ, the required power from the aircraft engine as a function of speed V can be obtained as

$$P(V) = FV = DV \tag{12.3}$$

$$= \underbrace{c_1 V^3}_{\text{parasitic}} + \underbrace{\frac{c_2}{V}}_{\text{induced}}, \tag{12.4}$$

where we have used the fact that the required power from the aircraft's engine is equal to its generated thrust multiplied by the aircraft speed.

It is observed from Eq. (12.4) that the propulsion power consumption of a fixed-wing UAV contains two components: the *parasitic power* that is needed to overcome the parasitic drag caused by the motion of the aircraft in the air, and the *induced power* needed to overcome the induced drag resulting from the lift force to maintain the aircraft airborne. While the parasitic power increases with the cube of the aircraft speed V, the induced power decreases inversely proportionally with V. A typical plot of $P(V)$ versus speed V is given in Figure 12.2(a). It is observed that, for $V = 0$, we have $P(0) \to \infty$, which reflects the well-known fact that fixed-wing UAVs must maintain a minimum forward speed to remain airborne.

Two particular UAV speeds that are of great practical interest are the *maximum-endurance (ME) speed* and the *maximum-range (MR) speed*, which are denoted as V_{me} and V_{mr}, respectively.

- *ME speed*: By definition, the ME speed V_{me} is the optimal UAV speed that maximizes the UAV endurance for any given on-board energy, which is

$$V_{\text{me}} = \arg \min_{V \geq 0} P(V). \tag{12.5}$$

For a fixed-wing UAV, as $P(V)$ in Eq. (12.4) is a convex function, by setting its first-order derivative to zero, it can be obtained that the ME speed is given by $V_{\text{me}} = [c_2/(3c_1)]^{1/4}$, and the corresponding minimum power consumption is $P(V_{\text{me}}) = 1.75 c_1^{1/4} c_2^{3/4}$.

- *MR speed*: On the other hand, the MR speed V_{mr} is the optimal UAV speed that maximizes the total traveling distance with any given on-board energy. Mathematically, it is given by

$$V_{\text{mr}} = \arg \min_{V \geq 0} E_0(V) \triangleq \frac{P(V)}{V}. \tag{12.6}$$

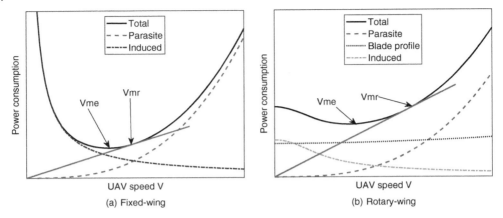

Figure 12.2 Typical plot for UAV propulsion power consumption versus speed. Source: From Zeng et al. [9].

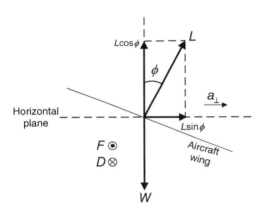

Figure 12.3 A schematic of the forces on an aircraft in banked level turn. The aircraft moves normal to the page. The horizontal component of the lift causes the centrifugal acceleration to change the heading. Source: From Zeng and Zhang [6].

Note that the function $E_0(V)$ has the unit joules per meter (J m^{-1}), which signifies the UAV energy consumption per unit traveling distance. For fixed-wing UAVs, it can be obtained based on Eq. (12.4) that $V_{mr} = (c_2/c_1)^{1/4} = 1.32 V_{me}$, and the corresponding optimal energy consumption per unit distance is $E_0(V_{mr}) = 2\sqrt{c_1 c_2}$. Note that we have $V_{mr} > V_{me}$ in general.

12.1.1.3 Circular Flight

For a fixed-wing aircraft to change its heading, it must roll to a banked position so that the lift L generates a lateral component to support the centrifugal acceleration, i.e., the acceleration component normal to the flying direction. This is illustrated in Figure 12.3, with ϕ denoting the bank angle, which is the angle between the vertical plane and the aircraft's symmetry plane. The forces exerted on the aircraft are also shown in the figure. For a UAV following a circular path on a certain horizontal plane with radius r and constant speed V, the centrifugal acceleration is

$$a_\perp = \frac{V^2}{r}, \tag{12.7}$$

which is supported by the horizontal component of the lift L. Thus, we should have

$$L \sin \phi = m a_\perp = \frac{m V^2}{r}. \tag{12.8}$$

Furthermore, to maintain a constant altitude without descending or ascending, we should have

$$L \cos \phi = W. \tag{12.9}$$

It then follows from Eqs. (12.8) and (12.9) that the loading factor κ is given by

$$\kappa = \sqrt{1 + \frac{V^4}{g^2 r^2}}. \tag{12.10}$$

Moreover, a constant speed V implies zero tangential acceleration. Thus, we should have

$$F = D \tag{12.11}$$

$$= \left(c_1 + \frac{c_2}{g^2 r^2} \right) V^2 + \frac{c_2}{V^2}, \tag{12.12}$$

where the second equality follows from Eqs. (12.1) and (12.10). As a result, the required propulsion power for circular flight, which is a function of radius r and speed V, can be obtained as

$$\overline{P}_{\text{cir}}(V, r) = FV = \left(c_1 + \frac{c_2}{g^2 r^2} \right) V^3 + \frac{c_2}{V}. \tag{12.13}$$

It is observed that, as $r \to \infty$, the expression (12.13) reduces to that for straight flight in Eq. (12.4).

12.1.1.4 Arbitrary Level Flight

For a fixed-wing aircraft with arbitrary level flight with constant altitude, denote its 2D trajectory by $\boldsymbol{q}(t) \in \mathbb{R}^{2 \times 1}$, $0 \leq t \leq T$. Then $\boldsymbol{v}(t) \triangleq \dot{\boldsymbol{q}}(t)$ and $\boldsymbol{a}(t) \triangleq \ddot{\boldsymbol{q}}(t)$ are the instantaneous velocity and acceleration, respectively. It is derived in [6] that the propulsion energy consumption can be modeled as

$$\bar{E}(\boldsymbol{q}(t)) = \underbrace{\int_0^T \left[c_1 \| \boldsymbol{v}(t) \|^3 + \frac{c_2}{\| \boldsymbol{v}(t) \|} \left(1 + \frac{\| \boldsymbol{a}(t) \|^2 - (\boldsymbol{a}^{\mathrm{T}}(t) \boldsymbol{v}(t))^2 / \| \boldsymbol{v}(t) \|^2}{g^2} \right) \right] \mathrm{d}t}_{\text{work to overcome air resistance}}$$

$$\underbrace{+ \frac{1}{2} m (\| \boldsymbol{v}(T) \|^2 - \| \boldsymbol{v}(0) \|^2)}_{\Delta_K, \text{ change in kinetic energy}}. \tag{12.14}$$

The above expression shows that, for level flight with fixed altitude, the UAV's energy consumption depends only on its speed $\| \boldsymbol{v}(t) \|$ and centrifugal acceleration

$$a_\perp(t) \triangleq \sqrt{\| \boldsymbol{a}(t) \|^2 - \frac{(\boldsymbol{a}^{\mathrm{T}}(t) \boldsymbol{v}(t))^2}{\| \boldsymbol{v}(t) \|^2}}$$

(i.e., the acceleration component that is normal to the UAV velocity vector and accounts for heading change yet without altering the UAV speed; see Figure 12.4), rather than depending

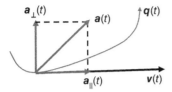

Figure 12.4 Decomposition of the acceleration vector into tangential and centrifugal components.

on its actual location $\mathbf{q}(t)$ or the tangential acceleration $a_{\parallel}(t)$ (i.e., the acceleration component parallel to the UAV's velocity vector). Such a result can be interpreted based on the well-known work–energy principle. The integral term in Eq. (12.14), which is guaranteed to be positive, is the work required from the aircraft's engine to overcome the air resistance force. The second term in Eq. (12.14), denoted as Δ_K, represents the change in the UAV's kinetic energy, which is an aggregated effect of the UAV's tangential acceleration component over time. Thus, Δ_K depends only on the initial and final speeds $\| \mathbf{v}(T) \|$ and $\| \mathbf{v}(0) \|$, rather than the intermediate UAV state. It can be verified that, for the special case of straight or circular flight, Eq. (12.14) reduces to Eq. (12.4) and Eq. (12.13), respectively.

The energy expression (12.14) is usually difficult to optimize directly due to the complicated expression for $a_{\perp}(t)$. One useful upper bound can be obtained by replacing $|a_{\perp}(t)|^2$ with $\| \mathbf{a}(t) \|^2$, which yields [6]

$$\bar{E}(\mathbf{q}(t)) \leq \bar{E}_{\mathrm{ub}}(\mathbf{q}(t)) \triangleq \int_0^T \left[c_1 \| \mathbf{v}(t) \|^3 + \frac{c_2}{\| \mathbf{v}(t) \|} \left(1 + \frac{\| \mathbf{a}(t) \|^2}{g^2} \right) \right] dt + \Delta_K. \quad (12.15)$$

Note that the above upper bound in Eq. (12.15) is tight for constant-speed flight, in which case $\mathbf{a}^{\mathrm{T}}(t)\mathbf{v}(t) = 0$ for all t.

12.1.1.5 Arbitrary 3D Flight

The derivation of the energy consumption model for a fixed-wing UAV following an arbitrary 3D trajectory, i.e., with ascending and descending, is quite challenging. While there are rich results in the aircraft design literature [8], to the best of our knowledge, no closed-form expression in terms of the UAV's 3D flying velocity and acceleration has been reported. On the other hand, as the UAV ascending/descending mainly introduces a change in its potential energy, one heuristic model might be

$$E_{\mathrm{3D}}(\mathbf{q}(t)) = E(\mathbf{q}(t)) + mg([\mathbf{q}(T)]_3 - [\mathbf{q}(0)]_3), \quad (12.16)$$

where $\mathbf{q}(t) \in \mathbb{R}^{3 \times 1}$ represents the 3D trajectory, with $[\mathbf{q}(t)]_3$ denoting its height at time t. Note that the first term in Eq. (12.16) is obtained by directly applying the energy consumption model in Eq. (12.14) to the 3D trajectory, and the second term is the change of the UAV's potential energy. However, proper care should be taken while using Eq. (12.16), since it ignores the effect of the additional lift and drag when the UAV moves vertically. More research endeavour is thus needed to rigorously derive the propulsion energy consumption of a fixed-wing UAV with arbitrary 3D trajectory.

12.1.2 Rotary-Wing Energy Model

Due to the fundamentally different flying mechanisms, the above energy consumption model for fixed-wing UAVs is no longer applicable for rotary-wing UAVs. In [7], it is derived that, for a rotary-wing UAV in straight and level flight with speed V, the power consumption can be expressed as

$$P(V) = P_0\left(1 + \frac{3V^2}{U_{\text{tip}}^2}\right) + P_i\left(\sqrt{1 + \frac{V^4}{4v_0^4}} - \frac{V^2}{2v_0^2}\right)^{1/2} + \frac{1}{2}d_0\rho sAV^3, \tag{12.17}$$

$$\underbrace{\phantom{P_0\left(1 + \frac{3V^2}{U_{\text{tip}}^2}\right)}}_{\text{blade profile}} \quad \underbrace{\phantom{P_i\left(\sqrt{1 + \frac{V^4}{4v_0^4}} - \frac{V^2}{2v_0^2}\right)^{1/2}}}_{\text{induced}} \quad \underbrace{\phantom{\frac{1}{2}d_0\rho sAV^3}}_{\text{parasitic}}$$

where P_0 and P_i, respectively, represent the *blade profile power* and *induced power* in hovering status, which depend on the aircraft's weight, air density ρ, rotor disk area A, etc., U_{tip} denotes the tip speed of the rotor blade, v_0 is known as the mean rotor induced velocity in hovering, and d_0 and s are the fuselage drag ratio and rotor solidity, respectively.

It is observed from Eq. (12.17) that similar to fixed-wing UAV, the propulsion power consumption of rotary-wing UAVs contains both the parasite power and the induced power, which increases and decreases with the speed V, respectively. Moreover, $P(V)$ in Eq. (12.17) contains a new component, namely the blade profile power, which is required to overcome the profile drag caused by blade rotation and increases quadratically with V. Another difference from the fixed-wing UAV model is that, at $V = 0$, the power consumption in Eq. (12.17) is a finite value, which corroborates the well-known fact that rotary-wing UAVs have the hovering capability. As V increases, it can be verified that $P(V)$ in Eq. (12.17) first decreases and then increases with V, i.e., hovering is in general not the most power-conserving status. It can be verified that the power function $P(V)$ in Eq. (12.17) is neither convex nor concave with respect to V, which is much more involved than that for fixed-wing UAVs in Eq. (12.4).

A typical plot of $P(V)$ versus UAV speed V is shown in Figure 12.2(b). For scenarios when $V \gg v_0$, by applying the Taylor approximation $(1+x)^{1/2} \approx 1 + \frac{1}{2}x$ for $|x| \ll 1$, Eq. (12.17) can be approximated by a convex function as

$$P(V) \approx P_0\left(1 + \frac{3V^2}{U_{\text{tip}}^2}\right) + \frac{P_iv_0}{V} + \frac{1}{2}d_0\rho sAV^3. \tag{12.18}$$

Similar to a fixed-wing UAV, the ME and MR speeds for rotary-wing UAV can be obtained based on Eq. (12.17). Though closed-form expressions for V_{me} and V_{mr} are difficult to obtain due to the complicated expression in Eq. (12.17), their numerical values can be easily obtained. Alternatively, V_{me} and V_{mr} can also be obtained graphically based on the power–speed curve $P(V)$. Specifically, while V_{me} corresponds to the minimum value of $P(V)$, V_{mr} can be obtained by drawing a tangential line from the origin to the power curve that corresponds to the minimum slope, as illustrated in Figure 12.2(b).

For a rotary-wing aircraft with an arbitrary level flight following a 2D trajectory $\boldsymbol{q}(t) \in \mathbb{R}^{2\times1}$, a similar acceleration decomposition technique as that for fixed-wing UAVs can be applied to obtain the general energy consumption model, which yields

$$E(\boldsymbol{q}(t)) = \underbrace{\int_0^T P_i\sqrt{1 + \frac{a_\perp^2(t)}{g^2}}\left(\sqrt{1 + \frac{a_\perp^2(t)}{g^2} + \frac{\|\boldsymbol{v}(t)\|^4}{4v_0^4}} - \frac{\|\boldsymbol{v}(t)\|^2}{2v_0^2}\right)^{1/2} dt}_{\text{induced}}$$

$$+ \underbrace{\int_0^T P_0\left(1 + \frac{3\|\boldsymbol{v}(t)\|^2}{U_{\text{tip}}^2}\right)dt}_{\text{blade profile}} + \underbrace{\int_0^T \frac{1}{2}d_0\rho sA\|\boldsymbol{v}(t)\|^3\,dt}_{\text{parasitic}} + \Delta_K, \tag{12.19}$$

where

$$a_\perp^2(t) = \|\boldsymbol{a}(t)\|^2 - \frac{(\boldsymbol{a}^\mathrm{T}(t)\boldsymbol{v}(t))^2}{\|\boldsymbol{v}(t)\|^2}$$

is the square of the centrifugal acceleration, and

$$\Delta_K \triangleq \frac{1}{2}m(\|\boldsymbol{v}(T)\|^2 - \|\boldsymbol{v}(0)\|^2)$$

is the change in kinetic energy. For an arbitrary 3D trajectory, a similar heuristic model as Eq. (12.16) can be obtained as well.

With the established mathematical models for UAV energy consumption, there are various ways to formulate the optimization problem to achieve energy-efficient UAV communications. Typical examples include maximizing the energy efficiency for a given finite time horizon, minimizing the UAV energy consumption subject to the communication requirement, and maximizing the communication performance with the given energy budget. In the following, we discuss the first two problems mentioned above, respectively.

12.2 Energy Efficiency Maximization

Energy efficiency is one of the most important performance metrics for wireless communication systems. It is measured by the number of information bits that can be successfully communicated between the transmitter and receiver per unit energy consumption. For UAV communication systems, the new consideration of the UAV's propulsion energy consumption renders the study on energy-efficient UAV communication significantly different from that in conventional terrestrial communication systems. In particular, as both UAV energy consumption and communication rate depend on the UAV's trajectory $\boldsymbol{q}(t)$ and communication resource allocation $\boldsymbol{r}(t)$, a generic expression for UAV energy efficiency can be written as

$$\mathrm{EE}(\boldsymbol{q}(t), \boldsymbol{r}(t)) = \frac{\overline{R}(\boldsymbol{q}(t), \boldsymbol{r}(t))}{E(\boldsymbol{q}(t)) + E_{\mathrm{com}}(\boldsymbol{r}(t))}. \tag{12.20}$$

Here the numerator is the aggregated communication throughput achieved by the UAV over a certain time horizon T, and the denominator is the UAV's total energy consumption, which includes both its propulsion energy consumption, which depends on its trajectory $\boldsymbol{q}(t)$ as discussed in Section 12.1, and communication-related energy consumption, which in general depends on the communication resource allocation $\boldsymbol{r}(t)$.

Then a generic problem formulation for energy efficiency maximization can be written as

$$\max_{\{\boldsymbol{q}(t)\},\{\boldsymbol{r}(t)\}} \mathrm{EE}(\boldsymbol{q}(t), \boldsymbol{r}(t)) \tag{12.21}$$

$$\text{s.t. } f_i(\{\boldsymbol{q}(t)\}) \geq 0, \quad i = 1, \dots, I_1, \tag{12.22}$$

$$g_i(\{\boldsymbol{r}(t)\}) \geq 0, \quad i = 1, \dots, I_2, \tag{12.23}$$

$$h_i(\{\boldsymbol{q}(t)\}, \{\boldsymbol{r}(t)\}) \geq 0, \quad i = 1, \dots, I_3, \tag{12.24}$$

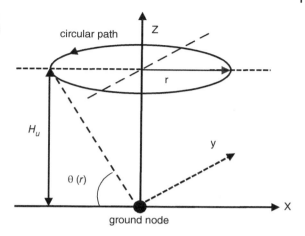

Figure 12.5 Wireless communication enabled by a fixed-wing UAV following a circular path with radius r.

where the $f_i(\cdot)$ include all constraints related to UAV mobility, the $g_i(\cdot)$ denote the communication resource constraints, and the $h_i(\cdot, \cdot)$ specify the coupled constraints involving both UAV trajectory and communication variables.

It is noted that the above energy efficiency maximization problem corresponds to one instance of the generic problem formulation (11.1)–(11.4) for joint UAV trajectory and communication resource optimization. Therefore, the techniques discussed in Chapter 11, such as TSP/PDP-based path initialization (TSP = traveling salesman problem; PDP = pick-up-and-delivery problem), block coordinate descent (BCD), and successive convex approximation (SCA), can also be applied for solving problem (12.21)–(12.24), except that now the objective function is more complicated due to the involved energy consumption models, as discussed in Section 12.1.

To gain useful insights for energy-efficient UAV communications, we consider the simple scenario where a fixed-wing UAV communicates with a single ground node. For simplicity, assume that the UAV follows a circular trajectory at a constant altitude H_U, with radius r and speed V, and the projection of the circle center on the ground coincides with the ground node, as shown in Figure 12.5. Then the elevation angle between the UAV and the ground node is a function of r given by $\theta(r) = \tan^{-1}(H_U/r)$. We consider the challenging urban environment where the line-of-sight (LoS) link may be occasionally blocked by buildings. With the elevation-angle-dependent probabilistic LoS channel model [9] and extending the result presented in [6] based on Jensen's inequality, the expected communication rate between the UAV and the ground node can be approximated as

$$\overline{R}(r) = \log_2\left(1 + \frac{\hat{P}_{\mathrm{LoS}}(r)\gamma_0}{(H_U^2 + r^2)^{\alpha/2}}\right). \tag{12.25}$$

Here $\gamma_0 = P_t\beta_0/\sigma^2$ is the received signal-to-noise ratio (SNR) at the reference distance of 1 m, with P_t denoting the constant transmit power, α is the path-loss exponent, and $\hat{P}_{\mathrm{LoS}}(r)$ is a decreasing function of r, which can be interpreted as the regularized LoS probability that takes into account both the LoS probability and the non-line-of-sight (NLoS) probability with the additional signal attenuation.

Then with the energy model in Eq. (12.13), the energy efficiency for a circular trajectory can be written as

$$
EE(r, V) = \frac{\overline{R}(r)}{\overline{P}_{cir}(V, r) + P_{com}} = \frac{\log_2 \left(1 + \dfrac{\hat{P}_{LoS}(r)\gamma_0}{(H_U^2 + r^2)^{\alpha/2}} \right)}{\left(c_1 + \dfrac{c_2}{g^2 r^2} \right) V^3 + \dfrac{c_2}{V} + P_{com}},
$$
(12.26)

where P_{com} is the communication-related power consumption of the UAV, which is assumed to be a constant in the considered scenario. Note that, since the numerator of Eq. (12.26) is independent of the UAV speed V, for any given radius r, the optimal speed V should minimize the energy consumption in the denominator.

By setting the first-order derivative equal to zero, the optimal UAV speed $V_{cir}^\star(r)$ for any given r can be readily obtained as

$$
V_{cir}^\star(r) = \left(\frac{c_2}{3(c_1 + c_2/(g^2 r^2))} \right)^{1/4}.
$$
(12.27)

The corresponding minimum UAV propulsion power consumption reduces to a univariate function of r as

$$
P_{cir}^\star(r) = A \left(c_1 + \frac{c_2}{g^2 r^2} \right)^{1/4},
$$
(12.28)

where $A \triangleq (3^{-3/4} + 3^{1/4})c_2^{3/4}$. As a result, the UAV's energy efficiency reduces to a univariate function with respect to radius r, i.e.,

$$
EE(r) = \frac{\log_2 \left(1 + \dfrac{\hat{P}_{LoS}(r)\gamma_0}{(H_U^2 + r^2)^{\alpha/2}} \right)}{A \left(c_1 + \dfrac{c_2}{g^2 r^2} \right)^{1/4} + P_{com}}.
$$
(12.29)

It is observed that, as r increases, both the denominator and numerator in Eq. (12.29) decrease. Thus, there exists a trade-off between minimizing the UAV's energy consumption and maximizing the communication throughput; hence there must exist a non-trivial optimal value $r^\star$ that maximizes $EE(r)$ in Eq. (12.29). This is validated by Figure 12.6, showing one typical plot of $EE(r)$ against r.

In general, a circular trajectory may not be optimal for energy efficiency maximization. In [6], a more general solution is obtained for fixed-wing UAV energy efficiency maximization by applying the SCA technique. The obtained trajectory of one example setup is shown in Figure 12.7. It is observed that an energy-efficient UAV trajectory for a fixed-wing UAV follows an interesting "figure-eight" shape, with the horizontal center of the shape coinciding with the location of the ground user. This is expected, since such a trajectory enables the UAV to hover around the ground user with smooth turning, so as to maintain good communication link yet without consuming excessive propulsion power.

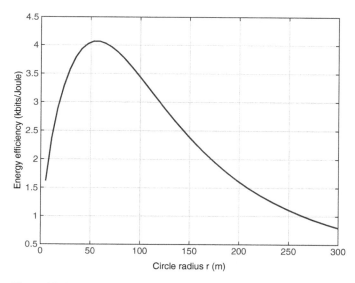

Figure 12.6 A typical plot of energy efficiency versus circle radius *r*. Source: From Zeng et al. [9].

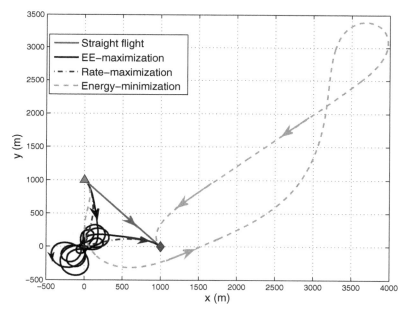

Figure 12.7 Energy-efficient trajectory for a fixed-wing UAV. The triangle and diamond denote the initial and final UAV locations, respectively. Source: From Zeng and Zhang [6].

12.3 Energy Minimization with Communication Requirement

Another important formulation for energy-efficient UAV communications is to minimize the UAV energy consumption, while satisfying the targeting communication performance. Such a formulation is relevant for certain scenarios when the amount of data to be transmitted/received by the UAV is fixed, e.g., for UAV-enabled data collection or information dissemination. Mathematically, such problems can be formulated as

$$\min_{\{q(t)\},\{r(t)\},T} E(q(t)) + E_{\text{com}}(r(t)) \tag{12.30}$$

$$\text{s.t. } \overline{R}(q(t), r(t)) \geq Q, \tag{12.31}$$

$$(12.22)-(12.24),$$

where Q denotes the targeting communication throughput requirement. Note that, different from the energy efficiency maximization problem (12.21)–(12.24), where the mission completion time T is given, in the above energy-minimization problem, T is also one of the optimization variables. This makes the energy-minimization problem in general more challenging to solve, since T appears in the integral upper bound of both the energy and communication throughput expressions, and there is no monotonic relationship for efficient search (e.g., bisection search) over T, except for a prohibitive exhaustive search. In [10], by applying the path discretization technique discussed in Section 11.3, an SCA-based algorithm is proposed for energy minimization for rotary-wing UAVs.

To further reduce the complexity and to gain useful insights, a fly–hover–communicate protocol can be applied for rotary-wing UAVs. Consider a UAV-assisted communication system with K ground users. With a fly–hover–communicate protocol, instead of continuously communicating with the ground users, the UAV successively visits K optimized hovering locations, and only communicates with the ground users when it is in the corresponding hovering locations, as illustrated in Figure 12.8. As a result, the trajectory optimization problem reduces to finding the optimal hovering locations and hovering time allocations for each of the K ground users, as well as the optimal flying path and speed between these hovering locations.

Assume that the UAV flies at a constant altitude H_U. Let $w_k \in \mathbb{R}^{2 \times 1}$ denote the horizontal coordinate of user k, and $\tilde{q}_k$ denote the horizontal coordinate of the UAV hovering location

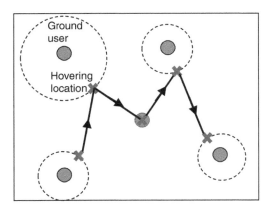

Figure 12.8 An illustration of the fly–hover–communicate protocol.

when it communicates with user k. Further denote by Q_k the bandwidth-normalized data requirement for user k. Then the total required communication time (or the hovering time at location $\tilde{\boldsymbol{q}}_k$) to meet the communication requirement is

$$T_k(\tilde{\boldsymbol{q}}_k) = \frac{Q_k}{\log_2\left(1 + \dfrac{\gamma_k}{(H_U^2 + \|\tilde{\boldsymbol{q}}_k - \boldsymbol{w}_k\|^2)^{\tilde{\alpha}}}\right)}, \tag{12.32}$$

where γ_k is the received SNR at the reference distance of 1 m, and $\tilde{\alpha}$ is equal to half of the path-loss exponent. Thus, the total required hovering and communication energy at the K locations is a function of $\{\tilde{\boldsymbol{q}}_k\}$, which can be expressed as

$$\begin{aligned}E_{\mathrm{hc}}(\{\tilde{\boldsymbol{q}}_k\}) &= (P_h + P_{\mathrm{com}})\sum_{k=1}^{K} T_k(\tilde{\boldsymbol{q}}_k) \\ &= \sum_{k=1}^{K}\frac{(P_h + P_{\mathrm{com}})Q_k}{\log_2\left(1 + \dfrac{\gamma_k}{(H_U^2 + \|\tilde{\boldsymbol{q}}_k - \boldsymbol{w}_k\|^2)^{\tilde{\alpha}}}\right)},\end{aligned} \tag{12.33}$$

where $P_h = P(0) = P_0 + P_i$ is the propulsion power consumption of a rotary-wing UAV at hovering status, which can be readily obtained via Eq. (12.17), and P_{com} is the communication-related power consumption.

On the other hand, the total required flying energy depends on the total traveling distance D_{tr} to visit all the K hovering locations $\{\tilde{\boldsymbol{q}}_k\}$, as well as the traveling speed $V(t)$ among them. It can be shown that, with the fly–hover–communicate protocol, the UAV should always travel with the MR speed V_{mr} defined in Eq. (12.6). Furthermore, for any given set of hovering locations $\{\tilde{\boldsymbol{q}}_k\}$ and initial/final locations $\boldsymbol{q}_I$ and $\boldsymbol{q}_F$, the total traveling distance D_{tr} depends on the visiting order of all the K locations, which can be represented by the permutation variables $\pi(k) \in \{1, \ldots, K\}$. Specifically, $\pi(k)$ gives the index of the kth ground user served by the UAV. Therefore, we have

$$D_{\mathrm{tr}}(\{\tilde{\boldsymbol{q}}_k\}, \{\pi(k)\}) = \sum_{k=0}^{K}\|\tilde{\boldsymbol{q}}_{\pi(k+1)} - \tilde{\boldsymbol{q}}_{\pi(k)}\|, \tag{12.34}$$

where $\tilde{\boldsymbol{q}}_{\pi(0)} = \boldsymbol{q}_I$ and $\tilde{\boldsymbol{q}}_{\pi(K+1)} = \boldsymbol{q}_F$.

Thus, the total required UAV traveling energy consumption with the optimal traveling speed V_{mr} can be written as

$$E_{\mathrm{tr}}(\{\tilde{\boldsymbol{q}}_k\}, \{\pi(k)\}) = E_0^\star D_{\mathrm{tr}}(\{\tilde{\boldsymbol{q}}_k\}, \{\pi(k)\}), \tag{12.35}$$

where $E_0^\star = E_0(V_{\mathrm{mr}})$ is the minimum energy consumption per unit traveling distance as defined in Eq. (12.6). The total UAV energy consumption is thus given by

$$E_{\mathrm{tot}}(\{\tilde{\boldsymbol{q}}_k\}, \{\pi(k)\}) = E_{\mathrm{hc}}(\{\tilde{\boldsymbol{q}}_k\}) + E_{\mathrm{tr}}(\{\tilde{\boldsymbol{q}}_k\}, \{\pi(k)\}). \tag{12.36}$$

As a result, the energy-minimization problem with the fly–hover–communicate protocol reduces to

$$\min_{\{\tilde{\boldsymbol{q}}_k\},\{\pi(k)\}} E_{\mathrm{tot}}(\{\tilde{\boldsymbol{q}}_k\}, \{\pi(k)\})$$

$$\text{s.t. } [\pi(1), \ldots, \pi(K)] \in \mathcal{P}, \tag{12.37}$$

where $\mathcal{P}$ represents the set of all the $K!$ possible permutations for the K ground users. The above formulation greatly simplifies the general energy-minimization problem, but it is still a non-convex optimization problem. Fortunately, by applying the TSPN (traveling salesman problem with neighborhood) and SCA techniques discussed in Chapter 11, an efficient solution is proposed in [10].

12.4 UAV–Ground Energy Trade-off

While most existing works study the energy consumption of either ground users or UAVs, there exists an interesting trade-off for the energy consumption between these two types of nodes for UAV-assisted wireless communications. Take UAV-assisted data collection as an example [11, 12]. Intuitively, the closer the UAV flies to the ground nodes, the less energy is needed for the ground nodes to complete its transmission of a given amount of data. However, this usually comes at the cost of more UAV energy consumption. Such a trade-off has been rigorously characterized in [11] for a fixed-wing UAV system via jointly optimizing the transmit power of ground users, the mission completion time, and the UAV flying speed.

For example, for the particular circular path shown in Figure 12.5, given the target number of information bits Q to be collected, the required energy consumption of the ground node, denoted by E_1, can be expressed in closed form with respect to the energy consumption of the UAV, denoted by E_2, as [11]

$$E_1 = T\left[\left(H_U^2 + \frac{1}{g^2}\frac{c_2}{\dfrac{E_2^4}{T^4(3^{-3/4}+3^{1/3})^4 c_2^3} - c_1}\right)\frac{2^{Q/BT}-1}{\tilde{\gamma}_0} + P_c\right], \tag{12.38}$$

where T is the mission completion time, B is the bandwidth, $\tilde{\gamma}_0 = \beta_0/\sigma^2$ represents the channel-power-to-noise ratio at the reference distance of 1 m, and P_c denotes the constant circuit power of the ground node. Note that, for simplicity, Eq. (12.38) assumes the free-space propagation environment and ignores the communication-related energy consumption by the UAV.

For any fixed mission completion time T, the function E_1 in Eq. (12.38) monotonically decreases with E_2, which clearly demonstrates the trade-off between E_1 and E_2. Time T can be further optimized to improve such a trade-off. An example plot of E_1 versus E_2 is given in Figure 12.9. Note that, while the absolute energy consumptions for the UAV and the ground user are on different orders, the changes in terms of the percentage of energy consumption along the trade-off curve are of similar levels. This validates the exploitation of such a trade-off to save the energy of one entity while compromising that for the other.

12.5 Chapter Summary

In this chapter, we consider energy-efficient communications for wireless systems enabled by UAV, which is a new and important line of research for UAV-assisted communications.

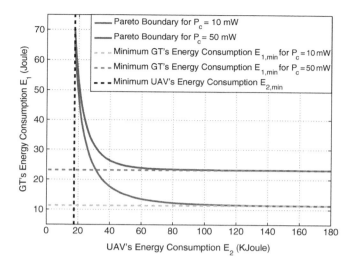

Figure 12.9 Energy trade-off between UAV and ground users. Source: From Yang et al. [11].

On the one hand, UAVs have very limited on-board energy, which makes it a necessity to efficiently expend their available energy. On the other hand, different from the conventional terrestrial communication systems, UAVs have one additional energy consumption component, namely, the propulsion energy to maintain them airborne and support their mobility. This renders the existing techniques for energy-efficient communications developed for terrestrial systems no longer applicable. To fill the gap, this chapter introduces rigorous mathematical models for the propulsion energy consumption of both fixed-wing and rotary-wing UAVs, and then presents the problem formulation and solution to maximize the energy efficiency with a given time horizon, and minimize the total energy consumption while satisfying the communication performance requirement.

Note that research for energy-efficient UAV communication is still in its infancy, with many issues remaining unsolved, including the derivation of more general energy consumption models with 3D trajectory, the development of low-complexity energy-efficient trajectory design techniques, the need to relax the assumption of perfect knowledge of channel and energy information, and so on.

References

1 C. D. Franco and G. Buttazzo, "Energy-aware coverage path planning of UAVs," in *Proc. IEEE International Conference on Autonomous Robot Systems and Competitions*, Apr. 2015.

2 A. Richards and J. P. How, "Aircraft trajectory planning with collision avoidance using mixed integer linear programming," in *Proc. IEEE American Control Conference*, May 2002.

3 C. S. Ma and R. H. Miller, "MILP optimal path planning for real-time applications," in *Proc. IEEE American Control Conference*, Jun. 2006.

4 E. I. Grotli and T. A. Johansen, "Path planning for UAVs under communication constraints using SPLAT! and MILP," *J. Intell. Robot Syst*, vol. 65, pp. 265–282, 2012.

5 Y. Mei, Y.-H. Lu, Y. C. Hu, and C. S. G. Lee, "Energy-efficient motion planning for mobile robots," in *Proc. IEEE International Conference on Robotics and Automation*, Apr. 2004.

6 Y. Zeng and R. Zhang, "Energy-efficient UAV communication with trajectory optimization," *IEEE Trans. Wireless Commun.*, vol. 16, no. 6, pp. 3747–3760, Jun. 2017.

7 Y. Zeng, J. Xu, and R. Zhang, "Energy minimization for wireless communication with rotary-wing UAV," *IEEE Trans. Wireless Commun.*, vol. 18, no. 4, pp. 2329–2345, Apr. 2019.

8 A. Filippone, *Flight performance of fixed and rotary wing aircraft*. American Institute of Aeronautics & Ast (AIAA), 2006.

9 Y. Zeng, Q. Wu, and R. Zhang, "Accessing from the sky: a tutorial on UAV communications for 5G and beyond," *Proc. of the IEEE*, vol. 107, no. 12, pp. 2327–2375, Dec. 2019.

10 Y. Zeng, J. Xu, and R. Zhang, "Rotary-wing UAV enabled wireless network: trajectory design and resource allocation," in *IEEE Global Communications Conference (GLOBECOM)*, 2018.

11 D. Yang, Q. Wu, Y. Zeng, and R. Zhang, "Energy trade-off in ground-to-UAV communication via trajectory design," *IEEE Trans. Veh. Technol.*, vol. 67, no. 7, pp. 6721–6726, Jul. 2018.

12 C. Zhan, Y. Zeng, and R. Zhang, "Energy-efficient data collection in UAV enabled wireless sensor network,"*IEEE Wireless Commun. Lett.*, vol. 7, no. 3, pp. 328–331, Jun. 2018.

13

Fundamental Trade-Offs for UAV Communications

Qingqing Wu[1], Liang Liu[2], Yong Zeng[3], and Rui Zhang[4]

[1]*State Key Laboratory of Internet of Things for Smart City, University of Macau, China*
[2]*Department of Electronic and Information Engineering, The Hong Kong Polytechnic University, Hong Kong*
[3]*National Mobile Communications Research Laboratory, Southeast University, China*
[4]*Department of Electrical and Computer Engineering, National University of Singapore, Singapore*

13.1 Introduction

Due to their prominent features of high mobility and flexible deployment, unmanned aerial vehicles (UAVs) will find many promising usages in the future Fifth-Generation (5G) networks and beyond, as shown in Figure 13.1. Particularly, UAVs can be used cost-effectively as on-demand aerial platforms to provide or enhance the communication services for terrestrial terminals in a multitude of applications, including aerial base stations (BSs)/relays in scenarios without terrestrial cellular coverage [1, 2], aerial helpers for providing new services such as data backhaul/offloading, cached-content multicasting and edge computing for terrestrial BSs/users, as well as mobile hubs for energy-efficient data collection [3] and wireless power transfer [4] for low-power Internet-of-Things (IoT) devices such as sensors and tags. On the other hand, UAVs in many civilian applications, such as cargo delivery and aerial video surveillance, may become new aerial users in the cellular network, which need to have high-performance two-way communications with the ground BSs to receive control signals and upload application data in real time.

Despite the above promising UAV applications, their future success critically depends on the development of new and effective UAV-to-ground communication technologies. Compared to the conventional terrestrial communications, UAV-to-ground communications enjoy the following two main advantages that can be exploited for throughput enhancement, namely, line-of-sight (LoS) dominated UAV-to-ground channels, and UAV's controllable high mobility in three-dimensional (3D) space.

On the one hand, thanks to the high altitude of UAVs, the probability of LoS channels between UAVs and the ground users/BSs is in general quite high, and thus UAV-to-ground communications are significantly less affected by channel impairments such as shadowing and fading as compared to terrestrial communications. On the other hand, thanks to the high mobility, swift 3D deployment or even dynamic movement of UAVs becomes feasible so that they can adjust their locations/trajectories based on the locations and/or movement

UAV Communications for 5G and Beyond, First Edition.
Edited by Yong Zeng, Ismail Guvenc, Rui Zhang, Giovanni Geraci, and David W. Matolak.
© 2021 John Wiley & Sons Ltd. Published 2021 by John Wiley & Sons Ltd.

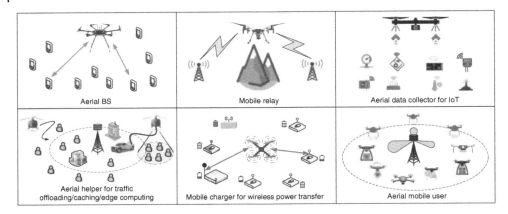

Figure 13.1 Typical UAV applications in 5G and beyond. From Wu et al. [5].

of the ground BSs/users to maintain favorable LoS channels with them. It is worth noting that the LoS channels enable UAVs to have their signal coverage over a much larger number of ground users or BSs as compared to the BSs/users in terrestrial communications. Consequently, to achieve optimal communication and trajectory design, each UAV should not only maintain strong channels to its served users or connected BSs via flying in proximity of them, but also control its interferences to other UAVs as well as ground users/BSs so as to achieve the maximum network throughput.

Besides throughput, two important factors also need to be considered in UAV communication and trajectory design, namely, delay and energy. First, to maximize throughput, each UAV should communicate with a ground user/BS when flying sufficiently close to it so as to reduce their distance and hence improve the link capacity. However, this inevitably incurs more delay in communication due to the UAV movement. Thus, there is an interesting throughput–delay trade-off in UAV-to-ground communication, as shown in Figure 13.5(a).

Second, there also exists a new trade-off between throughput and energy in UAV-enabled communication, as shown in Figure 13.5(b), since the UAV generally needs to consume more propulsion energy to move closer to the ground users/BSs in order to gain higher throughput. As commercial UAVs usually have limited on-board energy, more propulsion energy consumption leads to shorter endurance of UAVs, thus imposing a critical constraint on their practical applications. Last, the above two trade-offs naturally imply a delay–energy trade-off, as shown in Figure 13.5(c), as delay in UAV-to-ground communication can be reduced if more propulsion energy is consumed by the UAV to move faster to the ground users/BSs it is designated to communicate with.

Motivated by the above new and interesting trade-offs among the throughput, delay, and (propulsion) energy consumption in UAV communication and trajectory design, this chapter aims to provide an overview on the state-of-the-art results on them. In particular, we will focus on the use of UAVs as communication platforms (e.g., aerial BSs/relays) to serve terrestrial users, although such fundamental trade-offs also exist similarly in the other paradigm with UAVs as new aerial users to be served by the ground BSs in the cellular network [1].

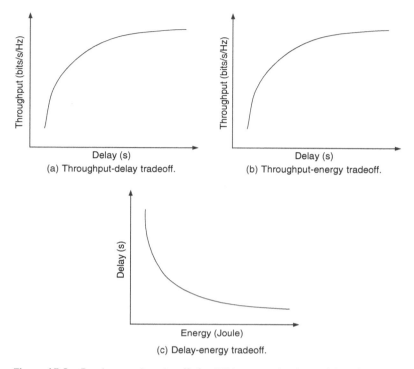

(a) Throughput-delay tradeoff.

(b) Throughput-energy tradeoff.

(c) Delay-energy tradeoff.

Figure 13.2 Fundamental trade-offs for UAV communications with trajectory design.

13.2 Fundamental Trade-offs

In this section, we discuss the above fundamental trade-offs in UAV-enabled communication and highlight the main differences with their counterparts in traditional terrestrial communication.

13.2.1 Throughput–Delay Trade-Off

The throughput–delay trade-off has been extensively studied for terrestrial wireless communications. For a basic point-to-point wireless communication link, the maximum achievable rate over fading channels, defined as the ergodic capacity, is achieved by coding over a sufficiently large number of channel coherence intervals to fully exploit the ergodicity of fading channels [6]. However, this comes at the cost of long transmission delay, which may not be tolerable for applications with stringent latency requirement. On the other hand, channel coding can be performed over each coherence interval to reduce the delay, resulting in the so-called delay-limited capacity [6]. However, the delay-limited capacity is in general smaller than the ergodic capacity for a given fading channel, and outage is usually inevitable in deep fading [6]. For general multiuser communication, the multiuser diversity gain can be attained to improve the network throughput by scheduling the user with the best channel among all users to communicate in each coherence interval, whereas this inevitably leads to more significant delay for each user as the number of users increases [6].

The above results show that there is a general throughput–delay trade-off for communication over fading channels. Moreover, it is shown in [7] that there is another trade-off between the total throughput of a mobile ad-hoc network (MANET) and the average delay tolerable by the users in the network due to the random user movement, as each user needs to wait before communicating with each other until they become sufficiently close.

By contrast, in UAV-enabled communication, channel fading is no longer a key factor contributing to the throughput–delay trade-off thanks to the LoS-dominant channels. Instead, the mobility of UAVs plays the decisive role in such a trade-off, as the UAV-to-ground LoS channels are solely determined by the distances between the UAV and ground users, which critically depend on the UAV's location. However, in sharp contrast to the random user movement in a MANET, where the delay is random and difficult to predict [7], the delay in UAV-enabled communication can be properly controlled via a joint UAV trajectory and communication scheduling design. Moreover, another key difference lies in the time scale of the delay between the terrestrial communication and UAV-to-ground communication: in the former case, the delay is measured in terms of channel coherence time, e.g., in milliseconds, while in the latter case, the delay is mainly due to the UAV flying time (distance divided by speed), e.g., in seconds or minutes. As a result, in order to fully exploit the throughput–delay trade-off via trajectory design in UAV-enabled communication, the application needs to be more delay-tolerant as compared to that in terrestrial communication [8].

13.2.2 Throughput–Energy Trade-Off

The throughput–energy trade-off in the traditional wireless communication is fundamentally rooted in the Shannon capacity formula, which explicitly suggests that the achievable rate increases monotonically with the transmit power [6]. One useful performance metric stemming from this trade-off is "energy efficiency," which measures the number of information bits that can be successfully communicated per unit energy consumption. If only the transmit energy is considered, it is well known that the energy efficiency monotonically increases with the decrease of the transmit rate/power [6], while if the circuit power at the transmitter is considered as well, it is shown in [9] that the energy efficiency first increases and then decreases with the transmit rate/power.

In UAV-enabled communication, the propulsion energy (usually in the order of kilowatts (kW)) required to maintain the UAVs airborne and support their high mobility is generally orders of magnitude higher than the transmit and circuit energy for communication (usually in the order of watts (W) or even smaller). As a result, the effect of propulsion energy on the UAV trajectory is the dominant factor determining the throughput–energy trade-off in UAV-enabled communication. For example, to enhance the throughput, each UAV needs to fly over a longer distance with a faster speed so that it can reach each of its served ground users as close as possible and stay near them as long as possible, given a finite flight duration, in order to exploit better LoS channels with them. Moreover, each UAV may also need to adjust its altitude and/or make sharp turns to avoid blockages in the directions of its served ground users. All these can lead to more significant propulsion energy consumption. As a result, for UAV-enabled communication, the energy efficiency is more appropriately defined in terms of information bits per joule (J) of propulsion energy, rather than that of transmit/circuit energy in traditional wireless communication. Such a new metric has a high practical significance, as it indicates the maximum number of information bits that can be communicated with a finite amount of the UAV's on-board energy.

13.2.3 Delay–Energy Trade-Off

As discussed in the above two subsections, the throughput–delay and throughput–energy trade-offs in UAV-enabled communication exhibit interesting new aspects compared to their traditional counterparts in terrestrial communication. As a result, their corresponding delay–energy trade-offs are also drastically different due to the new UAV trajectory design and the high UAV propulsion energy consumption. For example, to reduce the delay in movement and transmission, each UAV should fly between its served ground users with its maximum speed, but remain at its minimum speed (e.g., hovering) when serving them in its proximity, both resulting in more propulsion energy consumption in general.

In the following two sections, we will focus on examining the throughput–delay and throughput–energy trade-offs, respectively. Since the delay–energy trade-off becomes straightforward given the above two trade-offs, it is omitted for brevity. We will provide concrete examples to illustrate them more clearly, provide overviews on their state-of-the-art results, and also point out promising directions for future research.

13.3 Throughput–Delay Trade-Off

In this section, we investigate the joint UAV trajectory and communication design to characterize the throughput–delay trade-off. Specifically, we first consider a simple setup with one UAV serving two ground users (GUs) to draw useful insights. Then, we extend our study to the general case with multiple UAVs serving multiple GUs, followed by further discussions on related/future work.

13.3.1 Single-UAV-Enabled Wireless Network

As shown in Figure 13.3(a), we consider a UAV-enabled downlink communication system where one UAV is employed to serve two GUs in a finite period of T seconds. The UAV is assumed to fly at a constant altitude of H in meters with the maximum allowable speed denoted by V_{max} in meters per second (m s^{-1}). The air-to-ground channels from the UAV to the GUs are assumed to be dominated by the LoS links. As such, it is preferable to let the UAV fly as low as possible in order to reduce the signal path loss with the GUs. However, the minimum value of H is practically limited for terrain or building avoidance. The two GUs are assumed to be quasi-stationary with a distance of D meters between their nominal locations, as shown in Figure 13.3(a), where we assume that their maximum movement distances from their respective nominal locations within the given period T are negligible compared to D and the UAV altitude H; thus, their effects on the corresponding LoS channel gains are ignored. We consider that the UAV communicates with GUs via time-division multiple access (TDMA), i.e., only one GU is scheduled for communication at any time instant. To serve GUs continuously in a periodic manner, we assume that the UAV needs

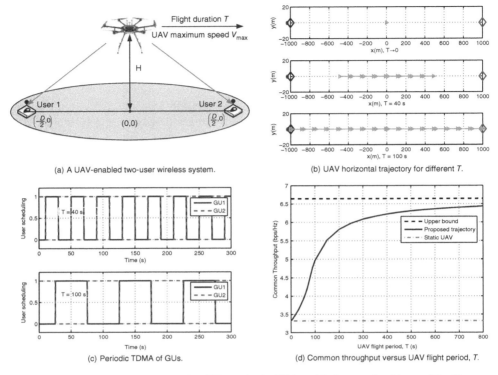

(a) A UAV-enabled two-user wireless system.

(b) UAV horizontal trajectory for different T.

(c) Periodic TDMA of GUs.

(d) Common throughput versus UAV flight period, T.

Figure 13.3 Throughput–delay trade-off for a single-UAV-enabled network with two GUs. The GUs' nominal locations are marked by '◇'s and the UAV trajectories are marked by '▷'s. The maximum transmit power and the receiver noise power are set as 20 dBm (0.1 W) and −110 dBm, respectively, and the channel power gain at the reference distance of 1 m is set as −50 dB. Other required parameters are set as follows: $V_{max} = 50$ m s^{-1}, $H = 100$ m, and $D = 2000$ m.

to return to its initial location by the end of each flight period T while the initial location can be optimized for maximizing the throughput. To ensure fairness among GUs, we aim to maximize the common (minimum) throughput among the GUs via jointly optimizing the UAV trajectory and communication scheduling.

In Figure 13.3(b), we show the UAV's optimal trajectories projected onto the ground plane under different flight periods, T. It is observed that, as T increases, the UAV tends to fly closer to the two GUs, while when T is sufficiently large (e.g., $T = 100$ s), the UAV flies between the two GUs with its maximum speed to save more time for hovering right above each of them to maintain the best channel for communication.

Furthermore, at any time instant, to maximize the throughput, the GU that is closer to the UAV (thus with a better channel) is scheduled for communication, while the other GU has to wait until the UAV flies closer to it again. As such, each GU will experience a waiting time of $T/2$ for communicating with the UAV periodically. This is illustrated in Figure 13.3(c), where the user scheduling is plotted over time. It is observed that a larger T leads to a longer waiting time for each GU.

Finally, we show in Figure 13.3(d) the achievable common throughput in bits per second per Hertz (bps/Hz) versus T. Note that the throughput upper bound is obtained by ignoring the time spent on traveling between the two GUs, which holds when T goes to infinity. In addition, the throughput of a static UAV is obtained by fixing the UAV at the middle location between the two GUs at all times. One can observe that, compared to the case of a static UAV, the common throughput can be significantly improved as T increases with a mobile UAV. However, such a throughput gain is at the cost of increasing the user delay (or larger T), which thus reveals a new throughput–delay trade-off in UAV-enabled wireless network.

13.3.2 Multi-UAV-Enabled Wireless Network

The use of multiple UAVs for cooperatively serving the GUs is an effective solution to improve the throughput–delay trade-off over the single-UAV-enabled network, by dividing the GUs into smaller-size groups, each served by one of the UAVs.

To demonstrate this, we consider a multi-UAV-enabled downlink transmission system as shown in Figure 13.4(a), where two UAVs are employed to serve a group of K GUs in a finite period of duration T. To achieve high spectral efficiency, we consider a spectrum sharing system, where the UAVs share the same frequency band for communication and each of the UAVs serves its associated GUs via the periodic TDMA. As such, each GU suffers from severe interference from other non-associated UAVs due to the LoS channel, which needs to be effectively mitigated by employing inter-UAV interference coordination (IUIC) via jointly designing the UAV trajectories, transmit power and user associations.

Similar to Section 13.3.1, we aim to maximize the common throughput of all GUs with optimally designed IUIC. However, this problem is a non-convex optimization problem involving infinite variables due to the continuous UAV trajectory. To tackle this problem, we first apply time discretization to divide the UAV flight period into a finite number of equal-time slots, each with a nominal location of the UAV. Then, we apply the block coordinate descent (BCD) and successive convex approximation (SCA) optimization techniques of Chapter 11 to obtain a suboptimal solution to the IUIC design [10]. As an initial UAV trajectory is needed for our algorithm, we adopt a simple and yet practical circular UAV trajectory for initialization [10].

For the purpose of illustration, we consider a setup with $K = 6$ GUs. Specifically, we show the optimized UAV trajectories without and with power control in Figure 13.4(b) and (c), respectively, for $T = 120$ s. In the former case, both UAVs transmit with their maximum power at all times. It is observed from Figure 13.4(b) that the optimized UAV trajectories tend not only to shorten the communication distances between the UAVs and their associated GUs (e.g., from $t = 0$ to $t = 20$ s), but also to enlarge the separations of the two UAVs to help alleviate the co-channel interference (e.g., from $t = 40$ s to $t = 60$ s), in the case without power control. However, at certain pairs of UAV locations, enlarging the UAVs' separation is achieved at the cost of compromising direct link gains, especially when the UAVs are flying on their ways to serve two GUs (e.g., the two nearby GUs around the center in Figure 13.4(b)) that are close to each other.

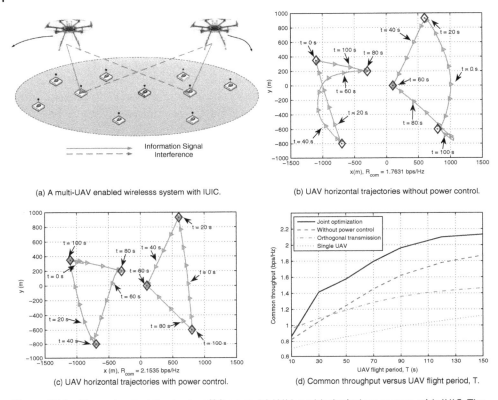

(a) A multi-UAV enabled wirelesss system with IUIC.

(b) UAV horizontal trajectories without power control.

(c) UAV horizontal trajectories with power control.

(d) Common throughput versus UAV flight period, T.

Figure 13.4 Throughput–delay trade-off for a multi-UAV-enabled wireless system with IUIC. The GUs' nominal locations are marked by '◊'s and the UAV trajectories are marked by '▷'s. The simulation parameters are set to be the same as those in Figure 13.3. The user common throughput is denoted by R_{com} in bps/Hz.

In contrast, for the case with power control, it is observed from Figure 13.4(c) that the optimized UAV trajectories do not tend to compromise the direct link gains in return for large distance separation. This is because power control can help avoid strong interference even when the two UAVs have to be close to each other (e.g., when serving the two nearby GUs around the center).

As a result, the common throughput is substantially improved over the case without power control, as shown in Figure 13.4(d). In addition, an orthogonal UAV transmission scheme is adopted for comparison where the two UAVs take turns to transmit information to serve GUs over orthogonal time slots, and the system is then interference-free. One can observe that, for short flying time T, which implies limited UAV flying ranges, the orthogonal transmission even achieves higher throughput than those non-orthogonal schemes, since the latter suffers from severe interference between the UAVs. However, as T increases, the proposed joint design significantly outperforms the orthogonal transmission, since the UAVs' trajectories can be more flexibly designed to enlarge the inter-UAV distance such that the spectrum can be better reused by the two UAVs with small interference. Finally, it is also observed that the user throughput in the multi-UAV network is significantly improved over the single-UAV network at the same delay, thus verifying the improved throughput–delay trade-off via effective multi-UAV cooperation with optimized IUIC.

13.4 Throughput–Energy Trade-Off

In this section, we investigate further the throughput–energy trade-off in UAV-enabled communication and trajectory design. First, we discuss the energy consumption models of UAVs. Then, we revisit the single-UAV-enabled system described in Section 13.3.1 by taking into account the UAV's propulsion energy consumption, followed by discussions on other related work and future research directions.

13.4.1 UAV Propulsion Energy Consumption Model

Fixed-wing and rotary-wing UAVs are the two main types of UAVs that have been widely used in practice. Both of them possess respective unique sets of advantages and limitations that render them more or less suitable for different applications.

To investigate the throughput–energy trade-off in UAV-enabled communication, the UAV's propulsion energy consumption needs to be properly modeled first. Towards this end, two analytical propulsion power models have been presented for fixed-wing and rotary-wing UAVs in [11] and [10], respectively. In general, the propulsion power required for the UAV depends on its velocity (including both the flying speed and direction) as well as the acceleration. In Figure 13.5, the typical propulsion power consumption versus the UAV's flying speed is illustrated for both fixed-wing and rotary-wing UAVs. In both cases, it is observed that, as the UAV's flying speed increases, the corresponding propulsion power required first decreases and then increases, which implies that flying at too high or too low speeds is not energy-efficient. Furthermore, flying at a very low speed is extremely energy-consuming and even impossible for fixed-wing UAVs in practice, which renders them very difficult to hover over a small geographical area to serve GUs, while this is

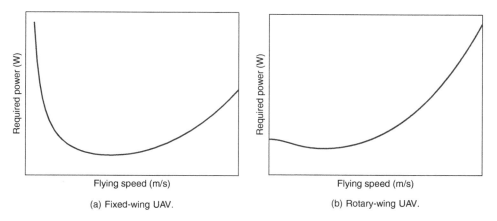

(a) Fixed-wing UAV.   (b) Rotary-wing UAV.

Figure 13.5 Typical propulsion power consumption versus the UAV's flying speed.

not an issue for rotary-wing UAVs. However, rotary-wing UAVs suffer from consuming excessive propulsion power when the UAV's flying speed is very high, which makes them inefficient for tasks over a wide geographical area. In practice, fixed-wing and rotary-wing UAVs can be both leveraged simultaneously to enhance the communication efficiency. For example, a promising UAV-enabled networking architecture is to deploy rotary-wing UAV-enabled BSs hovering at well-selected locations for establishing signal hotspots and at the same time to despatch fixed-wing UAV-enabled BSs flying around periodically for wider coverage and higher throughput.

13.4.2   Energy-Constrained Trajectory Optimization

As shown in Figure 13.6(a), we consider the same UAV-enabled two-user system as in Section 13.3.1 for a given UAV flight period T where the UAV has a limited on-board energy, and thus the maximum propulsion energy that can be consumed during this period is denoted by E_{max}. For the purpose of exposition, we consider a fixed-wing UAV with the minimum speed and maximum acceleration denoted by V_{min} in m s^{-1} and a_{max} in m s^{-2},

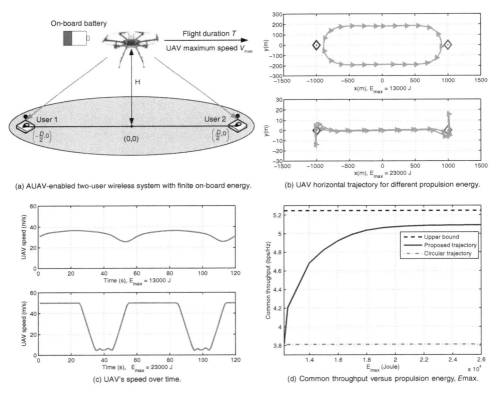

Figure 13.6   Throughput–energy trade-off for a single-UAV-enabled network with two GUs. The GUs' nominal locations are marked by '◇'s and the UAV trajectories are marked by '▷'s. For the propulsion power consumption model in [10], the constants c_1 and c_2 are set as 9.26×10^{-4} and 2250, respectively. The simulation parameters are set as follows: $V_{max} = 50$ m s^{-1}, $V_{min} = 5$ m s^{-1}, $a_{max} = 5$ m s^{-2}, and $T = 120$ s. Other parameters are set to be the same as those in Section 13.3.1.

respectively. Similarly as in Section 13.3.1, we consider the common throughput maximization for the two GUs via jointly optimizing the UAV trajectory as well as the user scheduling, and subject to the new UAV's total energy constraint and the mobility constraints (on its speed and acceleration).

In Figure 13.6(b), we plot the UAV's optimized trajectories under different constraints on the propulsion energy. It is observed that the UAV flies close to the two GUs by following a smooth trajectory with relatively large turning radii when $E_{max} = 13$ kJ; whereas when E_{max} is increased to 23 kJ, the UAV's trajectory tends to approach that without the propulsion energy constraint shown in Figure 13.3(b). This is because, in the latter case, sharp turning in the flight direction to quickly shorten the UAV–GU distance requires more propulsion energy consumption.

Furthermore, the UAV's flying speeds over time in the above two cases are illustrated in Figure 13.6(c). It is observed that, in the first case, the UAV's flying speed does not vary much around 30 m s^{-1} during the total period due to the limited propulsion energy; while in the latter case, with more available energy, the UAV first flies at the maximum speed (50 m s^{-1}) to get close to each of the GUs and then hovers around the GU at the minimum speed (5 m s^{-1}), so as to maximize the throughput.

Finally, the achievable throughput versus the propulsion energy is plotted in Figure 13.6(d). The throughput upper bound is obtained by ignoring the propulsion energy constraint, which is the same as that in Figure 13.3(d) under the same T. The throughput lower bound is achieved by the initial circular trajectory [12] with the UAV's speed equal to 30 m s^{-1}. One can observe that the common throughput can be significantly improved at the cost of more propulsion energy consumption. In particular, as the propulsion energy increases, the common throughput first increases rapidly and then approaches a constant that is strictly lower than the throughput upper bound. This is because, in addition to the propulsion energy constraint, the practically achievable throughput is also subjected to the UAV's mobility constraints on the minimum speed and maximum acceleration.

13.5   Further Discussions and Future Work

Besides orthogonal multiple access schemes such as TDMA considered for multiuser communications, non-orthogonal multiple access schemes based on superposition coding (SC) or dirty paper coding [6] can be jointly designed with the UAV trajectory to further improve the throughput–delay trade-off and achieve the capacity limits of UAV-enabled wireless networks [13]. For example, a two-user broadcast channel (BC) is studied in [13], where it is shown that a simple and practical "hover–fly–hover" (HFH) trajectory with SC achieves the capacity region. However, whether similar results hold for a UAV-enabled BC with more than two users or other multiuser channel models still remains an open problem that is worth investigating in future work.

Furthermore, in our study above, the user delay is roughly measured in terms of the UAV flight period. However, the delay requirements in 5G networks may vary dramatically in time scale, from milliseconds (e.g., for online gaming/video streaming) to seconds or even minutes (e.g., for large file sharing/sensor data collection). Thus, how to model such heterogeneous delay requirements and design the joint UAV trajectory and communication resource allocation to efficiently meet them is also an important problem for future research.

For the multi-UAV-enabled network, we propose the IUIC as an effective technique to mitigate the strong LoS interference by exploiting the coordinated multi-UAV trajectory design. Alternatively, motivated by the rapid advance of the wireless backhaul technologies, the UAVs can share messages and perform cooperative beamforming for more efficient interference mitigation, a technique called coordinate multi-point (CoMP) in the sky [14]. It is worth noting that the methodology for designing the optimal UAV trajectories for CoMP is generally different from that for IUIC. For example, to maximize the cooperative beamforming gain in CoMP, it may be desirable to let some UAVs form a fleet to serve the GUs along the same trajectory, while this is apparently undesired in the IUIC case due to the inter-UAV interference. Another important issue worthy of further investigation is how to dynamically adjust the UAV trajectories according to the GUs' movements to improve their throughput and/or delay performances [14].

The throughput–energy trade-off can be further extended by taking the GUs' energy consumption into account, e.g., in the application of UAV-enabled data collection in IoT networks [15]. Since IoT devices are generally of low power and limited battery life, how to prolong their lifetime is critical for the sustainability and proliferation of future IoT ecosystem. Thanks to the controllable mobility, a UAV-enabled mobile data collector can move sufficiently close to the IoT devices, such as sensors or tags, to collect their data with minimum transmit energy. However, this will incur more propulsion energy consumption of UAVs, which implies an interesting new perspective in the throughput–energy trade-off in UAV-enabled communication [15].

On the other hand, the UAVs' energy supply can also be provisioned by means of other technologies such as solar energy harvesting and laser-beamed wireless power transfer by ground chargers. However, these technologies generally bring new design considerations that need to be further studied. For example, for solar-powered UAVs, while increasing the flying altitude will lead to higher path loss, it helps harvest more solar energy to support more flexible trajectory design to adapt to the GUs' dynamic locations and communication requirements. As such, the throughput–energy trade-off in UAV-enabled communication needs to be revised with carefully designed altitude control. Furthermore, in the case of multiple UAVs cooperatively serving the GUs, besides their communication cooperation through IUIC or CoMP, the design of multi-UAV trajectories also needs to consider their individual energy availability. For example, the propulsion energy consumptions of different UAVs should be balanced via cooperative trajectory design to maximize their endurance from a UAV network lifetime maximization perspective.

It is worth pointing out that, besides the three trade-offs considered in this chapter, there exist other important design considerations in UAV-enabled communication, which have not been fully explored yet and thus require further investigations. These may include, for example, the deployment cost of mobile UAVs, their wireless backhaul constraints, as well as the severe air-to-ground interference issue due to the LoS-dominant channels. For example, using multiple collaborative UAVs, each equipped with multiple antennas/full-duplex functionality, can largely improve the system throughput and/or reduce the user delay, while the system complexity and cost are also inevitably increased, which leads to the complexity/cost–throughput/delay trade-off.

On the other hand, the UAV-to-ground LoS channel model is only appropriate for rural or suburban areas or when the UAV altitude is sufficiently high. However, for other cases, such as in urban environments, other air-to-ground channel models, such as probabilistic LoS model and Ricean fading model, are more suitable. It is worth noting that such non-LoS channel models may have significant impacts on the optimal UAV trajectory design in UAV-enabled wireless networks. For instance, lowering the UAV's flying altitude under the probabilistic LoS channel model generally decreases the probability of having LoS links with GUs, while it is always beneficial under the LoS model. As a result, a more complex 3D trajectory optimization problem (as compared to the 2D design in our previous examples under the LoS model) needs to be investigated. Moreover, although the presence of LoS links makes the UAVs well suitable for 5G technologies such as millimeter wave (mmWave) and massive multiple input–multiple output (M-MIMO) communications, the severe air-to-ground interference issue and 3D mobility-induced Doppler effect deserve more investigations in the future.

13.6 Chapter Summary

In this chapter, we revisit the fundamental throughput, delay, and energy trade-offs in UAV-enabled wireless communication. In particular, we show that the communication throughput, delay, and UAV's propulsion energy can be optimally traded off between each other, via judiciously optimizing the UAV's trajectory with communication resource allocation jointly, for both single-UAV- and multi-UAV-enabled networks. While we focus on employing UAVs as aerial BSs in this chapter, the discussed new trade-offs are general and also applicable to other UAV-mounted platforms [16] or cellular-connected UAV users [1]. Finally, some relevant topics are also highlighted to motivate future research. It is hoped that the new design trade-offs and useful insights unveiled in this chapter will be helpful in the practical design of UAV-enabled communication systems in the future.

References

1 Y. Zeng, Q. Wu, and R. Zhang, "Accessing from the sky: a tutorial on UAV communications for 5G and beyond," *Proc. of the IEEE*, vol. 107, no. 12, pp. 2327–2375, Dec. 2019.
2 J. Chen and D. Gesbert, "Optimal positioning of flying relays for wireless networks: a LOS map approach," in *Proc. IEEE International Conference on Communications (ICC)*, May 2017.
3 M. Mozaffari, W. Saad, M. Bennis, and M. Debbah, "Mobile unmanned aerial vehicles (UAVs) for energy-efficient Internet of things communications," *IEEE Trans. Wireless Commun.*, vol. 16, no. 11, pp. 7574–7589, Nov. 2017.
4 J. Xu, Y. Zeng, and R. Zhang, "UAV-enabled wireless power transfer: trajectory design and energy optimization," *IEEE Trans. Wireless Commun.*, vol. 17, no. 8, pp. 5092–5106, Aug. 2018.

5 Q. Wu, L. Liu, and R. Zhang, "Fundamental tradeoffs in communication and trajectory design for UAV-enabled wireless network," *IEEE Wireless Commun.*, vol. 26, no. 1, pp. 36–44, Feb. 2019.

6 D. Tse and P. Viswanath, *Fundamentals of Wireless Communication*. Cambridge University Press, 2005.

7 M. Grossglauser and D. N. C. Tse, "Mobility increases the capacity of ad hoc wireless networks," *IEEE/ACM Trans. Networking*, vol. 10, no. 4, pp. 477–486, Aug. 2002.

8 Q. Wu and R. Zhang, "Common throughput maximization in UAV-enabled OFDMA systems with delay consideration," *IEEE Trans. Commun.*, vol. 66, no. 12, pp. 6614–6627, Dec. 2018.

9 Q. Wu, G. Y. Li, W. Chen, D. W. K. Ng, and R. Schober, "An overview of sustainable green 5G networks," *IEEE Wireless Commun.*, vol. 24, no. 4, pp. 72–80, Aug. 2017.

10 Y. Zeng, J. Xu, and R. Zhang, "Energy minimization for wireless communication with rotary-wing UAV," *IEEE Transactions on Wireless Communications*, vol. 18, no. 4, pp. 2329–2345, Apr. 2019.

11 Y. Zeng and R. Zhang, "Energy-efficient UAV communication with trajectory optimization," *IEEE Trans. Wireless Commun.*, vol. 16, no. 6, pp. 3747–3760, Jun. 2017.

12 Q. Wu, Y. Zeng, and R. Zhang, "Joint trajectory and communication design for multi-UAV enabled wireless networks," *IEEE Trans. Wireless Commun.*, vol. 17, no. 3, pp. 2109–2121, Mar. 2018.

13 Q. Wu, J. Xu, and R. Zhang, "Capacity characterization of UAV-enabled two-user broadcast channel," *IEEE J. Sel. Areas Commun.*, vol. 36, no. 9, pp. 1955–1971, Sep. 2018.

14 S. Z. L. Liu and R. Zhang, "CoMP in the sky: UAV placement and movement optimization for multi-user communications," to appear in *IEEE Trans. Commun.*, arXiv preprint arXiv:1802.10371,2019.

15 D. Yang, Q. Wu, Y. Zeng, and R. Zhang, "Energy trade-off in ground-to-UAV communication via trajectory design," *IEEE Trans. Veh. Technol.*, vol. 67, no. 7, pp. 6721–6726, Jul. 2018.

16 Y. Zeng, R. Zhang, and T. J. Lim, "Wireless communications with unmanned aerial vehicles: opportunities and challenges," *IEEE Commun. Mag.*, vol. 54, no. 5, pp. 36–42, May 2016.

14

UAV–Cellular Spectrum Sharing

Chiya Zhang[1,2] and Wei Zhang[3]

[1]*School of Electronic and Information Engineering, Harbin Institute of Technology, Shenzhen, China*
[2]*Peng Cheng Laboratory (PCL), Shenzhen, China*
[3]*School of Electrical Engineering and Telecommunications, University of New South Wales, Sydney, Australia*

14.1 Introduction

14.1.1 Cognitive Radio

The growing massive data traffic and demand for better and faster broadband services require additional radio spectrum. Radio spectrum has become a vital limited resource regulated by governments and agencies. Cognitive radio is the technology enabling mobile devices to access and share frequency bands, when other services are not using them at the moment. The idea of cognitive radio was proposed in [1–3]. In a typical cognitive radio network, there are two types of users, primary users (PUs) and secondary users (SUs). A PU is a licensed user of a certain spectrum band and has priority to access the spectrum. An SU is an unlicensed user that has limited authority to reuse the unused licensed band in an opportunistic manner. Cognitive radio technology enables SUs to adaptively obtain dynamic access to the spectrum of PUs. It is an intelligent radio and network technology enabling more communications to operate at the same time as well as having a more efficient overall radio operating behavior. The licensed PU frequency band that is not being used for some time period in a specific area is referred to as the spectrum hole or white space.

14.1.1.1 Overlay Spectrum Sharing

In overlay spectrum sharing [4–6], an SU has the knowledge of the codebooks and messages from the PU, and then mitigates interference by assisting the primary transmission and canceling the interference from PU to SU. It has the following pros and cons:
   Advantages

- Transmission has mutual benefits, as both PU and SU send the incumbent messages.
- The network does not necessarily require the presence of spectrum holes.
- The SU's throughput is guaranteed by the interference cancellation.

UAV Communications for 5G and Beyond, First Edition.
Edited by Yong Zeng, Ismail Guvenc, Rui Zhang, Giovanni Geraci, and David W. Matolak.
© 2021 John Wiley & Sons Ltd. Published 2021 by John Wiley & Sons Ltd.

Challenges

- The global channel state information (CSI) is assumed to be available at the SU, which may not be practicable.
- The network needs advanced transmission and coding schemes to activate the certain level of coordination.
- Security issues need to be considered at both SU and PU sides.

14.1.1.2 Underlay Spectrum Sharing

In underlay spectrum sharing [7–12], PU and SU transmit on the same frequency band concurrently. The underlay spectrum sharing explores spatial spectrum holes by restricting SU interference to the primary receivers (PRs). It has the following pros and cons:

Advantages

- Spectrum is utilized efficiently.
- The network does not necessarily require the presence of spectrum holes during the transmission phase.
- Existing interference mitigation techniques, such as beamforming, spread spectrum-based techniques, and power control techniques, can be straightforwardly applied.

Challenges

- CSI and interference threshold are assumed to be available at SU, which may not be practicable.
- It is difficult to restrict the interference within regulatory limits.
- It is not suitable for high-interference regions.

The primary exclusive regions (PERs) centered at every primary receiver are designed in [13]. PER is a fixed region centered at each primary receiver in which no secondary transmitter is allowed to transmit signals. Applying PERs provides an extra level of control to the network performance. It is often used to further guarantee the primary performance, e.g., when the primary user requires a very small outage. The relationship between shadowing with path-loss exponent and PERs is further analyzed in [14]. As the idea of PER effectively guarantees the primary service under limited spectrum resources, research efforts have been made to implement PERs into the optimization of different spectrum sharing schemes [15, 16]. Assuming all transmitters are 2D Poisson-distributed nodes with limited feedback of channel quality information from their local receivers, two limited feedback-based underlay spectrum sharing schemes were proposed in [15], with and without PERs, respectively.

14.1.2 Drone Communication

An unmanned aerial vehicle (UAV), commonly known as a drone, is an aircraft without a human pilot aboard. Drones have been applied in many applications, such as policing and surveillance, and scientific research. As illustrated in Figure 14.1, UAV aerial base stations can establish communication connection between ground nodes A and B. UAVs can also act as aerial sensors to collect data and transmit data to the control station. UAVs can act as aerial base stations or relays, namely drone small cells (DSCs), to provide communication services to areas after natural disasters. DSCs can be deployed in a high-altitude platform (HAP), which is above 10 km height, or in a low-altitude platform (LAP), below 10 km [17].

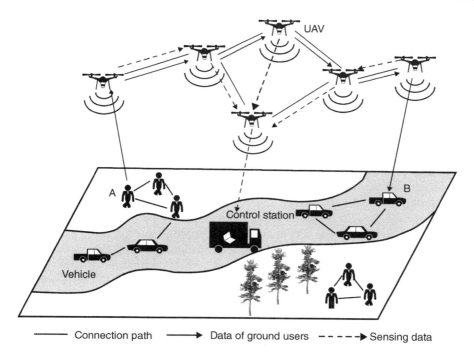

Figure 14.1 Application scenario of a UAV network.

A DSC network has several advantages over conventional cellular networks. The Fifth-Generation (5G) communication utilizing mmWave signals suffers many propagation-related shortcomings, such as relatively short range and vulnerability to blockage. UAVs can exploit its mobility to avoid blockage and provide seamless coverage. Further, aerial base stations are robust to environmental changes. Another advantage of UAV networks is the flexibility of reconfiguration. For example, UAVs can be deployed to help offloading cellular networks in wireless congestion events with low cost. In [18], the optimal deployment altitude of a drone providing maximum coverage is discussed. The deployment design and performance analysis of DSCs at LAP are further studied in [19], in which the optimal altitude maximizing ground coverage and minimizing required transmit power for a single DSC is derived. Their result shows that the optimal deployment altitude varies according to different environment changes.

Drones play an important role in distributing information not only by extending the coverage of the network, but also by providing extra capacity for deploying cellular networks in remote locations. Many recent studies have been conducted on various aspects of drone communications. The ultimate goal is the development of high-capacity, low-latency, and robust UAV communication systems. We can expect that the research field of drone communication and networking will keep growing in the future and drones will be used ever more widely over the skies.

14.1.2.1 UAV Spectrum Sharing

The Third Generation Partnership (3GPP) recently proposed using drones as user equipment (UE) in the sky, helping the conventional cellular network [20]. The survey shows that the air-to-ground link improves the performance of the desired connection, but it also increases the interference. Hence, interference management and control is one of the critical issues to build up drone cellular networks. As illustrated in Figure 14.1, in a UAV network, there is a control station which is mainly responsible for dispatching, coordinating, charging, and collecting of UAVs [21]. When UAVs act as aerial sensors, the traffic is generated in UAVs and transmitted to the control station via multi-hop transmissions. When UAVs act as base stations, the traffic is generated from ground nodes and is forwarded to other ground nodes via multiple UAVs.

In these two scenarios, repeated data transmissions in multi-hop transmission manner will consume radio resources and decrease network capacity. Thus, environment cognition has been proposed to improve the capacity of a UAV network. Also, a broadband UAV network would need to be constructed to support massive data transmission. According to the Shannon–Hartley theorem [22], the capacity increases logarithmically with the transmit power. However, the capacity increases linearly with the bandwidth of the spectrum. In a UAV network, due to the fact that UAVs have limited carrying capacity and energy supply, the transmit power of the UAV cannot be very large. Thus, exploiting more spectrum is a practical way to increase the capacity of a UAV network. With multiple sensors such as radar, Global Positioning System (GPS), and camera implemented in UAVs, environment cognition including traditional spectrum sensing can be realized to further improve the capacity and adaptability of the UAV network.

14.1.2.2 UAV Spectrum Sharing with Exclusive Regions

It is known that there exist temporal spectrum opportunities in PER. The architecture of three regions, including black, gray, and white regions, has been proposed [23]. The black region is surrounded by the gray region, which in turn is surrounded by the white region [23]. In the black region, SUs are not allowed to transmit. In the gray region, SUs can exploit temporal spectrum opportunities. Namely, SUs can use the spectrum that is not used by PUs. In the white region, because SUs are far away from PUs, SUs can transmit all the time with maximum power.

When primary receivers are densely deployed, the union of PERs of all primary receivers forms a layer. When a UAV is located below the layer, it is not allowed to transmit. Moreover, the architecture of the three regions can be applied in the UAV network. As illustrated in Figure 14.2, layer 1 and layer 2 divide the 3D space into three regions. In region 1, since UAVs are close to the ground users, UAVs are not allowed to transmit. In region 2, UAVs are

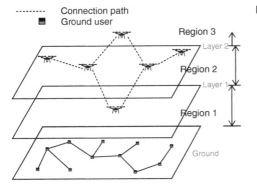

Figure 14.2 PER layers of UAV network.

far from the ground users. Thus, UAVs can exploit the temporal spectrum opportunities. In region 3, since UAVs are sufficiently high above the ground, UAVs can transmit all the time. Regions 1, 2, and 3 are similar to the black, gray, and white regions, respectively, in [24]. It is verified in [24] that the capacity of a secondary network with three regions is larger than that with PER. Hence the architecture of three regions can also be applied in a cognitive UAV network.

The traditional cognitive radio technology focuses on spectrum sensing, spectrum decisions, and spectrum sharing, etc. However, the mobility of UAVs can be controlled to improve network capacity. Besides, the environment information collected by multiple sensors of UAVs can be used to determine the optimal resource allocation scheme.

14.1.3 Chapter Overview

The aim of this chapter is to present an efficient transmission scheme and to optimally allocate resources in an underlay spectrum sharing between drone networks and cellular networks. In Section 14.2 the aim is to study the effect of the unique characteristics of an elevated base station on the drone network distribution. Taking advantage of the stochastic geometry, the 3D blockage effect by random shape theory is determined. Under a blockage model with volume, the exact expression of the signal-to-noise ratio (SNR) meta-distribution is derived. The result relaxes the assumption of conventional 2D Poisson network studies.

In Section 14.3 the focus is to study the deployment of aerial base stations in DSCs while sharing the spectrum with cellular networks. Specifically, the optimal DSC base station density in optimizing the downlink throughput is studied in two scenarios: a spectrum sharing of single-tier DSCs in a 3D network, and a spectrum sharing between the DSC network and the traditional 2D cellular network. The optimal DSC base station density by maximizing the DSC network throughput is derived. Then, the underlay spectrum sharing between the 3D DSC network and traditional 2D cellular networks is analyzed, and the optimal density of DSC aerial base stations that maximizes the DSC network throughput while satisfying the cellular network efficiency constraint is derived.

14.2 SNR Meta-Distribution of Drone Networks

14.2.1 Stochastic Geometry Analysis

The random and irregular nature of the locations of nodes in the wireless network has led to a growing interest in the use of stochastic geometry and Poisson point processes (PPPs) for accurate, flexible, and tractable spatial modeling and analysis. We take advantage of Slivnyak's theorem [25], which states that, for a PPP, the distribution of the original process is equal to its reduced Palm distribution. The overall network performance analysis is conducted by considering the typical user performance metrics. The most representative performance metrics are the signal-to-interference-plus-noise ratio (SINR) and signal-to-interference ratio (SIR), which are defined as

$$\text{SINR} = \frac{S}{I+N}, \tag{14.1}$$

$$\mathrm{SIR} = \frac{S}{I}. \tag{14.2}$$

Here S is the received power at the receiver, namely the incoming signal from the desired transmitter, I is the combined power of all the other (interfering) transmitters in the network, and N is the power from some thermal noise term.

The aim of stochastic geometry wireless network models is to derive closed-form expressions for SINR or SIR determining the successful probability, which is defined as

$$P_c = \mathbb{P}(\mathrm{SIR} > \theta),$$

where θ is the predefined outage constraint. The calculation of the SIR and SINR distributions at the typical receiver or, equivalently, the success probability of transmissions over the typical link in Poisson bipolar and cellular networks under the Rayleigh fading assumption, is relatively straightforward [26].

However, the relative research based on the successful probability only provides limited information on the success probabilities of the individual links. The typical successful probability derived by stochastic geometry is found by taking the average of a group of probabilities. Practically, network operators would be more interested in looking at the distribution of such a group of probabilities. The SIR meta-distribution [27] is defined as the distribution of the conditional success probability, given the point processes. The meta-distribution provides fine-grained information on the SIR. It answers questions such as: "What fraction of users in a network can achieve a desired link reliability given the required SIR threshold?"

Formally, the downlink SIR distribution is defined as

$$F(\theta, x) = \mathbb{P}^o(P_s(\theta) > x), \quad \theta \in \mathbb{R}^+, \ x \in [0, 1],$$

where $P_s(\theta) = \mathbb{P}(\mathrm{SIR} > \theta | \Phi)$ is the successful probability conditioned on a realization of the base station point process Φ, $\mathbb{P}^o$ is the Palm measure of the point process, given an active receiver at the origin, and the SIR is measured at that receiver. Since it is infeasible to directly derive the tractable exact expression, researchers study the nth moments of such a distribution [27], which is defined as

$$M_n(\theta) = \mathbb{E}(P_s(\theta)^n).$$

The first moment is just the typical successful probability by stochastic geometry. It has been shown [27] that approximating the distribution by a beta distribution using the first and second moments results in a very good match.

This section aims to study the effect of the unique characteristics of an elevated base station on the distribution, e.g., a DSC network [28]. We need to consider the 3D blockage effect by random shape theory [29] to determine the line-of-sight probabilities for the received power. The result in this section provides answers to questions such as: "What fraction of users in a network can achieve a desired link reliability given the required SNR threshold with aerial base stations 150 m above the ground in an urban area?"

14.2.2 Characteristic Function of the SNR Meta-Distribution

As shown in Figure 14.3, a drone downlink model is considered. The active elevated base stations are modeled as PPP Φ in a 2D plane with height H_a above the ground with density

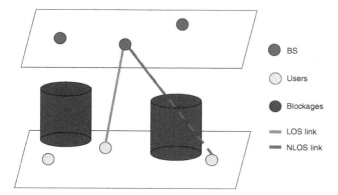

Figure 14.3 System model of the elevated base stations with blockages.

λ_a. The user is connected to its closest base station. Base stations transmit with unit power. Blockages are modeled as cylinders with the same height H_b and radius R. The center of blockages on the ground are modeled as another PPP Φ_b.

For the line-of-sight (LOS) links, the small-scale fading is ignored. Denote the path-loss exponent by α and attenuation factor due to the non-line-of-sight (NLOS) connection by η. Assume the noise N_0 is a constant. The received SNR of a typical receiver at the origin x away from its associated transmitter is given by

$$\text{SNR}_L = \frac{x^{-\alpha}}{N_0}.$$

For NLOS links, the small-scale fading channel is modeled as an exponentially distributed variable with unit mean, $h \sim \exp(1)$. If the link is NLOS

$$\text{SNR}_N = \frac{|h|x^{-\alpha_N}}{N_0}.$$

Denote $a \in \{L, N\}$ as the subscript indicating that the link is LOS or NLOS, and

$$P_a(\theta) = \mathbb{P}(\text{SNR} > \theta \mid \Phi, \Phi_b).$$

Since $\mathbb{P}(P_a > \tau) = \mathbb{P}(\log(P_a) > \log(\tau))$, we look at the characteristic function of $\log(P_a)$:

$$\begin{aligned}
\phi(t) &= \mathbb{E}(\exp(jt \log(P_a))) \\
&= \mathbb{E}^{\Lambda_a}(\mathbb{E}^{\Lambda_b}(P_a^{it})) \\
&= \mathbb{E}^{\Lambda_a}\left(\mathbb{E}^N \int_{\mathbb{R}^2} \cdots \int_{\mathbb{R}^2} \lambda_b(P_a^{it}) \, db_1 \cdots db_N\right).
\end{aligned} \tag{14.3}$$

The events of whether the typical link is LOS or NLOS are mutually exclusive. We define spaces $\mathbb{L}, \mathbb{N} \subset \mathbb{R}^{2N}$ that represent the patterns of drones and blockages resulting in the link being LOS or NLOS.

Then we can further derive the above equation as

$$\begin{aligned}
\phi(t) &= \mathbb{E}^{\Lambda_a}\left(\mathbb{E}^N\left(\int_{\mathbb{L}} \lambda_b(P_L^{it}) \, db_1 \cdots db_N + \int_{\mathbb{N}} \lambda_b(P_N^{it}) \, db_1 \cdots db_N\right)\right) \\
&= \mathbb{E}^{\Lambda_a}(\mathbb{E}^{\Lambda_b, \mathbb{L}}(P_L^{it}) + \mathbb{E}^{\Lambda_b, \mathbb{N}}(P_N^{it}))
\end{aligned}$$

$$
= \mathbb{E}^{\Lambda_a}(\mathbb{E}^{\Lambda_b,\mathbb{L}}(P_L^{it})) + \mathbb{E}^{\Lambda_a}(\mathbb{E}^{\Lambda_b,\mathbb{N}}(P_N^{it}))
$$

$$
= \mathbb{E}^{\Lambda_a}(\mathbb{E}^{\Lambda_b,\mathbb{L}}(1)(\mathbb{P}(\mathrm{SNR}_L > \theta \mid \Phi)^{it})) + \mathbb{E}^{\Lambda_a}(\mathbb{E}^{\Lambda_b,\mathbb{N}}(1)(\mathbb{P}(\mathrm{SNR}_N > \theta \mid \Phi)^{it})),
$$

(14.4)

where $\mathbb{E}^{\Lambda_b,\mathbb{L}}(1)$ is the probabilistic LOS probability, which is derived in the next subsection:

$$
\mathbb{E}^{\Lambda_b,\mathbb{L}}(1) = 1 - \mathbb{E}^{\Lambda_b,\mathbb{N}}(1) = \exp\left(-\lambda_b\left(\pi R^2 + 2R\frac{H_b}{H_a}x\right)\right).
$$

(14.5)

We also have

$$
\mathbb{P}(\mathrm{SNR}_L > \theta \mid \Phi) = \mathbb{1}_{\mathrm{SNR}_L > \theta} = \mathbb{1}_{x < (\theta N_0)^{-1/\alpha}}.
$$

(14.6)

To average over the node distribution, the probability density function (pdf) $f(x)$ of the distribution of the distance from the origin to the closest base station is derived in the next section. We have

$$
\mathbb{E}^{\Lambda_a}(\mathbb{E}^{\Lambda_b,\mathbb{L}}(1)(\mathbb{P}(\mathrm{SNR}_L > \theta \mid \Phi)^{it}))
$$

$$
= \int_{H_a}^{\infty} \mathbb{1}_{x < (\theta N_0)^{-1/\alpha}}^{it}\left(\exp\left(-\lambda_b\left(\pi R^2 + 2R\frac{H_b}{H_a}x\right)\right)\right)f(x)\,\mathrm{d}x
$$

$$
= \int_{H_a}^{(\theta N_0)^{-1/\alpha}}\left(\exp\left(-\lambda_b\left(\pi R^2 + 2R\frac{H_b}{H_a}x\right)\right)\right)f(x)\,\mathrm{d}x,
$$

(14.7)

where

$$
f(x) = 2\pi\lambda_a x\exp(-\pi\lambda_a(x^2 - H_a^2)).
$$

(14.8)

When $\alpha = 2$, Eq. (14.7) can be further evaluated as

$$
\mathbb{E}^{\Lambda_a}(\mathbb{E}^{\Lambda_b,\mathbb{L}}(1)(\mathbb{P}(\mathrm{SNR}_L > \theta \mid \Phi)^{it})) = \left(\frac{L_1(L_2 + L_3 + L_4)}{L_5}\right)\Big|_{x = H_a}^{x = (\theta N_0)^{-1/\alpha}},
$$

(14.9)

where

$$
L_1 = \exp\left(\frac{H_a^4\lambda_a\pi^2 + H_b^2 R^2\lambda_b^2 - \pi^2 R^2 H_a^2\lambda_a\lambda_b}{\pi H_a^2\lambda_a}\right),
$$

(14.10)

$$
L_2 = \Gamma\left(1, \frac{\pi^2 H_a^2\lambda_a^2 x^2 + 2\pi H_a H_b R\lambda_a\lambda_b x + H_b^2 R^2\lambda_b^2}{\pi H_a^2\lambda_a}\right)H_a\sqrt{\pi\lambda_a}\,|2\pi H_a\lambda_a x + 2H_b\lambda_b R|,
$$

(14.11)

$$
L_3 = -2\Gamma\left(\frac{1}{2}, \frac{\pi^2 H_a^2\lambda_a^2 x^2 + 2\pi H_a H_b R\lambda_a\lambda_b x + H_b^2 R^2\lambda_b^2}{\pi H_a^2\lambda_a}\right)\pi H_a H_b R\lambda_a\lambda_b x,
$$

(14.12)

$$
L_4 = -2\Gamma\left(\frac{1}{2}, \frac{\pi^2 H_a^2\lambda_a^2 x^2 + 2\pi H_a H_b R\lambda_a\lambda_b x + H_b^2 R^2\lambda_b^2}{\pi H_a^2\lambda_a}\right)H_b^2 R^2\lambda_b^2,
$$

(14.13)

$$
L_5 = H_a\sqrt{\pi\lambda_a}\,|2H_a\lambda_a\pi x + 2H_b R\lambda_b|.
$$

(14.14)

Similarly, for the NLOS component, we have

$$
\mathbb{E}^{\Lambda_a}(\mathbb{E}^{\Lambda_b,\mathbb{N}}(1)(\mathbb{P}(\mathrm{SNR}_N > \theta \mid \Phi)^{it}))
$$

$$
= \int_{H_a}^{\infty}\exp(-\theta N_0 x^{\alpha_N})^{it}\left[1 - \left(\exp\left(-\lambda_b\left(\pi R^2 + 2R\frac{H_b}{H_a}x\right)\right)\right)\right]f(x)\,\mathrm{d}x.
$$

(14.15)

There is no closed-form expression for the integral above. One way to approximate the distribution is to find the moments from the characteristic function and form a beta distribution approximation (since the LOS part is independent of t, the moments of the distribution depend on the NLOS part):

$$\mathbb{E}(\log{(P_a)}^n) = i^{(-n)}\phi^{(n)}(0).$$ (14.16)

The first moment is calculated as

$$\mathbb{P}(\mathrm{SNR}_N > \theta) = \int_{H_a}^{\infty} \theta N_0 x^{\alpha_N} \left[1 - \left(\exp\left(-\lambda_b\left(\pi R^2 + 2R\frac{H_b}{H_a}x\right)\right)\right)\right] f(x)\, dx.$$ (14.17)

With the first and second moments M_1 and M_2, the beta distribution pdf is given by

$$f_B(X) = \frac{X^{[\mu(\beta+1)-1]/(1-\mu)}(1-X)^{\beta-1}}{B(\mu\beta/(1-\mu),\beta)},$$ (14.18)

where

$$\mu = M_1,$$ (14.19)

$$\beta = \frac{(\mu - M_2)(1-\mu)}{M_2 - \mu^2}.$$ (14.20)

Figure 14.4 shows the numerical evaluation of the approximation of the integrals above. The integral is evaluated by the trapezoidal method. One piece of the interesting work for the future would be to randomize the shape of the blockages.

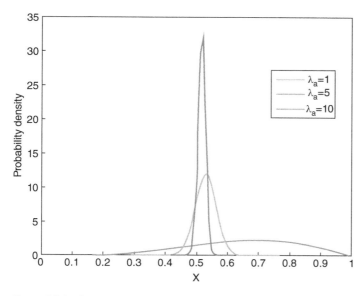

Figure 14.4 Probability density function of the approximated meta-distribution.

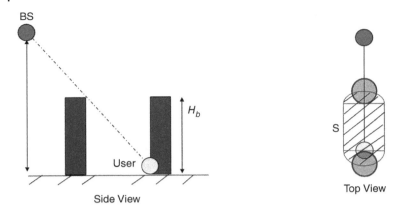

Figure 14.5 Illustration of when the blockage is at the edge of the link connection. The LOS probability is the Poisson null probability that no blockage center is in the shaded ground area S.

14.2.3 LOS Probability

In 2D stochastic geometry, the blockages are often modeled as line segments given a random direction and width [29]. Here we model 3D blockages as cylinders with radius R and height H_b, where the center of each blockage on the ground is distributed as a 2D PPP Φ_b with λ_b. For the simplest case where the radius and height of the blockage are fixed, the probability of having no blockage is the Poisson null probability that no blockage center is in the area S shown in Figure 14.5. Hence the LOS probability is given by

$$p_{\text{LOS}} = \exp(-S\lambda_b) = \exp\left(-\lambda_b\left(\pi R^2 + 2R\frac{H_b}{H_a}x\right)\right).$$

Note that this model does not consider the reflection of the signals. It is assumed that the user is connected to its closest base station; i.e., every base station has a service region regardless of the blockage. In a 2D network we have the distance distribution of the nearest neighbour as [30]

$$f_{\text{2D}}(r) = 2\pi\lambda_a r \exp(-\pi\lambda_a r^2). \tag{14.21}$$

The distance from a receiver at the origin to the base station H_a above the ground corresponding to the 2D distance to its projection $g(x)$ is

$$g(x) = \sqrt{x^2 - H_a^2}. \tag{14.22}$$

The distance distribution from the origin to the nearest neighbour H_a above the ground is derived as

$$f(x) = f_{\text{2D}}(g(x))\,|g^{-1}(x)|' = 2\pi\lambda_a x \exp(-\pi\lambda_a(x^2 - H_a^2)). \tag{14.23}$$

14.3 Spectrum Sharing of Drone Networks

The spectrum sharing of 2D cellular wireless networks has been well studied in recent years. Different interference management strategies in 2D Poisson cognitive radio networks have been developed [31–34]. The focus of this section is to study the aerial base station deployment in DSCs while sharing the spectrum with cellular networks. Specifically, the optimal DSC base station density in optimizing downlink throughput is studied in two scenarios: a spectrum sharing of single-tier DSCs in a 3D network, and a spectrum sharing between the DSC network and the traditional 2D cellular network. The optimal DSC base station density by maximizing the DSC network throughput is derived. Then, the underlay spectrum sharing between the 3D DSC network and traditional 2D cellular network is analyzed, and the optimal density of DSC aerial base stations that maximizes the DSC network throughput while satisfying the cellular network efficiency constraint is derived.

14.3.1 Spectrum Sharing in Single-Tier DSCs

As shown in Figure 14.6, we consider DSC networks in 3D space. We assume that DSC aerial base stations follow a 3D PPP $\{X_i \in \Phi_d\}$ with density λ_d in an infinite 3D space $\mathbb{V}$, but the height is limited to L, that is, $\mathbb{V} = \{(x, y, z) : x, y \in \mathbb{R}, z \in [0, L]\}$. The channel between any pair of DSC aerial base stations and the user here is assumed to undergo path loss and small-scale fading. The path loss is proportional to $x^{-\alpha}$, where x is the distance between the transmitting aerial base station and the typical user, and α is the mean path-loss exponent. The power gain of small-scale fading channel h_i is exponentially distributed with unit mean, and the noise N_0 is additive white Gaussian noise following the distribution $N_0 \sim N(0, N)$. We assume that DSCs only transmit while they are static, i.e., moving DSCs will perform transmission on certain "stop points."

We assume that all aerial base stations transmit at the same power level P_d. For a typical link, the received signal power is hence $h_0 D^{-\alpha}$, where D is the distance between a typical user and a typical aerial base station. We assume that the distance between a user and its serving base station is independent of the base station density.

The transmission is successful if the received SINR at a receiver is larger than a certain threshold. We set the SINR thresholds of DSC users to θ. A typical DSC user at the origin O will receive interference from other transmitting DSC base stations while receiving the

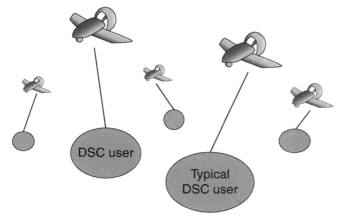

Figure 14.6 Drone small cell (DSC) networks.

desired signal. The received SINR of a typical user at the origin O is

$$\text{SINR}_d = \frac{P_d h_0 D^{-\alpha}}{N + \sum\limits_{x_i \in \Phi_d \backslash \{0\}} P_d h_i x_i^{-\alpha}}. \tag{14.24}$$

The transmission rate of a network is defined as [26]

$$T_d = \lambda_d P(\text{SINR}_d > \theta) \log(1 + \theta). \tag{14.25}$$

First, the transmission rate scales with increase of λ_d. However, with the increase of the density of DSC base stations λ_d, more interference will be generated, thereby resulting in smaller success probability $P(\text{SINR}_d > \theta)$.

Our first goal is to find the optimal DSC base station density λ_d^* that maximizes the throughput of the network. We assume that the DSC network has an outage probability constraint $\epsilon_d \ll 1$. Then the optimal DSC base station density can be obtained by solving the following optimization problem:

$$\max_{\lambda_d} \quad T_d \tag{G1}$$

$$\text{s.t.} \quad P(\text{SINR}_d > \theta) > 1 - \epsilon_d.$$

Next, we derive the optimal DSC base station density for a single-tier DSC downlink network in 3D space. Let the interference $I = \sum_{x_i \in \Phi_d \backslash \{0\}} h_i x_i^{-\alpha}$. From Eq. (14.24) we have the coverage probability as

$$P(\text{SINR}_d > \theta) = P\left(\frac{h_0 D^{-\alpha}}{N/P_d + I} > \theta\right) \tag{14.26}$$

$$= \mathbb{E}_I\left[P\left(h_0 > \theta D^{\alpha}\left(\frac{N}{P_d} + I\right) \mid I\right)\right]$$

$$= \mathbb{E}_I\left[\exp\left(-\theta D^{\alpha}\frac{N}{P_d}\right)\exp(-\theta D^{\alpha}I)\right]$$

$$= \exp\left(-\theta D^{\alpha}\frac{N}{P_d}\right)\mathcal{L}_I(\theta D^{\alpha}). \tag{14.27}$$

Here $\mathcal{L}_I(\theta D^{\alpha})$ is the Laplace transform of I, which can be further derived as

$$\mathcal{L}_I(\theta D^{\alpha}) = \mathbb{E}_I[\exp(-\theta D^{\alpha}I)]$$

$$= \mathbb{E}_{\Phi_d, h_i}\left[\prod_{x_i \in \Phi_d \backslash \{0\}} \exp(-\theta D^{\alpha}h_i x_i^{-\alpha})\right]$$

$$= \mathbb{E}_{\Phi_d}\left[\prod_{x_i \in \Phi_d \backslash \{0\}} \mathbb{E}_{h_i}[\exp(-\theta D^{\alpha}h_i x_i^{-\alpha})]\right] \tag{14.28}$$

$$= \mathbb{E}_{\Phi_d}\left[\prod_{x_i \in \Phi_d \backslash \{0\}} \frac{1}{1 + \theta D^{\alpha}x^{-\alpha}}\right]. \tag{14.29}$$

Equation (14.28) follows because of the independent and identically distributed (i.i.d.) h_i and its further independence from the point process Φ_d. Equation (14.29) follows because h_i is exponentially distributed with unit mean.

The probability generating functional of a set $\mathbb{V}$ is given by [35]

$$\mathbb{E}(\prod_{x_i \in \Phi} f(x)) = \exp\left(-\lambda_d \int_{\mathbb{V}} [1 - f(x)] \, dx\right). \tag{14.30}$$

Applying Eq. (14.30) to Eq. (14.29), we have

$$\begin{aligned}
\mathcal{L}_I(\theta D^\alpha) &= \exp\left(-\lambda_d \int_{\mathbb{V}} \left(1 - \frac{1}{1 + \theta D^\alpha x^{-\alpha}}\right) \, dx\right) \\
&= \exp\left(-\lambda_d \int_0^L \int_0^{2\pi} \int_0^\infty \left(1 - \frac{1}{1 + \theta D^\alpha(\sqrt{(r^2 + z^2)})^{-\alpha}} r\right) \, dr \, d\phi \, dz\right) \\
&= \exp\left(-2\pi\lambda_d \int_0^L \int_0^\infty \frac{\theta D^\alpha r(\sqrt{(r^2 + z^2)})^{-\alpha}}{1 + \theta D^\alpha(\sqrt{(r^2 + z^2)})^{-\alpha}} \, dr \, dz\right) \\
&= \exp(-\lambda_d H(L, \theta, D, \alpha)), \tag{14.31}
\end{aligned}$$

where

$$H(L, \theta, D, \alpha) = \int_0^L \int_0^\infty \frac{2\pi\theta D^\alpha r(\sqrt{(r^2 + z^2)})^{-\alpha}}{1 + \theta D^\alpha(\sqrt{(r^2 + z^2)})^{-\alpha}} \, dr \, dz. \tag{14.32}$$

With Eqs. (14.27)–(14.31), we are now able to solve the optimization problem (G1). Since T_d in Eq. (14.25) is unimodal in terms of λ_d and $P(\text{SINR}_d > \theta)$ is a decreasing function of λ_d, Karush–Kuhn–Tucker (KKT) conditions [32] are applicable. The Lagrange function of the above optimization problem is given by

$$\mathcal{L}(\lambda_d) = T_d + \mu(P(\text{SINR}_d > \theta) - 1 + \epsilon_d), \tag{14.33}$$

where μ is the Lagrange multiplier. Then, the KKT conditions are given as

$$\frac{d\mathcal{L}(\lambda_d)}{d\lambda_d} = \frac{dT_d}{d\lambda_d} + \mu\frac{dP(\text{SINR}_d > \theta)}{d\lambda_d}, \tag{14.34}$$

$$\mu(P(\text{SINR}_d > \theta) - 1 + \epsilon_d) = 0, \tag{14.35}$$

$$P(\text{SINR}_d > \theta) - 1 + \epsilon_d \geq 0, \tag{14.36}$$

$$\mu \geq 0 \quad \text{and} \quad \lambda_d \geq 0. \tag{14.37}$$

Solving Eqs. (14.34)–(14.37) yields the optimal DSC aerial base stations, as follows:

$$\lambda_d^* = \frac{\left[-\ln\left(\frac{1 - \epsilon_d}{\exp(-\theta D^\alpha N/P_d)}\right)\right]^+}{H(L, \theta, D, \alpha)} \tag{14.38}$$

where $[\cdot]^+$ denotes $\max(\cdot, 0)$. The maximum DSC network throughput is given by

$$T_d^* = \frac{\left[-\ln\left(\frac{1 - \epsilon_d}{\exp(-\theta D^\alpha N/P_d)}\right)\right]^+}{H(L, \theta, D, \alpha)} (1 - \epsilon_d) \log(1 + \theta). \tag{14.39}$$

It can be seen from Eq. (14.38) that the optimal primary density is obtained when the success probability just meets the outage constraint. Since $\epsilon_d \ll 1$, we have $\ln(1 - \epsilon_d) \approx -\epsilon_d$.

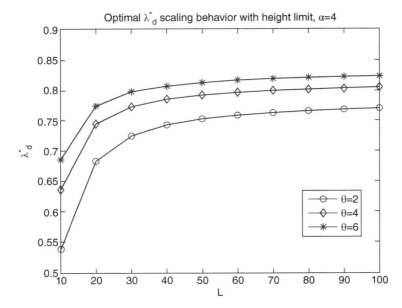

Figure 14.7 Optimal λ_d^* scaling behavior with height limit, $\alpha = 4$

Therefore, for small ϵ_d, both optimal DSC density and potential throughput are linear with the DSC network outage constraint ϵ_d.

Figure 14.7 plots the optimal λ_d^* as a function of L. From Figure 14.7 we can see that the optimal DSC density increases to a limit as the height limit L increases. This is because, as the height limit increases, more aerial base stations are allowed to transmit. Also, the optimal λ_d^* increases with the transmission threshold θ. As θ increases, the success probability decreases, and a larger node density is needed to reach the maximum throughput.

14.3.2 Spectrum Sharing with Cellular Network

In this subsection, a scenario in which the 3D DSC network coexists with a 2D downlink cellular network in underlay spectrum sharing is analyzed, and optimal deployment density of the DSCs is studied.

As shown in Figure 14.8, cellular base stations are distributed on the ground as a 2D PPP $\{Y_j \in \Phi_c\}$ with density λ_c. The distance between a typical cellular base station and its associated user is d. Similarly to the DSC network, we assume that all cellular base stations transmit with power P_c. The channel between any base station and the user undergoes path loss and small-scale fading. The power gain of the small-scale fading channel h_j is exponentially distributed with unit mean, and the noise is $N_0 \sim N(0, N)$. Each cellular transmitter decides to transmit if the received SINR is larger than a certain threshold θ_c with a one-bit feedback from the cellular receiver.

Consider a typical DSC user at the origin O, it will receive interference not only from transmitting DSC base stations but also from transmitting cellular base stations. The SINR

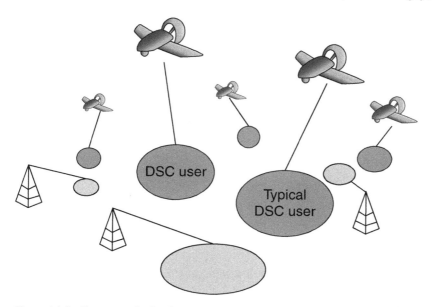

Figure 14.8 Spectrum sharing between drone small cell (DSC) networks and cellular networks.

of the typical DSC user is given by

$$\text{SINR}_{dc} = \frac{P_d h_0 D^{-\alpha}}{N + P_d \sum\limits_{i \in \Phi_d \backslash \{0\}} h_i x_i^{-\alpha} + P_c \sum\limits_{j \in \Phi_c} h_j y_j^{-\alpha}}. \tag{14.40}$$

Similarly, the SINR expressions of a typical cellular user with and without the DSC network are given, respectively, as follows:

$$\text{SINR}_c = \frac{P_c h_0 d^{-\alpha}}{N + P_d \sum\limits_{i \in \Phi_d} h_i x_i^{-\alpha} + P_c \sum\limits_{j \in \Phi_c \backslash \{0\}} h_j y_j^{-\alpha}}, \tag{14.41}$$

$$\text{SINR}_c' = \frac{P_c h_0 d^{-\alpha}}{N + P_c \sum\limits_{j \in \Phi_c \backslash \{0\}} h_j y_j^{-\alpha}}. \tag{14.42}$$

The throughputs of the DSC and cellular networks are

$$T_{dc} = \lambda_{dc} P(\text{SINR}_{dc} > \theta) \log(1 + \theta), \tag{14.43}$$

$$T_c = \lambda_c P(\text{SINR}_c > \theta_c) \log(1 + \theta_c), \tag{14.44}$$

$$T_c' = \lambda_c P(\text{SINR}_c' > \theta_c) \log(1 + \theta_c), \tag{14.45}$$

where T_c and T_c' are the cellular network throughputs with and without the DSC network, respectively. Define δ as the cellular efficiency loss ratio $\delta = (T_c - T_c')/T_c$ and r_{th} as the cellular efficiency loss constraint [15]. The optimal DSC base station density coexisting with a cellular network λ_{dc} can be obtained by solving the following optimization problem:

$$\max_{\lambda_{dc}} \quad T_{dc} \tag{G2}$$

s.t. $P(\text{SINR}_{dc} > \theta) > 1 - \epsilon_{dc}$,

$\delta \leq r_{th}$.

Next, we derive the optimal DSC base station density when coexisting with cellular networks.

Consider a DSC network that coexists with a 2D cellular network. Denote $I_c = \sum_{j \in \Phi_c} h_j y_j^{-\alpha}$. The success probability of the typical DSC user is given by

$$
P(\text{SINR}_{dc} > \theta) = P\left(\frac{h_0 D^{-\alpha}}{\dfrac{N}{P_d} + I + \dfrac{P_c}{P_d} I_c} > \theta \right)
$$

$$
= \exp\left(-\theta D^\alpha \frac{N}{P_d} \right) \mathbb{E}_I[\exp(-\theta D^\alpha I)] \mathbb{E}_{I_c}\left[\exp\left(-\theta D^\alpha \frac{P_d}{P_c} I_c \right) \right],
$$

(14.46)

where I was given just before Eq. (14.26). The second term in Eq. (14.46) has been evaluated by Eqs. (14.27)–(14.31). The third term in Eq. (14.46) is the interference from other cellular base stations following a 2D PPP and can be derived as [15]:

$$
\mathbb{E}_{I_c}\left[\exp\left(-\theta D^\alpha \frac{P_d}{P_c} I_c \right) \right] = \exp\left(-\lambda_c C D^2 \left(\frac{P_d}{P_c} \theta \right)^{2/\alpha} \exp\left(-d^\alpha \frac{N}{P_c} \theta_c \right) \right),
$$

(14.47)

where

$$
C = \frac{2\pi^2}{\alpha \sin(2\pi/\alpha)}.
$$

The success probability of a typical DSC user is

$$
P(\text{SINR}_{dc} > \theta) = A_1 \exp(-A_2 \lambda_d) \exp(-A_3 \lambda_c),
$$

(14.48)

where

$$
A_1 = \exp\left(-\theta D^\alpha \frac{N}{P_d} \right),
$$

$$
A_2 = H(L, \theta, D, \alpha),
$$

(14.49)

$$
A_3 = C D^2 \left(\frac{P_d}{P_c} \theta \right)^{2/\alpha} \exp\left(-d^\alpha \frac{N}{P_c} \theta_c \right).
$$

With the success probability expressions, we are able to solve the optimization problem (G2). Likewise, for deriving the success probability of a typical cellular user, along with Eqs. (14.44) and (14.45), the cellular efficiency loss ratio is first calculated:

$$
\delta = \frac{T_c - T_c'}{T_c}
$$

$$
= 1 - \exp\left(-H\left(L, \theta_c \frac{P_c}{P_d}, d, \alpha \right) \lambda_{dc} \right).
$$

(14.50)

The Lagrange function of the optimization problem (G2) is hence given by

$$
\mathcal{L}(\lambda_{dc}) = T_{dc} + \mu_1 (P(\text{SINR}_{dc} > \theta)) - \mu_2 (\delta - r_{th}).
$$

(14.51)

The KKT conditions are listed as follows:

$$\frac{\mathrm{d}\mathcal{L}(\lambda_{dc})}{\mathrm{d}\lambda_{dc}} = \frac{\mathrm{d}T_{dc}}{\mathrm{d}\lambda_{dc}} + \mu_1 \frac{\mathrm{d}P(\mathrm{SINR}_{dc} > \theta)}{\mathrm{d}\lambda_{dc}} - \mu_2 \frac{\mathrm{d}\delta}{\mathrm{d}\lambda_{dc}}, \tag{14.52}$$

$$\mu_1(P(\mathrm{SINR}_{dc} > \theta) - 1 + \epsilon_{dc}) = 0, \tag{14.53}$$

$$\mu_2(\delta - r_{th}) = 0, \tag{14.54}$$

$$(-1 + \epsilon_{dc}) \leq P(\mathrm{SINR}_{dc} > \theta), \tag{14.55}$$

$$\delta \leq r_{th}, \tag{14.56}$$

$$\mu_1 \geq 0, \quad \mu_2 \geq 0 \quad \text{and} \quad \lambda_{dc} \geq 0. \tag{14.57}$$

Solving the above equations yields the following:

$$\text{if } \epsilon_{dc} < \eta, \quad \lambda_{dc}^* = \frac{[-\ln(1 - \epsilon_{dc}) - D^\alpha(N/P_d)\theta - A_3\lambda_c]^+}{A_2}, \tag{14.58}$$

$$\text{if } \epsilon_{dc} \geq \eta, \quad \lambda_{dc}^* = \frac{\ln(1 - r_{th})}{H(L, \theta_c(P_c/P_d), d, \alpha)}, \tag{14.59}$$

where

$$\eta = 1 - \exp(-A_3\lambda_c)A_1(1 - r_{th})^{H(L,\theta,D,\alpha)/H(L,\theta_c(P_c/P_d),d,\alpha)}.$$

We have the maximized potential throughput if $\epsilon_{dc} < \eta$:

$$T_{dc}^* = \lambda_{dc}^* P(\mathrm{SINR}_{dc*} > \theta)\log(1 + \theta), \tag{14.60}$$

where $P(\mathrm{SINR}_{dc*} > \theta)$ can be evaluated by substituting λ_{dc}^* into Eq. (14.48).

For $\epsilon_{dc} < \eta$, since $1 - \epsilon_{dc} \approx 1$, $\ln(1 - \epsilon_{dc}) \approx -\epsilon_{dc}$. The approximated optimal DSC density and potential throughput are both linear with the DSC network outage constraint ϵ_{dc}. For $\epsilon_{dc} > \eta$, both optimal DSC density and potential throughput are constants which are independent of ϵ_{dc}.

14.4 Summary

Spectrum sharing is a promising technology that improves the spectral utilization efficiency and provides the foundation stone to build up future drone cellular networks. This chapter provides an analytical framework for drone network performance analysis. It also provides a design for an efficient transmission scheme to optimally allocate resources in an underlay spectrum sharing between drone networks and cellular networks, under primary cellular throughput protection, and maximized drone network throughput, and spectrum utilization efficiency.

In Section 14.2, the aim is to study the effect of the unique characteristics of an elevated base station on the drone network distribution. Taking advantage of the stochastic geometry, the 3D blockage effect is analyzed by random shape theory. Under a blockage model with volume, the exact expression of the SNR meta-distribution is presented. The focus of

Section 14.3 is to study the deployment of aerial base stations in DSCs while sharing the spectrum with cellular networks. Starting with a consideration of the spectrum sharing of single-tier DSCs in a 3D network, the optimal DSC base station density by maximizing the DSC network throughput is derived. Then, the underlay spectrum sharing between the 3D DSC network and traditional 2D cellular networks is studied.

References

1 J. Mitola and G. Q. Maguire, "Cognitive radio: An integrated agent architecture for software defined radio," in *IEEE Personal Commun.*, vol. 6, pp. 13–18, Aug 1999.

2 S. Haykin, "Cognitve radio: brain-empowered wireless communications," in *J. Sel. Areas Commun.*, vol. 23, pp. 201–220, Feb. 2005.

3 I. F. Akyildiz, W.-Y. Lee, M. C. Vuran, and S. Mohanty, "Next generation/dynamic spectrum access/cognitive radio wireless networks: a survey," in *Computer Networks*, vol. 50, pp. 2127–2159, Sep. 2006.

4 S. Srinivasa and S. A. Jafar, "The throughput potential of cognitive radio: a theoretical perspective," in *IEEE Commun. Mag.*, pp. 73–79, May 2007.

5 Y. Han, A. Pandharipande, and S. H. Ting, "Cooperative decode-andforward relaying for secondary spectrum access," in *IEEE Trans. Wireless Commun.*, vol. 9, pp. 2914–2923, Sep. 2010.

6 C. Zhai, W. Zhang, and P. C. Ching, "Cooperative spectrum sharing based on two-path successive relaying," in *IEEE Trans. Wireless Commun.*, vol. 61, pp. 2260–2270, Jun. 2013.

7 L. Gao, R. Zhang, C. Yin, and S. Cui, "Throughput and delay scaling in supportive two-tier networks," in *IEEE J. Sel. Areas Commun.*, vol. 30, pp. 415–424, Feb. 2012.

8 S.W. Jeon, N. Devroye, M. Vu, S.-Y. Chung, and V. Tarokh, "Cognitive networks achieve throughput scaling of a homogeneous network," in *IEEE Trans. Inf. Theory*, vol. 57, pp. 5103–5115, Nov. 2011.

9 X. Song, C. Yin, D. Liu, and R. Zhang, "Spatial throughput characterization in cognitive radio networks with threshold-based opportunistic spectrum access," in *IEEE J. Sel. Areas Commun.*, vol. 32, no. 11, pp. 2190–2204, November 2014

10 C. Lee and M. Haenggi, "Interference and outage Poisson cognitive networks," in *IEEE Trans. Wireless Commun.*, vol. 11, pp. 1392–1401, Apr. 2012.

11 L. Wang and V. Fodor, "On the gain of primary exclusion region and vertical cooperation in spectrum sharing wireless networks," in *IEEE Trans. Veh. Technol.*, vol. 61, pp. 3746–3758, Oct. 2012.

12 M. Vu and V. Tarokh, "Scaling laws of single-hop cognitive networks," in *IEEE Trans. Wireless Commun.*, vol. 8, pp. 4089–4097, Aug. 2009.

13 M. Vu, N. Devroye, and V. Tarokh, "On the primary exclusive region of cognitive networks," *IEEE Trans. Wireless Commun.*, vol. 8, pp. 3380–3385, Jul. 2009.

14 A. Bagayoko, P. Tortelier and I. Fijalkow, "Impact of shadowing on the primary exclusive region in cognitive networks," in *Wireless Conf. (EW)*, European, Lucca, 2010, pp. 105–110.

15 Z. Wang and W. Zhang, "Opportunistic spectrum sharing with limited feedback in Poisson cognitive radio networks," *IEEE Trans. Commun.*, vol. 13, no. 12, pp. 7098–7109, Dec. 2014.

16 R. Dahama, K. W. Sowerby and G. B. Rowe, "Estimating protection distances in spectrum sharing systems," *IEEE Trans. on Signal Process.*, vol. 61, no. 17, pp. 4284–4295, Sep. 2013.

17 A. Al-Hourani, S. Kandeepan and A. Jamalipour, "Modeling air-to-ground path loss for low altitude platforms in urban environments," in *IEEE Global Commun. Conf. (GLOBECOM)*, Austin, TX, 2014, pp. 2898–2904.

18 A. Al-Hourani, S. Kandeepan, and S. Lardner (2014). Optimal LAP altitude for maximum coverage. *IEEE Wireless Communications Letters*, vol. 3, no. 6, pp. 569–572, Dec. 2014. doi: 10.1109/LWC.2014.2342736

19 M. Mozaffari, W. Saad, M. Bennis and M. Debbah, "Drone small cells in the clouds: design, deployment and performance analysis," in *IEEE Global Commun. Conf. (GLOBECOM)*, San Diego, CA, 2015, pp. 1–6.

20 3GPP Technical Report 36.777. Technical specification group radioaccess network. Study on enhanced LTE support for aerial vehicles(Release 15). Dec. 2017

21 Z. Wei, H. Wu, S. Huang and Z. Feng (2017). Scaling Laws of Unmanned Aerial Vehicle Network with Mobility Pattern Information, *IEEE Communications Letters*, vol. 21, no. 6, pp. 1389–1392, Jun. 2017. doi: 10.1109/LCOMM.2017.2671861

22 T. Cover and J. Thomas(2012). *Elements of Informantion Theory*, John Wiley and Sons.

23 Z. Wei, Z. Feng, Q. Zhang and W. Li, Three Regions for Space-Time Spectrum Sensing and Access in Cognitive Radio Networks, *IEEE Transactions on Vehicular Technology*, vol. 64, no. 6, pp. 2448–2462, Jun. 2015. doi: 10.1109/GLOCOM.2012.6503290

24 Z. Wei, Z. Feng, Q. Zhang and W. Li (2015). Three Regions for Space-Time Spectrum Sensing and Access in Cognitive Radio Networks. *IEEE Transactions on Vehicular Technology*, vol. 64, no. 6, pp. 2448–2462, June 2015. doi: 10.1109/TVT.2014.2342612

25 Coeurjolly J. F., Moller J., Waagepetersen R., "A tutorial on Palm distributions for spatial point processes", *International Statistical Review*, 2017, vol. 85, no.3, pp. 404–420

26 J. G. Andrews, F. Baccelli and R. K. Ganti, "A tractable approach to coverage and rate in cellular networks," *IEEE Trans. Commun.*, vol. 59, no. 11, pp. 3122–3134, Nov. 2011.

27 M. Haenggi, "The meta distribution of the SIR in Poisson bipolar and cellular networks," *IEEE Trans. Wireless Commun.*, vol. 15, no. 44, pp. 2577C2589, Apr. 2016.

28 L. Yang and W. Zhang, "Hierarchical codebook and beam alignment for UAV communications", in *Proc. IEEE Global Communications Conference (GLOBECOM 2018) Workshop on Wireless Networking and Control for UAV*, Abu Dhabi, UAE, Dec. 9–13, 2018.

29 T. Bai, R. Vaze and R. W. Heath, "Analysis of Blockage Effects on Urban Cellular Networks," in *IEEE Transactions on Wireless Communications*, vol. 13, no. 9, pp. 5070–5083, Sept. 2014.

30 D. Moltchanov, "Survey paper: Distance distributions in random networks," *Ad Hoc Networks*, vol. 10, no. 6, pp. 1146–1166, August 2012

31 A. K. Gupta, X. Zhang and J. G. Andrews, "SINR and throughput scaling in ultradense urban cellular networks," *IEEE Wireless Commun. Lett.*, vol. 4, no. 6, pp. 605–608,

32 Z. Wang and W. Zhang, "Spectrum sharing with limited feedback in Poisson cognitive network." in *IEEE Int. Conf. Commun. (ICC)*, Sydney, NSW, 2014, pp. 1441–1446.

33 Z. Wang and W. Zhang, "Exploiting Multiuser Diversity with 1-bit Feedback for Spectrum Sharing," *IEEE Trans. on Commun.*, vol. 62, no. 1, pp. 29–40, Jan. 2014.

34 C. Zhang and W. Zhang, "Spectrum sharing for drone networks," *IEEE J. Sel. Areas Commun.*, vol. 35, no. 1, Jan. 2017.

35 R. L. Streit, "Probability generating functional," in *Poisson point processes: imaging, tracking, and sensing*, Springer Science & Business Media, 2010, pp. 27.

Part IV

Other Advanced Technologies for UAV Communications

15

Non-Orthogonal Multiple Access for UAV Communications

Tianwei Hou[1], Yuanwei Liu[2,*], and Xin Sun[1]

[1]School of Electronic and Information Engineering, Beijing Jiaotong University, PR China
[2]School of Electronic Engineering and Computer Science, Queen Mary University of London, UK

15.1 Introduction

In the past decades, much research effort has been directed towards developing remotely operated unmanned aerial vehicles (UAVs), which stand as a potential candidate for aerial base stations (BSs) to provide access services to wireless devices located on the ground [1] or in the sky [2]. The air-to-air channel characterization in [3] studied the influence of the altitude-dependent Ricean K-factor. This work indicated that the impact of ground-reflected multipath fading reduces with increasing UAV altitude. A UAV-assisted ground-to-air network, where Ricean channels are used for evaluating strong line-of-sight (LoS) links between the UAV and ground users, was studied in [4]. It is indicated that the Rayleigh fading channel, which is a well-known model in scattering environments, can also be used to model the UAV channel characteristics in the case of large elevation angles in the mixed urban environment. A UAV-assisted underlaid device-to-device (D2D) network with probabilistic LoS channels was proposed in [2], which depends on the height of the UAV, the horizontal distance between the UAV and users, the carrier frequency, and the type of environment. In the case that a LoS link exists, a fixed LoS coefficient, e.g., an extra 20 dB attenuation, is the dominant component of small-scale fading channels.

In order to provide tractable analytical results, the performance of a downlink UAV network over Nakagami-m fading channels was evaluated in [5], where UAVs are distributed in a finite 3D network. An uniform binomial point process was invoked to model the proposed network. A UAV-assisted cellular hotspot scenario was investigated in [6], where the UAV flies cyclically along the cell edge for offloading actions. A multiple input–multiple output (MIMO) assisted UAV for uplink transmission in a cellular network with air-to-ground interference was investigated by utilizing multi-beam techniques [7]. In order to improve the spectrum efficiency and energy efficiency of UAV networks, new research on UAVs under emerging next-generation network architectures is needed.

In UAV-enabled wireless communications, the total UAV energy is limited, which includes propulsion energy and communication-related energy [8]. Therefore, integrating

*Corresponding Author: Yuanwei Liu

UAV Communications for 5G and Beyond, First Edition.
Edited by Yong Zeng, Ismail Guvenc, Rui Zhang, Giovanni Geraci, and David W. Matolak.
© 2021 John Wiley & Sons Ltd. Published 2021 by John Wiley & Sons Ltd.

UAVs and non-orthogonal multiple access (NOMA) into cellular networks is considered to be a promising technique to significantly enhance the performance of terrestrial users in the next-generation wireless system and beyond, where energy efficiency and spectrum efficiency can be greatly enhanced in downlink transmission to minimize communication-related energy [9]. A general introduction to UAV communications has been proposed in [10]. Three case studies, i.e., performance evaluation, joint trajectory design, and machine-learning-assisted UAV deployment [11], were carried out in order to better understand NOMA-enabled UAV networks. A fixed number of UAVs used as flying relays to support wireless backhaul networks was considered in [12]. The impact of system performance of MIMO-NOMA-assisted UAV networks was proposed in [13], where closed-form expressions of outage performance and ergodic rate were evaluated in the downlink scenario. In [14], a NOMA-assisted uplink scenario of UAV-assisted cellular communication was investigated, where two special strategies, i.e., egoistic and altruistic transmission strategies, were considered to derive the optimized solutions.

15.1.1 Motivation

The aforementioned two communication concepts, UAV and NOMA, can be naturally linked together as a new spectrum and energy-efficient wireless transmission technique, which is the focus of this chapter. In NOMA-assisted cellular networks, a natural question arises: "How to implement the NOMA technique for multiple randomly roaming users?" In this chapter, by proposing two potential association strategies, namely user-centric strategy and UAV-centric strategy, the effect of a NOMA-assisted UAV network is evaluated. Since NOMA is co-channel interference-limited, it is realistic to implement NOMA with conventional orthogonal multiple access (OMA) technologies. For example, we can first schedule users in pairs to perform NOMA, and then use conventional time/frequency/code-division multiple access to serve the different user pairs.

In this chapter, users and UAVs are spatially randomly deployed on the ground and in the sky via a homogeneous Poisson point process (HPPP), respectively. Here, two user selection strategies are proposed, based on the serving purposes to perform NOMA: (i) the user-centric strategy is a promising solution for providing access services after disasters in remote areas, where all the terrestrial users located in the Voronoi cell can be served by UAVs; and (ii) the UAV-centric strategy can be perfectly deployed in dense networks, i.e., concerts or football matches, to provide supplementary access services for offloading actions, where terrestrial users are located in a regular disk. Note that a non-negligible difference is that user association is decided by an individual user or a UAV for the user-centric strategy or for the UAV-centric strategy, respectively.

15.2 User-Centric Strategy for Emergency Communications

We first focus on the scenario that all the terrestrial users need to be served equally for emergency communications, e.g., after disasters, in remote areas or in rural areas [15]. Motivated by this purpose, we propose the user-centric strategy for providing emergency access services to all the terrestrial users.

Focusing on downlink transmission scenarios, we consider the user-centric strategy as shown in Figure 15.1(a). In this chapter, the UAV equipped with a single antenna communicates with multiple users equipped with a single antenna each. In the user-centric strategy, the locations of terrestrial users are random for emergency services, and there is no further information for UAVs to properly plan their trajectories. In order to serve all the terrestrial users equally, multiple UAVs should be distributed uniformly, which conforms to the definition of HPPP. Thereby, the UAVs are distributed according to an HPPP Ψ with density λ.

For the simplicity of theoretical analysis, as shown in Figure 15.1(b), a user is located at the origin point in the user-centric strategy, which becomes the typical user. The user-centric strategy is a useful model for large-scale networks, i.e., a rural area, where users are uniformly located in the Voronoi cell according to an HPPP Φ_u with density λ_u. It is worth mentioning that, in the case that the density of users λ_u is low, the user-centric strategy performs much better than the UAV-centric strategy.

Without loss of generality, we consider that there is one user, namely a fixed user, already connected to the UAV in the previous round of the user association process. In practice, multiple users are connected to the transmitter (UAV) one by one. For simplicity, we assume that the horizontal distance between the fixed user and the connected UAV is r_k, which can be any arbitrary value, and the horizontal distance between the typical user and the connected UAV is random, denoted by r. In the user-centric strategy, we consider that two users, the fixed user and the typical user, are paired to perform the NOMA technique, where paired NOMA users share the same frequency, time, and code resource blocks.

Consider the use of a composite channel model with two parts, large-scale fading and small-scale fading. It is assumed that the horizontal distance r is independently and identically distributed (i.i.d.). In this chapter, large-scale fading represents the path loss between the UAV and users. A log-normal distributed random variable for shadowing on both the desired and interference signals was considered in [16], but it is mathematically intractable while not changing the main trends, i.e., the diversity orders and high signal-to-noise ratio (SNR) slopes are the same. Thus, we neglect it in this chapter for simplicity.

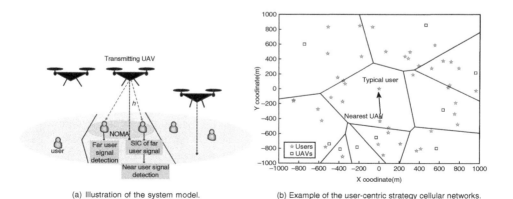

(a) Illustration of the system model. (b) Example of the user-centric strategy cellular networks.

Figure 15.1 Illustration of NOMA-assisted user-centric strategy model.

15.2.1 System Model

In order to illustrate the LoS link between the UAV and user, the small-scale fading is defined by Nakagami fading, and the probability density function (PDF) can be expressed as

$$f(x) = \frac{m^m x^{m-1}}{\Gamma(m)} e^{-mx}, \tag{15.1}$$

where m denotes the fading parameter, and $\Gamma(m)$ represents the Gamma function. Note that $\Gamma(m) = (m-1)!$ when m is an integer. The serving distance between the connected UAV to the typical user can be written as

$$r_t = \sqrt{h^2 + r^2}, \tag{15.2}$$

where r is the nearest horizontal distance allowed between the typical user and its connected UAV, and h denotes the height of the UAV.

Therefore, the large-scale fading can be expressed as

$$L_t = r_t^{-\alpha}, \tag{15.3}$$

where α denotes the path-loss exponent between the typical user and the connected UAV. Thus, the received power from the associated UAV for the user at the origin is given by

$$P_t = P_u L_t |h_t|^2, \tag{15.4}$$

where P_u denotes the transmit power of the UAV, and h_t denotes the channel coefficients for the typical user and its associated UAV.

In downlink transmission, paired NOMA users also detect interference from neighboring UAVs. Therefore, the co-channel interference I can be further expressed as follows:

$$I \triangleq \sum_{j \in \Psi, \, d_j > r_t} |g_j|^2 P_u d_j^{-\alpha_I}, \tag{15.5}$$

where d_j and $|g_j|^2$ denote the distance and the small-scale fading between the user and the jth interfering UAV, and α_I denotes the path-loss exponent between the interfering UAV and the typical user.

Besides, in practical wireless communication systems, obtaining the channel state information (CSI) at the transmitter or receiver is a non-trivial problem, which requires the classic pilot-based training process. Therefore, in order to provide more engineering insights, it is assumed that the CSI of UAVs is partly known at the typical user, where only distance information between UAVs and the typical user is required. The signal-to-interference-plus-noise ratio (SINR) of the user-centric strategy will be derived in the following subsection.

For the user-centric strategy, since the distance of the typical user is not predetermined, there are two potential cases for the typical user, namely far user case and near user case. We then turn our attention to the SINR analysis of these two potential cases.

15.2.1.1 Far user case

For the far user case, where the serving distance of the typical user is greater than that of the fixed user, i.e., $r > r_k$, the typical user treats the signal from the fixed user as interference, and thus the SINR can be expressed as

$$\text{SINR}_{t,\text{far}} = \frac{|h_t|^2 r_t^{-\alpha} P_u \alpha_v^2}{\sigma^2 + |h_t|^2 r_t^{-\alpha} P_u \alpha_w^2 + \sum_{j \in \Psi, \, d_j > r_t} |g_j|^2 P_u d_j^{-\alpha_l}}, \tag{15.6}$$

where σ^2 denotes the additive white Gaussian noise (AWGN) power, and α_v^2 and α_w^2 denote the power allocation factors for the far user and the near user, respectively. Note that $\alpha_v^2 + \alpha_w^2 = 1$ in NOMA communication.

For the far user case, the successive interference cancellation (SIC) technique is deployed at the fixed user. Thereby the fixed user needs to decode the information from the typical user with the following SINR:

$$\text{SINR}_{f \to t,\text{far}} = \frac{|h_f|^2 R_k^{-\alpha} P_u \alpha_v^2}{\sigma^2 + |h_f|^2 R_k^{-\alpha} P_u \alpha_w^2 + \sum_{j \in \Psi, \, d_j > r_t} |g_j|^2 P_u d_j^{-\alpha_l}}, \tag{15.7}$$

where $R_k = \sqrt{r_k^2 + h^2}$, and h_f denotes the channel coefficients for the fixed user.

Once it is decoded successfully, the fixed user will decode its own signal with imperfect SIC coefficient, and the SINR can be expressed as

$$\text{SINR}_{f,\text{far}} = \frac{|h_f|^2 R_k^{-\alpha} P_u \alpha_w^2}{\sigma^2 + \beta |h_f|^2 R_k^{-\alpha} P_u \alpha_v^2 + \sum_{j \in \Psi, \, d_j > r_t} |g_j|^2 P_u d_j^{-\alpha_l}}, \tag{15.8}$$

where β denotes the imperfect SIC coefficient. Since in practice SIC is not perfect, a fraction $0 < \beta < 1$ is considered in our model for the user with better channel gain. On the one hand, $\beta = 0$ when perfect SIC is assumed, and the near user can perfectly decode the signal intended for the far user. On the other hand, when SIC fails or there is no corresponding SIC, we have $\beta = 1$.

15.2.1.2 Near user case

For the near user case, when the typical user has smaller serving distance to the UAV than does the fixed user, i.e., $r < r_k$, the signal of the typical user can be treated as interference at the fixed user, and thus the SINR of the fixed user can be expressed as

$$\text{SINR}_{f,\text{near}} = \frac{|h_f|^2 R_k^{-\alpha} P_u \alpha_v^2}{\sigma^2 + |h_f|^2 R_k^{-\alpha} P_u \alpha_w^2 + \sum_{j \in \Psi, \, d_j > r_t} |g_j|^2 P_u d_j^{-\alpha_l}}. \tag{15.9}$$

The SIC technique can be deployed at the typical user for decoding the signal from the fixed user, and the SINR at the typical user in the near user case can be expressed as

$$\text{SINR}_{t \to f,\text{near}} = \frac{|h_t|^2 r_t^{-\alpha} P_u \alpha_v^2}{\sigma^2 + |h_t|^2 r_t^{-\alpha} P_u \alpha_w^2 + \sum_{j \in \Psi, \, d_j > r_t} |g_j|^2 P_u d_j^{-\alpha_l}}. \tag{15.10}$$

Once the typical user decodes the information from the fixed user successfully, it can decode its own signal with the SINR as follows:

$$\text{SINR}_{t,\text{near}} = \frac{|h_t|^2 r_t^{-\alpha} P_u \alpha_w^2}{\sigma^2 + \beta |h_t|^2 r_t^{-\alpha} P_u \alpha_v^2 + \sum_{j \in \Psi, \, d_j > r_t} |g_j|^2 P_u d_j^{-\alpha_l}}. \tag{15.11}$$

15.2.2 Coverage Probability of the User-Centric Strategy

In the networks considered, we first focus on analyzing the PDF of user distance distributions for paired NOMA users, which will be used for both user-centric strategy and UAV-centric strategy.

Lemma 15.1 *We assume that UAVs are distributed according to an HPPP with density λ, and an extra user is located at the origin of the disk in the user-centric strategy, or an extra UAV is located at the origin of the disk in the UAV-centric strategy, which is under the expectation over HPPP. Thus, in the user-centric strategy, the horizontal distance r between the typical user and UAVs follows the distribution*

$$f_r(r) = 2\pi \lambda r \, e^{-\pi \lambda r^2}, \quad r \geq 0. \tag{15.12}$$

Then, we focus on analyzing the user-centric strategy of the proposed network in order to increase the system fairness. In the user-centric strategy, the user association is based on connecting the nearest UAV to the typical user. As such, the first step is to derive the Laplace transform of interference for the typical user.

Lemma 15.2 *For the user-centric strategy, and based on the characteristic of stochastic geometry, the interference received at both the typical user and fixed user can be recognized as the same. Therefore, based on [17], the Laplace transform of the interference distribution for the paired NOMA users is given by*

$$\mathcal{L}_t(s) = \exp\left(-\frac{2\pi\lambda}{\alpha_I} \sum_{i=1}^{m_I} \binom{m_I}{i} \left(\frac{sP_u}{m_I}\right)^{\delta_I} (-1)^{\delta_I - i} B\left(\frac{-sP_u}{m_I r_t^{\alpha_I}}; \, i - \delta_I, \, 1 - m_I\right)\right), \tag{15.13}$$

where $\delta_I = 2/\alpha_I$, m_I denotes the fading parameters between the typical user and interfering UAVs, and $B(\cdot\,; \cdot, \cdot)$ represents the incomplete Beta function.

Proof: Please refer to appendix A of [18]. ■

In the case of large-scale networks, the existence of a LoS link between the interfering UAVs at infinity and users is unlikely. Thus, the minimum received power of inter-cell interference for cellular UAV networks is worth estimating, where the fading parameters m_I between ground users and interfering UAVs are equal to one. It is also assumed that the path-loss exponent $\alpha_I = 4$ because that path-loss exponent is normally in the range of 2–4, where 2 is for propagation in free space, and 4 is for relatively lossy environments and in the case of full specular reflection from the Earth's surface.

Corollary 15.1 *For the special case that the small-scale fading channels between interference sources and users follow Rayleigh fading, thereby $m_I = 1$ and $\alpha_I = 4$ for the user-centric strategy, the Laplace transform of the interference distribution for both paired NOMA users can be transformed into*

$$\mathcal{L}_t(s) \overset{(a)}{=} \exp\left(-\frac{2\pi\lambda P_u r_t^{2-\alpha_I}}{\alpha_I(1-\delta_I)} \,_2F_1(1, \ 1-\delta_I; \ 2-\delta_I; -sP_u r_t^{-\alpha_I})\right)$$

$$\overset{(b)}{=} \exp\left(-\pi\lambda\sqrt{sP_u}\tan^{-1}\left(\frac{\sqrt{sP_u}}{r_t^2}\right)\right), \tag{15.14}$$

where (a) is obtained by applying $m_I = 1$, (b) is obtained by substituting $\alpha_I = 4$, and $_2F_1(\cdot, \cdot; \cdot; \cdot)$ represents the hypergeometric function.

Then, we focus on the coverage behavior of the user-centric strategy. The fixed power allocation strategy is deployed at the UAV, where the power allocation factors α_w^2 and α_v^2 are constant during transmission. It is assumed that the target rates of the typical user and the fixed user are R_t and R_f, respectively. Based on SINR analysis in Eqs. (15.6), (15.10) and (15.11), the coverage probability of the typical user can be expressed as follows:

$$P_t(r) = P_{t,\text{near}}(r) + P_{t,\text{far}}(r)$$

$$= \Pr\left(\text{SINR}_{t\to f,\text{near}} > \varepsilon_f, \ \text{SINR}_{t,\text{near}} > \varepsilon_t\right) + \Pr\left(\text{SINR}_{t,\text{far}} > \varepsilon_t\right), \tag{15.15}$$

where $\varepsilon_t = 2^{R_t} - 1$, $\varepsilon_f = 2^{R_f} - 1$, and $P_{t,\text{near}}(r)$ and $P_{t,\text{far}}(r)$ denote the coverage probability of the typical user for the near user case and far user case, respectively. Therefore, the coverage probabilities of the typical user for the near user case and far user case are given in the following two lemmas.

Lemma 15.3 *The coverage probability conditioned on the serving distance of the typical user for the near user case in the user-centric strategy is expressed in closed form as*

$$P_{t,\text{near}}(r) = \sum_{n=0}^{m-1}\sum_{p=0}^{n}\binom{n}{p}\frac{(-1)^n}{n!}\Lambda_4^n\Lambda_5^n \exp(-mM_{t*}\sigma^2 r_t^\alpha - \Lambda_3 r_t^{2+(\alpha-\alpha_I)(i+a)})$$

$$\times r_t^{\alpha(1-j)q_j+(2+(\alpha-\alpha_I)(i+a)-\alpha b)q_b+\alpha n}, \tag{15.16}$$

where

$$M_t^n = \frac{\varepsilon_t}{P_u(\alpha_w^2 - \beta\varepsilon_t\alpha_v^2)}, \qquad M_{t\to f} = \frac{\varepsilon_f}{P_u(\alpha_v^2 - \varepsilon_f\alpha_w^2)}, \qquad M_{t*} = \max\{M_t^n, M_{t\to f}\},$$

$$\Lambda_3 = \frac{2\pi m\lambda}{\alpha_I}\sum_{a=0}^{\infty}\frac{(m_I)_a}{a!(i-\delta_I+a)}\sum_{i=1}^{m_I}\binom{m_I}{i}\left(\frac{M_{t*}P_u}{m_I}\right)^{i+a}(-1)^a,$$

$$\Lambda_4^n = \sum p!\prod_{j=1}^{p}\frac{\left[(-mM_{t*}\sigma^2)\prod_{k=0}^{j-1}(1-k)\right]^{q_j}}{q_j!(j!)^{q_j}},$$

and

$$\Lambda_5^n = \sum(n-p)!\prod_{b=1}^{n-p}\frac{\left[(-\Lambda_3)\prod_{k=0}^{b-1}(\delta_I-k)\right]^{q_b}}{q_b!(b!)^{q_b}}.$$

Proof: Please refer to appendix B of [18]. ∎

For the far user case, note that decoding will succeed if the typical user can decode its own message by treating the signal from the fixed user as interference. The coverage probability conditioned on the serving distance of the typical user for the far user case is calculated in the following lemma.

Lemma 15.4 *The coverage probability conditioned on the serving distance of the typical user for the far user case in the user-centric strategy is expressed in closed form as*

$$
P_{t,\text{far}}(r) = \sum_{n=0}^{m-1} \sum_{p=0}^{n} \binom{n}{p} \frac{(-1)^n}{n!} \Lambda_4^f \Lambda_5^f \exp(-mM_t^f \sigma^2 r_t^\alpha - \Lambda_3^f r_t^{2+(\alpha-\alpha_I)(i+a)})
$$
$$
\times r_t^{\alpha(1-j)q_j + (2+(\alpha-\alpha_I)(i+a) - \alpha b)q_b + \alpha n}, \tag{15.17}
$$

where

$$
M_t^f = \frac{\varepsilon_t}{P_u(\alpha_v^2 - \varepsilon_t \alpha_w^2)},
$$

$$
\Lambda_3^f = \frac{2\pi m \lambda}{\alpha_I} \sum_{a=0}^{\infty} \frac{(m_I)_a}{a!(i - \delta_I + a)} \sum_{i=1}^{m_I} \binom{m_I}{i} \left(\frac{M_t^f P_u}{m_I} \right)^{i+a} (-1)^a,
$$

$$
\Lambda_4^f = \sum p! \prod_{j=1}^{p} \frac{\left[(-mM_t^f \sigma^2) \prod_{k=0}^{j-1} (1-k) \right]^{q_j}}{q_j! (j!)^{q_j}},
$$

and

$$
\Lambda_5^f = \sum (n-p)! \prod_{b=1}^{n-p} \frac{\left[(-\Lambda_3^f) \prod_{k=0}^{b-1} (\delta - k) \right]^{q_b}}{q_b! (b!)^{q_b}}.
$$

Proof: Based on the SINR analysis in Eq. (15.6), and following a procedure similar to that in appendix B of [18], with interchanging M_{t*} with M_t^f, we can obtain the desired result in Eq. (15.17). ∎

Remark 15.1 The results in Eqs. (15.16) and (15.17) demonstrate that the coverage probability of the typical user is determined by imperfect SIC coefficient, its own target rate, the fading parameter m of the small-scale fading channels, and the distance of the fixed user served by the same UAV.

Remark 15.2 Inappropriate power allocation, such as $\alpha_v^2 - \varepsilon_t \alpha_w^2 < 0$ and $\alpha_w^2 - \beta \varepsilon_t \alpha_v^2 < 0$, will lead to the coverage probability always being zero.

Based on Lemmas 15.3 and 15.4, the coverage probability of the typical user in the user-centric strategy can be calculated in the following theorem.

Theorem 15.1 *The exact expression of the coverage probability for the typical user is expressed as*

$$
P_t = \int_0^{r_k} P_{t,\text{near}}(r) f_r(r) \, dr + \int_{r_k}^{\infty} P_{t,\text{far}}(r) f_r(r) \, dr, \tag{15.18}
$$

where $P_{t,\text{near}}(r)$ is given in Eq. (15.16), $P_{t,\text{far}}(r)$ is given in Eq. (15.17), and $f_r(r)$ is given in Eq. (15.12).

Proof: Considering the distance distributions of the typical user at the origin associated with the UAV at the horizontal serving distance r_k, which is the horizontal distance of the fixed user in the previous round, based on the definition of coverage probability in [19, 20], we can readily obtain the desired results in Eq. (15.18). ∎

Remark 15.3 Based on the result in Eq. (15.18), the coverage probability of the typical user is dependent on the distance of the fixed user in the user-centric strategy.

In order to provide more insights for UAV-assisted cellular networks, the coverage probability of the typical user is also derived for the OMA-assisted UAV cellular networks, i.e., time-division multiple access (TDMA). The typical user and fixed user follow the same distance distributions and small-scale fading channels in the OMA-assisted cellular UAV networks. The OMA benchmark adopted in this treatise is that by dividing the two users in equal time/frequency slots.

Corollary 15.2 *The coverage probability conditioned on the serving distance of the typical user for the OMA-assisted UAV cellular networks in the user-centric strategy is expressed in closed form as*

$$P_{\text{cov},t,o}(x) = \sum_{n=0}^{m-1}\sum_{p=0}^{n} \binom{n}{p} \frac{(-1)^n}{n!} \Lambda_4^o \Lambda_5^o \exp(-mM_t^o \sigma^2 r_t^\alpha - \Lambda_3^o r_t^{2+(\alpha-\alpha_I)(i+a)})$$

$$\times r_t^{\alpha(1-j)q_j+(2+(\alpha-\alpha_I)(i+a)-\alpha b)q_b+\alpha n}, \tag{15.19}$$

where

$$M_t^o = \frac{\varepsilon_t^o}{P_u}, \qquad \varepsilon_t^o = 2^{2R_t} - 1,$$

$$\Lambda_3^o = \frac{2\pi m\lambda}{\alpha_I}\sum_{a=0}^{\infty}\frac{(m_I)_a}{a!(i-\delta_I+a)}\sum_{i=1}^{m_I}\binom{m_I}{i}\left(\frac{M_t^o P_u}{m_I}\right)^{i+a}(-1)^a,$$

$$\Lambda_4^o = \sum p!\prod_{j=1}^{p}\frac{\left[(-mM_t^o\sigma^2)\prod_{k=0}^{j-1}(1-k)\right]^{q_j}}{q_j!(j!)^{q_j}},$$

and

$$\Lambda_5^o = \sum(n-p)!\prod_{b=1}^{n-p}\frac{\left[(-\Lambda_3)\prod_{k=0}^{b-1}(\delta_I-k)\right]^{q_b}}{q_b!(b!)^{q_b}}.$$

Proof: Following a procedure similar to that in appendix B of [18], with interchanging M_t^f with M_t^o, we can obtain the desired result in Eq. (15.19). ∎

15.3 UAV-Centric Strategy for Offloading Actions

In a conventional BS communication system, the BSs are distributed in order to cover all the ground, whereas UAV communications mainly focus on providing access services for

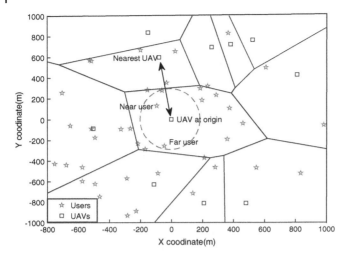

Figure 15.2 Example of the UAV-centric strategy cellular networks.

supporting BSs in the hotspot areas of the dense networks, i.e., airports or resorts, where most users are located in the lounge [6]. Based on the insights of [21], where the serving area can be considered as a regular disk, another strategy considered in this chapter is the UAV-centric strategy, where paired NOMA users are located inside the coverage disk as shown in Figure 15.2. It is also worth noting that the locations of UAVs are properly selected to serve terrestrial users in the hotspot areas based on user density in the UAV-centric strategy. Based on the insights of a Poisson cluster process (PCP), the users are located in multiple small clusters in practice.

For the UAV-centric strategy, a UAV is located at the origin point, which becomes the typical UAV serving users in the typical cell. Therefore, it is assumed that the distance between the UAV at the origin and the nearest UAV is R, and the potential paired NOMA users are located in the coverage area within radius $R/2$. In the UAV-centric strategy, the user pairing strategy is determined by the connected UAV, where all the users in the coverage disk are connected to the UAV. In the user association, for simplicity we assume that there are two users, near user w and far user v, that have access to the UAV at the origin to perform NOMA. It is assumed that the users are uniformly located, which is according to the HPPP, denoted by Ψ_u, and it is associated with the density λ_u, within a large ring and small disk with radii $R/2$ and $R/4$, respectively. By doing so, the NOMA technique can be performed without accurate CSI.

15.3.1 SINR Analysis

For the UAV-centric strategy, the distances between the interfering UAVs and the users are more complicated. For notational simplicity, the location of the jth interfering UAV is denoted by y_j, where $y_j \in \Psi$. The locations of the users are conditioned on the locations of their cluster heads (UAVs). As such, the SINR of the far user v can be derived as

$$\text{SINR}_v = \frac{|h_v|^2 d_v^{-\alpha} P_u \alpha_v^2}{\sigma^2 + |h_v|^2 d_v^{-\alpha} P_u \alpha_w^2 + \sum_{j \in \Psi} |g_j|^2 P_u d_j^{-\alpha_I}}, \tag{15.20}$$

where $|h_v|^2$ and d_v denote the small-scale channel gain and the distance between the far user and the UAV, and $|g_j|^2$ and d_j denote the small-scale channel gain and the distance between the jth interfering UAV and the user, respectively.

The near user w will first decode the signal of the far user v with the following SINR:

$$\text{SINR}_{w \to v} = \frac{|h_w|^2 d_w^{-\alpha} P_u \alpha_v^2}{\sigma^2 + |h_w|^2 d_w^{-\alpha} P_u \alpha_w^2 + \sum_{j \in \Psi} |g_j|^2 P_u d_j^{-\alpha_I}}, \tag{15.21}$$

where $|h_w|^2$ and d_w denote the small-scale channel gain and the distance between the near user and the UAV. If the signal of the vth user can be decoded successfully, the wth user then decodes its own signal. As such, the SINR at the wth user can be expressed as

$$\text{SINR}_w = \frac{|h_w|^2 d_w^{-\alpha} P_u \alpha_w^2}{\sigma^2 + \beta |h_w|^2 d_w^{-\alpha} P_u \alpha_v^2 + \sum_{j \in \Psi} |g_j|^2 P_u d_j^{-\alpha_I}}. \tag{15.22}$$

15.3.2 Coverage Probability of the UAV-Centric Strategy

Consider a disk centered at the origin with radius $R/2$, which is shown in Figure 15.2. In order to implement the NOMA protocol, we separate the disk into two power zones equally, a small disk with radius $R/4$ and a ring with radius from $R/4$ to $R/2$, to serve paired NOMA users. It is assumed that the near users and the far users are located in the small disk and large ring, respectively.

Focusing on the typical cell, which is located at the origin, the PDF of the distance for the near users conditioned on serving distance R follows

$$f_w(r \mid R) = \frac{32r}{R^2}, \quad 0 \le r \le l_1, \tag{15.23}$$

where $l_1 = R/4$.

The PDF of far users can be obtained by

$$f_v(r \mid R) = \frac{32r}{3R^2}, \quad l_1 \le r \le l_2, \tag{15.24}$$

where $l_2 = R/2$.

In order to derive the system performance, the Laplace transform of interfering UAVs needs to be derived. We calculate the Laplace transform of inter-cell interference for the paired users in the following lemma.

Lemma 15.5 *For the UAV-centric strategy, the Laplace transform of interference distribution conditioned on the serving distance R for the paired NOMA users is given by*

$$\mathcal{L}_U(s \mid R) = \exp\left(-\frac{l_I}{R}\left(1 - \left(1 + \frac{SP_u}{m_I l_I^{\alpha_I}}\right)^{-m_I}\right)\right)$$

$$\times \exp\left(-\frac{2\pi\lambda}{\alpha_I} \sum_{i=1}^{m_I} \binom{m_I}{i} \left(\frac{sP_u}{m_I}\right)^{\delta_I} (-1)^{(\delta_I - i)}\right.$$

$$\times \left. B\left(\frac{-sP_u l_I^{-\alpha_I}}{m_I}; i - \delta_I, 1 - m_I\right)\right), \tag{15.25}$$

where $l_I = \sqrt{R^2 + h^2}$.

Proof: Please refer to appendix C of [18]. ∎

It is also worth noting that, for the non-line-of-sight (NLoS) case, the small-scale fading between users and interfering UAVs can be considered as Rayleigh fading. Thus, the Laplace transform can be further obtained in the following corollary.

Corollary 15.3 *For the NLoS link, the Laplace transform of the interference distribution conditioned on the serving distance R is given by*

$$
\mathcal{L}_U(s \mid R) = \exp\left(-\frac{l_I}{R}\left(\frac{SP_u}{l_I^{\alpha_I} + SP_u}\right)\right)
$$
$$
\times \exp\left(-\frac{2\pi \lambda P_u l_I^{2-\alpha_I}}{\alpha_I(1-\delta_I)} \, {}_2F_1(1, \ 1-\delta_I; \ 2-\delta_I; -sP_u l_I^{-\alpha_I})\right). \tag{15.26}
$$

Then, we focus on the coverage behavior of paired NOMA users in the UAV-centric strategy. In the UAV-centric strategy, the coverage probability is more complicated than the user-centric strategy due to the fact that it is necessary to evaluate separately the interfering UAV located at distance R. It is assumed that the target rates of user w and user v are R_w and R_v, respectively. Therefore, the coverage probability of the wth user is given in the following lemma.

Lemma 15.6 *The closed-form expression of the coverage probability conditioned on serving distance for the near user is expressed as*

$$
P_{\text{cov},w}(r \mid R) = \sum_{n=0}^{m-1}\sum_{k=0}^{n}\sum_{l=0}^{k} \frac{(-1)^n r_w^{an}}{l!(k-l)!(n-k)!} \, \Theta_3\Theta_4\Theta_5
$$
$$
\times \exp\left(-mM_{w*}\sigma^2 r_w^\alpha - \Theta_1 r_w^{\alpha(i+a)} - \frac{ml_I}{R} + \Theta_2 r_w^{\alpha U}\right)
$$
$$
\times r_w^{\alpha(1-j)q_j + \alpha(i+a-g)q_g + \alpha n + \alpha(U-b)q_u}, \tag{15.27}
$$

where

$$
M_w = \frac{\varepsilon_w}{P_u(\alpha_w^2 - \beta\varepsilon_w\alpha_v^2)}, \qquad M_v = \frac{\varepsilon_v}{P_u(\alpha_v^2 - \varepsilon_v\alpha_w^2)},
$$
$$
\varepsilon_w = 2^{R_w} - 1, \qquad \varepsilon_v = 2^{R_v} - 1,
$$
$$
M_{w*} = \max\{M_w, M_v\}, \qquad r_w = \sqrt{r^2 + h^2},
$$
$$
\Theta_1 = \pi m\delta_I \lambda \sum_{i=1}^{m_I}\binom{m_I}{i}(-1)^{\delta_I-1}\sum_{a=0}^{\infty}\frac{(m_I)_a}{a!(i-\delta_I+a)}\left(\frac{M_{w*}P_u}{m_I}\right)^{i+a} l_I^{-\alpha_I(i-\delta_I+a)},
$$
$$
\Theta_2 = \frac{ml_I}{R}\sum_{U=0}^{\infty}(-1)^U C_{m_I+U+1}^U\left(\frac{M_{w*}P_u}{l_I^{\alpha_I}m_I}\right)^U,
$$

$$\Theta_3 = \sum (n-k)! \prod_{j=1}^{n-k} \frac{[(-mM_{w*}\sigma^2)\prod_{p=0}^{j-1}(1-p)]^{q_j}}{q_j!(j!)^{q_j}},$$

$$\Theta_4 = \sum (k-l)! \prod_{b=1}^{k-l} \frac{[(-\Theta_2)\prod_{p=0}^{b-1}(U-p)]^{q_u}}{q_u!(j!)^{q_u}},$$

and

$$\Theta_5 = \sum l! \prod_{g=1}^{l} \frac{[(-\Theta_1)\prod_{p=0}^{g-1}(i+a-g)]^{q_g}}{q_g!(j!)^{q_g}}.$$

Proof: Please refer to appendix D of [18]. ∎

Similar to Lemma 15.6, the coverage probability of the far user can be derived in the following lemma.

Lemma 15.7 *The closed-form expression of the coverage probability conditioned on serving distance for the far user is expressed as*

$$P_{\text{cov},v}(r \mid R) = \sum_{n=0}^{m-1}\sum_{k=0}^{n}\sum_{l=0}^{k} \frac{(-1)^n r_v^{\alpha n}}{l!(k-l)!(n-k)!} \, \Theta_{3,v}\Theta_{4,v}\Theta_{5,v}$$

$$\times \exp\left(-mM_v\sigma^2 r_v^\alpha - \Theta_{1,v} r_v^{\alpha(i+a)} - \frac{ml_I}{R} + \Theta_{2,v} r_v^{\alpha U}\right)$$

$$\times r_v^{\alpha(1-j)q_j + \alpha(i+a-g)q_g + \alpha n + \alpha(U-b)q_u}, \tag{15.28}$$

where

$$r_v = \sqrt{r^2 + h^2},$$

$$\Theta_{1,v} = \pi m \delta_I \lambda \sum_{i=1}^{m_I} \binom{m_I}{i} (-1)^{\delta_I - 1} \sum_{a=0}^{\infty} \frac{(m_I)_a}{a!(i - \delta_I + a)} \left(\frac{M_v P_u}{m_I}\right)^{i+a} l_I^{-\alpha_I(i-\delta_I+a)},$$

$$\Theta_{2,v} = \frac{ml_I}{R} \sum_{U=0}^{\infty} (-1)^U C_{m_I + U + 1}^U \left(\frac{M_v P_u}{l_I^{\alpha_I} m_I}\right)^U,$$

$$\Theta_{3,v} = \sum (n-k)! \prod_{j=1}^{n-k} \frac{[(-mM_v\sigma^2)\prod_{p=0}^{j-1}(1-p)]^{q_j}}{q_j!(j!)^{q_j}},$$

$$\Theta_{4,v} = \sum (k-l)! \prod_{b=1}^{k-l} \frac{[(-\Theta_{2,v})\prod_{p=0}^{b-1}(U-p)]^{q_u}}{q_u!(j!)^{q_u}},$$

and

$$\Theta_{5,v} = \sum l! \prod_{g=1}^{l} \frac{[(-\Theta_{1,v})\prod_{p=0}^{g-1}(i+a-g)]^{q_g}}{q_g!(j!)^{q_g}}.$$

Proof: Based on [22], the derivation in Eq. (15.28) can be readily proved. ∎

Then, the coverage probability of the paired NOMA users in the UAV-centric strategy can be derived in the following theorem.

Theorem 15.2 *Based on Lemmas 15.6 and 15.7, the exact expressions of the coverage probability for the paired NOMA users can be expressed as*

$$P_{\text{cov},w} = \int_0^\infty \int_0^{l_1} P_{\text{cov},w}(r \mid R) f_w(r \mid R) \, dr \, f_r(R) \, dR \tag{15.29}$$

and

$$P_{\text{cov},v} = \int_0^\infty \int_{l_1}^{l_2} P_{\text{cov},v}(r \mid R) f_v(r \mid R) \, dr \, f_r(R) \, dR, \tag{15.30}$$

where $l_1 = R/4$, $l_2 = R/2$, $P_{\text{cov},w}(r \mid R)$ is given in Eq. (15.27), $P_{\text{cov},w}(r \mid R)$ is given in Eq. (15.28), $f_w(r \mid R)$ is given in Eq. (15.23), $f_v(r \mid R)$ is given in Eq. (15.24), and $f_r(R)$ is given in Eq. (15.12).

Proof: By utilizing the PDF in Eq. (15.23), the coverage probability of the near user conditioned on the serving distance can be obtained as

$$P_{\text{cov},w}(R) = \int_0^{l_1} P_{\text{cov},w}(r \mid R) f_w(r \mid R) \, dr, \tag{15.31}$$

The overall coverage probability can be derived by the serving distance of UAV-assisted cellular networks, which can be expressed as

$$P_{\text{cov},w} = \int_0^\infty P_{\text{cov},w}(R) f_r(R) \, dR, \tag{15.32}$$

Plugging Eq. (15.12) in to Eq. (15.32), and after some mathematical manipulations, the coverage probability of the near user can be obtained. Thus, the proof is complete. ∎

15.4 Numerical Results

In this section, numerical results are provided to facilitate the performance evaluation of NOMA-assisted UAV cellular networks. Monte Carlo simulations are conducted to verify our analytical results. In the considered network, it is assumed that the power allocation factors are $\alpha_v^2 = 0.6$ for the far user and $\alpha_w^2 = 0.4$ for the near user. The path-loss exponent of interference α_I is set to 4, and the path-loss exponent of the desired transmission is smaller than 4. The height of the UAV is fixed to 100 m. In the Monte Carlo simulations, it is not possible to simulate a real infinite distribution for UAVs. Hence, the UAVs are distributed in a disk, and the radius of the disk is 10 000 m. The bandwidth of the downlink transmission is set to BW = 300 kHz, and the power of the AWGN is set to $\sigma^2 = -174 + 10\log_{10}(\text{BW})$ dBm. The UAV density $\lambda = 1/(500^2\pi)$. It is also worth noting that LoS and NLoS scenarios are

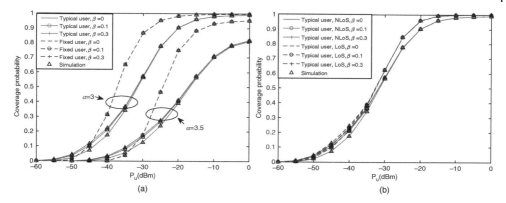

Figure 15.3 Coverage probability of the paired NOMA users versus the power of the UAV in the user-centric strategy, with target rate $R_t = 1$ BPCU and $R_f = 0.5$ BPCU. The horizontal distance of the fixed user is 300 m. The exact results of NOMA are calculated from Eq. (15.18). (a) Coverage probability in NLoS scenario with different path-loss exponents, where the fading parameters are set to $m = 1$ and $m_l = 1$. (b) Coverage probability in both the NLoS and LoS scenarios, where the fading parameters are $m = 2$ and $m_l = 1$. The path-loss exponent of the desired link is set to $\alpha = 3$.

indicated by the Nakagami fading parameter m, where $m = 1$ for NLoS scenarios (Rayleigh fading) and $m > 1$ for LoS scenarios. Without loss of generality, we use $m = 2$ to represent the LoS scenario throughout Section 15.4.

15.4.1 User-Centric Strategy

First, we evaluate the coverage performance of the downlink NOMA users in the user-centric strategy. In Figure 15.3(a), for a given set of the distance of fixed users, the solid curves and dashed curves are the coverage probability for typical users and fixed users, respectively. We can see that, as the power of the UAV increases, coverage ceilings, which are the maximum coverage probability for the proposed networks, of both typical and fixed NOMA users occur. This is due to the fact that, as the higher power level of interfering UAVs is deployed, the received SINR decreases dramatically. It is observed that, as the imperfect SIC coefficient β increases, the coverage probability of typical users decreases, which indicates that the performance of NOMA-assisted UAV communication can be effectively improved by decreasing the imperfect SIC coefficient. For example, in the case of $\beta = \frac{2}{3}$, the power residual from imperfect SIC is higher than the power of near users, i.e., $\alpha_w^2 < \alpha_v^2 \beta$.

We can also see that, in the case of $\beta = 0, 0.1,$ and 0.3, the coverage probabilities of fixed users are the same. This is due to the fact that the imperfect SIC is the critical component of typical users, whereas the imperfect SIC has no effect for fixed users in the case $R_f = 0.5$ BPCU (bits per channel use). As we can see in the figure, the outage of typical users occurs more frequently than for fixed users. This is due to the choice of the power allocation factors and the distance of the fixed user. Note that the simulation results and analytical results match perfectly in Figure 15.3(a), which demonstrates the accuracy of the developed analytical results.

Figure 15.3(b) shows the coverage probability achieved by typical users in both the NLoS and LoS scenarios. In order to illustrate the performance affected by the LoS transmission, the NLoS case is also shown in the figure as a benchmark for comparison. In Figure 15.3(b), we can see that higher fading parameter m would result in reduced outage probability for different UAV power levels and different imperfect SIC coefficients. This is because the LoS link between the UAV and users provides higher received power level. It is also worth noting that the proposed network does not require a large transmit power to increase the coverage probability due to the fact that the coverage ceiling occurs in the high-SNR regime.

In Figure 15.4, the impact of different choices of UAV density and the distance of fixed users is studied. As can be observed from the figure, increasing the distance of fixed users will decrease the coverage probability for fixed users, whereas the coverage probability of typical users increases. This is due to the fact that the distance of fixed users has an effect on user association for typical users. For fixed users, the received power decreases dramatically when the distance increases. On the other hand, for the dashed and starred curves, where the density of UAVs is 10 times higher than for the solid and dotted curves, the coverage probability of typical NOMA users in the case of high UAV density is much higher than the case of low UAV density. This is because the number of UAVs is increased, which leads to

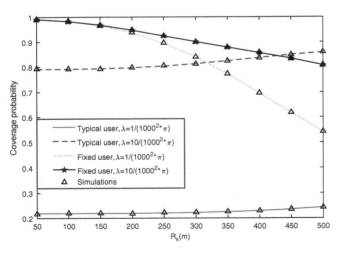

Figure 15.4 Coverage probability of the user-centric strategy versus the distance of fixed users, with target rate $R_t = 1$ BPCU and $R_f = 0.5$ BPCU. The path-loss exponent is fixed to $\alpha = 3$, and the power of UAV is set to -30 dBm.

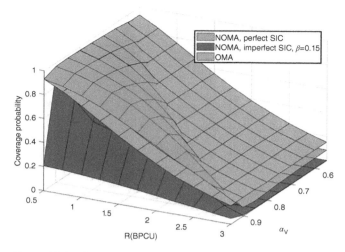

Figure 15.5 Coverage probability of typical users versus the target rate $R_t = R$ BPCU and power allocation factor α_v, with the imperfect SIC coefficient $\beta = 0$ and 0.15. The target rate of fixed users $R_f = 0.5$ BPCU, and the horizontal distance of the fixed user is 300 m. The transmit power of UAVs is fixed to -30 dBm with path-loss exponent $\alpha = 3$. The fading parameters are set to $m = 3$ and $m_l = 2$.

the decrease of the distance of connected UAVs. It is also worth noting that there are two crossings for fixed users, which mean that there exists an optimal distance of fixed users for given UAV density.

Next, Figure 15.5 plots the coverage probability of the paired NOMA users in the user-centric strategy versus the target rate R and power allocation factor α_v. It is observed that the coverage probability is zero in the case of inappropriate target rates and power allocation factors, which verifies the insights in Remark 15.2. The coverage probability of typical users in OMA is also plotted, which indicates that NOMA is able to outperform OMA for the appropriate power allocation factors and target rates of the paired NOMA users. One can also observe that NOMA cannot outperform OMA in the case of $\beta = 0.15$ for the user-centric strategy. This indicates that a hybrid NOMA/OMA-assisted UAV network may be a good solution in the case of poor SIC quality. The UAV could intelligently choose the access techniques for improving the system coverage probability.

15.4.2 UAV-Centric Strategy

In the UAV-centric strategy, $\varepsilon = 0.1$ m is used to evaluate the interference received from the UAV located at distance R. Then, we evaluate the performance of the downlink users in the UAV-centric strategy. In Figure 15.6(a) and 15.6(b), the impact of the NOMA-assisted UAV-centric strategy on the coverage probability is studied. The target rates of near users and far users are set to $R_w = 1.5$ BPCU and $R_v = 1$ BPCU, respectively. The solid and dashed

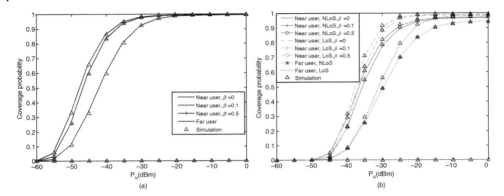

Figure 15.6 Coverage probability of the paired NOMA users in the UAV-centric strategy versus the transmit power, with target rates $R_w = 1.5$ BPCU and $R_v = 1$ BPCU, respectively. The exact results of NOMA are calculated from Eq. (15.29). (a) Coverage probability in NLoS scenario, where the fading parameters are set to $m = 1$ and $m_l = 1$. The path-loss exponent is set to $\alpha = 3$. (b) Coverage probability in both NLoS and LoS scenarios with path-loss exponent $\alpha = 3.5$, where the fading parameters are set to $m = 2$ and $m_l = 1$.

curves are the coverage probability of near users and far users, respectively. An interesting phenomenon occurs in the UAV-centric strategy, namely that, in the case of $\beta = 0.5$, the coverage probabilities of near users are all zero, which indicates that the transmission has failed. This is again due to the fact that $\alpha_w^2 - \beta \alpha_v^2 \varepsilon_w < 0$, which verifies our obtained insights in Remark 15.2.

Comparing Figure 15.6(a) and 15.6(b), one can observe that the impact of fading parameter m on the coverage probability is also significant, which is due to the fact that the received power level is greater in the case of larger m. Again, we can see that the coverage probability is also one for near users in the case of $\beta = 0.5$, which indicates that the LoS link has no effect on Remark 15.2. It is also worth noting that the coverage probability of the user-centric strategy is much greater than for the UAV-centric strategy in the case of $\beta = 0.5$, which indicates that the UAV-centric strategy is more susceptible to an imperfect SIC factor than the user-centric strategy.

Figure 15.7 plots the coverage probability for near users in the UAV-centric strategy in the cases of $\beta = 0$, $\beta = 0.1$, and $\beta = 0.5$. One can obtain that, on the one hand, inappropriate power allocation will lead to the coverage probability always being zero, which also verifies Remark 15.2. On the other hand, we can see that, in the case of $\beta > 0$, the coverage probability decreases dramatically when increasing target rate, which verifies that the SIC residue is the dominant interference in NOMA. In order to provide more insights, the coverage performance of OMA in the UAV-centric strategy is also provided. We can see that, in the case of $\beta = 0$, NOMA performs better than OMA, which indicates that the proposed networks are analytically shown to be applicable for UAV communications. We can also see that, in the case of $\beta = 0.15$, the coverage performance of NOMA- and OMA-assisted UAV cellular networks show close agreement, which also indicates that a hybrid NOMA/OMA-assisted UAV network may be a good solution for the UAV-centric strategy.

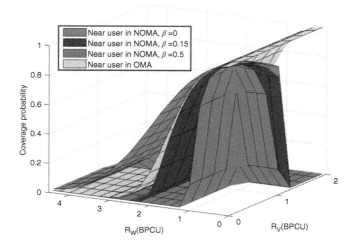

Figure 15.7 Coverage probability of the near user versus the target rate. The transmit power of UAVs is fixed to -40 dBm. The fading parameters are set to $m = 3$ and $m_l = 2$.

15.5 Conclusions

In this chapter, the application of NOMA to UAV-assisted cellular communications has been considered. A pair of association strategies, namely, user-centric strategy and UAV-centric strategy, have been proposed. The user-centric strategy is applicable when all the users located in the Voronoi cell need to be served by the UAV simultaneously. The derived results provide the benchmark for NOMA-assisted UAV cellular networks. The UAV-centric strategy is motivated by the fact that, in practice, it is more applicable to serve users in dense networks. The key idea of the UAV-centric strategy is to provide services for hotspot areas only, i.e., airports or resorts. Then, the performances of the proposed network were evaluated, where multiple UAVs are deployed in the sky to serve multiple users on the ground. Additionally, new analytical expressions for the interference and coverage probability were derived for characterizing the performance in NOMA-assisted UAV cellular networks.

References

1 Y. Zeng, R. Zhang and T. J. Lim, "Wireless communications with unmanned aerial vehicles: opportunities and challenges", *IEEE Commun. Mag.*, vol. 54, no. 5, pp. 36–42, May 2016.

2 M. Mozaffari, W. Saad, M. Bennis and M. Debbah, "Unmanned aerial vehicle with underlaid device-to-device communications: Performance and tradeoffs", *IEEE Trans. Wireless Commun.*, vol. 15, no. 6, pp. 3949–3963, Jun. 2016.

3 N. Goddemeier and C. Wietfeld, "Investigation of air-to-air channel characteristics and a UAV specific extension to the Rice model," *2015 IEEE Globecom Workshops (GC Wkshps)*, pp. 1–5, Dec. 2015.

4 F. Jiang and A. L. Swindlehurst, "Optimization of UAV heading for the ground-to-air uplink," *IEEE J. Sel. Areas Commun.*, vol. 30, no. 5, pp. 993–1005, Jun. 2012.

5 V. V. Chetlur and H. S. Dhillon, "Downlink coverage analysis for a finite 3-D wireless network of unmanned aerial vehicles," *IEEE Trans. Commun.*, vol. 65, no. 10, pp. 4543–4558, Oct. 2017.

6 J. Lyu, Y. Zeng and R. Zhang, "UAV-aided offloading for cellular hotspot," *IEEE Trans. Wireless Commun.*, vol. 17, no. 6, pp. 3988–4001, Jun. 2018.

7 L. Liu and S. Zhang and R. Zhang, "Multi-beam UAV communication in cellular uplink: Cooperative interference cancellation and sum-rate maximization," *IEEE Trans. Wireless Commun.*, vol. 18, no. 10, pp. 4679–4691, Oct. 2019.

8 Y. Zeng, J. Xu and R. Zhang, "Energy minimization for wireless communication with rotary-wing UAV", *IEEE Trans. Wireless Commun.*, vol. 18, no. 4, pp. 2329–2345, Apr. 2019.

9 Y. Zeng, J. Lyu and R. Zhang, "Cellular-connected UAV: Potential, challenges, and promising technologies," *IEEE Wireless Commun.*, vol. 26, no. 1, pp. 120–127, Feb. 2019.

10 Y. Liu, Z. Qin, Y. Cai, Y. Gao, G. Y. Li and A. Nallanathan, "UAV communications based on non-orthogonal multiple access," *IEEE Wireless Commun.*, vol. 26, no. 1, pp. 52–57, Feb. 2019.

11 X. Liu, Y. Liu, Y. Chen and L. Hanzo, "Trajectory design and power control for multi-UAV assisted wireless networks: A machine learning approach," *IEEE Trans. Veh. Technol.*, vol. 68, no. 8, pp. 7957–7969, Aug. 2019.

12 T. M. Nguyen, W. Ajib and C. Assi, "A novel cooperative NOMA for designing UAV-assisted wireless backhaul networks," *IEEE J. Sel. Areas Commun.*, vol. 36, no. 11, pp. 2497–2507, Nov. 2018.

13 T. Hou, Y. Liu, Z. Song, X. Sun and Y. Chen, "Multiple antenna aided NOMA in UAV networks: A stochastic geometry approach," *IEEE Trans. Commun.*, vol. 67, no. 2, pp. 1031–1044, Feb. 2019.

14 W. Mei and R. Zhang, "Uplink cooperative NOMA for cellular-connected UAV," *IEEE Journal of Selected Topics in Signal Processing*, vol. 13, no. 3, pp. 644–656, Jun. 2019.

15 N. Zhao, W. Lu, M. Sheng, Y. Chen, J. Tang, F. R. Yu and K. Wong, "UAV-assisted emergency networks in disasters," *IEEE Wireless Commun.*, vol. 26, no. 1, pp. 45–51, Feb. 2019.

16 J. G. Andrews, F. Baccelli and R. K. Ganti, "A tractable approach to coverage and rate in cellular networks," *IEEE Trans. Commun.*, vol. 59, no. 11, pp. 3122–3134, Nov. 2011.

17 I. S. Gradshteyn and I. M. Ryzhik, "Table of integrals, series and products", 6th ed. New York, NY, USA: Academic Press, 2000.

18 T. Hou, Y. Liu, Z. Song, X. Sun and Y. Chen, "Exploiting NOMA for UAV communications in large-scale cellular networks," *IEEE Trans. Commun.*, vol. 67, no. 10, pp. 6897–6911, Oct. 2019.

19 T. M. Cover and J. A. Thomas, "Elements of information theory," 6th ed, Wiley and Sons, New York, 1991.

20 E. Hildebrand, "Introduction to numerical analysis," New York, USA: Dover,1987.

21 K. S. Ali, M. Haenggi, H. ElSawy, A. Chaaban and M. Alouini, "Downlink non-orthogonal multiple access (NOMA) in poisson networks," in *IEEE Trans. Commun.*, vol. 67, no. 2, pp. 1613–1628, Feb. 2019.

22 N. Bourbaki, "Elements of the history of mathematics paperback", 22nd ed. Berlin, Germany: Springer Berlin Heidelberg, 2008.

16

Physical Layer Security for UAV Communications

Nadisanka Rupasinghe[1,3,], Yavuz Yapici[1], Ismail Guvenc[1], Huaiyu Dai[1], and Arupjyoti Bhuyan[2]*

[1]*Department of Electrical and Computer Engineering, North Carolina State University, Raleigh, NC, USA*
[2]*Idaho National Laboratory, Idaho Falls, ID, USA*
[3]*DOCOMO Innovations, Inc., Palo Alto, CA, USA*

16.1   Introduction

Secure communications in an unmanned aerial vehicle (UAV) based wireless network is of significant importance for the vast majority of practical applications. Along with the confidentiality of user messages, sophisticated transmission technologies are required to improve the secure data rates (i.e., secrecy rates). Although there are various ways for each layer of the network stack to make the underlying communications more secure, the physical layer techniques stand out from the others by providing robust and unbreakable secrecy rates.

In this chapter, we study how to enhance physical layer security (PLS) of UAV-based communications networks. To this end, we first briefly look at possible security attacks on wireless networks, which is followed by describing the wireless network security requirements. We subsequently discuss the importance of PLS for wireless networks, and present some interesting PLS techniques from the existing literature. After describing the ways to achieve secure wireless communications in general, we focus on UAV-based wireless networks. We study in detail some interesting PLS techniques proposed for enhancing security. Further, we discuss a case study where a UAV equipped with a base station (BS) is deployed to provide broadband connectivity over a densely packed user area, which well represents most of the temporary events (e.g., a football match in a stadium).

The rest of this chapter is organized as follows: In Section 16.2, we discuss possible security attacks in wireless networks, followed by security requirements for wireless networks in Section 16.3. Then, in Section 16.4, we discuss in detail how PLS can enhance secure communication within wireless networks. A detailed study of different PLS techniques to enhance security in UAV-based communication networks is provided in Section 16.5, followed by an interesting case study in Section 16.6.

* Corresponding Author: Nadisanka Rupasinghe; nrupasinghe@docomoinnovations.com

UAV Communications for 5G and Beyond, First Edition.
Edited by Yong Zeng, Ismail Guvenc, Rui Zhang, Giovanni Geraci, and David W. Matolak.
© 2021 John Wiley & Sons Ltd. Published 2021 by John Wiley & Sons Ltd.

16.2   Breaching Security in Wireless Networks

Wireless networks are highly prone to the attacks of various adversaries. Unlike the case for conventional wired networks, the broadcast nature of the wireless medium makes undesired breaches coming from illegitimate users more probable. Along with the new wireless technologies deployed, the number of people relying on wireless services increases day by day. For instance, nowadays, online banking via smart mobile devices is very popular for many people. However, during such communication, important/sensitive information is transmitted through wireless connections. The ability to share data reliably while preserving confidentiality is therefore extremely important in a wireless network, especially in the presence of adversaries. Considering the growing dissemination of wireless networks in this information era, the threat of security breaches in wireless communications therefore has to be taken care of carefully.

Security breaches in a wireless network are the result of either active or passive attacks [38, 39]. Active attacks can have a significant impact on the information exchange in a wireless network, since attackers often try to alter the network data. The most common forms of active attacks include denial-of-service (DoS) attacks, masquerade and replay attacks, and information disclosure and message modification attacks. Passive attacks, on the other hand, do not disrupt network operation, where the objective is to steal information passing through the wireless network. Two types of passive attacks are often encountered: eavesdropping intrusion and traffic analysis [39]. In the following, we briefly look at some of these network attacks, followed by security requirements to be considered for wireless networks.

16.2.1   Denial-of-Service Attacks

In a DoS attack, the attacker tries to disrupt the communication within a network. This can be achieved either by disabling the network or by exhausting the resources available to legitimate users [30, 39]. Radio-frequency (RF) jamming is one widely known DoS attack type [38]. In this category, an adversary can utilize jamming signals to disrupt the on-going communication such that a network suffers from DoS in a specific region.

16.2.2   Masquerade Attacks

In a masquerade attack, the attacker pretends to be a legitimate user, and breaches into the network through falsifying the network authentication system. A masquerade attack usually involves another form of active attack. For instance, an authentication sequence communicating through the network can be captured by an attacker, which is subsequently used to obtain privileges to access certain information in an illegal fashion [38].

16.2.3   Message Modification Attacks

In a message modification attack, an attacker alters some portion of the messages which belong to the legitimate users of a network. The modification can be an addition, deletion, or reordering of a portion of an original message which belongs to a legitimate user in the network [28].

16.2.4 Eavesdropping Intruders

Eavesdropping is a well-known type of passive attack targeting a confidential data exchange in a communication network [52]. In this security breach, a malicious receiver, referred to as eavesdropper, intercepts a message illegally. Assuming a mobile communications session with an exchange of confidential data, the ultimate goal of the network design is to prevent eavesdroppers from either detecting the message and/or learning the content. Encryption is the most commonly used technique for masking the important contents. In this strategy, eavesdroppers might be able to intercept the user messages, but cannot reveal any critical information due to the underlying encryption. Note that PLS eliminates any intrusion of the illegitimate receiver as much as possible.

16.2.5 Traffic Analysis

The network traffic can also be analyzed by a passive malicious receiver to determine the location and/or identities of the communicating parties. To this end, the attacker intercepts the transmitted messages, and tries to reveal as much information as possible. The traffic information itself may be useful for the attacker, which enables the tracking of the communication patterns of any two parties. This type of attack can happen even if the messages are encrypted, and hence the malicious receivers can use the information gleaned from this type of attack for other forms of attack. PLS is also very useful in this type of passive attack, since it prevents an attacker from intercepting the confidential data.

16.3 Wireless Network Security Requirements

Wireless networks require special attention to their security measures, which are not completely aligned with their counterparts associated with wired networks. In particular, the MAC (medium access control) and PHY (physical) layers of wireless networks are vulnerable to different type of attacks as compared to those encountered in wired networks [60]. Hence, the security requirements of wireless networks are specified focusing on protecting the wireless transmissions against specialized attack types involving eavesdropping attacks, jamming attacks, DoS attacks, data falsification attacks, and node compromise attacks. In order to prevent malicious receivers from intercepting confidential messages and revealing the associated content, the overall network should satisfy the requirements of *authenticity*, *confidentiality*, *integrity*, and *availability* [38], which we discuss briefly in the following.

16.3.1 Authenticity

The authenticity helps in confirming the identity of an authorized user, so that the network can distinguish between authorized and unauthorized users. In particular, communicating nodes in a wireless network should first perform a mutual authentication before starting data transmission [18]. Typically, a network node is equipped with a wireless network interface card, and has a unique MAC address, which can be used for authentication purposes. In addition, there are other wireless authentication methods happening in the upper layer of the protocol stack (i.e., network-layer authentication, transport-layer authentication, and application-layer authentication).

16.3.2 Confidentiality

The confidentiality ensures the data accessibility only by the intended nodes, and therefore no other unauthorized party can access the confidential data [39]. For instance, in the symmetric key encryption technique, the source node first encrypts the original data using an encryption algorithm with the aid of a secret key, which is shared with the intended destination only. The encrypted data is then transmitted to the destination node where the received data is decrypted using the secret key, which is shared previously. Since the malicious receiver has no knowledge of the secret key, it is unable to interpret the encrypted data. Traditionally, the classic Diffie–Hellman key agreement protocol is used to achieve the key exchange between the source and destination and requires a trusted key management center [46]. Recently, PLS has emerged as a means of protecting the confidentiality of wireless transmission against eavesdropping attacks [5, 33].

16.3.3 Integrity

The integrity of information flows within a wireless network is yet another important security requirement. In particular, to ensure integrity, the information should be accurate and reliable during its entire life-cycle without any falsification and modification by unauthorized users. The data integrity may be damaged by so-called insider attacks (e.g., node compromise attacks) [9, 22]. Here, a legitimate node that is altered and compromised by an adversary is termed as a compromised node. The compromised node may inflict damage upon the data integrity by launching malicious attacks, which includes message injection, false reporting, and data modification. In general, it is quite challenging to detect attacks by compromised nodes, since these compromised nodes running malicious codes still have valid identities.

16.3.4 Availability

The availability guarantees that the authorized users are capable of accessing a wireless network anytime and anywhere whenever required. As discussed in Section 16.2.1, DoS attacks make a wireless network unavailable to authorized users, which in turn results in unsatisfactory user experience [17, 47]. As an example, any unauthorized node can launch DoS attacks at the physical layer by maliciously generating interference, which disrupts the desired communications between legitimate users, and is known also known as a *jamming attack*.

16.4 Physical Layer Security

Passive attackers, or eavesdroppers to use their common name, are the primary hurdle towards achieving a secured data transmission in a wireless network. Owing to the broadcast nature of wireless transmission, any eavesdropper is very likely to intercept a sufficient amount of a confidential message exchanged by legitimate users. The ability of the network to meet the expected level of security (i.e., authenticity, confidentiality, integrity, and availability) requires that adequate security measures should be taken especially for eavesdroppers within the transmission range of the transmitter (i.e., base station in the downlink, and legitimate users in the uplink).

16.4.1 Physical Layer versus Upper Layers

Encryption techniques have been widely adopted in the upper layer of the protocol stack to guarantee communications security. These techniques, however, become insufficient or even not suitable along with growing mobile connections. One particular reason, among many others, for that is the requirement of a secure channel to exchange private keys during encryption. As an example, the symmetric cryptographic techniques, such as Data Encryption Standard (DES), require a common private key to be shared between the source and the destination nodes [38]. There should be a secure channel even for the encryption itself, which is not readily available for mobile wireless connections.

The PLS techniques, on the other hand, stand out as a self-sufficient communications security strategy. These techniques basically exploit the characteristics of wireless communications channels (e.g., fading, noise, interference) so as to avoid the use of extra spectral resources, and to reduce signalling overhead [37]. As a result, both the industry and academia have steered their attention to PLS techniques very recently [15, 24, 31, 32, 38, 56]. Note that PLS techniques can be employed both to improve the security of the overall network transmission, and to securely distribute the private keys of the desired encryption strategies.

PLS techniques have even been shown to provide perfect secrecy as long as the channel state information (CSI) is not available to unauthorized users, or the transmission links towards unauthorized users are noisier than those of authorized users [50]. From this perspective, the PLS strategy actually tries to increase the performance gap of the link quality between the legitimate user and eavesdropper through exploiting the physical properties of the wireless channels [15]. The respective techniques are most of the time energy-efficient and not complicated, which makes them appropriate for Internet-of-Things (IoT) devices and UAV systems with low processing capability and limited power resource.

16.4.2 Physical Layer Security Techniques

There are various techniques proposed in the literature to improve the security in the physical layer of the network stack [15, 24, 31, 32, 56]. Most of these techniques focus on enhancing the *secrecy capacity*, which is the maximum data rate achievable between the legitimate ends of the transmission subject to the constraints on information attainable by an unauthorized receiver [4, 37, 38]. In the pioneering work of Wyner [50], which constitutes the foundation and starting point of research on PLS, it was shown for a discrete memoryless channel that confidential communication between legitimate users is possible without sharing a secret key if the eavesdropper's channel is a degraded (much noisier) version of the intended user's channel. Accordingly, by using channel-dependent stochastic encoders, which generate random secrecy codes, confidentiality is achieved by exploiting the channel without using shared secret keys. A similar result has also been generalized to Gaussian channels in [14]. It is therefore very common to consider secrecy capacity (or, equivalently, achievable secrecy rates) while developing and/or measuring the effectiveness of PLS techniques. In the following, we briefly overview various PLS techniques that are widely used in the literature.

16.4.2.1 Artificial Noise

Artificial noise (AN) is an effective way of improving the secrecy capacity even when the eavesdroppers have better channel conditions than the legitimate users [13, 40]. The main objective of the AN approach is to purposefully create non-resolvable interference – only – at the eavesdropper, which degrades the respective channel quality [13]. It is worth remarking that this approach does not affect the channel quality of the legitimate users since AN is intentionally steered towards eavesdroppers [12, 38].

As an example, a multi-antenna transmission strategy is proposed in [56] along with AN-based beamforming and sectoring techniques. Considering scenarios with perfect and imperfect CSI at both ends of the transceiver, an AN transmission scheme is used to enhance secrecy capacity in [58]. In particular, the transmit power allocation between the information signal and the AN is studied to maximize achievable secrecy rates. It is also shown that creating more AN (to confuse eavesdroppers) is superior to allocating more power to the information transmission (to increase signal-to-noise ratio) as long as the CSI is imperfect.

An interesting security breach is considered in [49] for a time-division duplex (TDD) transmission in a massive multiple input–multiple output (MIMO) setting. An active eavesdropper attacks the uplink training phase to cause pilot contamination at the transmitter, which will result in more signal transmitted towards the eavesdropper (due to the inadequate downlink precoder). It is shown that matched filter (MF) precoding along with AN transmission is a powerful way to improve the asymptotic achievable secrecy rate when the number of transmit antennas is sufficiently large. The trade-off between maximizing power allocated to AN transmission (to hide information from a potential eavesdropper) and maintaining a predefined signal-to-interference-plus-noise ratio (SINR) for the desired link quality has been studied [27].

16.4.2.2 Cooperative Jamming

Cooperative jamming is another technique aimed at degrading the channel quality at the eavesdropper by creating intentional interference [16, 43]. In this category, cooperative relays are studied in [11] for AN generation. In particular, a secret transmission protocol involving a messaging relay, which assists the data exchange between the transmitter and the legitimate user, and a set of intervening relays are employed to generate AN targeting the eavesdroppers. It is also quantitatively analyzed how many eavesdroppers can be tolerated without affecting the communications secrecy in a wireless network supporting a certain number of legitimate users.

16.4.2.3 Protected Zone

In addition to AN transmission and cooperative jamming, protected zone is yet another interesting technique to enhance security in the physical layer. A protected zone is defined surrounding the transmitter in [31] along with beamforming and AN transmission to enhance the PLS in a multi-antenna communications system. In this implementation, the protected zone guarantees that no eavesdropper can exist within the specified (i.e., protected) region. The work of [24] takes into account both single- and multi-antenna scenarios along with non-orthogonal multiple access (NOMA) transmission within the context of a hybrid PLS strategy. The work proposes an eavesdropper-exclusion area to enhance secrecy rates in a single-antenna setting, and considers AN generation towards undesired directions to enhance PLS when the use of multiple antennas is possible.

16.5 Physical Layer Security for UAVs

UAVs are attracting significant interest in areas related to search and rescue, inspection and surveillance, cargo transportation, etc., mainly due to their controllable mobility, low cost, and on-demand rapid deployment. Recently, UAVs have also found increasingly more substantial applications in wireless communication [53], and are expected to play a significant role in the next-generation communication networks [48, 55]. In particular, UAV-based communication networks are receiving a great deal of attention due to their flexibility in rapid deployment of a wireless network infrastructure during temporary events [1–3, 26, 35, 53]. Such temporary events include natural disasters, where the existing network infrastructure is destroyed [2, 26], or sports events in stadiums, where there may be thousands of mobile users straining the available communication resources [3, 35]. On the other hand, the other paradigm where UAVs act as new "sky" users in the cellular networks enables communication to be achieved with UAV users even beyond the range of their operators' visual line-of-sight (LoS). In fact, the next-generation communication systems are specifically focusing on enhancing such long-range real-time communication with UAVs [23, 42].

Ensuring a secure communication in a UAV-based communication network is in fact more important than a fixed infrastructure communication network. This is because, since these networks are deployed during temporary events, as mentioned previously, and legitimate users may be deprived of any other means to achieve their communication, these networks can carry some sensitive information. Further, due to the rapid deployment of these networks, there may be less attention on securing the on-going communication within the network. As a result, these communication networks are highly vulnerable to eavesdropper attacks, as shown in Figure 16.1.

As discussed in Section 16.4, PLS can be a smart solution to ensure security in wireless communication systems with limited resources and to provide additional security on top of the state-of-the-art cryptographic techniques. Subsequently, we discuss a few interesting PLS techniques proposed for UAV-based communication networks, followed by an interesting case study from [33].

16.5.1 UAV Trajectory Design to Enhance PLS

In this section, we discuss how the mobility of UAVs can be effectively utilized for enhancing PLS in a UAV-based communication network. As shown in Figure 16.2, the core objective here is to design the UAV trajectory so as to establish a stronger channel to the legitimate user compared to that of the eavesdropper. In fact, by flying the UAV closer to (farther

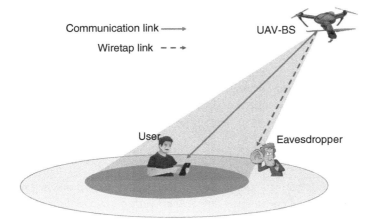

Figure 16.1 A scenario where an eavesdropper breaches the on-going communication between a legitimate user and the UAV base station.

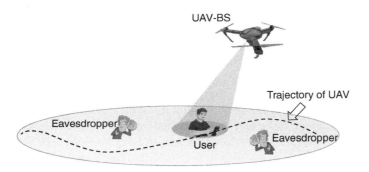

Figure 16.2 The trajectory of the UAV is optimized to maximize secrecy rates.

from) the legitimate user (eavesdropper) via proper trajectory design, this objective can be achieved [54].

In order to achieve a satisfactory level of secrecy performance with this solution, having accurate location information of legitimate users and eavesdroppers at the UAV is important. However, if there is ambiguity in the exact location information, the expected level of performance may not be achievable. There are some works in the literature explicitly taking this fact into consideration within the design process [7].

Another important aspect to be considered while designing the UAV trajectory to enhance PLS is no-fly zones (NFZs), as discussed in [10]. For instance, there are certain restrictions enforced by the US Federal Aviation Administration (FAA) on UAV operations around airports. In fact, consideration of NFZs within the trajectory design problem is important for practical deployments. In [10], the joint UAV trajectory and power optimization problem is formulated in order to maximize the achievable average secrecy rate over a finite flight period, while guaranteeing the UAV's maximum speed, the initial and final locations, the NFZ constraints, as well as the transmit power constraints.

Deploying UAVs as relays to connect transmitter–receiver pairs especially with no LoS conditions is yet another interesting and important application of UAVs. However, when a UAV is acting as a relay, it may not be feasible to employ some advance cryptographic techniques, since transmitter and receiver do not have a direct communication link and communication needs to go via a UAV. Hence, these type of networks are highly vulnerable to eavesdropper attacks. As discussed, trajectory design of a UAV relay can be a smart solution here to enhance secrecy performance. In fact, there are some works in the literature investigating how to enhance secrecy rates in such a UAV-aided mobile relaying system, consisting of a source, a destination, a UAV relay, and an eavesdropper, by designing the UAV trajectory [45].

16.5.2 Cooperative Jamming to Enhance PLS

Cooperative jamming is yet another interesting solution proposed for enhancing PLS in UAV-based communication networks [19, 21, 25, 59]. Since there are multiple eavesdroppers and their exact locations are unknown, it is difficult for the source UAV to make sure that eavesdroppers are not in the proximity of a legitimate user. As a result, there is a considerable possibility that the channel between the source UAV and the legitimate user is worse than that between the source UAV and the eavesdropper(s). This then leads to a higher vulnerability for secrecy breaching.

In order to overcome this, a cooperative *jammer* UAV can be deployed to enhance the secrecy performance of the UAV-based communication system, as captured in Figure 16.3. In particular, the jammer UAV can be scheduled to fly as close as possible to the ground eavesdroppers, while transmitting jamming signals towards eavesdroppers to compromise the wiretap channel [21]. In fact, as discussed in [19], a cooperative jammer UAV can be deployed to support confidential data transmission to a scheduled user by transmitting jamming signals towards unscheduled users, since those unscheduled users can be potential eavesdroppers. Further, as proposed in [25], cooperative jammer UAV(s) can be a handy solution to improve secrecy performance while UAV relays are deployed for connecting source and destination nodes.

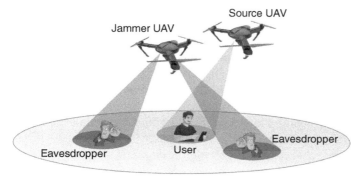

Figure 16.3 An example cooperative jamming scenario with UAVs: jamming signals are transmitted towards eavesdroppers by the jammer UAV while the source UAV serves the legitimate user.

Note that, in addition to deploying a cooperative jammer UAV, joint optimization of transmit power and flying trajectories for both UAVs (i.e., source UAV and jammer UAV) can also be considered for enhancing PLS further [19, 21].

16.5.3 Spectral- and Energy-Efficient PLS Techniques

Due to the limited power resources on-board a UAV BS, it is of utmost importance to achieve energy-efficient communication in a UAV-based communication network while ensuring confidentiality for on-going communications. As discussed in [20], the flight duration and secrecy performance of a UAV-based communication network are fundamentally restricted by the on-board battery power, while the total energy of the UAV is mainly consumed by the wireless communication part, compared with the propulsion and computational energy. Hence, enhancing energy and spectral efficiencies is essential in UAV-based communication networks. To that end, NOMA has been identified as a promising technology for the next-generation wireless communication systems for enhancing energy and spectral efficiencies [36, 41]. NOMA with multi-antenna transmission can be used for further improving the spectral efficiency. In fact, NOMA can be an effective solution to enhance spectral and energy efficiencies in power-limited UAV-based communication networks as well [34, 35].

However, achieving a secure communication is indeed pivotal with NOMA transmission. There are some studies in the literature on how to enhance secrecy performance when NOMA transmission is considered [24, 29, 44, 57]. In particular, [44] studies the secrecy sum rate optimization problem for a MIMO-NOMA system. Considering single-antenna and multi-antenna scenarios, PLS with NOMA transmission in large-scale networks is investigated in [24]. In particular, for the single-antenna scenario, an eavesdropper exclusion area is proposed, while for the multi-antenna scenario, AN generation towards undesired directions is introduced to enhance PLS. The secrecy performance of NOMA with randomly deployed users and eavesdroppers is analyzed in [29] considering large-scale networks. In order to improve the secrecy performance of such networks, a protected zone around the BS is introduced within which it is guaranteed that no eavesdropper exists.

Interestingly, inter-user interference (as a result of superposing of user messages in NOMA) in NOMA transmission can effectively be utilized for confusing the eavesdropper who is trying to intercept the communication between the BS and legitimate NOMA users. In particular, as shown in Figure 16.4, NOMA transmission is deployed in [6] along with massive MIMO techniques and non-orthogonal training in the uplink. User clusters are identified based on spatial information utilizing antenna arrays.

Inter-user interference in the downlink due to NOMA transmission is then effectively utilized to degrade the signal received at the eavesdropper. In addition, with the help of non-orthogonal training, the possibility of interception of a legitimate user transmission by the eavesdropper is further degraded. Specifically, the inter-user interference here is harnessed and served as an AN signal to confuse the eavesdroppers.

Enhancing PLS in a UAV-based communication network with NOMA transmission is even more important than that in a fixed infrastructure communication network. As discussed previously, this is mainly due to the rapid deployment of such networks and to the fact that users may be deprived of any other means to achieve their communication. In

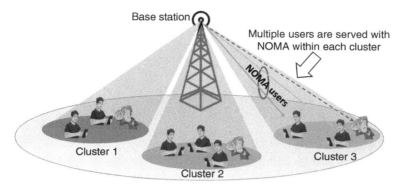

Figure 16.4 Multiple users are served within each cluster using NOMA. Non-orthogonal training is considered in the uplink to further degrade reception at the eavesdropper in addition to inherent inter-user interference in NOMA. Source: Modified from Chen et al. [6].

[33], an interesting solution is proposed to enhance PLS in a UAV–NOMA network which is deployed to provide broadband connectivity over a densely packed user area, which well represents most temporary events (e.g., football match in a stadium). In the next section, we discuss in detail the solution proposed in [33] for enhancing PLS in a UAV–NOMA network as a case study to better educate the interested reader.

16.6 A Case Study: Secure UAV Transmission

In this section, we provide an interesting case study on how PLS techniques can be effectively utilized to enhance security in a UAV–MIMO communication network. As shown in Figure 16.5, a UAV BS is deployed to provide coverage over a densely packed user region. However, as can be observed, there are eavesdroppers outside of the user region (known as the *Eve region*) trying to breach communications going on between legitimate users and the UAV BS. Note here that it is assumed that there are no eavesdroppers within the user region.

In order to enhance the PLS, a protected zone is introduced around the user region. Note that it can be guaranteed that the protected zone is free of eavesdroppers. For instance, a UAV BS can transmit a jamming signal towards the protected zone such that eavesdroppers within the protected zone cannot intercept communication going on between the UAV BS and legitimate users. However, due to physical constraints, such as limited power resources on-board the UAV, the protected zone may not be able to cover the entire Eve region. Hence, an approach is proposed in [33] to optimize the protected zone shape based on UAV BS hovering altitude such that the achievable NOMA sum secrecy rates are maximized. Subsequently, we capture details about the proposed solution.

16.6.1 System Model

We consider a mmWave–NOMA transmission scenario where a single UAV BS equipped with an M-element uniform linear array (ULA) is serving single-antenna users in the downlink. We assume that all the users lie inside a specific user region, as shown in Figure 16.5.

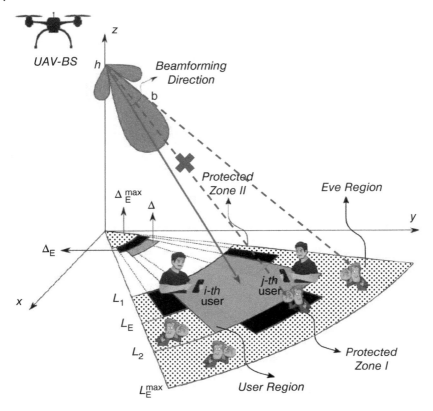

Figure 16.5 System scenario where NOMA transmission serves multiple users simultaneously in a single downlink beam.

A three-dimensional (3D) beam is generated by the UAV BS which entirely covers the user region. We assume that there are K users in total, and the users can be represented by the set $\mathcal{N}_U = \{1, 2, \ldots, K\}$. The user region is identified by an inner radius L_1, an outer radius L_2, and the angle Δ, which is the fixed angle within the projection of the horizontal propagation pattern of the UAV BS on the xy-plane. Note that it is possible to reasonably model various different hotspot scenarios, such as a stadium, concert hall, traffic jam, and urban canyon, by modifying these control parameters.

We assume that, although the user region is free from eavesdroppers, the surrounding region includes eavesdroppers trying to intercept the transmission between the UAV BS and the legitimate users. We designate the bounded region around the user region, which includes eavesdroppers, as the Eve region. Similar to the user region, we identify the Eve region by the same inner radius L_1, but an outer radius $L_E^{\max}$ (greater than L_2), and angle $\Delta_E^{\max}$ (greater than Δ), as shown in Figure 16.5. We assume K_E eavesdroppers in total, which are represented by the set $\mathcal{N}_E = \{1, 2, \ldots, K_E\}$. Note that the horizontal footprint of the UAV BS beam pattern covers the Eve region (so that any eavesdropper has nonzero channel to the UAV BS), as well, but the coverage over the Eve region might be provided by the side lobes depending on the specific radiation pattern.

16.6.1.1 Location Distribution and mmWave Channel Model

We assume that users and eavesdroppers are uniformly and randomly distributed within their specified regions following a homogeneous Poisson point process (HPPP) with the densities λ and λ_E, respectively. The number of users (eavesdroppers) in the user (Eve) region is therefore Poisson distributed, i.e.,

$$P(k \text{ users in the user region}) = \frac{\mu^k e^{-\mu}}{k!} \quad \text{with} \quad \mu = (L_2^2 - L_1^2)\frac{\Delta}{2}\lambda.$$

We assume that all the users have LoS paths since (i) the UAV BS is hovering at relatively high altitudes, and (ii) the LoS path is much stronger than the non-line-of-sight (NLoS) paths in the mmWave frequency band [8, 35]. The channel $\boldsymbol{h}_k$ between the kth user and the UAV BS is therefore given as

$$\boldsymbol{h}_k = \sqrt{M}\frac{\alpha_k \boldsymbol{a}(\theta_k)}{[\text{PL}(\sqrt{d_k^2 + h^2})]^{1/2}}, \tag{16.1}$$

where h, d_k, α_k, and θ_k represent UAV BS hovering altitude, horizontal distance between kth user and UAV BS, small-scale fading gain (i.e., complex Gaussian with $\mathcal{CN}(0, 1)$), and angle of departure (AoD), respectively. In addition, $\boldsymbol{a}(\theta_k)$ is the steering vector associated with AoD θ_k, and PL(x) represents the path loss (PL) over the distance x. Note that the channel between the ℓth eavesdropper in the Eve region (i.e., $\ell \in \mathcal{N}_E$) and the UAV BS can also be given using Eq. (16.1).

16.6.2 Protected Zone Approach for Enhancing PLS

The overall transmission scheme between the UAV BS and legitimate users presented in Figure 16.5 is highly prone to eavesdropper attacks, and the PLS is accordingly impaired. In this study, we consider the protected zone approach to enhance the secrecy rates of the network [24, 31]. In the proposed approach, an additional area (i.e., protected zone) around the user region (and inside the Eve region) has been cleared from eavesdroppers by means of some measures, as shown in Figure 16.6. This protected area is actually a fraction of the complete Eve region, and we denote this fraction by q with $q \leq 1$. Note that, since clearing eavesdroppers in the protected zone requires certain resources being spent on the ground, our goal is to keep this area as small as possible. In addition, we consider optimizing the shape of the protected zone to enhance secrecy rates while keeping its area the same, which is the main problem we tackle in this study.

The protected zone can be represented by an angle–distance (radius) pair (Δ_E, L_E) with $\Delta_E^{\min} \leq \Delta_E \leq \Delta_E^{\max}$ and $L_1 \leq L_E \leq L_E^{\max}$. Note that $\Delta_E^{\min}$ is the minimum angle value which occurs when $L_E = L_E^{\max}$. We can therefore represent $\Delta_E^{\min}$ as follows:

$$\Delta_E^{\min} = \frac{q[((L_E^{\max})^2 - L_1^2)\Delta_E^{\max} - (L_2^2 - L_1^2)\Delta]}{(L_E^{\max})^2 - L_2^2}. \tag{16.2}$$

As sketched in Figure 16.6, it is possible to have different shapes for the protected zone for a fixed q value. Note that, whenever we have $\Delta_E \leq \Delta$, L_E should be sufficiently greater than L_2 (e.g., "Protected Zone I" in Figure 16.6) to have a nonzero protected zone. When $\Delta \leq \Delta_E \leq \Delta_E^{\max}$, L_E might, however, be smaller (e.g., "Protected Zone II" in Figure 16.6)

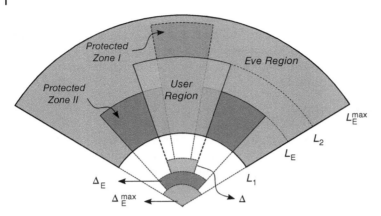

Figure 16.6 Footprint of the protected zone represented by angle–distance pair (Δ_{E}, L_{E}), which is free from any eavesdroppers.

or greater than L_2 depending on the area of the user region and the particular q choice. Specifically, L_{E} can be parametrically expressed as follows:

$$L_{\mathrm{E}}^2 = L_2^2 + \frac{q}{\Delta_{\mathrm{E}}}[((L_{\mathrm{E}}^{\max})^2 - L_1^2)\Delta_{\mathrm{E}}^{\max} - (L_2^2 - L_1^2)\Delta], \tag{16.3}$$

for $\Delta_{\mathrm{E}}^{\min} \leq \Delta_{\mathrm{E}} \leq \Delta$. Whenever we have $\Delta < \Delta_{\mathrm{E}} \leq \Delta_{\mathrm{E}}^{\max}$, then

$$L_{\mathrm{E}}^2 = L_1^2 + \frac{q}{\Delta_{\mathrm{E}}}\left[((L_{\mathrm{E}}^{\max})^2 - L_1^2)\Delta_{\mathrm{E}}^{\max} + \frac{1-q}{q}(L_2^2 - L_1^2)\Delta\right], \tag{16.4}$$

provided $L_{\mathrm{E}}^2 \geq L_2^2$, and L_{E} is otherwise expressed as

$$L_{\mathrm{E}}^2 = L_1^2 + \frac{q}{\Delta_{\mathrm{E}} - \Delta}[((L_{\mathrm{E}}^{\max})^2 - L_1^2)\Delta_{\mathrm{E}}^{\max} - (L_2^2 - L_1^2)\Delta]. \tag{16.5}$$

16.6.3 Secure NOMA for UAV BS Downlink

In this section, we consider NOMA transmission in the UAV BS downlink to enhance the spectral efficiency, and evaluate the associated secrecy rates in the presence of a protected zone.

16.6.3.1 Secrecy Outage and Sum Secrecy Rates

We assume that the UAV BS generates a beam $\boldsymbol{b}$ where the respective projection in the azimuth domain is in the direction of $\bar{\theta}$ with $\bar{\theta} \in [0, 2\pi]$ [35]. Assuming a critically spaced array, the effective channel gain of user $k \in \mathcal{N}_{\mathrm{U}}$ for beamforming direction $\bar{\theta}$ can be given using Eq. (16.1) as follows [35]:

$$
\begin{aligned}
|\boldsymbol{h}_k^{\mathrm{H}}\boldsymbol{b}|^2 &\approx \frac{|\alpha_k|^2}{M \times \mathrm{PL}(\sqrt{d_k^2 + h^2})}\left|\frac{\sin(\pi M(\bar{\theta} - \theta_k)/2)}{\sin(\pi(\bar{\theta} - \theta_k)/2)}\right|^2 \\
&= \frac{|\alpha_k|^2}{\mathrm{PL}(\sqrt{d_k^2 + h^2})}F_M(\pi[\bar{\theta} - \theta_k]),
\end{aligned} \tag{16.6}
$$

where $F_M(\cdot)$ is the Fejér kernel. Similarly, the effective channel gain of the most detrimental eavesdropper, g_E, is given as

$$g_E = \max_{k_E \in \mathcal{N}_E} |\boldsymbol{h}_{k_E}^H \boldsymbol{b}|^2, \tag{16.7}$$

where $\boldsymbol{h}_{k_E}$ is the channel gain of the k_Eth eavesdropper.

When deriving secrecy rates in NOMA transmission, we assume that the UAV BS knows the effective channel gains of desired users, while those of eavesdroppers are unknown. Without any loss of generality, we also assume that the users in set $\mathcal{N}_U$ are already indexed from the best to the worst with respect to their effective channel gains as represented by Eq. (16.6). Defining β_k to be the power allocation coefficient of the kth user, we therefore have $\beta_1 \leq \cdots \leq \beta_K$ such that $\sum_{k=1}^{K} \beta_k^2 = 1$. The transmitted signal is generated by superposition coding as

$$\boldsymbol{x} = \sqrt{P_{Tx}} \, \boldsymbol{b} \sum_{k=1}^{K} \beta_k s_k, \tag{16.8}$$

where P_{Tx} and s_k are the total downlink transmit power and the kth user's message, respectively. The received signal at the kth user is then given as

$$y_k = \boldsymbol{h}_k^H \boldsymbol{x} + v_k = \sqrt{P_{Tx}} \, \boldsymbol{h}_k^H \boldsymbol{b} \sum_{k=1}^{K} \beta_k s_k + v_k, \tag{16.9}$$

where v_k is a zero-mean complex additive white Gaussian noise with variance N_0.

With the received signal as in Eq. (16.9) in hand, each user first decodes messages of all weaker users (allocated with larger power) sequentially in the presence of the stronger users' messages (allocated with smaller power). Those decoded messages are then subtracted from the received signal in Eq. (16.9), and each user decodes its own message, treating the stronger users' messages as noise. This overall decoding process is known as successive interference cancellation (SIC), and the kth user decodes its own message after SIC with the following SINR:

$$\text{SINR}_k = \frac{P_{Tx} |\boldsymbol{h}_k^H \boldsymbol{b}|^2 \beta_k^2}{(1 - \delta_{k1}) P_{Tx} \sum_{l=1}^{k-1} |\boldsymbol{h}_k^H \boldsymbol{b}|^2 \beta_l^2 + N_0}, \tag{16.10}$$

where δ_{k1} is the Kronecker delta function taking value 1 if $k = 1$, and 0 otherwise. Assuming that eavesdroppers have powerful detection capability [24, 56], the most detrimental eavesdropper decodes the kth user message with the SINR given as

$$\text{SINR}_k^E = \frac{P_{Tx} \beta_k^2 g_E}{(1 - \delta_{k1}) P_{Tx} \sum_{l=1}^{k-1} \beta_l^2 g_E + N_0^E}, \tag{16.11}$$

where N_0^E is the associated noise variance.

Considering the SINR in Eq. (16.10), the instantaneous rate at the kth user is $R_k^{NOMA} = \log_2(1 + \text{SINR}_k)$. Similarly, considering Eq. (16.11), the instantaneous rate at the most detrimental eavesdropper for decoding the kth user's message is given as

$R_{k,\mathrm{E}}^{\mathrm{NOMA}} = \log_2(1 + \mathrm{SINR}_k^{\mathrm{E}})$. The secrecy rate for the kth legitimate user can therefore be given as [4, 24]

$$C_k^{\mathrm{NOMA}} = [R_k^{\mathrm{NOMA}} - R_{k,\mathrm{E}}^{\mathrm{NOMA}}]^+, \tag{16.12}$$

where $[x]^+ = \max\{x, 0\}$. As Eq. (16.12) implies, the secrecy rates are always strictly positive [51]. Assuming that $\overline{R}_k$ denotes the desired secrecy rate for the user $k \in \mathcal{N}_{\mathrm{U}}$, we define the secrecy outage event as occurring whenever $C_k^{\mathrm{NOMA}} < \overline{R}_k$ with the respective secrecy outage probability $\mathrm{P}_k^o = \mathrm{P}\{C_k^{\mathrm{NOMA}} < \overline{R}_k\}$. As a result, the outage sum secrecy rate with NOMA transmission can be given as

$$R^{\mathrm{NOMA}} = \sum_{k=1}^{K}(1 - \mathrm{P}_k^o)\overline{R}_k. \tag{16.13}$$

For performance comparison, we also consider the outage sum secrecy rate with orthogonal multiple access (OMA) transmission.

16.6.3.2 Shape Optimization for Protected Zone

In this section, we discuss optimization of the protected zone shape to enhance the secrecy rates while keeping its area (i.e., q) the same. We note that any particular subregion within the Eve region does not equally impair the achievable secrecy rates even if the subregion areas are the same and the eavesdroppers are equally capable. This is basically due to the varying effective channel gain between the UAV BS and an eavesdropper in different sub-regions, which is a function not only of the distance but also of the relative angle (i.e., angle offset from the beamforming direction) associated with each eavesdropper.

Considering Eq. (16.12), the subregion involving the most detrimental eavesdropper has the largest impact on the secrecy rates. Hence, instead of choosing the subregions arbitrarily to form the protected zone, it is more meaningful to include (i.e., protect) subregions that result in better effective channel gain for potential eavesdroppers, and hence is likely to involve the most detrimental eavesdropper.

As we will show in Section 16.6.3.4, the location distribution of the most detrimental eavesdropper depends also on the hovering altitude of the UAV BS. In particular, the most detrimental eavesdroppers are likely to be present in a subregion where $\Delta_{\mathrm{E}} \geq \Delta$ and $L_{\mathrm{E}} \leq L_2$, which is represented by "Protected Zone II" in Figure 16.6, when the altitude is low. In contrast, the region including the most detrimental eavesdroppers becomes closer to "Protected Zone I" of Figure 16.6 with $\Delta_{\mathrm{E}} \leq \Delta$ and $L_{\mathrm{E}} \geq L_2$ when the altitude is high. We therefore conclude that the shape of the protected zone should be optimized to take into account the UAV BS hovering altitude. Hence, at a particular altitude and for a given q, the optimal shape of the protected zone can be identified as

$$\Delta_{\mathrm{E}}^*, \ L_{\mathrm{E}}^* = \mathrm{argmax}_{\Delta_{\mathrm{E}}, L_{\mathrm{E}}} \ R^{\mathrm{NOMA}}$$
$$\text{s.t. } \Delta_{\mathrm{E}}^{\min} \leq \Delta_{\mathrm{E}} \leq \Delta_{\mathrm{E}}^{\max},$$
$$L_{\mathrm{E}} \text{ is computed by Eqs. (16.3)–(16.5)} \tag{16.14}$$

where R^{NOMA} is given in Eq. (16.13).

16.6.3.3 Numerical Results

In this section, we present numerical results to show the importance of shape optimization of the protected zone and its impact on the achievable sum secrecy rates with varying UAV BS hovering altitudes. Considering Figure 16.5, we assume that $L_2 = 100$ m, $L_1 = 25$ m, $L_E^{max} = 1.5L_2$ m, $\Delta = 0.02$ rad $(1.145°)$, $\Delta_E^{max} = 2\Delta$, $\overline{\theta} = 0°$, and $M = 100$. The distribution of users is based on HPPP with $\lambda = 1$, and user target secrecy rates are $\overline{R}_j = 4$ BPCU (bits per channel use) and $\overline{R}_i = 1$ BPCU, respectively. The power allocation ratios are $\beta_j^2 = 0.25$ and $\beta_i^2 = 0.75$, while $P_{Tx} = 10$ dBm and $N_0 = -35$ dBm. We assume two-user NOMA transmission with $j = 1$ and $i = 20$ after ordering users with respect to their effective channel gains. The path-loss model is assumed to be $\mathrm{PL}(\sqrt{d_k^2 + h^2}) = 1 + (\sqrt{d_k^2 + h^2})^\gamma$ with $\gamma = 2$ [8], and the UAV BS altitude is $h \in [10,150]$ m.

16.6.3.4 Location of the Most Detrimental Eavesdropper

We present the angle and distance distributions of the location of the most detrimental eavesdropper in Figures 16.7 and 16.8, respectively, for two different altitudes of $h = \{10,100\}$ m, and HPPP densities of $\lambda_E = \{0.1, 1\}$. In Figure 16.7, we observe that the most detrimental Eve is very likely to have a relative angle that is greater than $\Delta/2$ at a lower altitude of $h = 10$ m. In particular, the relative angle of the most detrimental eavesdropper exceeds $\Delta/2$ all the time for $\lambda_E = 1$ while it drops to approximately 70% of the time for $\lambda_E = 0.1$. When the altitude becomes higher (i.e., $h = 100$ m), the relative angle of the most detrimental eavesdropper becomes smaller than $\Delta/2$.

In Figure 16.8, we observe that the PL distance of the most detrimental eavesdropper is smaller (greater) than 100 m at lower (higher) altitudes, i.e., $h = 10$ m ($h = 100$ m). We therefore conclude that the most detrimental eavesdropper tends to have larger relative angles and smaller PL distances at lower altitudes in comparison to those at higher altitudes.

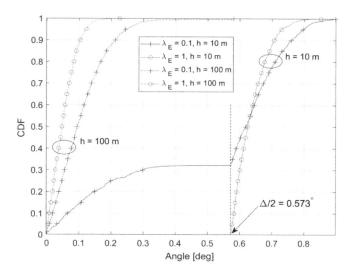

Figure 16.7 Angle distribution of the most detrimental eavesdropper (CDF = cumulative distribution function).

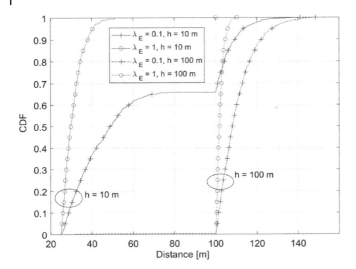

Figure 16.8 Distance distribution of the most detrimental eavesdropper (CDF = cumulative distribution function).

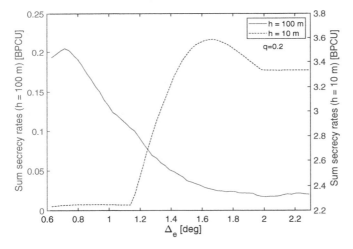

Figure 16.9 Sum secrecy rates of NOMA along with the protected zone angle (i.e., Δ_E) for $h = \{10,100\}$ m, $q = 0.2$, and $\lambda_E = 0.1$.

16.6.3.5 Impact of the Protected Zone Shape on Secrecy Rates

In Figure 16.9, we depict the sum secrecy rates along with the protected zone angle (i.e., Δ_E) at altitudes of $h = \{10,100\}$ m assuming $q = 0.2$. We observe that, while the secrecy rates get maximized at $\Delta_E \approx 1.7°$ ($> \Delta$) for $h = 10$ m, the optimal angle turns out to be $\Delta_E \approx 0.7°$ ($< \Delta$) at $h = 100$ m. This observation is consistent with the discussion in Section 16.6.3.4 in the sense that the most detrimental eavesdropper has a relative angle greater (smaller) than $\Delta/2$ at low (high) altitudes.

Similarly, Figure 16.10 presents the secrecy rates along with the protected zone distance (i.e., L_E) for the same settings as in Figure 16.9. We observe that, while the optimal distance

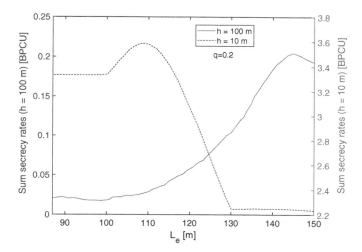

Figure 16.10 Sum secrecy rates of NOMA along with the protected zone distance (i.e., L_E) for $h = \{10,100\}$ m, $q = 0.2$, and $\lambda_E = 0.1$.

maximizing the secrecy rates is $L_E \approx 110$ m at $h = 10$ m, it turns out to be $L_E \approx 145$ m at $h = 100$ m. As before, this observation also agrees nicely with our discussions in Section 16.6.3.4 regarding the distance distribution of the most detrimental eavesdropper. This shows the importance of optimizing the protected zone shape at different hovering altitudes to maximize sum secrecy rates.

16.6.3.6 Variation of Secrecy Rates with Altitude

In Figure 16.11, we present the sum secrecy rates of NOMA and OMA transmission along with varying altitude of $h \in [10,150]$ m and for different protected zone sizes (i.e., $q \in \{0, 0.2, 0.5\}$). For a nonzero protected zone (i.e., $q \neq 0$), considering shape optimization as discussed in Section 16.6.3.2, sum secrecy rates are identified. In addition, Figure 16.11 also captures the sum secrecy rate variation with $q = 0.2$ for a fixed shape (optimal shape at $h = 10$ m). As can be observed, the fixed protected zone shape yields sum secrecy rates comparable to that of the optimized protected zone shape only around $h = 10$ m and performs worse at all the other altitudes. Further, we observe that the secrecy rates improve if a large portion of the Eve region can be covered by the protected zone (i.e., q increases). Based on the target sum secrecy rate and the operational altitude, the smallest q can also be determined. In this way, the desired secrecy rates can be achieved optimally by designating less area as protected zone, which would relieve the burden of clearing any unnecessary region free from eavesdroppers. Note also that the secrecy rate associated with NOMA is much larger than those of OMA, especially at lower altitudes.

In Figures 16.12 and 16.13, the variations of the optimal shape of the protected zone (angle and distance, respectively) are captured for $q = 0.2$ and 0.5. In particular, Figure 16.12 shows the optimal angle Δ_E^* variation whereas Figure 16.13 depicts the optimal distance L_E^* variation with UAV BS hovering altitude. As can be observed, Δ_E^* decreases with altitude (see Figure 16.12) while L_E^* increases with altitude (see Figure 16.13). This observation aligns nicely with the discussion in Section 16.6.3.4 which tells us that, at

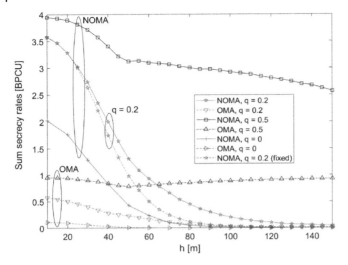

Figure 16.11 Sum secrecy rates for NOMA and OMA along with UAV BS hovering altitude, where $q \in \{0, 0.2, 0.5\}$, and $\lambda_E = 0.1$.

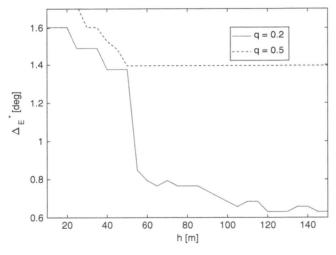

Figure 16.12 Optimal angle Δ_E^* variation with varying UAV BS hovering altitudes. Here $q = 0.2, 0.5$.

lower altitudes, the most detrimental eavesdropper tends to have a larger relative angle and smaller distance, whereas at higher altitudes, this is vice versa.

Summary

Wireless communication networks are highly prone to security attacks from malicious users. This is mainly due to the wireless channel used for achieving communication

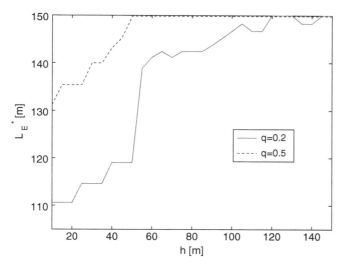

Figure 16.13 Optimal distance L_E^* variation with varying UAV BS hovering altitudes. Here $q = 0.2, 0.5$.

between the transmitter and legitimate users. In this chapter, we first discuss possible security attacks in a wireless communication network followed by the state-of-the-art PLS techniques to prevent such attacks of malicious users. In particular, we discuss extensively how PLS techniques can ensure secure communication within UAV-based wireless communication networks. To that end, the importance of UAV trajectory design, cooperative jamming, and introducing protected zones is discussed extensively.

An interesting case study is presented towards the end of the chapter, which investigates achievable secrecy rates of a UAV-based mmWave communication network considering PLS techniques along with NOMA transmission. In particular, a resource constrained protected zone approach is introduced to enhance the PLS. First, the location distribution of the most detrimental eavesdropper, which impairs secrecy rates the most, is investigated at different UAV altitudes. Based on the location distribution of the most detrimental eavesdropper, the associated optimal shape of the protected zone is determined for a given hovering altitude of the UAV in order to enhance secrecy rates.

References

1 Alex Fitzpatrick. Drones are here to stay. Get used to it. *TIME*, May. 2018. URL http://time.com/longform/time-the-drone-age/.

2 AT&T. Flying COW connects Puerto Rico, Nov. 2017. URL http://about.att.com/inside.connections.blog/flying.cow.puertori.

3 BBC. Drones to the rescue, May 2018. URL http://www.bbc.com/news/business-43906846.

4 M. Bloch, J. Barros, M. R. D. Rodrigues, and S. W. McLaughlin. Wireless information-theoretic security. *IEEE Trans. Inf. Theory*, 54 (6): 2515–2534, June 2008. ISSN 0018-9448. doi: 10.1109/TIT.2008.921908.

5 X. Chen, K. Makki, K. Yen, and N. Pissinou. Sensor network security: a survey. *IEEE Commun. Surv. Tut.*, 11 (2): 52–73, Apr. 2009. ISSN 1553-877X. doi: 10.1109/SURV.2009.090205.

6 X. Chen, Z. Zhang, C. Zhong, D. W. K. Ng, and R. Jia. Exploiting inter-user interference for secure massive non-orthogonal multiple access. *IEEE J. Sel. Areas in Communi.*, 36 (4): 788–801, Apr. 2018. doi: 10.1109/JSAC.2018.2825058.

7 M. Cui, G. Zhang, Q. Wu, and D. W. K. Ng. Robust trajectory and transmit power design for secure UAV communications. *IEEE Trans. on Vehic. Technol.*, 67 (9): 9042–9046, Sep. 2018. doi: 10.1109/TVT.2018.2849644.

8 Z. Ding, P. Fan, and H. V. Poor. Random beamforming in millimeter-wave NOMA networks. *IEEE Access*, (99): 1–1, 2017. ISSN 2169-3536. doi: 10.1109/ACCESS.2017.2673248.

9 D. Dzung, M. Naedele, T. P. Von Hoff, and M. Crevatin. Security for industrial communication systems. *Proc. of the IEEE*, 93 (6): 1152–1177, June 2005. ISSN 0018-9219. doi: 10.1109/JPROC.2005.849714.

10 Y. Gao, H. Tang, B. Li, and X. Yuan. Joint trajectory and power design for uav-enabled secure communications with no-fly zone constraints. *IEEE Access*, 7: 44459–44470, 2019. doi: 10.1109/ACCESS.2019.2908407.

11 D. Goeckel, S. Vasudevan, D. Towsley, S. Adams, Z. Ding, and K. Leung. Artificial noise generation from cooperative relays for everlasting secrecy in two-hop wireless networks. *IEEE J. Sel. Areas in Communi.*, 29 (10): 2067–2076, Dec. 2011. ISSN 0733-8716. doi: 10.1109/JSAC.2011.111216.

12 S. Goel and R. Negi. Secret communication in presence of colluding eavesdroppers. In *IEEE Military Communi. Conf. (MILICOM)*, pages 1501–1506 Vol. 3, Oct 2005. doi: 10.1109/MILCOM.2005.1605889.

13 S. Goel and R. Negi. Guaranteeing secrecy using artificial noise. *IEEE Trans. on Wireless Communi.*, 7 (6): 2180–2189, June 2008. ISSN 1536-1276. doi: 10.1109/TWC.2008.060848.

14 P. K. Gopala, L. Lai, and H. El Gamal. On the secrecy capacity of fading channels. *IEEE Trans. on Info. Theory*, 54 (10): 4687–4698, Oct 2008. ISSN 0018-9448. doi: 10.1109/TIT.2008.928990.

15 J. M. Hamamreh, H. M. Furqan, and H. Arslan. Classifications and applications of physical layer security techniques for confidentiality: A comprehensive survey. *IEEE Commun. Surveys Tuts*, pages 1–1, 2018. ISSN 1553-877X. doi: 10.1109/COMST.2018.2878035.

16 X. He and A. Yener. *Cooperative Jamming: The Tale of Friendly Interference for Secrecy.* Securing Wireless Communications at the Physical Layer, Springer US, 2010.

17 H. Huang, N. Ahmed, and P. Karthik. On a new type of denial of service attack in wireless networks: The distributed jammer network. *IEEE Trans. Wireless Communi.*, 10 (7): 2316–2324, July 2011. ISSN 1536-1276. doi: 10.1109/TWC.2011.052311.101613.

18 Y. Jiang, C. Lin, X. Shen, and M. Shi. Mutual authentication and key exchange protocols for roaming services in wireless mobile networks. *IEEE Trans. on Wireless Commun.*, 5 (9): 2569–2577, Sep. 2006. ISSN 1536-1276. doi: 10.1109/TWC.2006.05063.

19 H. Lee, S. Eom, J. Park, and I. Lee. UAV-aided secure communications with cooperative jamming. *IEEE Trans. on Vehic. Technol.*, 67 (10): 9385–9392, Oct. 2018. doi: 10.1109/TVT.2018.2853723.

20 B. Li, Z. Fei, Y. Zhang, and M. Guizani. Secure UAV communication networks over 5G. *IEEE Wireless Communi.*, pages 1–7, 2019a. doi: 10.1109/MWC.2019.1800458.

21 Y. Li, R. Zhang, J. Zhang, S. Gao, and L. Yang. Cooperative jamming for secure UAV communications with partial eavesdropper information. *IEEE Access*, 7: 94593–94603, 2019b. doi: 10.1109/ACCESS.2019.2926741.

22 X. Lin. Cat: Building couples to early detect node compromise attack in wireless sensor networks. In *Proc. IEEE Global Telecommun. Conf.*, pages 1–6, Nov. 2009. doi: 10.1109/GLOCOM.2009.5425922.

23 X. Lin, R. Wiren, S. Euler, A. Sadam, H. Maattanen, S. Muruganathan, S. Gao, Y. . E. Wang, J. Kauppi, Z. Zou, and V. Yajnanarayana. Mobile network-connected drones: Field trials, simulations, and design insights. *IEEE Vehic. Technol. Magazine*, 14 (3): 115–125, Sep. 2019. doi: 10.1109/MVT.2019.2917363.

24 Y. Liu, Z. Qin, M. Elkashlan, Y. Gao, and L. Hanzo. Enhancing the physical layer security of non-orthogonal multiple access in large-scale networks. *IEEE Trans. Wireless Commun.*, 16 (3): 1656–1672, Mar. 2017. ISSN 1536-1276. doi: 10.1109/TWC.2017.2650987.

25 R. Ma, W. Yang, Y. Zhang, J. Liu, and H. Shi. Secure mmwave communication using UAV-enabled relay and cooperative jammer. *IEEE Access*, 7: 119729–119741, 2019. doi: 10.1109/ACCESS.2019.2933231.

26 Arvind Merwaday, Adem Tuncer, Abhaykumar Kumbhar, and Ismail Guvenc. Improved throughput coverage in natural disasters: Unmanned aerial base stations for public-safety communications. *IEEE Veh. Technol. Mag.*, 11 (4): 53–60, 2016.

27 A. Mukherjee and A. L. Swindlehurst. Robust beamforming for security in mimo wiretap channels with imperfect csi. *IEEE Trans. on Signal Processing*, 59 (1): 351–361, Jan 2011. ISSN 1053-587X. doi: 10.1109/TSP.2010.2078810.

28 T. Ohigashi and M. Morii. A practical message falsification attack on WPA. In *Proc. Joint Workshop Inf. Security*, pages 1–12, Aug. 2009.

29 Z. Qin, Y. Liu, Z. Ding, Y. Gao, and M. Elkashlan. Physical layer security for 5G non-orthogonal multiple access in large-scale networks. In *IEEE Int. Conf. on Communi. (ICC)*, pages 1–6, May 2016. doi: 10.1109/ICC.2016.7510755.

30 D. R. Raymond and S. F. Midkiff. Denial-of-service in wireless sensor networks: Attacks and defenses. *IEEE Perv. Comput.*, 7 (1): 74–81, Jan 2008. ISSN 1536-1268. doi: 10.1109/MPRV.2008.6.

31 N. Romero-Zurita, D. McLernon, M. Ghogho, and A. Swami. Phy layer security based on protected zone and artificial noise. *IEEE Sig. Process. Lett.*, 20 (5): 487–490, May 2013. ISSN 1070-9908. doi: 10.1109/LSP.2013.2252898.

32 N. Rupasinghe, Y. Yapici, I. Guvenc, and Y. Kakishima. Non-orthogonal multiple access for mmWave drones with multi-antenna transmission. In *Proc. IEEE Asilomar Conf. Sig., Sys. Comp.*, 2017.

33 N. Rupasinghe, Y. Yapici, I. Guvenc, H. Dai, and A. Bhuyan. Enhancing physical layer security for NOMA transmission in mmwave drone networks. In *2018 52nd Asilomar Conf. Signal, Syst., Computers*, pages 729–733, Oct. 2018. doi: 10.1109/AC-SSC.2018.8645326.

34 N. Rupasinghe, Y. Yapici, I. Guvenc, M. Ghosh, and Y. Kakishima. Angle feedback for NOMA transmission in mmwave drone networks. *IEEE J. Sel. Topics in Signal Process.*, 13 (3): 628–643, June 2019. doi: 10.1109/JSTSP.2019.2905226.

35 N. Rupasinghe, Y. Yapici, I. Guvenc, and Y. Kakishima. Non-orthogonal multiple access for mmWave drone networks with limited feedback. *IEEE Trans. Commun.*, pages 1–1, Jan. 2019. ISSN 0090-6778. doi: 10.1109/TCOMM.2018.2867465.

36 Y. Saito, Y. Kishiyama, A. Benjebbour, T. Nakamura, A. Li, and K. Higuchi. Non-orthogonal multiple access (NOMA) for cellular future radio access. In *Proc. IEEE Veh. Technol. Conf. (VTC)*, pages 1–5, Jun. 2013. doi: 10.1109/VTCSpring.2013.6692652.

37 C. E. Shannon. Communication theory of secrecy systems. *The Bell System Technical Journal*, 28 (4): 656–715, Oct. 1949. ISSN 0005-8580. doi: 10.1002/j.1538-7305.1949.tb00928.x.

38 Y. Shiu, S. Y. Chang, H. Wu, S. C. Huang, and H. Chen. Physical layer security in wireless networks: a tutorial. *IEEE Wireless Commun.*, 18 (2): 66–74, Apr. 2011. ISSN 1536-1284. doi: 10.1109/MWC.2011.5751298.

39 W. Stallings. *Cryptography and Network Security Principles and Practices*. Prentice Hall PTR, 2006.

40 A. L. Swindlehurst. Fixed sinr solutions for the mimo wiretap channel. In *Proc. IEEE Int. Conf. on Acoustics, Speech and Signal Process.*, pages 2437–2440, April 2009. doi: 10.1109/ICASSP.2009.4960114.

41 Technical Specification Group Radio Access Network. Study on downlink multiuser superposition transmission (MUST) for LTE. Technical Report 3GPP TR36.859 v13.0.0, 3rd Generation Partnership Project (3GPP), Dec. 2015.

42 Technical Specification Group Radio Access Network. Enhanced LTE support for aerial vehicles. Technical Report 3GPP TR36.777, 3rd Generation Partnership Project (3GPP), Dec. 2017.

43 E. Tekin and A. Yener. The general gaussian multiple-access and two-way wiretap channels: Achievable rates and cooperative jamming. *IEEE Trans. on Info. Theory*, 54 (6): 2735–2751, June 2008. ISSN 0018-9448. doi: 10.1109/TIT.2008.921680.

44 M. Tian, Q. Zhang, S. Zhao, Q. Li, and J. Qin. Secrecy sum rate optimization for downlink MIMO nonorthogonal multiple access systems. *IEEE Signal Process. Lett.*, 24 (8): 1113–1117, Aug. 2017. doi: 10.1109/LSP.2017.2711022.

45 Q. Wang, Z. Chen, H. Li, and S. Li. Joint power and trajectory design for physical-layer secrecy in the uav-aided mobile relaying system. *IEEE Access*, 6: 62849–62855, 2018. doi: 10.1109/ACCESS.2018.2877210.

46 Y. Wei, K. Zeng, and P. Mohapatra. Adaptive wireless channel probing for shared key generation. In *Proc. IEEE INFOCOM*, pages 2165–2173, Apr. 2011. doi: 10.1109/INFCOM.2011.5935028.

47 A. D. Wood and J. A. Stankovic. Denial of service in sensor networks. *IEEE Computer*, 35 (10): 54–62, Oct. 2002. ISSN 0018-9162. doi: 10.1109/MC.2002.1039518.

48 Q. Wu, G. Y. Li, W. Chen, D. W. K. Ng, and R. Schober. An overview of sustainable green 5G networks. *IEEE Wireless Communi.*, 24 (4): 72–80, Aug. 2017. doi: 10.1109/MWC.2017.1600343.

49 Y. Wu, R. Schober, D. W. K. Ng, C. Xiao, and G. Caire. Secure massive mimo transmission with an active eavesdropper. *IEEE Trans. on Info. Theory*, 62 (7): 3880–3900, July 2016. ISSN 0018-9448. doi: 10.1109/TIT.2016.2569118.

50 A. D. Wyner. The wire-tap channel. *The Bell System Tech. J.*, 54 (8): 1355–1387, Oct. 1975. ISSN 0005-8580. doi: 10.1002/j.1538-7305.1975.tb02040.x.

51 N. Yang, H. A. Suraweera, I. B. Collings, and C. Yuen. Physical layer security of tas/mrc with antenna correlation. *IEEE Trans. Inf. Forensics Security*, 8 (1): 254–259, Jan. 2013. ISSN 1556-6013. doi: 10.1109/TIFS.2012.2223681.

52 Y. Zeng and R. Zhang. Active eavesdropping via spoofing relay attack. In *Proc. IEEE Int. Conf. on Acoust., Speech and Signal Process. (ICASSP)*, pages 2159–2163, Mar. 2016. doi: 10.1109/ICASSP.2016.7472059.

53 Y. Zeng, R. Zhang, and T. J. Lim. Wireless communications with unmanned aerial vehicles: opportunities and challenges. *IEEE Commun. Mag.*, 54 (5): 36–42, May 2016. ISSN 0163-6804. doi: 10.1109/MCOM.2016.7470933.

54 G. Zhang, Q. Wu, M. Cui, and R. Zhang. Securing uav communications via joint trajectory and power control. *IEEE Trans. on Wireless Communi.*, 18 (2): 1376–1389, Feb 2019. doi: 10.1109/TWC.2019.2892461.

55 S. Zhang, Q. Wu, S. Xu, and G. Y. Li. Fundamental green tradeoffs: Progresses, challenges, and impacts on 5G networks. *IEEE Communi. Surveys Tuts*, 19 (1): 33–56, Firstquarter 2017. doi: 10.1109/COMST.2016.2594120.

56 X. Zhang, X. Zhou, and M. R. McKay. Enhancing secrecy with multi-antenna transmission in wireless ad hoc networks. *IEEE Trans. Inf. Forensics Security*, 8 (11): 1802–1814, Nov. 2013. ISSN 1556-6013. doi: 10.1109/TIFS.2013.2279842.

57 N. Zhao, D. Li, M. Liu, Y. Cao, Y. Chen, Z. Ding, and X. Wang. Secure transmission via joint precoding optimization for downlink miso noma. *IEEE Trans. on Vehic. Technol.*, 68 (8): 7603–7615, Aug. 2019. doi: 10.1109/TVT.2019.2920144.

58 X. Zhou and M. R. McKay. Secure transmission with artificial noise over fading channels: Achievable rate and optimal power allocation. *IEEE Trans. on Vehic. Technol.*, 59 (8): 3831–3842, Oct 2010. ISSN 0018-9545. doi: 10.1109/TVT.2010.2059057.

59 Y. Zhou, P. L. Yeoh, H. Chen, Y. Li, R. Schober, L. Zhuo, and B. Vucetic. Improving physical layer security via a UAV friendly jammer for unknown eavesdropper location. *IEEE Trans. on Vehic. Technol.*, 67 (11): 11280–11284, Nov. 2018. doi: 10.1109/TVT.2018.2868944.

60 Y. Zou, J. Zhu, X. Wang, and L. Hanzo. A survey on wireless security: Technical challenges, recent advances, and future trends. *Proc. of the IEEE*, 104 (9): 1727–1765, Sep. 2016. ISSN 0018-9219. doi: 10.1109/JPROC.2016.2558521.

17

UAV-Enabled Wireless Power Transfer

Jie Xu[1], Yong Zeng[2,4,], and Rui Zhang[3]*

[1] *Future Network of Intelligence Institute (FNii) and School of Science and Engineering, The Chinese University of Hong Kong, Shenzhen, PR China*
[2] *National Mobile Communications Research Laboratory, School of Information Science and Engineering, Southeast University, PR China*
[3] *Department of Electrical and Computer Engineering, National University of Singapore, Singapore*
[4] *Purple Mountain Laboratories, Jiangsu, China*

17.1   Introduction

Besides wireless communications, radio-frequency (RF) signals have also enabled another emerging technology, namely, wireless power transfer (WPT), which is envisioned to be a promising solution to provide perpetual and cost-effective energy supplies to low-power electronic devices in future Internet-of-Things (IoT) wireless networks (see, e.g., [1, 2] and references therein). In conventional WPT systems, dedicated energy transmitters (ETs) are usually deployed at fixed locations to send RF signals to charge distributed energy receivers (ERs) such as low-power sensors and IoT devices. However, due to the severe propagation loss of RF signals over long distance, the performance of practical WPT systems for wide coverage range is fundamentally constrained by the low end-to-end power transmission efficiency. As a consequence, in order to provide ubiquitous wireless energy accessibility for massive low-power ERs scattered in a large area, ETs need to be deployed in an ultra-dense manner. This, however, would tremendously increase the cost, and thus hinder the large-scale implementation of future WPT systems. In the literature, various approaches have been proposed aiming to alleviate this issue by enhancing the WPT efficiency at the link level, via, for example, multi-antenna energy beamforming [3–5].

Different from these prior studies and motivated by the success of unmanned aerial vehicle (UAV) enabled wireless communications [6–8], in this chapter we resolve the above problem from a fundamentally new perspective at the system level, by proposing a radically novel architecture for WPT systems by utilizing UAVs as flying ETs. The proposed UAV-enabled WPT architecture is illustrated in Figure 17.1, in which a group of UAVs are dispatched as mobile ETs that fly above the serving area to cooperatively charge a set of distributed ERs on the ground. By exploiting the fully controllable mobility of the UAVs via joint trajectory design, the proposed system is expected to significantly boost the WPT

* Corresponding Author: Yong Zeng; yong_zeng@seu.edu.cn

UAV Communications for 5G and Beyond, First Edition.
Edited by Yong Zeng, Ismail Guvenc, Rui Zhang, Giovanni Geraci, and David W. Matolak.
© 2021 John Wiley & Sons Ltd. Published 2021 by John Wiley & Sons Ltd.

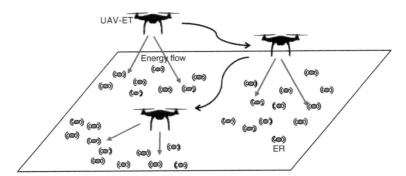

Figure 17.1 Illustration of a UAV-enabled WPT system. Source: From Xu et al. [9].

performance, while reducing the number of required ETs as compared to the conventional WPT systems with static ETs deployed at fixed locations.

A fundamental question to be addressed for UAV-enabled WPT systems is as follows: "How can one jointly optimize the trajectories of multiple UAVs so as to maximize the energy transferred to all ERs in a fair manner?" This question is non-trivial to be addressed, even for the simplest scenario with one UAV and two ERs. Notice that, in this basic setup, the power transferred from the UAV to the two ERs critically depends on the UAV's flying trajectory. For example, when the UAV moves from above one ER to the other, their received power will decrease and increase, respectively, thus resulting in an interesting power trade-off between the two ERs (see, e.g., [9–11]).

For the purpose of exposition, in this chapter we focus on the UAV-enabled multiuser WPT system with one UAV (ET) flying in a horizontal plane with a fixed altitude above the ground to wirelessly charge $K > 1$ ERs. Under the considered setup, we aim to find the optimal UAV trajectory to maximize the amount of energy transferred to the K ERs during a finite charging period, subject to the UAV's maximum speed constraint. First, we consider the UAV's trajectory optimization in the horizontal plane with a fixed altitude above the ground to maximize the sum received energy of all ERs. Despite the fact that this problem is non-convex and involves an infinite number of variables, we derive its optimal solution, which shows that the UAV should hover at one single fixed optimal location during the whole charging period, which can be obtained via a two-dimensional (2D) exhaustive search. In general, the sum-energy maximization with the optimal fixed hovering location incurs a severe "near–far" fairness issue, especially for a network spanning over a large area, as the near ERs in close proximity to the UAV can receive significantly more energy than the far-away ERs.

Next, to resolve the fairness issue, we consider an alternative problem to maximize the minimum received energy among all ERs via trajectory optimization. This problem is more challenging than the previous sum-energy maximization. To obtain useful insight and a performance upper bound, we first consider an ideal case by assuming that the UAV's maximum speed constraint can be ignored, which corresponds to the scenario when the charging duration is sufficiently large as compared to the UAV flying time. In this case, the problem is shown to satisfy the so-called time-sharing condition [12], and thus can be optimally solved via the Lagrange dual method. The obtained optimal solution reveals that the UAV should hover over an optimal set of locations, with optimal hovering time allocations among them.

Last, we consider the above min-energy maximization problem for the general case with the UAV's maximum speed constraint considered. Inspired by the optimal multi-location-hovering solution in the ideal case, we propose a successive hover-and-fly trajectory design, where the UAV successively hovers at a given set of hovering locations (e.g., using the optimal set of hovering locations obtained in the ideal case if the charging duration is sufficiently large) each for a certain duration, and flies with the maximum speed between these hovering locations. The total flying time is minimized by finding the path with the shortest traveling distance to visit all of these hovering locations. The proposed trajectory is asymptotically optimal if the charging duration is sufficiently large, so that the total flying time becomes asymptotically negligible. Furthermore, we also propose a successive convex approximation (SCA) based algorithm to obtain a locally optimal solution to the min-energy maximization problem. By employing the successive fly-and-hover trajectory as the initial input, the SCA-based algorithm iteratively refines the UAV trajectory to improve the min-energy of all ERs until convergence is reached.

The remainder of this chapter is organized as follows. Section 17.2 presents the UAV-enabled WPT system model. Section 17.3 presents the optimal solution to the sum-energy maximization problem. Sections 17.4 and 17.5 present the optimal solutions to the min-energy maximization problem for two cases without and with the UAV maximum speed constraint, respectively. Section 17.6 provides numerical results to validate the effectiveness of our proposed trajectory designs. Section 17.7 concludes this chapter and discusses interesting research directions that are worthy of future investigation.

17.2 System Model

We consider a UAV-enabled multiuser WPT system, where a UAV is dispatched to deliver wireless energy to $K \geq 2$ ERs located on the ground. Let $\mathcal{K} \triangleq \{1, \ldots, K\}$ denote the set of ERs. Each ER $k \in \mathcal{K}$ has a fixed location on the ground, denoted by $(x_k, y_k, 0)$ in three-dimensional (3D) Euclidean coordinates, which is assumed to be known to the UAV a priori for its trajectory design. We consider a finite charging period with duration T, denoted by $\mathcal{T} \triangleq [0, T]$. At each time instant $t \in \mathcal{T}$, the UAV is assumed to fly at a fixed altitude $H > 0$ above the ground, whose time-varying location is denoted as $(x(t), y(t), H)$. We assume that the initial and final UAV locations at time $t = 0$ and $t = T$ are not predetermined, but instead can be freely optimized. Denote by V in meters per second (m s^{-1}) the maximum possible speed of the UAV. We then have the maximum speed constraint at each time instant expressed as

$$\sqrt{\dot{x}^2(t) + \dot{y}^2(t)} \leq V, \quad \forall\, t \in \mathcal{T}, \tag{17.1}$$

with $\dot{x}(t)$ and $\dot{y}(t)$ denoting the time derivatives of $x(t)$ and $y(t)$, respectively.

In practice, the RF-signal-enabled WPT is implemented over a distance of several or tens of meters, and therefore the UAV needs to fly at a relatively low altitude generally with line-of-sight (LoS) links with ERs on the ground. In this case, we consider the free-space path-loss model for the wireless channel between the UAV and each ER, similarly as in prior works on UAV-enabled wireless communications [6]. At time $t \in \mathcal{T}$, the channel power gain from the UAV to ER $k \in \mathcal{K}$ is modeled as $h_k(t) = \beta_0 d_k^{-2}(t)$, where $d_k(t) = \sqrt{(x(t) - x_k)^2 + (y(t) - y_k)^2 + H^2}$ is their distance and β_0 denotes the channel power gain at a reference distance of $d_0 = 1$ m. Assuming that the UAV has a constant transmit power P, the RF power received by ER k at time t is thus given by

$$Q_k(x(t), y(t)) = h_k(t)P = \frac{\beta_0 P}{(x(t) - x_k)^2 + (y(t) - y_k)^2 + H^2}. \tag{17.2}$$

The total energy received by each ER $k \in \mathcal{K}$ over the whole charging period is a function of the UAV's trajectory $\{x(t), y(t)\}$, which can be written as

$$E_k(\{x(t), y(t)\}) = \int_0^T Q_k(x(t), y(t)) \, dt. \tag{17.3}$$

Note that, at each ER, the received RF signals are converted into direct current (DC) signals for energy harvesting via rectifiers [4]. In practice, the RF-to-DC conversion is generally nonlinear and the conversion efficiency critically depends on the received RF power and waveform at the ER (see, e.g., [2, 14, 15]). Since, in general, the harvested DC power increases monotonically with the received RF power, for simplicity, in this chapter, we consider the received RF power in Eq. (17.2) and the resultant received energy in Eq. (17.3) by ERs prior to RF-to-DC conversion as the performance metrics.

17.3 Sum-Energy Maximization

In this section, we consider the maximization of the sum received energy of all ERs over the charging period, by optimizing the UAV's trajectory $\{x(t), y(t)\}$ subject to the speed constraints in Eq. (17.1). The problem can be expressed as

$$\max_{\{x(t), y(t)\}} \sum_{k \in \mathcal{K}} E_k(\{x(t), y(t)\})$$

$$\text{s.t. Eq. (1.1).} \tag{P1}$$

Problem (P1) involves an infinite number of optimization variables, i.e., $x(t)$ and $y(t)$ values over continuous time t. Furthermore, (P1) is a non-convex optimization problem, as the objective function is a non-concave function with respect to the trajectory $\{x(t), y(t)\}$. Therefore, problem (P1) is generally difficult to be optimally solved directly. However, with a careful examination of the special structure of (P1), in the following, we present the optimal solution to (P1).

For ease of description, given the UAV's location $x(t)$ and $y(t)$ at a given time t, we define the sum-power received by all the K ERs as

$$\psi(x(t), y(t)) \triangleq \sum_{k \in \mathcal{K}} Q_k(x(t), y(t)) = \sum_{k \in \mathcal{K}} \frac{\beta_0 P}{(x(t) - x_k)^2 + (y(t) - y_k)^2 + H^2}. \tag{17.4}$$

Accordingly, the sum-energy received by the K ERs over the whole charging period is

$$\sum_{k \in \mathcal{K}} E_k(\{x(t), y(t)\}) = \int_0^T \psi(x(t), y(t)) \, dt. \tag{17.5}$$

Let $x^\star$ and $y^\star$ denote an optimal UAV location that maximizes the function $\psi(x, y)$, i.e.,

$$(x^\star, y^\star) = \arg\max_{x,y} \psi(x, y). \tag{17.6}$$

As the function $\psi(x,y)$ is non-concave with respect to x and y, it is generally difficult to find closed-form expressions for $x^\star$ and $y^\star$. Fortunately, $\psi(x,y)$ in Eq. (17.6) only has two variables x and y. Besides, it is not difficult to show that $x^\star$ and $y^\star$ should satisfy $\underline{x} \leq x^\star \leq \overline{x}$ and $\underline{y} \leq y^\star \leq \overline{y}$, respectively, where

$$\underline{x} = \min_{k \in \mathcal{K}} x_k, \quad \overline{x} = \max_{k \in \mathcal{K}} x_k, \quad \underline{y} = \min_{k \in \mathcal{K}} y_k, \quad \overline{y} = \max_{k \in \mathcal{K}} y_k. \tag{17.7}$$

This is because, if $(x^\star, y^\star)$ lies outside the box specified above, we can always increase the energy transferred to all the K ERs by moving $(x^\star, y^\star)$ into the box. As a result, we can adopt a 2D exhaustive search over the box region $[\underline{x}, \overline{x}] \times [\underline{y}, \overline{y}]$ to find $(x^\star, y^\star)$. Notice that the optimal solution of $x^\star$ and $y^\star$ to problem (17.6) is generally non-unique.

Given $x^\star$ and $y^\star$, we have the following proposition, for which the proof can be found in [9].

Proposition 17.1 The optimal trajectory solution to problem (P1) is given as

$$x^\star(t) = x^\star, \quad y^\star(t) = y^\star, \quad \forall t \in \mathcal{T}. \tag{17.8}$$

Proposition 17.1 indicates that the UAV should hover at one single fixed location $(x^\star, y^\star, H)$ during the whole charging period, referred to as *single-location hovering*. Due to the non-uniqueness of the optimal solution $x^\star$ and $y^\star$ to problem (17.6), such an optimal hovering location $(x^\star, y^\star, H)$ is non-unique in general. However, this single-location-hovering solution can lead to a severe "near–far" fairness issue in multiuser WPT, as the near ERs in close proximity to the optimal hovering location can receive significantly more energy than those far ERs, especially in a large network with many ERs that are sufficiently separated from each other.

17.4 Min-Energy Maximization under Infinite Charging Duration

To overcome the aforementioned fairness issue in the sum-energy maximization problem, we consider an alternative performance metric, namely the min-energy maximization. Specifically, we maximize the minimum received energy among all the K ERs via optimizing the UAV's trajectory $\{x(t), y(t)\}$, subject to the maximum UAV flying speed constraints in Eq. (17.1). In general, this problem is formulated as

$$\max_{\{x(t),y(t)\}} \min_{k \in \mathcal{K}} E_k(\{x(t), y(t)\})$$

$$\text{s.t. Eq. (1.1).} \tag{P2}$$

Problem (P2) is non-convex, and more difficult to solve than the sum-energy maximization problem (P1). In particular, the single-location-hovering optimal solution of (P1) given in Proposition 17.1 is no longer valid for (P2).

To tackle (P2), in this section we first consider an ideal case by ignoring the UAV speed constraints in Eq. (17.1) and solve the relaxed problem optimally. In practice, the speed constraints in Eq. (17.1) can be ignored if the charging duration T is sufficiently large (see Proposition 17.3 for a more rigorous argument). For ease of presentation, we rewrite problem (P2) without constraint (17.1) in the following problem:

$$\max_{\{x(t),y(t)\}} \min_{k \in \mathcal{K}} E_k(\{x(t),y(t)\}). \tag{P3}$$

In Section 17.5, we will consider the general case of (P2) with the UAV speed constraint (17.1) included, and propose efficient solutions to (P2) based on the optimal solution obtained for (P3).

17.4.1 Multi-Location-Hovering Solution

This subsection presents the optimal solution to the min-energy maximization problem (P3) when $T \to \infty$. In this case, problem (P3) can be equivalently expressed as the following problem by introducing an auxiliary variable E:

$$\max_{\{x(t),y(t)\},E} E \tag{P3.1}$$

$$\text{s.t. } E_k(\{x(t),y(t)\}) \geq E, \quad \forall k \in \mathcal{K}. \tag{17.9}$$

Though problem (P3.1) is non-convex, it can be shown that it satisfies the so-called time-sharing condition in [12]. Therefore, the strong duality holds between (P3.1) and its Lagrange dual problem. As a result, we can optimally solve (P3.1) by using the Lagrange dual method [13].

Let $\lambda_k \geq 0$, $k \in \mathcal{K}$, denote the dual variable associated with the constraint in Eq. (17.9) for the kth ER. Then the Lagrangian associated with (P3.1) is given as

$$\mathcal{L}(\{x(t),y(t)\},E,\{\lambda_k\}) = \left(1 - \sum_{k \in \mathcal{K}} \lambda_k\right) E + \int_0^T \sum_{k \in \mathcal{K}} \lambda_k Q_k(x(t),y(t)) \, dt. \tag{17.10}$$

Accordingly, the dual function of (P3.1) is given by

$$f(\{\lambda_k\}) = \max_{\{x(t),y(t)\},E} \mathcal{L}(\{x(t),y(t)\},E,\{\lambda_k\}). \tag{17.11}$$

Notice that in order for the dual function $f(\{\lambda_k\})$ to be upper-bounded from above (i.e., $f(\{\lambda_k\}) < \infty$), it must hold that $\sum_{k \in \mathcal{K}} \lambda_k = 1$. Accordingly, the dual problem of (P3.1) is given by

$$\min_{\{\lambda_k\}} f(\{\lambda_k\}) \tag{D3.1}$$

$$\text{s.t. } \sum_{k \in \mathcal{K}} \lambda_k = 1, \tag{17.12}$$

$$\lambda_k \geq 0, \quad \forall k \in \mathcal{K}. \tag{17.13}$$

As a result, we can solve problem (P3.1) by equivalently solving its dual problem (D3.1). Let the feasible set of $\{\lambda_k\}$ characterized by the constraints in Eqs. (17.12) and (17.13) be denoted as $\mathcal{X}$. In the following, we first solve problem (17.11) to obtain $f(\{\lambda_k\})$ under any given feasible dual variables $\{\lambda_k\} \in \mathcal{X}$, then solve (D3.1) to find the optimal $\{\lambda_k\}$ to minimize $f(\{\lambda_k\})$, and finally construct the optimal primal solution to (P3.1).

First, we obtain the dual function $f(\{\lambda_k\})$ under any given $\{\lambda_k\} \in \mathcal{X}$ via solving problem (17.11). In this case, problem (17.11) can be decomposed into the following subproblems:

$$\max_E \left(1 - \sum_{k \in \mathcal{K}} \lambda_k\right) E, \tag{17.14}$$

$$\max_{x(t),y(t)} \tilde{\psi}^{\{\lambda_k\}}(x(t),y(t)) \triangleq \sum_{k \in \mathcal{K}} \lambda_k Q_k(x(t),y(t)), \quad \forall t \in \mathcal{T}. \tag{17.15}$$

In the above, Eq. (17.15) consists of an infinite number of subproblems, each corresponding to a time instant t. Let the optimal solutions to Eqs. (17.14) and (17.15) be denoted by $E^{\{\lambda_k\}}$, as well as $x^{\{\lambda_k\}}(t)$ and $y^{\{\lambda_k\}}(t)$, for all $t \in \mathcal{T}$, respectively.

As for subproblem (17.14), since $1 - \sum_{k \in \mathcal{K}} \lambda_k = 0$ holds for any feasible $\{\lambda_k\} \in \mathcal{X}$, its objective value is always zero. In this case, we can choose any arbitrary real number as the optimal solution $E^{(\{\lambda_k\})}$ for the purpose of obtaining the dual function $f(\{\lambda_k\})$.

On the other hand, note that the subproblems in Eq. (17.15) are identical for all time instants $t \in \mathcal{T}$. Therefore, we can drop the time index t and re-express each problem in Eq. (17.15) as

$$\max_{x,y} \tilde{\psi}^{\{\lambda_k\}}(x,y). \tag{17.16}$$

Notice that problem (17.16) has two optimization variables, and the optimal solution of $x^{\{\lambda_k\}}$ and $y^{\{\lambda_k\}}$ satisfies $\underline{x} \leq x^{\{\lambda_k\}} \leq \overline{x}$ and $\underline{y} \leq y^{\{\lambda_k\}} \leq \overline{y}$, with $\underline{x}, \overline{x}, \underline{y}$, and $\overline{y}$ given in Eq. (17.7). As a result, we can adopt a 2D exhaustive search over the box region $[\underline{x}, \overline{x}] \times [\underline{y}, \overline{y}]$ to find the optimal $x^{\{\lambda_k\}}$ and $y^{\{\lambda_k\}}$. Accordingly, the optimal solution to problem (17.15) is given by

$$x^{\{\lambda_k\}}(t) = x^{\{\lambda_k\}}, \quad y^{\{\lambda_k\}}(t) = y^{\{\lambda_k\}}, \quad \forall t \in \mathcal{T}. \tag{17.17}$$

Note that the optimal solution of $x^{\{\lambda_k\}}$ and $y^{\{\lambda_k\}}$ to Eq. (17.16) is generally non-unique, and we can arbitrarily choose any one of them to obtain the dual function $f(\{\lambda_k\})$.[2] By substituting $E^{\{\lambda_k\}}, x^{\{\lambda_k\}}(t)$ and $y^{\{\lambda_k\}}(t)$ into problem (17.11), the dual function $f(\{\lambda_k\})$ is obtained.

Next, with $f(\{\lambda_k\})$ obtained, we then solve the dual problem (D3.1) to find the optimal dual solution $\{\lambda_k\}$ to minimize $f(\{\lambda_k\})$. Note that the dual function $f(\{\lambda_k\})$ is always convex but generally non-differentiable [13]. As a result, problem (D3.1) can be solved by subgradient-based methods such as the ellipsoid method [16]. Note that the subgradient of the objective function $f(\{\lambda_k\})$ is given by

$$s_0(\lambda_1, \ldots, \lambda_K) = [TQ_1(x^{(\{\lambda_k\})}, y^{(\{\lambda_k\})}), \ldots, TQ_K(x^{(\{\lambda_k\})}, y^{(\{\lambda_k\})})],$$

where $E^{(\{\lambda_k\})} = 0$ is chosen for simplicity. Furthermore, the equality constraint in Eq. (17.12) can be viewed as two inequality constraints, $1 - \sum_{k \in \mathcal{K}} \lambda_k \leq 0$ and $-1 + \sum_{k \in \mathcal{K}} \lambda_k \leq 0$, whose subgradients are given by $s_1(\lambda_1, \ldots, \lambda_K) = -e$ and $s_2(\lambda_1, \ldots, \lambda_K) = e$, respectively, where e denotes an all-ones vector. We denote the obtained optimal dual solution to (D3.1) as $\{\lambda_k^*\}$.

Finally, based on the optimal dual solution $\{\lambda_k^*\}$ to (D3.1), we need to obtain the optimal primal solution to (P3.1), denoted by $\{x^*(t)\}, \{y^*(t)\}$, and E^*. It is worth noting that, when

---

2 When the optimal solution $x^{\{\lambda_k\}}(t), y^{\{\lambda_k\}}(t)$, and $E^{(\{\lambda_k\})}$ are non-unique, they may not be optimal for the primal problem (P3.1) after the dual problem is solved. As a result, an additional step is required to obtain the optimal primal solution of $E, x(t)$, and $y(t)$ to (P3.1), as will be shown later.

using the Lagrange dual method to solve problem (P3.1) via the dual problem (D3.1), the optimal solution to problem (17.11) under the optimal dual solution $\{\lambda_k^*\}$ (i.e., the $x^{\{\lambda_k^*\}}(t)$, $y^{\{\lambda_k^*\}}(t)$, and $E^{\{\lambda_k^*\}}$) is the optimal primal solution to (P3.1), if such a solution is unique and primal feasible [13]. On the other hand, when the optimal $x^{\{\lambda_k^*\}}(t)$, $y^{\{\lambda_k^*\}}(t)$, and $E^{\{\lambda_k^*\}}$ to problem (17.11) are non-unique, they may not be feasible nor optimal to problem (P3.1) in general. In the latter case, we need to time-share among these non-unique optimal solutions as follows to construct the optimal primal solution $\{x^*(t)\}$, $\{y^*(t)\}$, and E^* to (P3.1).

Under the optimal dual solution $\{\lambda_k^*\}$, suppose that problem (17.16) under $\{\lambda_k^*\}$ has a total number of $\Gamma \geq 1$ optimal location solutions to maximize $\psi^{\{\lambda_k^*\}}(x, y)$, denoted by $(x_1^*, y_1^*), \ldots, (x_\Gamma^*, y_\Gamma^*)$, which are obtained via a 2D exhaustive search over the box region $[\underline{x}, \overline{x}] \times [\underline{y}, \overline{y}]$. Let $Q_k(x_\gamma^*, y_\gamma^*)$ denote the corresponding received power at each ER $k \in \mathcal{K}$ when the UAV hovers at the location $(x_\gamma^*, y_\gamma^*, H)$. Due to the zero duality gap between (P3.1) and (D3.1), it is evident that, for any time $t \in [0, T]$, the optimal primal solution of $(x^*(t), y^*(t))$ must be chosen from the Γ hovering locations. Notice that, when the UAV hovers at the same location at different times t, the ERs will receive equal power. Therefore, we only need to decide the time-sharing ratio among the Γ solutions for constructing the optimal primal solution to (P3.1). Here, time-sharing means that the UAV should hover at each of these different locations for a certain portion of the total duration T. Let τ_γ denote the optimal hovering duration at $(x_\gamma^*, y_\gamma^*, H)$. Then, the optimal τ_γ^*, together with the maximum min-energy E^*, can be obtained by solving the following problem:

$$\max_{\{\tau_\gamma \geq 0\}, E} \quad E$$

$$\text{s.t.} \quad \sum_{\gamma=1}^{\Gamma} \tau_\gamma Q_k(x_\gamma^*, y_\gamma^*) \geq E, \quad \forall k \in \mathcal{K},$$

$$\sum_{\gamma=1}^{\Gamma} \tau_\gamma = T. \tag{17.18}$$

Note that problem (17.18) is a linear program (LP), which can be efficiently solved by using standard convex optimization techniques [13]. As a result, the optimal solution of E^* to (P3.1) is found. Finally, we obtain the optimal trajectory solution of $\{x^*(t), y^*(t)\}$ to (P3.1) (and thus (P3)), which is given in the following proposition based on the above time-sharing property, with the proof omitted for brevity.

Proposition 17.2 Let the whole charging period be partitioned into Γ portions, denoted by $\mathcal{T}_1, \ldots, \mathcal{T}_\Gamma$, where $\mathcal{T}_\gamma = \left(\sum_{i=1}^{\gamma-1} \tau_i^*, \sum_{i=1}^{\gamma} \tau_i^* \right]$ with duration τ_γ^* for $\gamma \geq 1$. Then, the optimal trajectory solution $\{x^*(t), y^*(t)\}$ to (P3.1) or (P3) is given by

$$x^*(t) = x_\gamma^*, \quad y^*(t) = y_\gamma^*, \quad \forall t \in \mathcal{T}_\gamma, \gamma \in \{1, \ldots, \Gamma\}, \tag{17.19}$$

where $\mathcal{T}_\gamma \cap \mathcal{T}_\zeta = \emptyset$, for all $\gamma \neq \zeta$, and $\bigcup_{\gamma=1}^{\Gamma} \mathcal{T}_\gamma = \mathcal{T}$.

Remark 17.1 Note that Proposition 17.2 implies that, to maximize the min-energy transferred to the K ERs, the UAV should hover above a number of fixed locations during the charging period, and the optimal hovering locations (i.e., the x_γ^* and y_γ^*) are generally different from the locations of the ERs (i.e., x_k and y_k). We refer to such a design as multi-location hovering. Different from the single-location hovering for sum-energy maximization, the result here shows that the UAV should in general hover over different locations so as to balance the energy transferred to all ERs.

17.5 Min-Energy Maximization Under Finite Charging Duration

In this section, we consider the general min-energy maximization problem (P2) by including the practical UAV's maximum speed constraints in Eq. (17.1). This problem is difficult to be solved globally optimally in general with $K > 2$. To tackle this problem, we propose two suboptimal solutions inspired by the optimal solution obtained above for problem (P3) in the ideal case without the UAV's maximum speed constraint.

17.5.1 Successive Hover-and-Fly Trajectory Design

In this subsection, we propose a successive hover-and-fly trajectory design to solve problem (P2) based on the optimal solution obtained for (P3) in the ideal case. Recall that the optimal solution to (P3) corresponds to Γ optimal hovering locations, i.e., $\{(x_\gamma^*, y_\gamma^*, H)\}_{\gamma=1}^{\Gamma}$. In the proposed trajectory design with the maximum speed constraint, the UAV sequentially hovers at each of these locations for a certain duration and flies from one location to another with the maximum speed V. As a result, to find the optimal successive hover-and-fly trajectory, we need to first determine the UAV's traveling path to visit all the Γ locations with the minimum flying distance so as to minimize the total flying time, and then optimize the hovering time at each of these locations for the remaining time in the charging duration.

17.5.1.1 Flying Distance Minimization to Visit Γ Hovering Locations

First, we determine the UAV's traveling path to visit all the Γ hovering locations with the minimum flying distance. For ease of description, let $d_{\gamma,\zeta} \triangleq \sqrt{(x_\gamma^* - x_\zeta^*)^2 + (y_\gamma^* - y_\zeta^*)^2}$ denote the distance between the γth hovering location $(x_\gamma^*, y_\gamma^*, H)$ and the ζth hovering location $(x_\zeta^*, y_\zeta^*, H)$. We define a binary variable $f_{\gamma,\zeta}$ for any $\gamma, \zeta \in \{1, \ldots, \Gamma\}$, $\gamma \neq \zeta$, where $f_{\gamma,\zeta} = 1$ indicates that the UAV should fly from the γth hovering location to the ζth hovering location, and $f_{\gamma,\zeta} = 0$ otherwise. The trajectory design problem thus becomes determining $\{f_{\gamma,\zeta}\}$ to minimize $\sum_{\gamma=1}^{\Gamma} \sum_{\zeta=1, \zeta \neq \gamma}^{\Gamma} f_{\gamma,\zeta} d_{\gamma,\zeta}$, provided that each of the Γ locations is visited exactly once.

The flying distance minimization problem considered here is reminiscent of the well-known traveling salesman problem (TSP) (see, e.g., [17, 18]), with the following difference. In the standard TSP, the salesman (or equivalently the UAV of our interest) needs to return to the origin city (the initial hovering location) after visiting all these cities (or hovering locations here); but our flying distance minimization problem does not have such a requirement, since the initial and final hovering locations can be optimized. Fortunately, it has been shown in [19] that our flying distance minimization problem can be transformed to the standard TSP as follows.

First, we add a dummy hovering location, namely the $(\Gamma + 1)$th hovering location, whose distances to all the existing Γ hovering locations are 0, i.e., $d_{\Gamma+1,\gamma} = d_{\gamma,\Gamma+1} = 0$, for all $\gamma \in$

$\{1, \dots, \Gamma\}$. Note that this dummy hovering location is a virtual node that does not exist physically. Then, we obtain the desirable traveling path by solving the standard TSP problem for the $\Gamma + 1$ hovering locations,[3] and then removing the two edges associated with the dummy location. For the obtained traveling path, we define the permutation $\pi(\cdot)$ over the set $\{1, \dots, \Gamma\}$, such that the UAV first visits the $\pi(1)$th hovering location, followed by the $\pi(2)$th, and so on, until the $\pi(\Gamma)$th hovering location at last. In this case, the resulting flying distance and flying duration with the maximum speed V are given as $D_{\text{fly}} = \sum_{\gamma=1}^{\Gamma-1} d_{\pi(\gamma), \pi(\gamma+1)}$, and $T_{\text{fly}} = D_{\text{fly}}/V$, respectively. We denote the corresponding trajectory as $\{\hat{x}(t), \hat{y}(t)\}_{t=0}^{T_{\text{fly}}}$.

It is worth noting that the above traveling path is only feasible when the charging duration T is no smaller than T_{fly}, i.e., $T \geq T_{\text{fly}}$, since otherwise the charging duration is insufficient for the UAV to visit all the Γ hovering locations. In the following, we first determine the hovering time allocation over different locations in the case with $T \geq T_{\text{fly}}$, and then refine the trajectory design in the case with $T < T_{\text{fly}}$.

17.5.1.2 Hovering Time Allocation When $T \geq T_{\text{fly}}$

First, we consider the case when $T \geq T_{\text{fly}}$. With the above traveling path $\{\hat{x}(t), \hat{y}(t)\}_{t=0}^{T_{\text{fly}}}$, the trajectory design problem remains to allocate the hovering duration $T - T_{\text{fly}}$ among the Γ locations to maximize the min-energy transferred to all the K ERs. Note that, based on the traveling path $\{\hat{x}(t), \hat{y}(t)\}_{t=0}^{T_{\text{fly}}}$, we can obtain the energy received by each ER $k \in \mathcal{K}$ during the UAV's flying time as $E_k^{\text{fly}} = \int_0^{T_{\text{fly}}} Q_k(\hat{x}(t), \hat{y}(t)) \, dt$, with $Q_k(\cdot, \cdot)$ given in Eq. (17.2). Also, recall that $Q_k(x_\gamma^*, y_\gamma^*)$ denotes the received power at ER $k \in \mathcal{K}$ when the UAV hovers at the location $(x_\gamma^*, y_\gamma^*, H)$. Then the optimal hovering durations, denoted as τ_γ^{**}, together with the corresponding maximum min-energy of the K ERs, denoted by E^{**}, can be obtained by solving the following LP:

$$\max_{\{\tau_\gamma \geq 0\}, E} E$$

$$\text{s.t.} \sum_{\gamma=1}^{\Gamma} \tau_\gamma Q_k(x_\gamma^*, y_\gamma^*) + E_k^{\text{fly}} \geq E, \quad \forall k \in \mathcal{K},$$

$$\sum_{\gamma=1}^{\Gamma} \tau_\gamma = T - T_{\text{fly}}. \tag{17.20}$$

With the optimal permutation $\pi(\cdot)$ and the optimal hovering durations $\{\tau_\gamma^{**}\}$ obtained, the successive hover-and-fly trajectory is finalized, which can be summarized as follows. Divide the charging period into $2\Gamma - 1$ slots; in the $(2\gamma - 1)$th slot with duration $\tau_{\pi(\gamma)}^{**}$, $\gamma \in \{1, \dots, \Gamma\}$, the UAV hovers at the $\pi(\gamma)$th hovering location $(x_{\pi(\gamma)}^*, y_{\pi(\gamma)}^*, H)$; and in the (2γ)th slot, $\gamma \in \{1, \dots, \Gamma - 1\}$, the UAV flies from the $\pi(\gamma)$th hovering location $(x_{\pi(\gamma)}^*, y_{\pi(\gamma)}^*, H)$ to the $\pi(\gamma + 1)$th hovering location $(x_{\pi(\gamma+1)}^*, y_{\pi(\gamma+1)}^*, H)$ with its maximum speed V.

---

3 Note that, although the TSP is an NP-hard problem in combinatorial optimization, various heuristic and approximation algorithms have been proposed to give efficient high-quality solutions for it (see, e.g., [17, 18]). In particular, it has been shown in [18] that the TSP problem can be formulated as a binary integer program, by incorporating a set of constraints to ensure that there is only a single tour connecting all visited locations. The binary integer program can be solved via CVX [20] by using the Mosek solver that supports the integer program (see https://mosek.com/ for details).

Proposition 17.3 When the charging duration T is sufficiently large such that $T \gg T_{\text{fly}}$, the successive hover-and-fly trajectory design is asymptotically optimal for problem (P2).

Proof: When $T \gg T_{\text{fly}}$, the flying time is negligible and thus the successive hover-and-fly trajectory is equivalent to the optimal multi-location-hovering solution to (P3). In this case, the objective value achieved by the successive hover-and-fly trajectory for (P2) asymptotically approaches the optimal value of (P3), which actually serves as the upper bound for that of (P2). Therefore, the proposed trajectory design is asymptotically optimal for (P2) when $T \gg T_{\text{fly}}$. ∎

17.5.1.3 Trajectory Refinement When $T < T_{\text{fly}}$

In this subsection, we consider the scenario when $T < T_{\text{fly}}$. In this case, the UAV traveling path $\{\hat{x}(t), \hat{y}(t)\}_{t=0}^{T_{\text{fly}}}$ based on the TSP solution is no longer feasible since the charging time is insufficient for the UAV to visit all the Γ hovering locations. To address this issue, we first find the solution to (P2) when T is sufficiently small (i.e., $T \to 0$) such that the UAV can only hover at one single location, and then reconstruct a modified successive hover-and-fly trajectory for the case of $T < T_{\text{fly}}$.

First, when $T \to 0$, the UAV should hover at one single fixed location, denoted by $(x_{\text{fix}}, y_{\text{fix}}, H)$, where x_{fix} and y_{fix} can be obtained by solving the following problem via a 2D exhaustive search over $(\underline{x}, \overline{x}) \times (\underline{y}, \overline{y})$:

$$(x_{\text{fix}}, y_{\text{fix}}) = \arg\max_{x,y} \ \min_{k \in \mathcal{K}} \ Q_k(x,y). \tag{17.21}$$

Next, we reconstruct the trajectory as follows, by down-scaling the previously obtained traveling path $\{(\hat{x}(t), \hat{y}(t), H)\}_{t=0}^{T_{\text{fly}}}$ for the case of $T = T_{\text{fly}}$ linearly towards the center point $(x_{\text{fix}}, y_{\text{fix}}, H)$, such that the resulting total flying distance equals VT:

$$
\begin{aligned}
x^{**}(t) &= \hat{x}(t/\kappa) + (1-\kappa)(x_{\text{fix}} - \hat{x}(t/\kappa)), \\
y^{**}(t) &= \hat{y}(t/\kappa) + (1-\kappa)(y_{\text{fix}} - \hat{y}(t/\kappa)), \quad \forall\, t \in [0, T],
\end{aligned}
\tag{17.22}
$$

where $\kappa = T/T_{\text{fly}} < 1$ denotes the linear scaling factor. Note that, when $T \to 0$, we have $\kappa \to 0$, and the above redesigned trajectory reduces to hovering at one single fixed location $(x_{\text{fix}}, y_{\text{fix}}, H)$; while, when $T \to T_{\text{fly}}$, we have $\kappa \to 1$, and the above redesigned trajectory becomes identical to the TSP-based trajectory $\{(\hat{x}(t), \hat{y}(t), H)\}_{t=0}^{T_{\text{fly}}}$.

17.5.2 SCA-Based Trajectory Design

In this subsection, we develop an alternative SCA-based algorithm to find a locally optimal solution to problem (P2). Note that the SCA-based trajectory design has been studied for throughput or energy-efficiency maximization for UAV-enabled wireless communication systems [7, 8], but the results cannot be directly applied for UAV-enabled WPT systems considered in this chapter.

With the SCA-based trajectory design, we first discretize the whole charging duration into a finite number of N time slots, each with duration $\Delta = T/N$. Note that the duration Δ is chosen to be sufficiently small, such that we can assume that the UAV's location is approximately unchanged during each slot n, which is denoted as $(x[n], y[n], H)$, $n \in \mathcal{N} \triangleq \{1, \ldots, N\}$. In this case, the energy received by each ER $k \in \mathcal{K}$ at slot n is given by

$$\hat{E}_k(x[n], y[n]) = \frac{\beta_0 P \Delta}{(x[n] - x_k)^2 + (y[n] - y_k)^2 + H^2}. \tag{17.23}$$

Accordingly, the min-energy maximization problem (P2) over the continuous trajectory $\{x(t), y(t)\}$ can be reformulated as follows over the discretized trajectory variables $\{x[n], y[n]\}_{n=1}^N$:

$$\max_{\{x[n], y[n]\}} \min_{k \in \mathcal{K}} \sum_{n=1}^N \hat{E}_k(x[n], y[n]) \tag{17.24}$$

$$\text{s.t. } (x[n] - x[n-1])^2 + (y[n] - y[n-1])^2 \leq V^2 \Delta^2,$$

$$\forall n \in \{2, \ldots, N\}, \tag{17.25}$$

where the constraints in Eq. (17.25) correspond to the discretized version of the maximum speed constraints in Eq. (17.1). Note that the constraints in Eq. (17.25) are all convex but the objective function in Eq. (17.24) is non-concave. Therefore, problem (17.24) is a non-convex optimization problem.

For the non-convex optimization problem (17.24), we obtain a locally optimal solution by proposing an SCA-based algorithm, which is operated in an iterative manner to successively maximize a lower bound of the objective function in Eq. (17.24) at each iteration. Particularly, let $\{x^{(0)}[n], y^{(0)}[n]\}$ denote the initial trajectory and $\{x^{(i)}[n], y^{(i)}[n]\}$ the obtained trajectory after iteration $i \geq 1$. We have the following lemma.

Lemma 17.1 *For any given $\{x^{(i)}[n], y^{(i)}[n]\}$, $i \geq 0$, it follows that*

$$\hat{E}_k(x[n], y[n]) \geq \hat{E}_k^{(i)}(x[n], y[n]), \quad \forall k \in \mathcal{K}, n \in \mathcal{N}, \tag{17.26}$$

where

$$\hat{E}_k^{(i)}(x[n], y[n]) \triangleq \frac{2\beta_0 P \Delta}{(x^{(i)}[n] - x_k)^2 + (y^{(i)}[n] - y_k)^2 + H^2}$$
$$- \frac{\beta_0 P \Delta ((x[n] - x_k)^2 + (y[n] - y_k)^2 + H^2)}{((x^{(i)}[n] - x_k)^2 + (y^{(i)}[n] - y_k)^2 + H^2)^2}. \tag{17.27}$$

The inequalities in Eq. (17.26) are tight for $x[n] = x^{(i)}[n]$ and $y[n] = y^{(i)}[n]$, i.e.,

$$\hat{E}_k(x^{(i)}[n], y^{(i)}[n]) = \hat{E}_k^{(i)}(x^{(i)}[n], y^{(i)}[n]), \quad \forall k \in \mathcal{K}, n \in \mathcal{N}. \tag{17.28}$$

Proof: See [9]. ∎

Based on Lemma 17.1, at each iteration $i + 1$, we optimize over $\{x[n], y[n]\}$ by replacing the $\hat{E}_k(x[n], y[n])$ in problem (17.24) with their respective lower bounds $\hat{E}_k^{(i)}(x[n], y[n])$ in Eq. (17.27). More specifically, the discretized trajectory is updated as

$$\{x^{(i+1)}[n], y^{(i+1)}[n]\} = \arg \max_{\{x[n], y[n]\}} \min_{k \in \mathcal{K}} \sum_{n=1}^N \hat{E}_k^{(i)}(x[n], y[n]),$$

$$\text{s.t. Eq. (1.25).} \tag{17.29}$$

Note that the function $\hat{E}_k^{(i)}(x[n], y[n])$ in Eq. (17.27) is jointly concave with respect to $x[n]$ and $y[n]$, and therefore the objective function in problem (17.29) is jointly concave with respect to $\{x[n], y[n]\}$. As a result, problem (17.29) is a convex optimization problem, and thus can be optimally solved by standard convex optimization techniques such as the interior point method [13]. Furthermore, due to Lemma 17.1, the objective function in problem (17.29) serves as a lower bound for that in problem (17.24). Therefore, after each iteration i, the objective function of problem (17.24) achieved by $\{x^{(i)}[n], y^{(i)}[n]\}$ monotonically increases [8]. As problem (17.24) has a finite optimal value, the SCA-based algorithm in Eq. (17.29) will lead to a converged solution to problem (17.24) in general.

It is worth noting that the performance of the SCA-based algorithm depends on the choice of the initial trajectory $\{x^{(0)}[n], y^{(0)}[n]\}$. Here, we choose the discretized version of the proposed successive hover-and-fly trajectory as $\{x^{(0)}[n], y^{(0)}[n]\}$. In this case, the SCA-based trajectory design can always achieve a performance that is at least no worse than the successive hover-and-fly trajectory design, as will be validated by the numerical results in the following section.

17.6 Numerical Results

In this section, we provide numerical results to evaluate the performance of our proposed trajectory designs. In the simulation, we set $\beta_0 = -30$ dB, $H = 5$ m, and $P = 40$ dBm. For all the simulations given below, we consider the average received power by the ER, which is obtained by normalizing the total received energy by the charging duration T. More specifically, we consider a UAV-enabled WPT system with $K = 10$ ERs, whose locations are shown in Figure 17.2.

First, we evaluate the performance of our proposed optimal solution for the sum-energy maximization problem (P1). In Figure 17.2, the green triangle denotes the optimal hovering

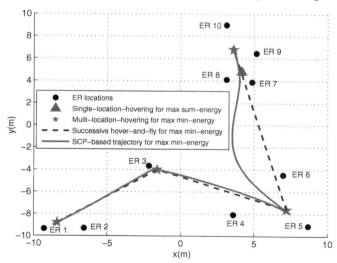

Figure 17.2 Trajectory designs for a UAV-enabled WPT system with $K = 10$ ERs. Source: From Xu et al. [9].

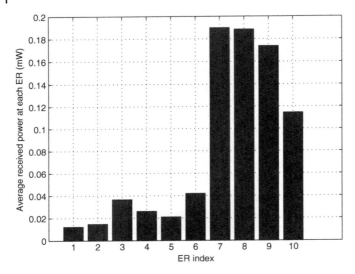

Figure 17.3 The average received power by different ERs with the optimal sum-energy maximization trajectory design. Source: From Xu et al. [9].

location for sum-energy maximization (i.e., $(x^\star, y^\star, H)$ given in Eq. (17.6)). It is observed that this hovering location is close to ERs 7–10, but far away from other ERs, especially ER 1. Figure 17.3 shows the corresponding average power received by each individual ER. It is observed that ERs 7–10 receive much more energy than the other ERs, which demonstrates the near–far fairness issue in the sum-energy maximization for this multiuser WPT system.

Next, we evaluate the performance of our proposed trajectory designs for the min-energy maximization problem (P2). Note that, besides the single-location-hovering solution for the sum-energy maximization problem (P1), Figure 17.2 also shows the multi-location-hovering solution for problem (P3) without the UAV speed constraint, as well as the proposed successive hover-and-fly trajectory design and the SCA-based trajectory design for problem (P2) with the UAV speed constraint considered, by assuming $T = 20$ s. First, it is observed that there are $\Gamma = 4$ optimal hovering locations for the min-energy maximization problem (P3), which are close to ERs 1–2, ER 3, ERs 4–6, and ERs 7–10, respectively. This clearly shows that, when the ERs are close to each other (e.g., ERs 7–10), then the UAV should hover at one single location above them to charge them more efficiently. It is also observed that the SCA-based trajectory design and the successive hover-and-fly trajectory design both visit the Γ optimal hovering locations. Furthermore, it is observed that the SCA-based trajectory in general deviates from the successive hover-and-fly trajectory when flying from one hovering location to another.

Figure 17.4 shows the max-min average power received by all ERs in the 10-ER WPT system in Figure 17.2 versus the charging duration T, for our proposed trajectory designs, as compared to the following two benchmark schemes.

- *Single-location hovering for minimum energy maximization*: During the whole charging period, the UAV hovers at a fixed location $(x_{\text{fix}}, y_{\text{fix}}, H)$ obtained in Eq. (17.21).

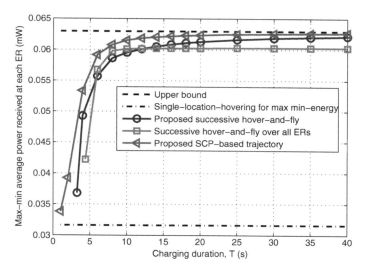

Figure 17.4 The max–min average received power at each ER versus the charging duration T with $K = 10$ ERs. Source: From Xu et al. [9].

- *Successive hover-and-fly over all ERs*: In this scheme, instead of optimizing the UAV's hovering locations, they are simply set as the K locations above the K ERs. This scheme can be implemented similarly as that in Section 17.5.1 via replacing $\{(x_\gamma^*, y_\gamma^*, H)\}_{\gamma=1}^\Gamma$ as $\{(x_k, y_k, H)\}_{k=1}^K$.

In Figure 17.4, the upper bound corresponds to the optimal value achieved by (P3) with the UAV's maximum speed constraints ignored. It is observed that the two proposed trajectory designs, namely the successive hover-and-fly and the SCA-based trajectory designs, outperform the single-location-hovering design, and achieve higher average max–min average power as T becomes large. When $T \geq 15$ s, the proposed successive hover-and-fly and the SCA-based trajectory designs also outperform the successive hover-and-fly trajectory over all the ERs. Furthermore, it is observed that the SCA-based trajectory achieves better performance than the successive hover-and-fly trajectory, and converges to the upper bound, when T becomes large.

17.7 Conclusion and Future Research Directions

This chapter presents a new UAV-enabled multiuser WPT system. We exploit the mobility of the UAV to maximize the energy transferred to all ERs over a given charging period by optimizing the UAV's trajectory under the practical flying speed constraint. First, we consider the sum-energy maximization of all ERs and obtain the optimal solution to this problem, which shows that the UAV should hover at only one single optimal location during the whole charging period. However, this single-location-hovering solution may lead to unfair performance among ERs due to their different distances from the UAV.

To address this issue, we further consider an alternative problem to maximize the minimum energy transferred to all ERs. We first consider the relaxed problem by ignoring the UAV speed constraints and derive the optimal solution, which shows that the UAV should hover over multiple fixed locations with optimal hovering time allocations among them. Based on this solution, we further propose two new trajectory designs for solving the min-energy maximization problem in the general case with the UAV speed constraint considered. Numerical results show that the proposed UAV-enabled WPT system with optimized UAV trajectory significantly enhances the WPT performance over the conventional WPT system with static ETs, and yet achieves fair energy delivery to distributed ERs.

It is our hope that this chapter can pave a new avenue for future research on UAV-enabled WPT and communication. In the following, we present some potential and promising extensions of the results in this chapter, and briefly discuss other important issues that remain unaddressed in this chapter to motivate future work.

(1) Efficient trajectory design in UAV-enabled WPT systems critically depends on the accurate modeling of the wireless channels from the UAV to the ERs. In this chapter, we assume that the UAV-to-ER channels are LoS-dominated, and thus use the free-space path-loss model, similarly as in prior works [6, 8]. In certain application scenarios, such as in forests or urban environments with high-rise buildings, however, there may exist obstacles and rich scatterers between the UAV and ERs, and thus different wireless channel models should apply. In such scenarios, we can use elevation-angle-dependent stochastic channel models such as probabilistic LoS models and Ricean fading models, where the probability of LoS occurrence and the Ricean K-factor depend on the elevation angle of the UAV [21]. However, how to design the optimal UAV trajectory under these new and more complex channel models is an interesting open problem. Alternatively, instead of using stochastic wireless channel models, the radio map technique [22, 23] can be employed for the UAV to acquire knowledge of channel propagation environments, for the purpose of facilitating the trajectory design. In general, how to optimize the UAV trajectory based on radio maps is a very interesting but challenging task, while preliminary work [24] shows that the multi-location-hovering solution is still applicable for maximizing the min-energy received among multiple ERs when the charging duration is large.

(2) In this chapter, to reveal the most essential design insights for UAV-enabled WPT, we focus on a basic system setup with one UAV (ET) and multiple ERs, by assuming free-space channel models and a priori known knowledge about the ERs' locations and CSI. Another interesting direction is to extend the UAV-enabled WPT to the general multi-UAV scenario. In this case, multiple UAVs should cooperate with each other to further improve the WPT performance. On the one hand, the UAVs can cooperatively design their trajectories subject to individual UAV speed and inter-UAV collision avoidance constraints; while, on the other hand, the UAVs can also cooperatively design their energy signals based on the technique of distributed energy beamforming [2]. How to jointly design multiple UAV trajectories and their distributed energy beamforming is a very interesting but non-trivial problem worthy of further investigation in the future. Furthermore, multi-UAV cooperation can be exploited to address the limited individual UAV endurance issue for achieving sustainable WPT. How to jointly control multiple UAV trajectories for WPT performance optimization with finite endurance of each individual UAV is also an interesting problem worth further investigation.

(3) It is worth pointing out that the trajectory design principles for energy maximization in UAV-enabled multiuser WPT systems are also applicable for rate maximization

and capacity region characterization in multiuser wireless communication systems, such as broadcast channels [25], multiple access channels [26], and multicast channels [27]. In general, for special cases when the communication duration becomes sufficiently long, the multi-location-hovering trajectory design, together with proper wireless resource allocation, have been shown to be optimal for maximizing the achievable data rates for certain UAV-enabled multiuser communication systems [25–27].

(4) Last, but not least, motivated by the UAV-enabled WPT, UAV-enabled wireless powered communication networks (WPCNs) have also recently emerged as an interesting direction, in which UAVs are utilized not only for charging ground users in the downlink, but also for collecting information from them in the uplink. In [28], the authors consider the case of one single-UAV-enabled WPCN, in which the time-division multiple access (TDMA) protocol is used for separating downlink WPT and uplink wireless communications to/from multiple users. Under this setup, the UAV trajectory design is optimized, jointly with the wireless resource allocation, to maximize the uplink common (minimum) throughput among all ground users over a finite UAV flight period, subject to its maximum speed constraint and a new set of energy neutrality constraints at users. How to use UAVs to realize a self-sustainable wireless network with unified wireless information and power transmission is an interesting future research direction.

References

1 S. Bi, C. K. Ho, and R. Zhang, "Wireless powered communication: Opportunities and challenges," *IEEE Commun. Mag.*, vol. 53, no. 4, pp. 117–125, Apr. 2015.

2 Y. Zeng, B. Clerckx, and R. Zhang, "Communications and signals design for wireless power transmission," *IEEE Trans. Commun.*, vol. 65, no. 5, pp. 2264–2290, May 2017.

3 J. Xu, L. Liu, and R. Zhang, "Multiuser MISO beamforming for simultaneous wireless information and power transfer," *IEEE Trans. Signal Process.*, vol. 62, no. 18, pp. 4798–4810, Sep. 2014.

4 J. Xu and R. Zhang, "Energy beamforming with one-bit feedback," *IEEE Trans. Signal Process.*, vol. 62, no. 20, pp. 5370–5381, Oct. 2014.

5 J. Xu and R. Zhang, "A general design framework for MIMO wireless energy transfer with limited feedback," *IEEE Trans. Signal Process.*, vol. 64, no. 10, pp. 2475–2488, May 2016.

6 Y. Zeng, R. Zhang, and T. J. Lim, "Wireless communications with unmanned aerial vehicles: Opportunities and challenges," *IEEE Commun. Mag.*, vol. 54, no. 5, pp. 36–42, May 2016.

7 Y. Zeng and R. Zhang, "Energy-efficient UAV communication with trajectory optimization," *IEEE Trans. Wireless Commun.*, vol. 16, no. 6, pp. 3747–3760, Jun., 2017.

8 Y. Zeng, R. Zhang, and T. J. Lim, "Throughput maximization for UAV-enabled mobile relaying systems," *IEEE Trans. Commun.*, vol. 64, no. 12, pp. 4983–4996, Dec. 2016.

9 J. Xu. Y. Zeng, and R. Zhang, "UAV-enabled wireless power transfer: Trajectory design and energy optimization," *IEEE Trans. Wireless Commun.*, vol. 17, no. 8, pp. 5092–5106, Aug. 2018.

10 J. Xu, Y. Zeng, and R. Zhang, "UAV-enabled wireless power transfer: trajectory design and energy region characterization," in *Proc. IEEE Globecom Workshop*, Dec. 2017, pp. 1–7.

11 Y. Hu, X. Yuan, J. Xu, and A. Schmeink, "Optimal 1D trajectory design for UAV-enabled multiuser wireless power transfer," *IEEE Trans. Commun.*, Early Access, 2019.

12 W. Yu and R. Lui, "Dual methods for nonconvex spectrum optimization of multicarrier systems," *IEEE Trans. Commun.*, vol. 54, no. 7, pp. 1310–1322, Jul. 2006.

13 S. Boyd and L. Vandenberghe, *Convex Optimization*. Cambridge, U.K.: Cambridge Univ. Press, 2004.

14 B. Clerckx and E. Bayguzina, "Waveform design for wireless power transfer," *IEEE Trans. Signal Process.*, vol. 64, no. 23, pp. 6313–6328, Dec. 2016.

15 M. R. V. Moghadam, Y. Zeng, and R. Zhang, "Waveform optimization for radio-frequency wireless power transfer," in *Proc. IEEE SPAWC*, 2017.

16 S. Boyd. *EE364b Convex Optimization II, Course Notes*, accessed on Jun. 29, 2017. [Online]. Available: http://www.stanford.edu/class/ee364b/

17 M. Padberg and G. Rinaldi, "A branch-and-cut algorithm for the resolution of large-scale symmetric traveling salesman problems," *SIAM Rev.*, vol. 33, no. 1, pp. 60–100, 1991.

18 C. E. Miller, A. W. Tucker, and R. A. Zemlin, "Integer programming formulation of traveling salesman problems," *J. ACM*, vol. 7, no. 4, pp. 326–329, Oct. 1960.

19 E. L. Lawler, J. K. Lenstra, A. H. G. R. Kan, and D. B. Shmoys, *The Traveling Salesman Problem: A Guided Tour of Combinatorial Optimization*, 1st ed. Wiley, 1985.

20 M. Grant and S. Boyd, CVX: Matlab software for disciplined convex programming, version 2.1, Mar. 2017. [Online]. Available: http://cvxr.com/cvx/.

21 Y. Zeng, Q. Wu, and R. Zhang, "Accessing from the sky: A tutorial on UAV communications for 5G and beyond," *Proc. IEEE*, vol. 107, no. 12, pp. 2327–2375, Dec. 2019.

22 O. Esrafilian, R. Gangula, and D. Gesbert, "Learning to communicate in UAV-aided wireless networks: Map-based approaches," *IEEE Internet Things J.*, vol. 6, no. 2, pp. 1791–1802, Apr. 2019.

23 S. Bi, J. Lyu, Z. Ding, and R. Zhang, "Engineering radio maps for wireless resource management," *IEEE Wireless Commun.*, vol. 26, no. 2, pp. 133–141, Apr. 2019.

24 X. Mo, Y. Huang, and J. Xu, "Radio-map-based robust positioning optimization for UAV-enabled wireless power transfer," *IEEE Wireless Commun. Letters*, vol. 9, no. 2, pp. 179–183, Feb. 2020.

25 Q. Wu, J. Xu, and R. Zhang, "Capacity characterization of UAV-enabled two-user broadcast channel," *IEEE J. Sel. Areas Commun.*, vol. 36, no. 9, pp. 1955–1971, Sep. 2018.

26 P. Li and J. Xu, "Fundamental rate limits of UAV-enabled multiple access channel with trajectory optimization," *IEEE Trans. Wireless Commun.*, vol. 19, no. 1, pp. 458–474, Jan. 2020.

27 Y. Wu, J. Xu, L. Qiu, and R. Zhang, "Capacity of UAV-enabled multicast channel: Joint trajectory design and power allocation," in *Proc. IEEE ICC*, 2018.

28 L. Xie, J. Xu, and R. Zhang, "Throughput maximization for UAV-enabled wireless powered communication networks," *IEEE Internet Things J.*, vol. 6, no. 2, pp. 1690–1703, Apr. 2019.

18

Ad-Hoc Networks in the Sky

Kamesh Namuduri

University of North Texas, USA

Ad-hoc networks provide a means for extending the range of wireless communications for unmanned aerial vehicles (UAVs). They provide a solution to beyond-radio-line-of-sight (BRLOS) communications among the UAVs as well as between a UAV and a ground control station (GCS). Although there are several ways to extend connectivity to BRLOS among UAVs, ad-hoc networking is the most efficient solution in terms of latency. However, it is extremely challenging to establish and maintain an ad-hoc network among UAVs, primarily due to the high mobility of UAVs. This chapter presents the fundamental concepts, challenges, and solutions to creating ad-hoc networks with UAVs as nodes. It discusses the mobility models, applicable standards, and products that are available today for creating mesh networking of UAVs.

18.1 Communication Support for UAVs

Traditionally, communication support for UAVs comes from satellites operating at the geostationary orbit level, as shown in Figure 18.1. However, two additional communication strategies are also being investigated for UAVs: 4G/5G cellular communications operating on the ground, as shown in Figure 18.2; and aerial networks and communications operating in the mid-air, as shown in Figure 18.3. These three strategies of communication support are independent, complementary, and may serve as backup to one another. They are applicable to manned and unmanned as well as small- and large-scale aircraft systems. The three modes of communication support differ in terms of performance aspects, such as latency and coverage among others. Satellite communication (SATCOM) offers unlimited coverage and large latency. Cellular communication offers medium coverage and medium latency. Direct communication via aerial ad-hoc networks provides limited coverage and low latency, but it comes with high Doppler effect as well as challenges arising from high mobility.

UAV Communications for 5G and Beyond, First Edition.
Edited by Yong Zeng, Ismail Guvenc, Rui Zhang, Giovanni Geraci, and David W. Matolak.
© 2021 John Wiley & Sons Ltd. Published 2021 by John Wiley & Sons Ltd.

Figure 18.1 UAV navigation and control using satellite communications.

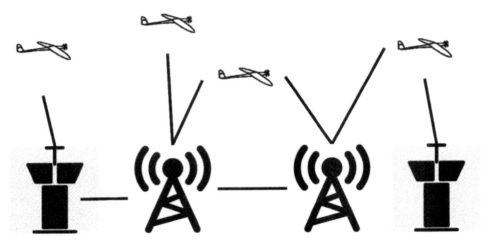

Figure 18.2 UAV navigation and control using cellular communications.

18.1.1 Satellite Connectivity

Satellite communication is the most widely used technology for commercial aircraft navigation (see Figure 18.1). Typically, geostationary satellites are used for aircraft navigation. The geostationary orbit has the advantage that the satellite remains in the same position throughout the day, and antennas can be directed towards the satellite and remain on track. Latency in SATCOM, the time it takes for a signal to make a round trip from the aircraft to the satellite, is about 638 ms (milliseconds), as the geostationary orbit is about 22,300 miles above the Earth.

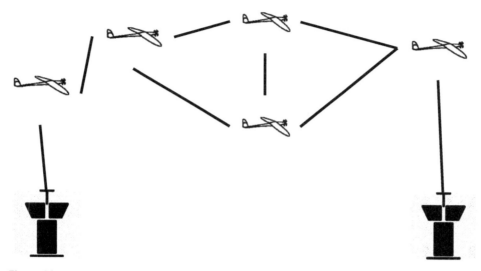

Figure 18.3   UAV navigation and control using aerial communications.

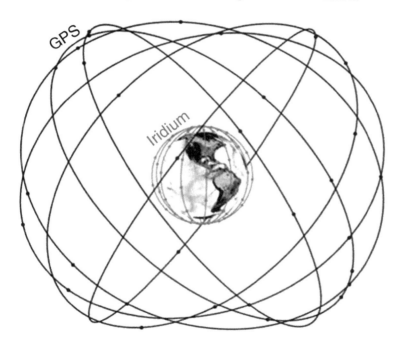

Figure 18.4   The 66-satellite Iridium constellation in low Earth orbit and 31-satellite GPS constellation in medium earth orbit. Sources: Tyler Reid [37]; Tyler Reid et al. [38].

Position, navigation, and timing (PNT) services from Low-Earth-Orbit (LEO) satellites are being used today. LEO satellites provide much stronger signals than Global Navigation Satellite Systems (GNSS) such as the Global Positioning System (GPS) in medium Earth orbit (MEO). Figure 18.4 shows the 31-satellite GPS constellation in contrast with the 66-satellite Iridium network [37, 38]. Iridium signals are 300–2400 times stronger than GNSS signals on the ground, making them attractive for PNT applications where GNSS signals are obstructed [23].

18.1.2 Cellular Connectivity

An alternative to SATCOM is cellular networking for unmanned aerial system (UAS) command, control, and navigation (C2N) services. Cellular connectivity to low-altitude UAVs has received significant interest recently, which has led to a 3GPP study on enhanced Long Term Evolution (LTE) support for connected UAVs in Release 15 [28]. The telecommunication industry is actively working towards this solution. In this strategy, a UAS is like user equipment (UE) or a mobile device. Just like the way a cellular network provides connectivity services for a mobile device, the same network can also provide C2N services for UAVs connected to it.

18.1.3 Aerial Connectivity

In a presentation made to the International Civil Aviation Organization (ICAO), the authors [10] pointed out that "the accident with MH370 flight shows the need for global tracking of airline flights." For ocean-crossing flight paths, such as that of flight MH370, and those remote from Air Traffic Control (ATC) surveillance facilities, the first thought that comes to mind is to use SATCOM. Indeed, SATCOM is the only solution for an aircraft to communicate with the ATC when it is beyond line of sight from ATC facilities, but this method has several disadvantages. It is a centralized solution that requires all aircraft to be equipped with expensive SATCOM equipment. Even if installed, it requires all aircraft to communicate with the ATC, and does not allow direct communication between aircraft, which may be more advantageous in many situations, such as close-proximity operations.

In the commercial aviation domain, aircraft-to-aircraft communication would facilitate global tracking of aircraft while in the air, thus avoiding incidents in which aircraft go missing. Automatic Dependent Surveillance Broadcast (ADS-B) transponders have been recommended by NextGen and the Single European Sky Air Traffic Management Research (SESAR) program as mandatory for the backbone of future air navigation systems. However, ADS-B is not technically capable of global monitoring of aircraft. Assume that airlines install the ADS-B transponders such that they cannot be switched off and their operation is foolproof against attacks (e.g., by hijacker, pilot, or crew). An aircraft responds to an ATC facility through a non-switchable secondary surveillance radar (SSR) transponder within its radio line of sight (RLOS). An aircraft equipped with a non-switchable ADS-B transponder can also respond to incoming messages from an ATC ADS-B ground station. However, SSR and ADS-B communication is only within RLOS. BRLOS communication is impossible within the SSR and ADS-B framework.

In the world of UAVs, aircraft-to-aircraft communication will play an even greater role, in the sense that it leads to information sharing among the UAVs, enhancing their situational awareness significantly compared to current systems. Amazon is suggesting segregation of airspace in which drones would follow set rules for flying below 500 feet. In this space, UAVs are expected to be connected to online networks and would directly communicate with each other, allowing for the automated control of flights in real time. Aerial connectivity is the key for the success of integrating UAVs into national air spaces. In order to realize applications such as package delivery, communication standards and protocols need to be in place. Vehicle-to-Vehicle (V2V) communications is the key to collision avoidance of UAVs, especially during close encounters. Innovations in geo-fencing [9, 24, 45, 48], conflict mitigation [12, 21], sense-and-avoid [26, 33], and protection of the airspace against rogue UAVs [35, 41] can be designed much more efficiently with aerial networking than by any other means that exist today. Direct over-the-air V2V communication is implemented in LTE through device-to-device (D2D) communications and is called "LTE sidelink." LTE sidelink is an adaptation of the core LTE standard that allows for communication between two or more nearby devices, using Evolved UMTS Terrestrial Radio Access Network (E-UTRAN) technology without the need for a base station [27].

18.2 The Mobility Challenge

Mobility poses a huge challenge for aircraft-to-aircraft communications. In order to understand this problem, imagine an aerial network established among a group of UAVs. This aerial network can be represented as a graph in which each vertex represents an aerial node and each edge represents a connection between a pair of aerial nodes. As the UAVs move around at high speeds, the connections among the nodes may change very frequently. A snapshot of the network taken at one instant may look quite different from the snapshot taken at a previous instant. At any time, some existing nodes may drop out of the network and some new nodes may join the network. The number of nodes and the number of connections in the network keep changing often depending on the speeds of the UAVs as well as the communication range of the radios being used. Thus, the graph representing the aerial network can be considered as a random graph. Topology changes created due to constantly moving UAVs lead to frequent loss of connectivity among UAVs. Loss of connectivity between nodes, in turn, leads to frequent changes in the routes established in the aerial network and ultimately to packet losses. Additional overhead traffic will be created to establish new routes that are needed for sharing information among nodes. While one can develop strategies to mitigate this effect, the impact of mobility on the overall network performance will be quite significant and cannot be completely eliminated or ignored.

18.2.1 UAS-to-UAS Communication

The high mobility of UAVs causes large Doppler spreads in received signals and time-varying communication channels between vehicles. Doppler shift (f_d), which is a function of carrier frequency (f_c) and relative velocity (V_r) of the transmitter and receiver, increases with the mobility of UAVs as

$$f_d = \frac{f_c V_r}{c} \cos(\theta), \tag{18.1}$$

where c is the speed of light and θ represents the angle of the relative velocity vector. To quantify the frequency offset caused by mobility, consider a simple scenario in which two UAVs are flying at the same altitude and maintaining a line of sight between them.

If the carrier frequency is assumed to be 5.9 GHz, as in dedicated short-range communication (DSRC) radios, even with a modest relative velocity of 30 m s^{-1} (corresponding to 108 km h^{-1}), the Doppler shift is about 590 Hz. Results of experiments conducted with DSRC radios on automobiles reported in [22] suggest that, in automobile V2V communications, Doppler shift reduces effective communication range and increases packet drops significantly.

In the case of V2V communications supported through an infrastructure such as a cellular network, network planning at the design stages is useful to combat the effect of Doppler shift. Some of the available solutions include configuration of the length of cyclic prefix and subcarrier spacing. However, accurate tracking of the fast time-varying channel caused by the high mobility of UAVs is a non-trivial problem [11].

18.2.2 Mobility Models

Airborne network connectivity depends on several physical, medium-access, and network-layer parameters including: (i) antenna type and its placement on the aircraft; (ii) type of radio used for communications; (iii) transmission range; (iv) number of aerial and ground vehicles present in the network; (v) mobility of nodes; and (vi) network and security protocols, among several others.

In order to analyze the network connectivity in different situations, one needs to consider a variety of flight test scenarios that involve: (i) two or more airborne nodes; (ii) ground vehicles and airborne vehicles; (iii) one or more relay nodes; (iv) aerial vehicles with different velocities; and (v) aerial vehicles with varying transmission range capabilities. The test scenarios reflect the mission- and application-specific characteristics of airborne networks.

A mobility model serves as a foundation of the analytical framework for network connectivity studies, network performance evaluation, and, eventually, the design of reliable network protocols [13]. In particular, mobility models capture the movement patterns of nodes from which the time-varying characteristics of the network, such as the node distribution and the statistics of link and path lifetimes, can be estimated. In order to provide accurate predictions that will help enhance airborne network connectivity, it is crucial to develop realistic and tractable mobility models for airborne networks. Some mobility models such as random direction (RD) and random waypoint (RWP) [6, 7, 19, 51] have been extensively investigated in the literature. The RWP model assumes that an agent chooses a random destination (waypoint) and traveling speed; upon arrival, it pauses before traveling to the next destination. The extended version of the RD model assumes that an agent chooses a speed and direction randomly after a randomly selected traveling time [14, 15]. The stochastic properties of these common models, such as their spatial distributions, can be found in, e.g., [6, 7, 19, 30].

The widely used RWP and RD models are well suited to describing the random activity of mobile users in mobile ad-hoc networks (MANETs); however, they lack the capability to describe features that are unique to airborne vehicles. For example, it is easy for mobile users/ground vehicles to slow down, make sharp turns, and travel in the opposite direction (see an enhanced random mobility model that captures such movement [5]). However, airborne vehicles, especially fixed-wing aircraft, tend to maintain the *same* heading and speed and change direction by making turns with a large radius. Rotary aircraft also require time and space to change their heading and direction. This unique feature stems from the mechanical and aerodynamic constraints for airborne vehicles and is reflected in the correlation in acceleration along spatial and temporal dimensions. Capturing this unique spatiotemporal correlation and incorporating it into mobility models can significantly improve path estimation and connectivity analysis for airborne networks and, in turn, facilitate the design of robust airborne networking. We propose to conduct a thorough investigation of realistic theoretical models that capture such features unique to airborne networks, but, simple and tractable enough to facilitate connectivity analysis and network design.

A basic smooth-turn (ST) mobility model that captures the preference of airborne vehicles toward making smooth trajectories (e.g., straight trajectories or turns with large radius), instead of sharp turns, is presented in [31, 49, 50]. The idea behind the smooth-turn random mobility model is simple. An airborne vehicle selects a point in the space along the line perpendicular to its heading direction and circles around it until the vehicle chooses another circle center. This perpendicularity is the key that ensures the smoothness of flight trajectories. In addition, since a vehicle commonly favors a straight trajectory and slight turns rather than very sharp turns, we model the inverse lengths of the circle radii using a Gaussian distribution. Furthermore, it is assumed that the waiting time for the next circle center change is memoryless, i.e., the timing of the change in circle center does not depend on the duration for which the UAV maintains its current circle center. The memoryless feature of waiting time is typically used to model the occurrence of random events, and brings in the nice features of renewable processes for tractable analysis [34]. For instance, connectivity analysis can be taken at any time instant without prior knowledge of the duration that a vehicle has kept its centripetal acceleration.

The ST mobility model naturally captures the movement of a UAV that makes smooth turns, and favors slight turns and straight trajectories over sharp turns. It is built based upon the rich literature on maneuver target tracking [18, 25, 29, 40]; these authors thoroughly studied the dynamics to capture the correlation of motion across both temporal and spatial coordinates. However, because these models are built for target tracking purposes, they focus on the high-precision prediction of the acceleration and paths for an individual aircraft and, therefore, their motion dynamics are too complex to be directly used for our purpose. Driven by the need to model the motion of airborne vehicles considering group statistics, the ST model captures the correlation across spatial coordinates through a simple random selected radius.

An analysis of mobility models may help mitigate the negative impact of mobility on network performance. Broadly, UAV networks may be classified based on the randomness of mobility patterns. Figure 18.5 illustrates four classes of mobility patterns: from completely random flight paths to well-defined planned flight paths. As the randomness is reduced, the path predictability increases. Path predictability can be utilized to plan the network design.

18.3 Establishing an Ad-Hoc Network

An ad-hoc network, by definition, lacks any infrastructure. It means neither cellular communication support on the ground nor satellite communication support from space is needed for a genuine ad-hoc network. Such a true ad-hoc network is extremely challenging

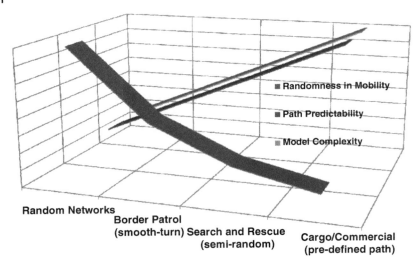

Figure 18.5 Mobility models with varying degrees of randomness, predictability, and complexity.

and nearly impossible to establish and maintain. The reasons for this difficulty will be explained later in this chapter. It also makes a lot of sense to take advantage of any available resources to enhance the performance of a UAV network. In this section, a strategy for establishing an ad-hoc network is discussed assuming that there is no infrastructure available to support the UAV network. This discussion serves two purposes: first, it explains the steps involved in establishing a network in detail; and second, it outlines the challenges involved in establishing an ad-hoc network of UAVs.

The process of establishing an ad-hoc network begins with a unique identification of each UAV. The aviation community is working towards remote identification (remote ID) specification for UAVs. A remote ID provides uniquely identifiable information for a UAS. Remote ID is needed for UAS traffic management (UTM), law enforcement, and security applications. For example, a law enforcement agency may want to know why a UAV is flying in a particular area by querying the UAV directly. The remote ID of a UAV will help verify the authorized use of the UAV and its intended flight path.

18.3.1 Network Addressing

While a remote ID is useful for validating and authorizing a UAS and its flight path, a network ID is essential to establish an ad-hoc mesh network. The purpose of a remote ID is different from that of a network ID.

The network ID allows a UAS to connect to its peers in a network and to share real-time mission-critical information in applications such as search and rescue. An Internet Protocol (IP) address, for example, identifies the host and its location in the network, and thus enables its peers to establish a path to that host. An ad-hoc network (or a mesh network) is highly suitable for UAS fleet and swarm applications. Assuming the use of IP for the mesh

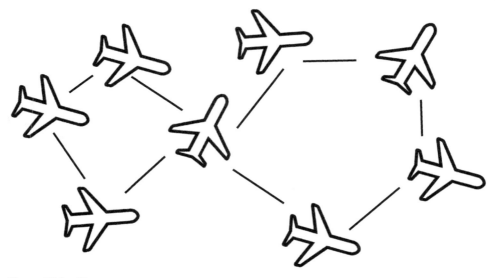

Figure 18.6 Illustration of an ad-hoc network of UAVs.

network, a uniquely identifiable IP address is required for each UAS in order to interconnect a fleet of UASs.

18.3.2 Routing

In an ad-hoc network, there is no central controller, so a UAS needs to find its own route to a destination UAS with which it wants to share real-time information. As a UAV moves, it makes connections with its immediate neighbors and establishes routes with peers which are farther away. Once a path is identified, a host can route its data packets through its connections on the path to its destination. Routing poses a big challenge in a UAV network because of the dynamics in network topology created by the high mobility of the UAVs. Consider, for example, a swarm consisting of eight UAVs forming an ad-hoc network, as illustrated in Figure 18.6. A route from one node to another node in this network may involve multiple hops. Any connection in the network, in particular, a connection that involves multiple links, may break instantly due to the high mobility of the UAVs. In a highly mobile network, a path may become invalid even before it is established! Due to this reason, it is nearly impossible to establish and maintain an ad-hoc network. Hence, it is logical to take advantage of any available infrastructure to enhance the connectivity of a UAS network. One such example is shown in Figure 18.3.

Routing protocols that are designed for traditional MANETs on the ground are not suitable for UAS networks. The same is true for security protocols that are developed for terrestrial networks. For example, certificate-based authentication protocols, primarily developed for terrestrial networks, will not be suitable for airborne networks because of the time that it takes to share the information back and forth with the authentication server. Hence, there is a need to develop light-weight security protocols.

18.4 Standards

Several international organizations are working towards developing standards that support UAS operations, particularly towards communications and traffic management. Some of the closely related and ongoing standards activities at various stages of progress are discussed below.

18.4.1 ASTM: Remote ID for UAS

ASTM International (known as the American Society for Testing and Materials), an international standards organization, is developing standards for remote ID and tracking of UASs. This task is led by ASTM Committee F38 and its scope is defined as [20]: "Any low altitude geographical area generally under 400 feet AGL [above ground level] both urban and rural (in and beyond cellular infrastructure range), regardless of airspace class. During operation of the UAS, a Unique ID (and possibly other registration numbers) along with location and vector (speed/direction) will be communicated at a regular interval such that a compliant receiver will be able to identify an aircraft that is within operating range of the receiver for broadcast mechanisms and network range for network mechanisms."

Remote ID enables a UAS to share its unique identification information with other parties. Remote ID will assist the aviation authorities, law enforcement, and security agencies to verify if a UAS is legally authorized to fly. Remote ID facilitates secure UAS traffic management and public safety.

18.4.2 EUROCAE: Safe, Secure, and Efficient UAS Operations

EUROCAE (the European Organisation for Civil Aviation Equipment) is the European leader in the development of worldwide-recognized industry standards for aviation. WG-105 is tasked to develop standards and guidance documents that will allow the safe operation of UASs in all types of airspace, at all times, and for all types of operations. The current focus areas are: UAS traffic management, Command, control, and communication, Detect and avoid, Design and airworthiness, Specific operations risk assessment, and Enhanced RPAS automation.

The Global UTM Association (GUTMA) is a non-profit consortium of worldwide UAS traffic management (UTM) stakeholders. Its purpose is to foster the safe, secure, and efficient integration of drones in national airspace systems with the mission to support and accelerate the transparent implementation of globally interoperable UTM systems.

18.4.3 3GPP: 4G LTE and 5G Support for Connected UAS Operations

Cellular networks can provide the desired communication infrastructure support for UAV operations, including reliable command and control. Cellular networks can match the reliability, latency and throughput requirements of UAS traffic management [47]. The 3GPP study on enhanced LTE support for connected UAVs in Release 15 addresses cellular connectivity to low-altitude UAVs. Release 16 includes support for remote identification of unmanned aerial systems. Release 17 includes work items that support UAS connectivity, identification, and tracking, 5G enhancements for UAS, and application-layer support for UAS [43].

18.4.4 IEEE P1920.1: Aerial Communications and Networking Standards

This standard defines a broad air-to-air communications framework for self-organized ad-hoc aerial networks. Aerial communications and networking standards are independent of any medium such as satellite or cellular communication systems. On the other hand, aerial communications can coexist and complement satellite and cellular communications. The IEEE P1920.1 standard includes use cases, network reference architecture, and security reference architecture. The IEEE P1920.1 is the first in a series of standards related to UAS communications and networking, and is expected to be released in 2020.

18.4.5 IEEE P1920.2: Vehicle-to-Vehicle Communications Standard for UAS

The V2V standard for UAS defines the protocol for exchanging information between vehicles. The information exchange will facilitate beyond-line-of-sight (BLOS) and BRLOS communications. The information exchanged between the aircraft may be for the purpose of command, control, and navigation, or for any application-specific purpose. The IEEE P1920.2 standard is its early stages of development.

18.5 Technologies and Products

Several technologies that can support ad-hoc mesh networking have been evolved in recent years. Some of these products have been field-tested for their applicability for UAV communications. A summary of commercial off-the-shelf products that can support ad-hoc networking of UAVs is provided below.

18.5.1 Silvus Streamcaster

Silvus Technologies introduced a multiple input–multiple output (MIMO) radio that is capable of supporting MANET with video and data communications capabilities for military, law enforcement, and public safety applications. Silvus Technologies' new product Streamcaster 4400 Enhanced (SC4400E) is a 4 × 4 MIMO radio that is convenient for use in portable and embedded applications where size, weight, and power need to be minimized. SC4400E can be used for ad-hoc mesh networking of UAVs [44].

18.5.2 goTenna

The product "goTenna" is a long-range lightweight device for mobile mesh networking. It is a solution for decentralized off-grid short-burst communications for public safety and other applications [36]. This product runs a mesh networking protocol that does not use any control packets for broadcasting and unicasting and builds state by observing the packet header

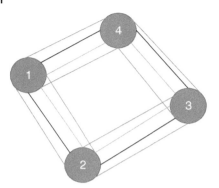

Figure 18.7 Rajant's Kinetic Mesh networking solution.

information. This is cooperatively a less-expensive solution for mesh networking. This product has been field-tested by the Colorado Center of Excellence for Advanced Technology Aerial Firefighting [32].

18.5.3 MPU5 and Wave Relay from Persistent Systems

The MPU5 radio system, built on 3×3 MIMO technology, allows for extended range and increased throughput in complex urban, subterranean, and maritime environments. The Wave Relay MANET is designed for mobility and is the core of the MPU5. Wave Relay is a scalable, peer-to-peer mesh networking solution for sharing data, video, and voice without any infrastructure support. This product has been field-tested for long-range (over 50 miles air-to-ground link) communications [46].

18.5.4 Kinetic Mesh Networks from Rajant

Rajant's Kinetic Mesh networking solution consists of BreadCrumb wireless nodes powered by the InstaMesh networking protocol. BreadCrumb nodes can be used to quickly deploy a wireless mesh network without the support of any infrastructure. One advantage of BreadCrumb radios is their capability of supporting multiple frequency bands, including 900 MHz, 2.4 GHz, 4.9 GHz, and 5 GHz [39].

Figure 18.7 illustrates how a multifrequency network adapts to changes caused by the movement of BreadCrumb node E. The dotted lines show locations where radio frequencies experience interference. In these situations, InstaMesh instantly reroutes traffic over a clear frequency. These capabilities greatly alleviate RF bottlenecks while maintaining robust fault tolerance, high throughput, and low latency.

18.6 Software-Defined Network as a Solution for UAV Networks

In an aerial network, UAVs cooperate with one another to keep the network connected even in the event of link or node failures. UAS networks are expected to be self-organizing, delay-tolerant, flexible, adaptable, and scalable. Such capabilities can be achieved through software-defined networking (SDN), which decouples the control and data planes of networking devices. SDN facilitates network management and control through a software abstraction layer provided by a logically centralized controller. New requirements have led to proposals to extend SDN to wireless networks. Software-defined wireless networks decouple control functions (radio control, spectrum management, mobility management, and interference management) from the radio data plane [4, 17, 42].

For aerospace networks, temporo-spatial SDN (TS-SDN), which takes advantage of the ability to predict the locations of many of the nodes in the network, has been proposed. The location and trajectory of a node along its flight path can be used to predict and control the network state, which can allow for better topology management, proactive packet routing, and radio resource management, along with the use of software-defined radios (SDRs), and scheduled operations [1–3]. The main advantage of SDN is evolvability and simplification [8]. SDN provides the flexibility to deploy new policies via software updates. SDRs extend this update flexibility to deploying new waveforms for use at the physical layer. As new algorithms and solutions are developed by the larger research community, software-defined aerial networks will be able to be updated without having to deploy new costly hardware [8].

In a distributed control paradigm, each UAS will control its own behavior [16]. Previous studies have demonstrated the feasibility of integrating OpenFlow with wireless mesh networks and the advantages of SDN-based control strategies over traditional strategies employed in MANETs.

18.7 Summary

This chapter outlined the three strategies for providing connectivity for UAVs: satellite, cellular, and aerial communications. It explained why mobility poses a huge challenge in establishing aerial connectivity. It presented the processes involved in establishing an ad-hoc mobile network of UAVs. It emphasized the near-impossibility of establishing and maintaining a true ad-hoc network of UAVs. On the other hand, with the support of available infrastructure, it is possible to establish a mesh network of UAVs. This chapter also discussed various related standards for remote ID that are currently in progress. Technologies and commercial products that are available today for creating mesh networks are discussed. Finally, SDN is discussed as a way to address the mobility challenge.

References

1 Brian Barritt and Wesley Eddy. Temporospatial sdn for aerospace communications. In *AIAA SPACE 2015 Conference and Exposition*, page 4656, 2015.

2 Brian Barritt, Tatiana Kichkaylo, Ketan Mandke, Adam Zalcman, and Victor Lin. Operating a uav mesh & internet backhaul network using temporospatial sdn. In *Aerospace Conference, 2017 IEEE*, pages 1–7. IEEE, 2017.

3 Brian James Barritt. *The Modeling, Simulation, and Operational Control of Aerospace Communication Networks*. PhD thesis, Case Western Reserve University, 2017.

4 Carlos J Bernardos, Antonio De La Oliva, Pablo Serrano, Albert Banchs, Luis M Contreras, Hao Jin, and Juan Carlos Zúñiga. An Architecture for Software Defined Wireless Networking. *IEEE Wireless Communications*, 21 (3): 52–61, June 2014.

5 C. Bettstetter. Smooth is better than sharp: a random mobility model for simulation of wireless networks. In *ACM International Workshop on Modeling, Analysis and Simulation of Wiresless and Mobile Systems*, Rome, Italy, July 2001.

6 C. Bettstetter, H. Hartenstein, and X. Peréz-costa. Stochastic properties of the random waypoint mobility model. *Wirelss Networks*, 10: 555–567, 2004.

7 J. Boudec and M. Vojnovic. Perfect simulation and stationarity of a class of mobility models. Technical report, Techinical Report IC/2004/59, 2004.

8 Salvatore Costanzo, Laura Galluccio, Giacomo Morabito, and Sergio Palazzo. Software defined wireless networks: Unbridling sdns. In *Software Defined Networking (EWSDN), 2012 European Workshop on*, pages 1–6. IEEE, 2012.

9 Evan T Dill, Kelly J Hayhurst, Steven D Young, and Anthony J Narkawicz. UAS hazard mitigation through assured compliance with conformance criteria. In *AIAA Information Systems-AIAA Infotech@ Aerospace*, pages 1218–1218. 2018.

10 Edward Falcov. Use of self-organizing airborne networks to monitor commercial aircraft globally. *Working Paper WP10, multidisciplinary meeting on global tracking*, 2014.

11 Pingzhi Fan, Jing Zhao, and I Chih-Lin. 5g high mobility wireless communications: Challenges and solutions. *China Communications*, 13 (2): 1–13, 2016.

12 Scott Xiang Fang, Siu O'Young, and Luc Rolland. Development of small uas beyond-visual-line-of-sight (bvlos) flight operations: System requirements and procedures. *Drones*, 2 (2): 13, 2018.

13 R. Ghanta and S. Suresh. Influence of mobility models on the performance of routing protocols in ad-hoc wireless networks. *IEEE 59th Vehicular Technology Conference*, pages 2185–2189, 2004.

14 B. Gloss, M. Scharf, and D. Neubauer. A more realistic random direction mobility model. In *4th Management Committee Meeting, Würzburg*, Germany, October 2005.

15 R. A. Guérin. Channel occupancy time distribution in a cellular radio system. *IEEE Transactions on Vehicular Technology*, 35 (3): 89–99, August 1987.

16 Lav Gupta, Raj Jain, and Gabor Vaszkun. Survey of important issues in uav communication networks. *IEEE Communications Surveys & Tutorials*, 18 (2): 1123–1152, Second Quarter 2015.

17 Tanzeena Haque and Nael Abu-Ghazaleh. Wireless software defined networking: A survey and taxonomy. *IEEE Communications Surveys & Tutorials*, 18 (4): 2713–2737, Fourth Quarter 2016.

18 J. P. Helferty. Improved tracking of maneuvering targets: the use of turn-rate distributions for acceleration modeling. *Proceedings of the IEEE International Conference on Multisensor Fusion and Integration for Intelligent Systems*, pages 515–520, October 1994.

19 E. Hyytiä, P. Lassila, and J. Virtamo. Spatial node distribution of the random waypoint mobility model with applications. *IEEE Transactions on mobile computing*, 5 (6): 680–694, June 2006.

20 ASTM International. Committee f38 on unmanned aircraft systems, 2019. URL https://www.astm.org/COMMIT/SUBCOMMIT/F38.htm.

21 Marcus Johnson. Unmanned aircraft systems traffic management: Conflict mitigation approach. 2018.

22 Dwayne Jordan, Nicholas Kyte, Scott Murray, Mohammad A Hoque, Md Salman Ahmed, and Asad Khattak. Poster: Investigating doppler effects on vehicle-to-vehicle communication: An experimental study. In *Proceedings of the 2nd ACM International Workshop on Smart, Autonomous, and Connected Vehicular Systems and Services*, pages 77–78. ACM, 2017.

23 D Lawrence, H Cobb, G Gutt, M OConnor, T Reid, T Walter, and D Whelan. Navigation from LEO: Current capability and future promise. *GPS World Magazine*, 28 (7): 42–48, 2017.

24 Jooyoung Lee, Inseok Hwang, and David Hyunchul Shim. Uas surveillance in low-altitude airspace with geofencing: Constrained stochastic linear hybrid systems approach. In *AIAA Information Systems-AIAA Infotech@ Aerospace*, page 0077. 2018.

25 X. R. Li and V. P. Jilkov. Survey of maneuvering target tracking: dynamic models. *Proceedings of SPIE Conference on Signal and Data Processing of Small Targets*, AES-6 (4048): 212–235, April 2000.

26 Zhaowei Ma, Chang Wang, Yifeng Niu, Xiangke Wang, and Lincheng Shen. A saliency-based reinforcement learning approach for a uav to avoid flying obstacles. *Robotics and Autonomous Systems*, 100: 108–118, 2018.

27 Rafael Molina-Masegosa and Javier Gozalvez. LTE-V for sidelink 5G V2X vehicular communications: A new 5g technology for short-range vehicle-to-everything communications. *IEEE Vehicular Technology Magazine*, 12 (4): 30–39, 2017.

28 Siva D Muruganathan, Xingqin Lin, Helka-Liina Maattanen, Zhenhua Zou, Wuri A Hapsari, and Shinpei Yasukawa. An overview of 3GPP release-15 study on enhanced LTE support for connected drones. *arXiv preprint arXiv:1805.00826*, 2018.

29 N. Nabaa and R. H. Bishop. Validation and comparison of coordinated turn aircraft maneuver models. *IEEE Transactions on aerospace and electronic systems*, 36 (1): 250–255, January 2000.

30 P. Nain, D. Towsley, B. Liu, and Z. Liu. Properties of random direction. In *IEEE INFOCOM*, pages 1897–1907, March 2005.

31 K. Namuduri, Y. Wan, Mahadevan Gomathisankaran, and Ravi Pendse. Airborne network: a cyber-physical system perspective. *ACM Mobihoc Conference*, June 2012.

32 Department of Public Safety. Report on TAK use for search and rescue. Technical report, Colorado Center of Excellence for Advanced Technology Aerial Firefighting, April 2019.

33 Edwin Ordoukhanian and Azad M Madni. Introducing resilience into multi-uav system-of-systems network. In *Disciplinary Convergence in Systems Engineering Research*, pages 27–40. Springer, 2018.

34 A. Papoulis and S. U. Pillai. *Probability, random variables and stochastic processes*. McGraw-Hill, 2002.

35 Alyssa Pierson, Zijian Wang, and Mac Schwager. Intercepting rogue robots: An algorithm for capturing multiple evaders with multiple pursuers. *IEEE Robotics and Automation Letters*, 2 (2): 530–537, April 2017.

36 Ram Ramanathan, Christophe Servaes, Warren Ramanathan Ayush Dusia, and Adarshpal Sethi. Long-range short-burst mobile mesh networking: Architecture and evaluation.

Proceedings of the IEEE International Conference on Sensing, Communication and Networking, June 2019.

37 Tyler Reid. *Orbital Diversity for Global Navigation Satellite Systems, PhD Dissertation*. PhD thesis, 2017.

38 Tyler GR Reid, Andrew M Neish, Todd Walter, and Per K Enge. Broadband LEO constellations for navigation. *Navigation: Journal of The Institute of Navigation*, 65 (2): 205–220, 2018.

39 Deepshikha Shukla. Controlling drones and uavs: Advancements in wireless technologies. *Electronics For You*, pages 70–71, December 2018.

40 R. A. Singer. Estimating optimal tracking filter performance for manned maneuvering targets. *IEEE Trans. Aerospace and Electronic Systems*, AES-6: 473–383, 1970.

41 Alexander Solodov, Adam Williams, Sara Al Hanaei, and Braden Goddard. Analyzing the threat of unmanned aerial vehicles (uav) to nuclear facilities. *Security Journal*, 31 (1): 305–324, 2018.

42 Keshav Sood, Shui Yu, and Yong Xiang. Software-defined wireless networking opportunities and challenges for internet-of-things: A review. *IEEE Internet of Things Journal*, 3 (4): 453–463, 2016.

43 The Mobile Broadband Standard. 3GPP list of work items, 2019a. URL https://www.3gpp.org/DynaReport/WI-List.htm.

44 The Mobile Broadband Standard. Streamcaster 4400 enhanced (sc4400e), 2019b. URL https://silvustechnologies.com/products/streamcaster-radios/.

45 Mia N Stevens and Ella M Atkins. Geofencing in immediate reaches airspace for unmanned aircraft system traffic management. In *2018 AIAA Information Systems-AIAA Infotech@ Aerospace*, page 2140. 2018.

46 Persistent Systems. MANET/mesh network for cooperative unmanned vehicles, 2013. URL https://www.persistentsystems.com/.

47 Attila Takacs, Xingqin Lin, Stephen Hayes, and Erika Tejedor. Drones and networks: Ensuring safe and secure operations. Technical report, Ericsson, 11 2018.

48 Ryan J Wallace, Jon M Loffi, Michael Quiroga, and Carlos Quiroga. Exploring commercial counter-uas operations: A case study of the 2017 dominican republic festival presidente. *International Journal of Aviation, Aeronautics, and Aerospace*, 5 (2): 8, 2018.

49 Y. Wan, K. Namuduri, Y. Zhou, D. He, and S. Fu. A smooth turn mobility model for airborne networks. *ACM Mobihoc Conference*, June 2012.

50 Yan Wan, Kamesh Namuduri, Yi Zhou, and Shengli Fu. A smooth-turn mobility model for airborne networks. *IEEE Transactions on Vehicular Technology*, 62 (7): 3359–3370, 2013.

51 J. Yoon, M. Liu, and B. Noble. Sound mobility models. *Proceedings of the Ninth Annual International conference on Mobile Computing and Networking*, pages 205–216, 2003.

Index

UAV Communications for 5G and Beyond, First Edition.
Edited by Yong Zeng, Ismail Guvenc, Rui Zhang, Giovanni Geraci, and David W. Matolak.
© 2021 John Wiley & Sons Ltd. Published 2021 by John Wiley & Sons Ltd.